About the Author

DR. KATHI J. KEMPER is the founding director of the Center for Integrative Health and Wellness at Ohio State University. She was recruited from Wake Forest University, where she held the Caryl J. Guth Chair for Complementary and Integrative Medicine and served as the founding director of the Center for Integrative Medicine.

Dr. Kemper was also the founding director of the Center for Holistic Pediatric Education and Research at Boston Children's Hospital and associate professor of pediatrics at Harvard Medical School. Dr. Kemper is past president of the Academic Pediatric Association and founder of the American Academy of Pediatrics Section on Integrative Medicine. She is recognized internationally as the leading authority on complementary therapies for children and is frequently consulted by media, including the *New York Times, Chicago Tribune, Newsweek,* ABC News, *Wall Street Journal, Reader's Digest, Redbook, First for Women,* and *USA Today.*

THE HOLISTIC PEDIATRICIAN

THE
HOLISTIC
PEDIATRICIAN

TWENTIETH ANNIVERSARY REVISED EDITION

A PEDIATRICIAN'S COMPREHENSIVE GUIDE TO SAFE AND EFFECTIVE THERAPIES FOR
THE 25 MOST COMMON AILMENTS OF INFANTS, CHILDREN, AND ADOLESCENTS

KATHI J. KEMPER, M.D., M.P.H.

Harper
New York • London • Toronto • Sydney

This book is intended to educate parents about a variety of approaches to children's health care needs. This book should not be a substitute for the personal care and treatment of a qualified physician, but, rather, should be used in conjunction with a physician's care in order to consider the full range of health care options available to your child. The author and publisher expressly disclaim responsibility for any adverse effects resulting from the information contained herein. Before giving your child any of the herbal treatments recommended in this book, consult a physician or pharmacist for possible drug interactions.

HarperCollins books may be purchased for educational, business, or sales promotional use. For information please e-mail the Special Markets Department at SPsales@harpercollins.com.

First Perennial edition published 1996.

Second edition published in Quill 2002.

Twentieth Anniversary Revised Edition published in Harper Paperbacks 2016.

Designed by Nancy Singer Olaguera

Library of Congress Cataloging-in-Publication Data has been applied for.

ISBN 978-0-06-256052-0

16 17 18 19 20   OV/RRD   10 9 8 7 6 5 4 3 2 1

To Daniel Alexander Kemper

CONTENTS

Contents

Foreword to the Twentieth Anniversary Edition of *The Holistic Pediatrician*

Andrew Weil, M.D.

It is my pleasure to write this Foreword for the twentieth anniversary edition of *The Holistic Pediatrician*. This newest edition marks a milestone in pediatric integrative medicine, a further indication of much progress in the field. As befits the changes we've seen over time, this edition contains new chapters on Anxiety, Autism, Fatigue, and Pain—conditions that have become all too common not just in adults, but in children and youth in twenty-first-century America.

So much has changed in twenty years. Back in the 1990s, when we began the physician fellowship training program at the University of Arizona and Dr. Kemper started writing the first edition of this book, pediatricians were still worried about chickenpox (now there's a vaccine, and rates have dropped over 90 percent), and email was a novelty. Google, WebMD, and Facebook did not exist. Physicians rarely learned much about nutrition or the importance of the environment to children's health. Parents who used therapies like herbs, meditation, and massage were considered eccentric. Instead of the prevailing dichotomy of conventional vs. alternative, Dr. Kemper promoted a synthesis or integration of therapies based on scientific evidence.

When *The Holistic Pediatrician* first appeared, it was heartily embraced, not only by parents, but also by physicians eager to fill the gaps in their education. Within two years of its publication, Dr. Kemper was recruited to my alma mater, Harvard Medical School, to found the first Center for Holistic Pediatrics in the United States. While at Harvard, Dr. Kemper developed the first pediatric training programs to help faculty physicians learn an integrative approach to health care. She and her team, the Longwood Herbal Task Force, created the first online interprofessional training program to enable health professionals to respond to families' questions about herbs and dietary supplements. She also founded the Academic Pediatric Association's Special Interest Group on Holistic and Complementary (now Integrative) Pediatrics to help innovators across the country learn from one another how to incorporate "alternative" care into academic settings that insist on the highest standards of evidence. These were the two most frequent questions colleagues asked: "Does it work?" and "Is it safe?"

As evidence grew that integrative approaches are indeed effective and safe, both families and pediatricians demanded more evidence-based, reliable information. In 2005, Dr. Kemper founded the American Academy of Pediatrics' Section on Integrative Medicine. In 2010, the Academy published her second book, *Mental Health Naturally*, to address America's growing health crisis with an integrative, comprehensive approach to promoting mental/emotional well-being. The same year, in response to a parent's request, she wrote a concise, practical summary for parents of children with attention-deficit hyperactivity disorder (ADHD), *Addressing ADD Naturally*.

Over the past decade, Dr. Kemper has not only pioneered the field of integrative pediatrics, she has also created two more integrative medicine centers at academic institutions—Wake Forest University Health Sciences in North Carolina and The Ohio State University. As the demand for integrative services expands, for adults as well as children, the main questions her colleagues now ask are not about safety and effectiveness but about how rapidly services can be delivered to patients who need them, and how they can be paid for.

This is where we need your help. I urge you to keep a copy of this book on your shelf for your own family's reference. In it, you will find reliable, practical advice and comfort for common childhood problems. Second, please encourage your physician to read it and refer to it. Thousands of pediatricians have already read, learned from, and changed their practices because of this book. That's a start. But in order for health care to truly change, your advocacy is needed to ensure that insurance plans cover integrative services—registered dietitians, acupuncturists, certified health coaches, and massage therapists—just as they do for physicians, chiropractors, psychologists, and physical therapists. Continue to work in your community and region to promote clean air (smoke free zones) and water and sustainable agriculture, energy, and transportation. And always keep in mind that children's healthy habits and habitats depend on the choices and actions of wise and loving adults.

ACKNOWLEDGMENTS

Thanks to the many parents and pediatric clinicians who provided constructive feedback and requests for more information in response to the first two editions. Thanks to visionary colleagues in the American Academy of Pediatrics who lead efforts to create our first truly integrative medical specialty—pediatrics! Thanks to many friends from the Academic Consortium for Integrative Medicine and Health and the Academy for Integrative Health and Medicine for organizing efforts to build collaborations that are helping to transform health care. Thanks to the many researchers worldwide who have done the hard work that makes it possible for us to provide evidence-informed care. Thanks to those who have gone before us and worked beside us to create a more humanistic health care that values clinicians as well as patient wellness. Thanks to those working to advocate for healthier habitats and policies that affect health through the environment, agriculture, transportation, education, housing, and justice. Thanks to the editing team at Harper Collins who has believed in this project for over 20 years. Deep gratitude goes to my friend and colleague, Andy Weil, for writing the Foreword and for being a gracious role model for many years. And most of all, thanks to Daniel, who has grown up as the son of the Holistic Pediatrician, and who makes her proud and grateful every day. You are my joy.

INTRODUCTION

Welcome to the twentieth anniversary edition of the classic
*The Holistic Pediatrician: A Pediatrician's Guide to Safe and Effective Therapies
for the 25 Most Common Ailments of Infants, Children, and Adolescents*!

"We heard you were open to holistic medicine, and we wanted you
to see our daughter, Shelly," Lisa and Larry Bradshaw began. "She's had
asthma since she was nine months old, and she just isn't getting over it.
We don't like the idea of giving her drugs every day, but we don't want
her to be sick either. We've treated ourselves with Chinese herbs and
homeopathic remedies, but we aren't sure what's safe for a two-year-
old like Shelly. Can you help?"

Lisa and Larry are typical of the groundswell of parents who have sought and used integrative care for themselves and who want to provide the safest, most effective care for their children. There has been a veritable explosion in the number of books on holistic medicine over the last twenty years, but few of these books are aimed at providing pediatric care.

This book was written to empower you to care for your child's health care needs, confidently armed with scientific information and grounded in a pediatrician's lifetime of practice experience.

As parents, you are the primary providers of your children's health care. Parents manage their children's home environment, diet, and medical care. Mothers characteristically cope with most minor illnesses by using home or folk remedies, and they often ask for advice from relatives and friends before seeking help from a health care professional. *Parents are generally very competent in caring for their children's illnesses.*

With commercial, government, professional organization, and academic information available over the Internet, parents have become inundated by a variety of goods and services. I see parents as the primary providers of health care who are trying their best, but often feeling overwhelmed and confused, looking for a knowledgeable, respectful coach who can help them navigate the minefields of misinformation to provide the safest, most effective care for their child. I'm here to serve as your guide and coach.

Many books written by health professionals discuss child development, behavior, health, and illness. However, none are written by an academic pediatrician, trained in the sciences, who has helped parents for over 20 years to integrate the best of modern medical science with proven therapies from healthy lifestyle, herbal medicine, homeopathy, and other healing techniques in a truly holistic approach to common and emerging childhood illnesses. This book does just that. I've updated it just for you! Diseases that barely registered on the radar screen 20 years ago (autism and fatigue) are grabbing front-page headlines as their rates soar. It's time for up-to-date, solid answers from a trusted source. Now you have it.

Who needs another book on holistic medicine?

Parents do! A 1993 study in the *New England Journal of Medicine* reported that nearly one out of every three American adults used alternative medical therapies in 1990; by 2007, this percentage had soared to over 40 percent. Many parents also seek alternative therapies for their children. When given an option between a non-drug home remedy and a drug, nearly 85 percent of parents prefer the home remedy. The families most likely to seek complementary care have children that suffer from a serious, chronic disease, such as rheumatoid arthritis, or that have ailments for which traditional medical care has not been of much help, such as recurrent ear infections or allergies; however, parents of healthy kids often use echinacea or vitamin C when their child has a cold, or try a back rub when the baby can't sleep. Parents want the best for their children—safe,

effective, personal care that is low in cost and side effects. Practitioners of "natural" therapies are often seen as providing more personalized care, listening to parents' concerns and preferences better than the average general practitioner. Most parents who seek alternative care are intelligent and well educated, and most take their children to regular medical doctors in addition to other types of health care providers.

What is holistic medicine? Holistic medicine is the foundation of good medicine. It promotes the child's joy and flourishing in the context of family, culture, and community. Holistic practitioners see the whole child—body, mind, emotions, spirit, and relationships with others. From a variety of potential treatments, holistic practitioners choose those that are best suited to the individual child and family and integrate those therapies into a unique plan.

One of the most holistic doctors I've ever met is Dr. David Heimbach, a burn surgeon at Harborview Medical Center in Seattle. You might wonder if a surgeon could really be holistic. Dr. Heimbach assembled a team of plastic surgeons, nurses, physical and occupational therapists, pediatricians, nutritionists, social workers, and psychologists to help meet the needs of children who have suffered severe burns. When he takes care of a burned child, he looks not only at the burn, but also at the whole child and the child's family. He makes sure that the out-of-state families have a place to stay while their child is being treated, and he implemented a fund to help pay for their housing. He asks about the child's school, friends, and church. On one occasion, he even made arrangements for a seriously burned child to be visited in the hospital by his puppy. (Yes, hygiene was maintained, and the visit was a rousing success for both boy and puppy.) The psychologists on the team use hypnosis and the latest electronic gaming technology to help children cope with the pain of the initial burn and the subsequent surgeries. Nutritionists help ensure that children are getting not only enough calories, but also additional vitamins and minerals to hasten healing. For children whose families are far away, Dr. Heimbach asks for volunteers from the community to play with, read to, and hold injured children. Yes, even surgeons in major academic medical centers can be holistic physicians.

On the other hand, practitioners who believe that all ailments can be traced to yeast infections or vitamin deficiencies are no more holistic than those who believe that all illness is due to germs. Some worrisome providers refuse to refer a sick feverish child to a medical doctor because they think the problem can be fixed with an alternative therapy alone. These practitioners may call themselves holistic, but in fact, they are ideologues with good marketing skills. Not all unconventional therapies are holistic, nor is mainstream medical practice necessarily NOT holistic. A single therapy, be it medication, surgery, nutrition, supplements, herbs, or massage, is not holistic unless it is done in the context of the whole child.

I try to avoid the terms "alternative" and "unconventional" medicine, despite their widespread use, because they are very difficult to define. One's definition of "alternative" depends very much on what one considers mainstream. For many Americans, Chinese medicine is an alternative. But for Chinese-Americans, it is mainstream. Other so-called alternatives such as chiropractic are as American as apple pie. Hypnosis and acupuncture used to be considered unconventional, but they are now used in major medical centers across America. Rather than try to define mainstream, alternative, and unconventional medicine, we'll look at the therapies themselves (Chapter 1) and how they can complement each other.

Illness versus disease. Disease is an abnormal condition in the body. Illness is one's experience of abnormal or suboptimal functioning. You can have a disease such as cancer for weeks or months before symptoms develop and you feel ill. In general we cure disease and heal illness.

Healing versus curing. In this book, curing means the elimination of symptoms or signs of a disease. For example, a child is cured of pneumonia when the fever and the cough are gone, and signs of infection disappear from the X-ray. There are no cures yet for many illnesses that affect children. The symptoms of cystic fibrosis may be minimized, but the underlying disease will not be eradicated until we come up with genetic therapies. Yet, even children with chronic or genetic diseases such as cystic fibrosis can be considered healed if they feel happy, loved, and able to function as well as they'd like. Being healed is a state of mind and spirit. Being cured is a physical phenomenon. We can compare the cure rates of different therapies, but we cannot accurately measure healing. When I describe the effectiveness of different therapies, I am talking about the effectiveness of those therapies in curing, not healing.

About the Author

I am a pediatrician, a medical doctor specializing in the care of children. My medical training has spanned many years at several institutions: the University of North Carolina (M.D., Preventive Medicine graduate training, and Master's in Public Health, specializing in maternal and child health); the University of Wisconsin (internship and residency in pediatrics); and Yale University (fellowship training in pediatric research). In 1988, I joined the faculty of the University of Washington, where I taught medical students and pediatric residents how to be good pediatricians. In 1994 I joined the staff of Swedish Medical Center in Seattle, helping to train family doctors in pediatrics. Shortly after the first edition of *The Holistic Pediatrician* was published, I was elected as President of the Academic Pediatric Association—an international academic organization of some of the top pediatricians in the world. That was the year I became a mom, and I can assure you that being a doctor is far easier than being a mother!

In 1998, I was recruited to Harvard Medical School to become the first director of the new Center for Holistic Pediatric Education and Research at The Children's Hospital in Boston—the first such center in the country. Since then, I've founded Centers for Integrative Medicine at Wake Forest and The Ohio State University. In 2005, I led the efforts to create the Section on Integrative Medicine in the American Academy of Pediatrics (AAP) while maintaining a practice and research career. In 2010, the AAP published my second book, *Mental Health Naturally*, and in response to a mother's request for something short and sweet on the topic, I published *Addressing ADD Naturally*. I have had over 170 scientific articles and chapters published in medical journals such as the *New England Journal of Medicine* and the *Journal of the American Medical Association*. I have also been a consultant and writer for *The New York Times, The Wall Street Journal, ABC News, Redbook,* and WebMD among many others.

Throughout my training and professional experience, it was my patients and their families who stimulated me the most to learn and do research. And now, of course, it is my son, who is the best teacher in the world. My colleagues throughout the United States have taught me an enormous amount about healing, about teaching, about writing grants and proposals, about starting and maintaining a new enterprise, and mostly about myself. To all of these teachers and mentors, I am thankful.

The book is organized for you to read selectively. You do not need to read it from cover to cover. I suggest that you read Chapters 1 and 2, and then skip to the chapters that are pertinent for your particular child. By reading Chapter 1, "The Therapeutic Mountain," first, you will be better able to understand all of the rest of the chapters. It describes a paradigm for truly integrative medicine that I developed in 1994 and since taught to thousands of pediatricians and parents to help us understand the relationship between different types of therapies. This chapter describes the therapies, the professionals who recommend them, and how to select a practitioner for your child. Chapter 2, "Trust Me, I'm a Doctor," is a short description of how you can evaluate claims that different treatments are effective for your child. Since the Internet has made it so easy to find the resources listed in the appendices of earlier editions, those appendices have been eliminated. That has already provided a bit of room for all the extra scientific information on new treatments that have been produced in the past twenty years.

To maintain a consistent, easy-to-follow approach, all of the chapters on a particular illness and conditions are organized in the same way. If you want a quick summary of how to treat a particular condition, see the last section of each chapter, "What I Recommend."

To avoid using cumbersome phrases such as he/she and him/her, males and females change chapter by chapter in the stories woven throughout the book. All of the case stories are composite descriptions of patient encounters; to preserve patient confidentiality, fictitious names, gender, and ethnicity are used.

THE HOLISTIC PEDIATRICIAN

1
THE THERAPEUTIC MOUNTAIN

"I've taken my child to so many doctors, I've lost count," Helen began. "The pediatrician put him on antibiotics to prevent any more ear infections, but the medicines gave him diarrhea and a yeast infection. The chiropractor said that adjusting his neck would help, but I didn't think it helped much, and I didn't like all the X-rays. The naturopath recommended some herbs and vitamins, but my insurance wouldn't pay for them. None of these doctors thought the other ones did any good. They all seemed more interested in promoting their own particular therapy than in working with each other to help my child. I'm frustrated and confused. How can the best, the safest, and most effective of all available treatments be combined for my child?"

Helen's story from 1994 epitomizes many families' complaints about the health care system. Back in the 1990s, my friend at Harvard, Dr. David Eisenberg, published research on the use of complementary and alternative medical (CAM) therapies in the prestigious *New England Journal of Medicine* and the *Journal of the American Medical Association*. His research showed that the percentage of Americans using CAM therapies increased from about 30 percent to over 40 percent in less than 10 years. And that was ignoring the use of prayer and multivitamins, which were used by a majority of Americans to improve health.

These publications grabbed the attention of leading academic medical centers, which started offering courses in CAM. A group of visionary philanthropists created the Bravewell Collaborative. The Collaborative funded leadership training and awards, education, clinical care, and research at major academic health centers, and then formed the Academic Consortium for Integrative Health and Medicine. By 2015, the Consortium boasted over 60 members across North America.

The Consortium collaborated with the International Society for Complementary Medicine Research to sponsor international research symposia from North America to Europe to Asia, building the groundwork for integrating complementary therapies into conventional settings internationally. There have been enormous growth and change in this field in the last twenty years.

Dr. John Astin published papers describing why CAM became so popular. He found that many people sought CAM care because it relied on values and worldviews that were more consistent with their personal beliefs than the disease-focused, high-tech, pharmaceutical industry–dominated world of modern medicine. These values include autonomy and interdependence, a respect for nature and the human capacity for health and healing, and a preference for humanistic, relationship-based care.

Dr. Wayne Jonas wrote about *salutogenesis* (the process of restoring health and well-being, which is the opposite of *pathogenesis,* the process of becoming sick) and optimal healing environments. He was appointed the first director of the National Institutes of Health's Office of Alternative Medicine, which is now the National Center for Complementary and Integrative Health.

In 2004, a small group of integrative pediatricians gathered in St. Paul, Minnesota, to discuss the future of pediatric integrative medicine. That meeting spawned a new Section on Integrative Medicine within the American Academy of Pediatrics, as well as an international online discussion group among pediatric professionals, a series of journal articles, a new textbook, and the emergence of residency

and fellowship training programs in integrative pediatrics. The group, which included several next-generation pediatricians, met again in 2015, seeking input from diverse non-physician clinicians and thousands of parents to pave the way for the next ten years of pediatric integrative medicine.

Over the years, the many different kinds of health professionals who care for children—whose backgrounds, theories, and therapies often differ—have advanced from hostility and competition to a greater emphasis on cooperation, teamwork, and patient-centered care. An increasing number of insurance companies cover therapies previously considered alternative, like acupuncture and massage. The U.S. Department of Defense created a program called Total Force Fitness, a care model for healthy warriors and their families. Today there are integrative pediatricians in every specialty, and the most common question is not, "Does this work?" but rather, "How do we get this paid for so our patients have access to it?" Integrative medicine has become mainstream.

Helen's experience from the 1990s has become much less common in the twenty-first century, as leading medical institutions have adopted a more holistic, integrative care model, focusing on patient-centered care. Integrative, holistic care has simply become the best kind of health care!

WHAT DO WE MEAN BY HOLISTIC HEALTH CARE?

In 1948, the World Health Organization defined health as "a state of complete physical, mental, spiritual, and social well-being and not merely the absence of disease or infirmity." Holistic care is based on this definition. It means caring for the whole person—body, mind, and spirit in the context of family, culture, and community.

WHAT DO WE MEAN BY INTEGRATIVE HEALTH CARE?

Integrative care is holistic care based on a respectful relationship between patients and clinicians. It is informed by evidence, and uses all appropriate therapeutic approaches, health care professionals, and disciplines to achieve optimal health and healing. It is not simply combining complementary and alternative medicine with conventional medications to treat diseases. Instead, it emphasizes articulating and achieving health goals, in the context of respectful relationships, and a common approach to health built on healthy habits in a healthy habitat. The reliance on respectful relationships also suggests the vital role of clinician compassion and loving-kindness—professionals' desire and behavior to relieve suffering and extend good will, not to make money or increase one's power over others or inflate one's ego, but simply to enrich the human community by serving others.

Ideally, both professionals and parents lay aside their personal concerns when faced with an ill child. The focus should be on the child's and family's goals for wellness and healing. Health goals can be thought of in several ways.

WHAT ARE HEALTH GOALS?

In conventional medicine, health goals are typically divided into curing, preventing, managing, or rehabilitating from a disease or injury. Table 1.1 shows examples of these different health goals.

In conventional care, we define or diagnose a condition, and then dispense a treatment (often, but not always a medication) to cure, prevent, manage, or rehabilitate it. I refer to this as the "diagnose and dispense" model of care.

When I began doing research on therapies like acupuncture, music, and Therapeutic Touch, I realized that many of these therapies

Table 1.1: Health Goals	
Goal	Examples
Cure	Penicillin for strep throat; surgery for appendicitis
Prevent	Polio vaccine to prevent polio; car seats to prevent injuries
Manage	Eye glasses to manage near-sightedness; insulin to manage diabetes
Rehabilitate	Physical or occupational therapy to recover from an injury

were not aimed at a specific disease or condition, but on the whole person. When the whole person felt better, she was more resilient and able to recover from and resist a variety of conditions. Integrative care means choosing the therapy that best helps the patient achieve his or her goal. It can include *both* models of care, depending on the patients' goals and needs.

MODELS OF CARE

Conventional Care (Diagnose and Dispense)

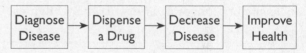

Holistic Care

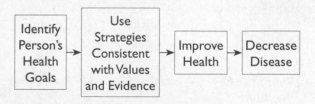

A person doesn't have to be sick to feel better after a massage, a loving dinner with family and friends, or a good night's sleep.

As I began writing *Mental Health Naturally,* I realized that there was a lot written on mental *disorders* like depression, but very little written on mental *health* and how to achieve health goals (as opposed to using antidepressant medications). So, I began asking my patients what their health goals were. This goal-oriented approach is most useful when considering an ongoing or chronic health situation.

Table 1.2 shows a few examples of common health goals.

An individual's health goals can change over time and with life circumstances. Goals set a positive direction. When we consider our health goals, rather than just our diagnoses, we often think more creatively and effectively about the strategies we can use to achieve those goals.

For example, if we just focus on a diagnosis like depression, the first remedy most physicians consider is an antidepressant medication (based on the "diagnose and dispense" model). But if we want to become more consistently cheerful, we may look at the literature on happiness and consider volunteering to help others; donating to a cause we believe in; caring for a pet or plant; changing our sleep, diet, or exercise habits; learning to practice mindfulness meditation; or a variety of other activities.

Focusing on goals rather than just diagnoses gives us many more options and strategies to achieve those goals.

STRATEGIES TO ACHIEVE HEALTH GOALS

When I began writing *The Holistic Pediatrician* back in 1994, I initially felt overwhelmed with all the therapies and strategies available to help cure, prevent, manage, and rehabilitate common childhood health problems. How could I remember all of them while keeping the focus on the patient and family?

At the time, I lived in Seattle, which has a skyline dominated by beautiful Mount Rainier. The image of the Therapeutic Mountain emerged to help me remember the four major categories of therapeutic strategies. I have now used the Therapeutic Mountain model for many years. It has held up well, and has been adopted by thousands of healers and researchers across the United States and around the world. It is simple, and it avoids polarizing treatments into "us vs. them" or "mainstream" and "alternative" medicine.

The mountain is an archetypal symbol of a high goal, achieved with dedication, preparation, persistence, and hard work. Such is the nature of healing. The goal is the well-being of

Table 1.2: Common Health Goals				
Physical	Emotional	Mental	Spiritual	Social
Comfort and ease	Calmness, confidence	Clarity, insight	Meaning and purpose	Supportive relationships
Vigor and vitality	Cheerfulness	Focus	Connection to something greater than oneself	Communication skills
Resistance to infections	Stability	Good memory	Compassion	Generosity
Strength and endurance	Resilience to stress	Problem-solving skills	Wisdom	Forgiveness
Flexibility	Peacefulness	Mindfulness	Gratitude	Harmony

the patient in the context of the family and community. Regardless of background, real healers are dedicated to this goal. They undergo extensive training and continue to learn from and listen to each patient, refining and enhancing their skills.

There are many sides to a mountain and many ways to reach the top. For the sake of simplicity, let's picture all of the primary healing modalities as occurring on one of the four sides of the Therapeutic Mountain (Figure 1.1).

THE THERAPEUTIC MOUNTAIN

1. South Side or Base: Healthy Habits in a Healthy Habitat (lifestyle therapies)
2. East Side: Biochemical Therapies (medications, vitamins, minerals, herbs, and other supplements)
3. West Side: Biomechanical Therapies (surgery, massage, bodywork, chiropractic, osteopathy)

4. North Side: Biofield or Bioenergetic (radiation therapy, acupuncture, Reiki, Therapeutic or Healing Touch, prayer, homeopathy)

Therapies on each side of the mountain are grouped together because of their functional similarities and because research on them uses similar strategies to test safety and effectiveness. Let's look at each side of the Therapeutic Mountain in more detail, starting with the south side, Healthy Habits in a Healthy Habitat).

SOUTH SIDE OF THE MOUNTAIN

The south side of the Therapeutic Mountain is the foundation of good health. We start here because a healthy lifestyle is the first and safest approach to improving health, regardless of specific health goals. The techniques on this side of the mountain are integral to healing traditions worldwide.

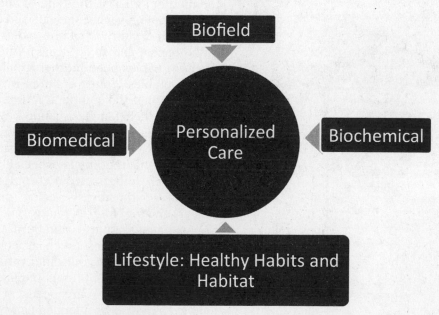

Figure 1.1: The Therapeutic Mountain

LIFESTYLE (HEALTHY HABITS IN A HEALTHY HABITAT) THERAPIES

These five factors are primarily regulated by the child and family with occasional professional advice. They are a part of daily life:

- Food: Nutrition
- Fitness: Exercise and Sleep
- Friendship with Self: Mindful Self-Care and Emotional Self-Regulation
- Fellowship and Meaning: Social Support and Connection
- Fields: Natural, Built, and Social Environments (Healthy Habitat)

As components of lifestyle, they are under our control, but paradoxically take more effort to change than using a medication. It's harder to transform a fast-food diet into a whole-foods regimen than to take vitamins. It's more challenging to develop and maintain an exercise program than to take high blood pressure

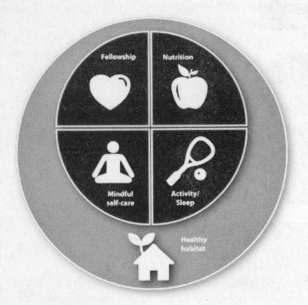

Figure 1.2: Five Elements of a Healthy Lifestyle

medicine. Despite the challenges involved in changing lifestyle and habits, parents who are interested in their child's lifelong health will focus on the fundamentals of a healthy lifestyle. Figure 1.2 is an image to help remember the five elements of healthy lifestyle.

The "glue" holding these five segments together is spirituality. For our purposes, spirituality does not refer to a specific religion or religious practice but to two ideas: first, a sense of connection to something or someone greater than one's individual self (God, Higher Self, Nature, Family, and/or Community); and, second, a sense of meaning and purpose in life (such as being loved and loving). Spirituality manifests itself in loving behaviors, generosity, kindness, wisdom, and serenity.

Food: Nutrition

"You are what you eat." Proper nutrition is the first fundamental of a healthy lifestyle (Table 1.3). Optimal nutrition for the child begins with the mother's diet at the time of conception. It includes not only what she eats and drinks, but, just as importantly, what she avoids (e.g., alcohol, tobacco, and toxic drugs). After delivery, newborn babies need mothers' milk, not fruits and vegetables. Young children increase their food vocabulary one food at a time and gradually adopt their families' eating habits. By the time children reach adolescence, the nutritional patterns for a lifetime have been established.

Nutritional therapy is especially important for children suffering from chronic conditions (such as cystic fibrosis, celiac disease, and cancer) and for children recovering from major trauma (such as injuries and burns). For everyday health problems (such as constipation), you can easily treat your child with diet: extra servings of fruit and bran muffins. Children with special health care needs or suspected food sensitivities or allergies can benefit from the help of a trained dietitian.

Table 1.3: Dr. Kemper's Nutrition Tips	
Do	Avoid
Drink pure, filtered water.	Carbonated or sweetened beverages
Eat at least 5–9 servings of fruits and veggies daily.	Fast food; junk food
Eat whole grains, seeds, nuts, and beans.	Processed foods and food or beverages containing high-fructose corn syrup
Eat foods rich in protein, such as beans, nuts, fish, low-fat dairy products, and/or sustainably raised, grass-fed meat.	Fried foods, foods with trans fats, and foods rich in saturated fats (whole milk, fatty meats). The wrong type of fatty foods can trigger inflammation and negatively change the balance of healthy bacteria that live in our intestines.
Eat foods that contain omega-three fatty acids or olive oil.	Food dyes and preservatives, pesticides, artificial sweeteners, and other chemical additives
Eat locally grown foods; consider a garden; visit a local farmer's market.	Food imported from developing countries; they have a higher rate of contamination than foods from developed countries.
Eat organic foods rich in essential micronutrients.	Genetically modified foods

Beware of non-credentialed, self-appointed diet gurus. Some Web sites, books, and "experts" are simply trying to sell products. Look for registered dietitians (R.D.s) who have advanced training in nutrition. Four of my favorite sources of nutritional advice are Drs. Walter Willett and David Ludwig from Harvard, Professor Marion Nestle from New York University, and writer Michael Pollan. All of them are wonderful writers and wise, effective advocates for sound nutrition. The advice from the Nutrition Action Newsletter from the Center for Science in the Public Interest is solid and unbiased. Check your sources and consult an expert before making radical changes in your child's diet.

Demonstrate your commitment to your child's healthy nutrition and the future health of the planet we share by planting an organic garden. Your child will be delighted to learn where real food comes from, have a better appreciation of the work involved in obtaining it, and perhaps develop a healthy hobby for the rest of his life! Advocate for community gardens. Shop at your local farmer's market and support restaurants that purchase from local growers. Fresh food is more nutritious than food that has spent days in shipping containers.

Fitness: Exercise and Sleep

A proper balance of exercise and rest is the second fundamental to maintaining health, and an important element of healing when we become ill. Exercise releases the body's own antidepressants and painkillers. It improves circulation, lung function, and brain functioning. Physical activity helps children deal with stress; maintain a healthy weight, strength, flexibility, coordination, balance, and endurance; and build self-esteem.

Whether it is organized soccer, a yoga class, or a backyard game, exercise is good for children. Other habits, such as excessive time in front of the television, promote a sedentary lifestyle and compete with more active pursuits. Developing a routine of regular physical activity during childhood is an antidote to the frightening national epidemics of obesity and diabetes. *Children should engage in at least an hour of physical activity daily.*

Most children suffering from a short-term illness such as a cold, flu, ear infection, or diarrhea will rest more and exercise less. Rest is as important as exercise in allowing the body's energy to be redirected to healing. Do not push your child to romp and play when he's acutely ill. Do encourage healthy sleep by setting a bedtime; engaging in a bedtime routine which includes reading before bed; having a dark, cool room; and using soothing music and aromas like lavender to help set the scene for peaceful slumber.

Friendship with Self: Emotional and Mental Self-Care

The third fundamental, friendship with self, encompasses a range of techniques from behavior management to self-hypnosis and professional psychological counseling. Even infants learn to calm and soothe themselves by sucking on their hands or a pacifier. Older children learn more complex techniques to calm themselves. Relaxation therapies have beneficial effects on conditions ranging from asthma to chronic diarrhea. They can also improve concentration for tests and sports.

The three most well-known techniques for self-regulation are the Relaxation Response, described by Harvard professor, Dr. Herbert Benson; Mindfulness Meditation, exemplified by another Massachusetts professor and author, Jon Kabat-Zinn; and Transcendental Meditation (TM), popularized in the 1960s by the Beatles. Meditation doesn't have to occur while sitting still; *yoga* and *tai chi* are popular forms of moving meditation. Around the world,

these meditation techniques are used to calm the mind and emotions of patients with high blood pressure, cancer, and other chronic illnesses. They can also help build focus, attention, clarity, and insight.

Despite its popular image as a stage technique, *hypnosis* is a safe, effective therapy used medically to help children. It has been endorsed by the American Psychiatric Association, the American Psychological Association, and the American Dental Association. It is widely used by pediatricians, pediatric psychologists, and behavioral therapists to assist children confronted by pain, chronic headaches, and behavioral problems such as bedwetting. Hypnosis is simply a state of focused, concentrated attention. Lamaze breathing to reduce the pain of childbirth is one type of hypnosis. These techniques can be as simple as focusing on breathing, becoming absorbed in a story or fantasy, or counting sheep.

Remember mood rings? By turning a deep blue when your hands were relaxed and warm, they gave you *biofeedback* that blood flow was increased to your fingers. Every moment, your child regulates his temperature, heart rate, breathing, and muscle tone. If he receives help in focusing on one of these processes and gets feedback about it, he can begin consciously to regulate it. For example, children can easily learn to increase and decrease the temperature in their fingers if they are given feedback by a machine in the form of sound or lights. This technique has been very helpful in reducing the frequency and severity of migraine headaches.

Biofeedback is generally taught by psychologists, but formal training and licensure are not required. Before taking your child to a therapist for biofeedback, check references from other parents whose children were treated. Also ask about fees and how much your insurance might be expected to cover.

Many types of therapists provide *coun-*

seling and *psychotherapy* to parents and their children. Treatment by counselors, support groups, behavioral therapists, psychologists, and psychiatrists can be helpful in addressing the thoughts and emotions that affect children's health. Like adults, children experience stress and negative emotions. Children whose parents are going through a divorce, for example, are prone to behavior problems, school and learning problems, sleep problems, stomachaches, and headaches. Children who suffer from chronic illnesses such as cancer or asthma or who depend on wheelchairs to get around can benefit from *support groups* of similarly affected peers. Such groups offer kids the opportunity to share strategies for managing their common challenges.

If you seek professional assistance because of a serious behavioral or psychological problem, please ask about the therapist's training in pediatrics and family issues; ask for and check references. Licensing requirements for psychologists vary from state to state.

Fellowship with Others: Social Support and Connection

You may have heard the saying, "If momma ain't happy, ain't nobody happy." This common sense wisdom reflects the importance of families (particularly mothers and other primary caregivers) on children's health. Parental depression, alcoholism, and drug use adversely affect children. Growing up in a family free of domestic violence, child abuse, sexual abuse, and neglect should be every child's birthright, but all too often, these toxic stressful conditions damage children and challenge their ability to grow up healthy. Children need a supportive family to thrive. If you did not have optimal role models of nurturing parenting in your own life, please consult with a family counselor, psychologist, or one of the many fine books on parenting to help you learn new

skills to help you give your child the kind of childhood you missed.

Fields: Environment

Children are extraordinarily sensitive to their physical and emotional environments. Unhealthy environments, burdened with pollution, noise, violence, poverty, sexism, and racism, are leading causes of illness in children. When trying to help heal a child, paying attention to the environment pays off. *Environmental activism is health care on a larger scale.* Specific examples of environmental therapies for children are:

- Air filters to remove airborne allergens
- Tepid baths to reduce itching
- Phototherapy (light therapy) to treat newborn jaundice
- Soothing music to promote restful sleep
- White noise to soothe a colicky baby
- Ice packs for a sprained ankle
- Mist tents for croup
- Bike trails and parks to promote healthy physical activity

Never underestimate the therapeutic value of the aroma of home cooking or the presence of the child's favorite blanket or stuffed animal.

Because poverty is the #1 risk factor for poor health, wise parents and pediatricians advocate for children to have equal access to high-quality health care and education; jobs for family members to provide for children's needs; family sick leave; maternity leave; elimination of racism, sexism, and poverty; and policies that promote pediatric safety in our homes, schools, and communities. Pediatricians are also on the forefront of advocating for gun safety and taking steps to minimize the effects of climate change.

EAST SIDE: BIOCHEMICAL THERAPIES

All of the therapies on the east side of the Therapeutic Mountain work through biochemistry. The three primary techniques of this side of the mountain are medications, herbs, and other nutritional or dietary supplements.

Let's start with the most familiar to modern families: medications.

Medications

When most of us think of medical therapy, the first thing that comes to mind is taking *medication*. Medications have very specific legal meanings that have implications for the standards by which they are manufactured and marketed. Today, most medicines are chemically synthesized, but originally many were derived from plants.

Penicillin is a good example. The most common medicine used to treat children with strep throats is penicillin. In 1928, a Scottish microbiologist, Alexander Fleming, noticed that a blue *Penicillium* mold growing on his laboratory cultures was killing the bacteria. Just as he was about to throw the ruined cultures away, he realized that the mold's deadly effect on the bacteria might be important. He was right. Penicillin, derived from Fleming's mold, has saved millions of lives; however, its overuse has led to an alarming increase in bacterial resistance, limiting the power of these medications to save lives. Medications are highly regulated in the United States by the Food and Drug Administration to ensure their safety, purity, uniformity, and potency.

Medications are lifesaving when it comes to acute, severe illnesses such as meningitis and pneumonia. They are also highly effective in managing certain chronic illnesses such as asthma and in curing previously fatal diseases such as childhood leukemia. However, medications are practically useless in curing many common childhood illnesses such as colds and coughs. Even used properly, medications have side effects. Antibiotics, for example, commonly cause stomachaches, diarrhea, and diaper rashes; for one in 10,000 children they cause a severe allergic reaction that can lead to death.

Though many medications can be purchased by parents over-the-counter, prescription medication is available only with a physician's order. M.D.s (Doctors of Medicine), D.O.s (Doctors of Osteopathy), and dentists are fully licensed to prescribe medications. Nurse practitioners have master's degrees, and they are licensed to prescribe many common medications. Physician's assistants (P.A.s) prescribe only under physician supervision. Over a dozen states license naturopathic doctors (N.D.s); most allow naturopaths to prescribe from a list of certain antibiotics and immunizations, but do not allow N.D.s to prescribe many other kinds of medication. A growing number of states allow pharmacists to prescribe and administer influenza vaccines. Other health professionals are not licensed to prescribe regulated medications.

Always ask about the risks, benefits, side effects, and alternatives to prescription medications. Many pharmacies carry drug information sheets that describe specific medications and their effects in detail. In this book, each chapter on a particular illness will give you information on non-prescription and prescription medications used in treating that particular condition.

Dietary Supplements

Vitamins and minerals are essential for maintaining health. A deficiency of any of the essential vitamins or minerals causes illness. British sailors recognized and effectively prevented scurvy by bringing along a supply of vitamin C–rich limes—hence, their nickname, "limeys." Normally, the body's need for vitamins and minerals can be met adequately with a healthy diet emphasizing fruits, vegetables, nuts, seeds, legumes, and whole grains.

Table 1.4 shows examples of good food sources of essential vitamins and minerals.

Nutritional supplements are also used to treat a variety of ailments. Vitamin B2 can help reduce migraine headaches. Children with cystic fibrosis require extra doses of vitamins A, D, E, and K. Magnesium given by vein can help children hospitalized with asthma. Children who require daily seizure medicine may need nutritional supplements to counteract some of the medication's effects. Supplements may also be useful for children who do not eat a balanced, healthy diet. It turns out that only about 1 percent of American children meet their required

Table 1.4: Essential Nutrients and Food Sources	
Nutrient	Food Sources
Essential fatty acids (EPA, DHA, and linolenic acid)	Fish (salmon, sardines, and mackerel), fish oil
	Flax seeds, flax oil, walnut oil; dark green leafy vegetables
	Animals that have eaten omega-3 rich diets such as eggs from chickens fed flax seed; grass-fed beef
Amino acids such as tryptophan and tyrosine	Soy, tofu, beans, lentils
	Seeds: sesame, pumpkin, sunflower
	Nuts, peanuts, peanut butter
	Milk, cheese, eggs; turkey, chicken, fish
Vitamins	
Vitamin B1 (thiamine)	Beans, lentils; nuts, seeds
	Whole grains; meat; dairy
	Oranges; yeast, brewer's yeast
Vitamin B3 (niacin)	Meat, fish, poultry, milk, eggs
	Green vegetables; whole grains
	Yeast, brewer's yeast
Vitamin B6 (pyridoxine)	Beans, nuts, legumes, soy
	Eggs, meats, fish, poultry
	Whole grains and fortified breads and cereals
Vitamin B9 (folate)	Dark green leafy vegetables, asparagus
	Beans and legumes; citrus fruits and juices
	Bran and whole grains, yeast; poultry, pork, shellfish, liver
Vitamin B12	Fish, shellfish, meat, dairy products; fortified foods
Choline	Egg yolks, soy, lecithin

(continued)

Table 1.4: Essential Nutrients and Food Sources *(continued)*

Vitamin C	All fruits and vegetables contain some vitamin C. Best sources include: green and red peppers, citrus fruits and juices, strawberries, tomatoes, broccoli, turnip greens and other leafy greens, sweet and white potatoes, cantaloupe, papaya, mango, watermelon, Brussels sprouts, cauliflower, cabbage, winter squash, raspberries, blueberries, cranberries, and pineapples.
Vitamin D	Fish, fish oils, oysters; fortified foods such as orange juice, cow milk, soy milk, and rice milk, and some cereals
Minerals	
Calcium	Milk, yogurt, buttermilk, cheese; calcium-fortified orange juice Green leafy vegetables (broccoli, collards, kale, mustard greens, turnip greens, bok choy, Chinese cabbage) Canned salmon and sardines; shellfish Almonds, Brazil nuts; dried beans
Chromium	Brewer's yeast; beef, chicken, oysters; eggs; wheat germ Green peppers, spinach; apples, bananas
Iodine	Iodized salt; seafood
Iron	Meats such as liver, fish, poultry, egg yolks Peas and beans; raisins; whole grain bread
Magnesium	Fish such as halibut Nuts: almonds, cashews, peanuts; beans: soybeans, black-eyed peas; lentils Leafy green vegetables, spinach, avocado; Dairy: yogurt, milk Whole grains, wheat bran, wheat germ, oat bran, brown rice
Selenium	Brazil nuts; shellfish, salmon, halibut Vegetables grown on soil containing selenium Whole grains, brown rice
Zinc	Beef, pork, and lamb; oysters; dark meat of poultry Peanuts, peanut butter, nuts, and legumes (beans)

dietary allowance (RDA) of nutrients through their diet. I typically recommend a good daily multivitamin/multimineral to just about everybody. Because brands and formulations change so often, you may want to check with the Center for Science in the Public Interest, *Consumer Reports on Health,* or ConsumerLab for reviews on commonly available supplements.

As with medicine, high doses of certain vitamins and minerals can have dangerous side effects. Although vitamin A is important for normal vision, too much can cause problems in the brain, liver, bones, and skin. Excessive vitamin C causes diarrhea. Iron overdoses can be fatal. Nature designed us to be in balance. When we take super-high doses of any one substance, we run the risk of upsetting the balance. That which has the power to cure also has the power to harm.

No particular license is required to recommend dietary supplements, and there are a lot of deceptive claims (pro and con) about the need for supplements. Unfortunately, there is little standardization in nutrition education and far too little research on optimal use of supplements in children. Medical doctors and osteopaths receive more training than in the past, but most rely on trained nutritionists (registered dietitians) to help children who have complex nutritional needs. Naturopaths and chiropractors also receive nutrition education, but it may not be specifically geared to children's needs. Be sure to ask your health care practitioner about his or her training in pediatric nutrition because it is a specialized field.

Herbs

Herbal, botanical, or *phyto-therapeutic medicines* have been used since ancient times to prevent and cure disease. Herbal medicines contain a complex mixture of chemicals. Some people feel that Nature combined the ingredients in plants for good reason, more wisely than a chemist distilling out a single active ingredient. For many herbal remedies, the active ingredient has not yet been identified, extracted, or synthesized. For example, chamomile tea (prescribed by Peter Rabbit's mother) is used around the world to soothe distressed babies and children. Its therapeutic effect does not depend on a single ingredient, but on a complex mixture of different chemicals that have not been duplicated in any lab.

Regardless of whether you take the chemically isolated, active ingredient (medication) or an herbal compound, herbs and medicines work basically the same way. That is, the active ingredients interact with particular molecules in the body or bacteria to achieve their beneficial effects. Medication is more highly purified and herbs contain more of the original natural ingredients, but their effectiveness is due to the same chemical principles.

Herbs tend to be both more subtle and more variable in their effects than medications. Because herbs are natural products, their potency and purity vary. In fact, as studies commissioned by *The New York Times, Los Angeles Times, Boston Globe,* and others have pointed out, the concentration of active ingredients can vary tenfold to a hundredfold in products made by the same manufacturer. This is scary for products such as ephedra, in which the margin between help and harm is narrow. Some herbal products contain none of the active ingredients listed in the label. This creates a strong potential for being ripped off when buying herbal products.

Mercury, arsenic, and lead can contaminate herbal remedies imported from developing countries. About one-third of Asian patent medicines (herbs compounded and sold in pill form) are intentionally "spiked" with medications such as steroids and antibiotics to enhance their effectiveness; the labels typically do not provide this information in English.

We cared for a boy at Boston Children's Hospital whose seizures were being treated with both mainstream medications and a Chinese patent medicine; unfortunately, he became sleepier and sleepier, and when he ended up in the intensive care unit, the toxicologist determined that the "herbs" actually contained very high concentrations of phenobarbital and bromides, potent anticonvulsants that nearly caused a fatal interaction with his other medicines. Fortunately, the mystery was solved, and he recovered; but it was a powerful lesson for all of us regarding the dangers of using Asian patent medicines.

Currently, herbal products are not regulated by the FDA the same way as medications, so standardization and quality control are quite variable. Safety is only addressed *after* a product is on the market and the FDA begins to receive reports of serious side effects. This lack of regulation doesn't matter very much if you're treating a stomachache with home-grown peppermint tea, but it could cause problems if you are using more potent herbs or use adult-sized doses for a child.

The use of adult-sized doses of Chinese herbal medicine, Jin Bu Huan, resulted in the near-fatal poisoning of three children in 1993, and since then there have been numerous reports of serious toxicity associated with using herbal products.

You cannot assume that a product is safe simply because it is natural. Nor can you assume that the government regulations provide consumer protection for herbs the same way as for medications.

No one has to have a degree in botany or biochemistry to prescribe herbs. Think of it this way. No one has to have a degree to tell you how much coffee or tea will help you wake up in the morning; you figure it out on your own based on trial and error. Herbs and supplements are regulated more like food than like medications. There are few studies on use of supplements in children compared with adults, so oftentimes we are left to use our best judgment and intuition. More research is needed to help us make better-informed decisions for children's health.

Herbs are not a panacea. Practitioners should be aware of when a child's illness is amenable to herbal therapies and when another type of therapy (medication, vitamins, massage, etc.) would be more effective. Beware of anyone who claims that herbs can cure anything that ails your child. That is no more holistic than claiming that medications cure everything.

I do not recommend you give your child an herbal remedy unless you or someone you trust can recognize the herb and has overseen its growth and preparation, or the product has been tested and verified independently.

WEST SIDE: BIOMECHANICAL THERAPIES

All of the therapies on the west side of the Therapeutic Mountain work biomechanically. These therapies affect tissues and organs by stimulating, realigning, moving, or removing them. They can also have system-wide effects by promoting the circulation of lymph and blood, triggering hormone release, and alleviating the pressure on pinched nerves.

BIOMECHANICAL THERAPIES

- Massage and Physical Therapy
- Spinal Manipulation (chiropractic and osteopathic adjustments)
- Surgery

Massage and Physical Therapy

Massage is an ancient healing technique used around the world. Parents practice informal massage when they encourage burps by rubbing or patting babies' backs. Formal bodywork techniques range from Swedish massage to Rolfing to deep tissue massage and physical therapy to rehabilitate muscles and joints after an injury. The various techniques all contribute to relaxation and well-being by stimulating blood flow, calming nervous impulses, and stretching and relaxing the tendons and ligaments that hold the bones and joints together.

Massaging one part of the body can also help draw attention away from another painful part of the body, reducing discomfort. The close personal interaction during massage also enhances the bond between parents and children.

Many massage therapists use oils or lotions to help reduce friction between the therapist's hands and the patient's body. Oil lubricates the skin and makes massage more comfortable. By adding aromatic oils (such as eucalyptus) to the base oil (usually some sort of vegetable oil), the massage can have even more benefit. In my family, our parents massaged Vick's Vapo-Rub™ or Mentholatum™ into our necks and chests when we had colds. The camphor and eucalyptus oil helped unclog our noses while the massage itself comforted and reassured us that we were loved and cared for, no matter how miserable we felt.

Massage oils can be stimulating (such as bergamot, grapefruit, geranium, and cardamom), calming (such as sandalwood, sage, chamomile, and rose), and even sedating (such as vanilla, hyacinth, lavender, jasmine, and valerian).

Some essential oils have natural antibacterial effects. Tea tree oil, for example, was used by Australian medics during World War II to prevent wound infections. Today it is a common ingredient in natural dandruff shampoos and acne remedies.

Massage therapists must be licensed, and licensing requirements vary by state. Many children's hospitals hire massage therapists to provide massage to help premature babies grow and develop and to comfort children hospitalized for cancer treatment.

If you learn to give a good massage for your child, the cost of care goes way down. Daily massage has proven helpful for low birth weight babies, depressed teenagers, kids with asthma, children with painful arthritis, and a variety of other conditions, too. I recommend massage for just about every patient I see.

Spinal Manipulation

Chiropractic therapy was invented by an Iowa grocer, Daniel Palmer, in 1895. Palmer believed that all ailments were due to misalignment (subluxation) of the spine, and that therefore, all ailments could be cured by manually realigning the spine. Today there are two main types of chiropractors—those who still believe that manipulating the spine can cure all ills, and those who make more limited claims about treating neck, back, joints, and injury-related pain.

Many chiropractors also recommend nutritional and herbal therapies as well as spinal manipulation. Chiropractic treatment is used by about 10 percent of American adults and 5 percent of children.

Scientific studies have documented the effectiveness of chiropractic treatment in helping adults with back and neck pain. However, there are fewer studies of the effectiveness of chiropractic manipulation in treating children. Children rarely experience back pain; when they do, serious diseases such as spinal infections and cancer must be considered. Reliance on repeated X-rays to document spinal changes puts children at risk from exposure to excessive radiation.

Most chiropractors are not specially trained to recognize or treat serious childhood illnesses, and they are not licensed to prescribe medications or to perform surgery. They may be very helpful if your child has a wrenched back or stiff neck from an injury, but be wary about claims to cure serious problems such as diabetes or kidney infections.

Chiropractors are licensed to practice in all 50 states. Most insurances cover chiropractic treatments.

Osteopathic manipulation was invented by Dr. Andrew Taylor Still in the late 1800s. Dr. Still believed that manipulating the spine and other joints would improve circulation and lead to more balanced functioning of the nervous system. Although Dr. Still rejected the use of drugs, his professional descendants have adopted a more holistic approach. Osteopathic physicians (DOs or Doctors of Osteopathy) are licensed in all 50 states and have the same prescriptive and practice privileges as medical doctors (MDs).

Surgery

Surgeons set fractures, remove brain tumors, perform skin grafts, and deliver babies. Modern techniques have made surgery far safer than it was 50 years ago. Surgeons undergo at least five years of training after completing medical (or osteopathic) school. Becoming board-certified in pediatric surgery requires at least two years of additional training. Before you agree to a non-emergency surgical procedure for you child, ask for a second opinion. Ask the surgeon about his or her credentials, alternatives to surgical treatment, and references to parents of children who have undergone the procedure recommended for your child.

NORTH SIDE: BIOFIELD OR BIOENERGETIC THERAPIES

All of the therapies on the north side of the mountain are based on the principle of an invisible, vital energy, information, or spirit that animates, flows through, and surrounds the body. The aim of all of these techniques is to restore a harmonious balance of energy or information, which, in turn, improves the functioning of molecules, cells, tissues, and organs.

BIOENERGETIC OR BIOFIELD THERAPIES

- Radiation Therapy, Diagnostic Ultrasound and Magnetic Therapies
- Acupuncture
- Reiki/Therapeutic Touch/QiGong/ Healing Touch
- Prayer and Ritual
- Homeopathy

Some people think that the therapies on the north side of the mountain are nonsense because they do not believe in invisible energies. However, radiation therapy and diagnostic use of ultrasounds and magnetic fields are considered conventional twenty-first century care.

Despite their enigmatic nature, the other biofield therapies have proven effective in diverse circumstances. Acupuncture, for example, is so effective in reducing pain that acupuncturists have become part of the medical teams caring for horses participating in the Olympics; and acupuncturists work in over one-third of academic pediatric pain treatment centers in North America. Prayer and pastoral care are offered in hospitals across America. Homeopathy reduces diarrhea among infants too young to understand the power of placebo. These therapies work—at least for some children, some of the time. We may not understand in western biomedical terms how they work or the best way to elicit their effects on a consistent basis (yet), but growing research supports their use.

Acupuncture

Acupuncture has been practiced in China for over 2,000 years. It is based on the theory that illness is caused by an imbalance in the body's flow of energy, called *qi* or *chi*. Treatment is aimed at restoring balance of the yin and yang (negative and positive aspects of *qi*) along the meridians or vessels that carry *qi* throughout the body. In classical acupuncture, the points along the meridians are stimulated by needles or heat (moxibustion), but they can also be stimulated with vigorous massage (shiatsu), tiny hammers, lasers, or electrical currents. *The Web That Has No Weaver*, by Harvard faculty member, Ted Kaptchuk, OMD, is my favorite book about acupuncture and Chinese medicine.

Like surgeons' knives, acupuncture needles actually penetrate the body. But, unlike surgery, in which healing is effected through rearranging tissues, in acupuncture treatment the needles' purpose is to restore proper energy flow.

Most of the scientific studies of acupuncture's effectiveness have been performed on adults, but more studies are being done in children as well. In 1997, the National Institutes of Health (NIH) expert panel on acupuncture concluded that acupuncture is effective for a number of painful conditions and for treating certain kinds of nausea. In a survey we performed of pediatric patients at Children's Hospital in Boston who had undergone acupuncture therapy for severe pain, only 1 in 50 reported it was uncomfortable and most actually thought it helped, even when other therapies didn't. Our New England patients had quickly discovered what the Chinese have known for thousands of years—acupuncture works! And it's very safe, even for children, when proper hygiene is observed and disposable needles are used to reduce the risk of infection—the same kind of precautions necessary when drawing blood or giving an injection.

Proper acupuncture training is extensive. The Chinese, Japanese, French, Americans, and Koreans all have somewhat different styles of acupuncture, but there are no studies suggesting substantial differences in their effects. Choose a therapist who has a title, Lic.Ac. (Licensed Acupuncturist) or Dipl.Ac. (a Diplomate of Acupuncture from the National Commission for the Certification of Acupuncturists). If your child's physician practices acupuncture, ask whether he or she belongs to the American Academy of Medical Acupuncture. Traditional Chinese doctors carry the title, O.M.D. (Oriental Medical Doctor), indicating additional training in herbal medicine. These initials DO NOT indicate that the practitioner has any training as an M.D. or in pediatrics. Insurance, reimbursement, and licensing requirements for acupuncture therapy vary from state to state.

Therapeutic Touch, Reiki, Healing Touch, Laying on of Hands

The practice of healing by transmitting energy from the therapist's hands to the patient's body has been around since the beginning of time. The formal technique, Therapeutic Touch (TT), was developed by Dr. Dolores Krieger, a professor at the New York University School of Nursing in conjunction with my teacher, Dora Kunz.

The TT technique develops the intuitive senses, enabling practitioners to sense blockages in the energy flow in and around the patient and to help correct the energy flow. Therapeutic Touch is akin to massage, but its practitioners do not need to actually touch the body; rather, they work just on the surface or in the energy fields surrounding it.

Laying on of hands, the Chinese technique known as QiGong, Reiki, and Healing Touch are all variations of this same theme.

Therapeutic Touch has undergone scientific study, and has proven useful in the treatment of pain, high blood pressure, anxiety, headache, and wound healing. Therapeutic Touch is practiced primarily by nurses. It has been taught and practiced in over 80 countries. Although formal training is available, certification and licensure are not required to practice Therapeutic Touch. Ask your prospective practitioner about references and experiences in working with children.

Many children's hospitals have nurses on staff who are trained in Therapeutic or Healing Touch or Reiki, and the hospitals have developed policies about who can provide these services and the circumstances in which they may be used. In the last twenty years, my colleagues, MaryJane Ott, ARNP, MN, and Larraine Bossi, MN (both from Dana Farber), Deborah Larrimore, RN, LMT, CHTI, and I have trained dozens of nurses, social workers, and physicians to provide Therapeutic and Healing

Touch. Since I received the Reiki Master initiation from Roy Bauer in October 2000, I have incorporated Reiki into my practice as well. In fact, the most frequent reasons I have been consulted in the hospital are to provide Reiki and Therapeutic Touch for patients in the intensive care unit, those recovering from major surgery or children with cystic fibrosis or cancer, as well as for outpatients with abdominal pain, headaches, or fatigue.

Prayer

Intercessory *prayer* asks the Unseen Power (God, Spirit, Allah, or Great Absolute) to heal the patient. Prayer forms an invisible connection between the patient and the Universal Energy or Spirit, thus restoring balance, wholeness, and health. Prayer and spiritual rituals are as old as humanity. Effective prayer does not require formal training or licensure. One does not need to practice a particular religion. Prayer has proven effective for numerous conditions in numerous studies. *Healing Words* by Dr. Larry Dossey is my favorite book summarizing the numerous scientific studies demonstrating that prayer works. Another wonderful book summarizing years of research and common sense is *The Healing Power of Faith* by Dr. Harold Koenig from Duke University. Although prayer does not *cure* in every case, sincere prayer often brings a sense of peace and genuine *healing* to both the person praying and the patient. I recommend prayer as a therapy for all kinds of conditions in children whose family beliefs include this spiritual tradition.

Homeopathy

Homeopathy was founded by a German physicist and chemist, Samuel Hahnemann, at the end of the eighteenth century. This was an era when the other main type of medical practice (called *allopathy* by Hahnemann) had little to offer patients except bleeding, sweating, vomiting, and purging. Given the allopathic alternatives at the time, homeopathy was definitely a safer treatment! In the intervening years, allopathic medicine has developed advanced diagnostic and surgical techniques, antibiotics, and effective treatments for cancer and other killer diseases. Although homeopathy is disparaged by most American physicians, it is widely used in India and Europe.

Homeopathy is based on two principles. The first principle is "like cures like." Hahnemann said that a substance that produces symptoms in a healthy person cures those same symptoms in a sick person. For example, *belladonna* causes flushing, fever, and a rapid heart rate. Thus, it would be the remedy of choice for a child with a sudden onset of fever, flushing, and a fast heartbeat. The theory behind this principle is that the body will tend to react to the remedy, fighting the symptoms it provokes, and thereby fighting the symptoms the body is experiencing. By paying careful attention to the whole range of the child's experience, the homeopathic practitioner matches the remedy that comes closest to causing those symptoms, and therefore cures them. Homeopathic practitioners are very interested in the entire constellation of symptoms the child experiences; they ask a lot of questions and don't rely much on laboratory tests.

The second principle has to do with the bioenergetic or spiritual aspects of homeopathic remedies: "The more the remedy is diluted, the more potent it is." Homeopathic remedies are diluted anywhere from 1:10 to 1 in billions. The theory states that the more diluted remedies are more potent than the more concentrated remedies. During each dilution, the compound is vigorously shaken or succussed, which is thought to increase its potency. Of course, this principle violates most concepts of chemistry and physics, which is why homeopathy is disparaged by most medical doctors. Nevertheless, several scientific studies have demonstrated homeopathy's effectiveness in treating hay fever, childhood diarrhea, and even ear infections.

Bach flower remedies, invented in the 1930s by Dr. Edward Bach, are an offshoot of homeopathy. Dr. Bach was a British homeopathic physician who used his intuition to select trees, flowers, and other plants used in diluted doses to heal the negative emotions underlying physical illness. Perhaps the most famous of the 38 remedies is Rescue Remedy™, a combination of diluted plant extracts: cherry, plum, clematis, impatiens, rock rose, and star of Bethlehem. Rescue Remedy™ is said to be useful in treating the shock resulting from a startling experience or acute injury. The remedies are available without prescription, and are sold at many health food stores and by mail order from British distributors. There are no scientific studies demonstrating the effectiveness of Bach flower remedies in treating children, but they are quite safe.

Legally, homeopathy can only be practiced by health professionals who are licensed to prescribe medications. In most states this means that homeopathic practice is legally limited to M.Ds, D.O.s, dentists, nurse practitioners, naturopaths, and chiropractors. Despite these regulations, many people without formal training prescribe homeopathic remedies, and remedies can be purchased through mail order catalogs and at many grocery stores. Although the remedies themselves are probably safe, the danger is delaying use of a more effective therapy.

How to Choose?

Now that we've looked at all the therapeutic options and strategies from the four sides of the mountain, how do you know which one to use?

Some healthy lifestyle strategies are nearly always helpful. There is no contraindication to a good night's sleep or a healthy diet! On the other hand, when considering more specific therapies, evidence-based medicine suggests that we start by identifying the specific goal or target of the therapy, and then weigh the effectiveness and safety of the remedy. Table 1.5 shows how I decide whether to recommend, tolerate, monitor, or avoid a particular therapy.

First, if a treatment has been proven both safe and effective for a particular condition or goal, I recommend it. There are some gray areas. What's the criterion for deciding a treatment has been proven effective or safe? The standard scientific criterion is that the therapy was statistically effective to a significant degree in two or more randomized controlled clinical trials. Sometimes it's just not possible to randomize patients. You can't ethically randomize babies, for example, to breast-feed or be given artificial formula to determine which is better for preventing eczema. Nor can you ethically protect a random group of children from tobacco smoke while randomizing other children to tobacco exposure to determine the impact on intelligence, weight gain, or asthma. So, for some therapies, we simply have to use different research designs and common sense.

Table 1.5: Weighing the Remedy		EFFECTIVE	
		Yes	No
SAFE	Yes	Use	Tolerate
	No	Monitor	Avoid

Second, if a treatment is proven effective, but it is costly or has serious side effects, we may go ahead and use it, but we carefully monitor the child to avoid or catch problems early. For example, chemotherapy has improved life expectancy for children with leukemia from being a death sentence to having a greater than 90 percent cure, but it has many side effects, and children are closely monitored to minimize their suffering.

Third, if a treatment is safe, but its effectiveness is unknown, not yet tested, not proved in large trials, or disproved, I tolerate its use, but I don't actively recommend it. For example, chicken soup is a common home remedy for the common cold. Yet, there are no large randomized controlled trials demonstrating that it reduces the duration of cold symptoms. So, I don't recommend a certain dose or brand of chicken soup to treat your child's cold. However, since chicken soup is safe (for everyone except the chicken!), inexpensive, consistent with many families' cultures and traditions, and is a way of showing care and supporting social connections, I support or tolerate its use. Many complementary therapies for common childhood conditions fall into this category. They haven't undergone sufficient study to actively recommend them, but they are often safe, inexpensive, consistent with family values, and promote other health goals like connection and support, so I support or tolerate their use.

The only time I strongly urge families to avoid therapies is when they are both ineffective and unsafe or costly, *or* if they delay or interfere with other treatments that are effective in treating a serious condition. For example, I would avoid recommending surgical treatments for the common cold! Surgery is good for a lot of things, but treating colds isn't one of them. And I'd avoid using shark cartilage instead of chemotherapy for childhood leukemia.

SEEKING INTEGRATIVE CARE

You need to be able to *trust* your child's physician and other health professionals on the team. Get a recommendation from someone you trust. Even if someone is highly trained and competent, you may not "click" or be on the same communication wavelength. Ask questions. Don't be afraid to let a physician know that you are interviewing a number of physicians to find the best one for your family. For more information in contacting a holistic physician, see the Academy of Integrative Health and Medicine or graduates from an integrative residency or fellowship training program.

SYSTEMS OF HEALTH CARE

In addition to biomedical care and the growing field of integrative health care, there are several traditional systems of care. Let's review those now.

Naturopathy as a separate healing tradition was born in Germany in the early twentieth century. Early naturopaths emphasized the importance of mineral baths, steam baths, and fasts. Modern naturopathic physicians (N.D.s) complete a four-year curriculum covering physiology, anatomy, herbal and homeopathic remedies, nutrition, exercise and environment, massage, and spinal manipulation. Not all naturopaths use all of these therapies. Fewer than 20 states currently license naturopaths.

Traditional Chinese Medicine includes not only acupuncture and herbal therapies, but also diet, exercise (such as Tai Chi), meditation, massage, spinal manipulation, and QiGong (transfer of healing energy from the practitioner to the patient).

Ayurvedic medicine is an ancient system of medicine from India that includes herbs, proper nutrition, meditation, environmental changes (aromatherapy and sound therapy), and massage.

PAYING FOR HOLISTIC HEALTH CARE

The systems for financing health care are changing rapidly. Many more therapies are covered by insurance today than they were 20 years ago, but the situation is far from optimal. This means you could end up paying "out of pocket" for the therapy of your choice. Be sure to talk about payment and insurance coverage up front. Be sure to ask not only about professional fees, but also about the cost and coverage for treatments and devices that the provider might recommend (such as vitamins, herbs, psychotherapy, or biofeedback devices). It's also a good idea to check your insurance policy and call about coverage for specific services or treatments before you start a course of therapy for your child. Be sure to get the agent's statement about coverage in writing in case there are any questions later. Know that you can advocate for coverage for the services that benefit you and your child. Not much progress has been made without a few squeaky wheels.

GENERAL ADVICE

Practitioners work best when they know the whole story about your child. If you seek care from more than one practitioner, let everyone know what kind of care your child is receiving and from whom. Ask the different practitioners if they are willing to cooperate with each other and send each other copies of their records, recommendations, and contact information.

You can contact your state department of licensing regarding regulations for the various kinds of health professionals in your state. Ask questions. Ask for references.

Doubt claims about panaceas. If something sounds too good to be true, it probably is. How do you know what to believe? That's what Chapter 2 is all about.

SUMMARY

1. Consider your health goals as well as your child's official diagnoses before you start thinking about which treatment to use. The "diagnose and dispense" model is quick and is useful in emergencies, but may over-focus too early on medications rather than considering the entire range of therapies.
2. Start with healthy habits in a healthy habitat. Do not try to replace a healthy diet with an over-reliance on supplements or herbs. Don't rely on massage or acupuncture to replace a healthy sleep environment or adequate exercise. Advocate for public policies that promote pediatric health.
3. Consider the whole range of available therapies. Think about the four types of therapies on the Therapeutic Mountain. No one technique—no matter how organic or high tech, ancient or modern—is a panacea. Do you know the story of the carpenter whose only tool was a hammer? When he used his hammer to replace a light, the bulb shattered. Some problems are best treated with therapies from one side of the mountain and some respond better to others. Ask your therapist or physician about their experience in treating children and what your insurance will cover.
4. Weigh the benefits (effectiveness) and risks before you decide which therapies to use, monitor, or avoid.
5. Choose your professional health care team wisely. Advocate with your insurance carrier to cover the range of effective, safe integrative therapies.

2

"TRUST ME, I'M A DOCTOR"

"Doctor Kemper, I'm so glad you're here. I really wanted to talk with a doctor about natural therapies for cancer. I read somewhere that there was this boy in California who had cancer, and he was cured with cat's claw. My cousin's son was just diagnosed with cancer. Do you think he should take cat's claw, too?" *So began a conversation with my friend, Toni Lozano.*

When a child is ill, we all want to help. In general, we prefer remedies that are natural, safe, good for the environment, and in harmony with our beliefs about ourselves and the world. Stories about dramatic cures are inspiring, and they give us hope that we, too, can overcome painful challenges.

"I know that I can't trust everything I read," Toni continued. *"There are so many claims out there about certain foods or whatever being bad for you and other foods or vitamins being good for you. It's everywhere—magazines, billboards, TV, radio talk shows, the Internet, friends, and neighbors. And they're all saying different things. Sometimes I feel overwhelmed. How do I know* *what really works and what is safe for my family?"*

Several years ago, a colleague gave me an aphorism I adore:

"In God we trust; everyone else must have data."

This is the era of the skeptic. Many have lost faith in traditional authority figures. Before trusting their child to a surgeon's knife, many well-informed parents ask for a list of potential alternative treatments, possible side effects, and a second opinion. When it comes to your child's health, skepticism is your right as a prudent parent.

"In God We Trust . . ."

Most of us do have faith in something or someone. Many have complete faith in their own doctor, though they distrust others. Some believe in the power of prayer. If you have complete faith in something, you don't need any evidence to know what to do. You follow your faith. Faith alone can be one of the most healing powers on earth. I have seen miraculous healings based on faith.

More and more hospitals encourage spiritual healers to visit patients and perform healing rituals. On my very first day at Children's Hospital in Boston back in 1998, I was asked by a dying teenager and his family to find a spiritual healer. We found a wonderful shaman who visited the boy, performed a ceremony in his hospital room, and brought great comfort and peace to everyone involved. Numerous scientific studies have demonstrated the healing power of prayer. Prayer is effective whether the seeker is Christian, Jewish, Muslim, Buddhist or Hindu—the words and the form don't matter. It is the faith and the intent that count.

On the other hand, I have also witnessed the suffering of children whose parents slavishly adhered to religious doctrines—a child who died of meningitis because the parents saw the illness as a test of their faith rather than seeking "worldly" antibiotics. Another child suffered for days from a low blood count following a serious injury because the parents refused a transfusion. Children have been bruised and battered, permanently scarred, and brain damaged because parents felt they needed to "beat the devil" out of them.

Our cultural beliefs, our sense of who we are, our connectedness to family and community, and the meaning of health and illness are vital to the healing process. In many Eastern and Native American cultures, illness is seen as a lack of balance within a person or between a person and their environment. Western culture looks at disease in terms of cause and effect (e.g., germs cause illness). It is impossible for science to prove whether Eastern or Western philosophies are more "true." But science can compare the effectiveness of different kinds of therapies in healing and curing children with different kinds of disease.

Toni believed in natural remedies, but she wasn't sure if an herb was powerful enough for a disease like cancer. She wondered if natural methods are always better.

Given equal effectiveness, cost, and safety, most people would choose natural therapies over artificial ones. But keep in mind that many modern conveniences such as central heating, hot running water, and jet airplanes are not natural. The lowest infant mortality rates in the world are found in the technologically advanced nations. There are some lifesaving advantages to modern medicine. On the other hand, many times physicians and families take the easy way out and choose a drug when a change in nutrition, exercise, or environment would do more good in the long run. We all have to balance our values, priorities, and resources when choosing therapies.

The Old Paradigm: Theory, Traditon, and Personal Experience

Prior to this century, medical practice was based on tradition and theory. Unfortunately, practices based on *theory* alone, while they may be well intentioned, can be disastrous for the patient. For example, the practice of bloodletting using leeches was based on the reasonable-sounding theory that it would rid the body of evil humors. It is now believed that excessive bloodletting is what ultimately killed George Washington. Many therapies that sound good in theory turn out to be at best ineffective and at worst deadly in actual practice.

Despite spectacular advances in medicine, much of health care continues to be based on *tradition*. Health care practitioners, like all human beings, tend to do things the way we were taught. But traditions change. My early professors taught me to follow up children diagnosed with an ear infection two weeks after their diagnosis to ensure that the fluid had cleared; by the time I worked at Yale, the tradition had changed to three weeks. We now know that only 50 percent of children clear that fluid within four weeks of a diagnosis, and many pediatricians do not recommend ear recheck visits at all unless the child is still having a problem. Is it wrong to recommend things based on tradition and experience? Not necessarily. Some traditions are simple common sense, and experience can be our greatest teacher.

On the other hand, some traditions based on short-term successes turn out to be hazardous or costly. Though it is a traditional herbal remedy for coughs, coltsfoot has recently been shown to be a cancer-causing liver toxin, and it is no longer regarded as safe. Even 20 years after scientific evaluations indicated that aggressively treating mildly jaundiced babies was probably unnecessary, some physicians continued to order phototherapy. Why? In part because of tradition, and in part because we doctors fear that if we break with tradition and anything bad happens, the lawyers would have a field day, and we'd never forgive ourselves. Stepping outside of our traditions can be scary.

Our own *experience* shapes our attitudes and recommendations. A prominent pediatrician told me the story of his teenage patient who had very smelly feet. The pediatrician had counseled him extensively about hygiene to no avail. One day the boy showed up for a routine physical exam, and the physician braced himself for the overwhelming aroma before entering the exam room. To his surprise, the foot odor was completely gone. He began to congratulate the boy on his hygiene, but the boy told him it had nothing to do with washing his socks. "Zinc," he said. "I started taking some of my mom's zinc supplements and the smell went away." The pediatrician was dubious. The boy said he, too, thought it was just a fluke, so he stopped the zinc. Within a few days the odor returned, so he resumed the zinc. On the basis of this experience, this physician started recommending zinc supplements to his patients with smelly feet. While zinc is probably safe in low doses, there are no studies demonstrating its effectiveness in treating foot odor. It may not affect other boys the same way. The boy may have adopted other healthy habits such as a good diet or exercise changes at the same time. Even impressive experiences such as that of the boy with smelly feet are not necessarily a solid scientific foundation for making treatment suggestions to others.

If we can't trust authorities, theories, tradition, and our own experience, what can we trust?

The New Paradigm: Evidence-Based Medicine

If you're like most people, you care less about *theory and tradition* than about how well a treatment actually works. *Evidence-based medicine* means that you look at the scientific evidence that a treatment is effective rather than relying on tradition, theory, or anecdotal experience when making a medical decision. Evidence-based medicine relies on several *levels of evidence* to evaluate the effectiveness (and safety) of therapies. Let's look at these different levels of evidence.

The Lowest Form of Evidence—The Anecdote or Case Report

"Child with fatal cancer, cured with coffee enema!"

"Ancient Amazon ointment reverses ravages of acne!"

These stories usually begin, "I had a patient (friend, cousin, etc.) once who . . ." This is the kind of claim that caught Toni's eye and impressed my colleague with the power of zinc. The more exotic, mysterious, instant, miraculous, secret, or organic the treatment, the greater the appeal. Those who market "the cure" may cite impressive and rational-sounding philosophical theories to support their claims.

These stories are called *anecdotes* in the popular press and *case reports* in scientific journals. They usually take the form of a hopelessly ill child who is cured with a wondrous ancient/new, secret therapy. Anecdotes grab our attention, pique our curiosity, and give us hope. In the typical anecdote there is insufficient information to tell if the child in the story is similar to your child or if the illness is really like your child's illness. It is also difficult to tell what other therapies the child may have been receiving and exactly what is in the miracle cure.

Be very wary of people marketing miracle cures by phone, Internet, or direct mail. Many of these "cures" are available only with a substantial outlay of cash up front. Or they promise you untold, easy riches if you, too, join the marketing pyramid. Although case reports may be intriguing, they do NOT provide a sufficient basis for recommending a therapy.

The Case Series—A Collection of Anecdotes

"We treated 100 feverish children with Brand X, and they all felt better within hours."

"Colicky babies, treated with chiropractic, dramatically improve within six weeks."

"I have used Vitamin C with my patients for years, and they all have done very well."

A *case series* is simply a collection of anecdotes. Although the numbers make the claim sound scientific, a case series is also weak evidence that a treatment works. Consider the first claim: Children treated with Brand X felt better within hours. Why did they have a fever? What other therapies were they given? Would they have gotten better regardless of what kind of therapy they received or if they didn't receive any therapy? Case series provide promising leads for scientists to test. But a collection of hopeful anecdotes is *not* sufficient grounds for saying that a therapy has proven effectiveness. Colicky babies treated with chiropractic do improve over six weeks, but so do colicky babies who never visit a chiropractor!

Variations on the Theme of Case Series: The Twenty-First Doctor

"I took my child to twenty doctors, and none of them could figure out what was the matter with him. Nothing they recommended worked, and they pronounced him incurable. Just when I was about to give up, I took him to Doctor Jones. He put my baby on these food supplements, and now he's like a different child. He's active and full of life. I'm sure glad we took him to Dr. Jones. Maybe you should take your son to him, too. Or maybe you just want to try these supplements. . . ."

This scenario is the prototype for the "Twenty-First Doctor Syndrome." It is really just a variation on the anecdote or case report. The story starts off with a difficult illness, resistant to the ministrations of numerous healers. The parent is frustrated and about to give up, but tries one last doctor or therapy. The story is usually told by someone who is genuinely convinced that the final treatment caused the cure.

In some cases, it does take one unique healer or one unique treatment to cure a child. However, in many more cases, the marvelous cure is the result of the parents' own strong hope and faith or the illness having finally run its course (such as infant who outgrows colic). Was the child about to get better anyway? Perhaps. Does this make the cure less remarkable or less true? Of course not. These stories remind us that the human mind and spirit are among the most powerful healing tools available. But bear in mind that the visible trigger (the doctor or the diet) for this particular cure may not be the key to healing for your child.

"Well, this is all very disappointing," Toni complained. *"If I can't trust friends' stories, and if I can't trust doctor's theories, how do I know who to trust? My husband says that all this holistic stuff is nothing but a placebo anyway. Just what is a placebo?"*

The Placebo Effect—A Psycho-Neuro-Immunologic Modulator?

Anecdotes about miraculous cures, stories about spontaneous remission, and the Twenty-First Doctor syndrome all raise the issue of the *placebo effect*. A placebo is an inert or inactive substance, such as water or sugar pills. The word placebo comes from the Latin *placere,* "to please." Placebos were originally devised to satisfy demanding patients who did not need or would not benefit from real medicine. A placebo can be anything—a word, pill, diet, exercise, or operation. For example, a patient takes the inert sugar pill, believing it will help him, and he is healed.

You might think that placebos work only for the weak minded and suggestible, but you would be wrong. The placebo effect occurs in about 20–30 percent of *all* patients with every imaginable disease. It benefits the gifted as well as the gullible. It can reverse real disease as well as overcome imaginary symptoms. It illustrates the amazing capacity of the power of the mind to bring about healing.

Recently I have begun telling my medical students and residents that we should replace the word placebo with the phrase *Psycho-Neuro-Immunologic Modulator* (PNIM). Why should we use such a cumbersome phrase? Because many people associate the word "placebo" with something that is worthless. But placebos are far from inactive. After many years of research, we know that there are direct connections between the mind (Psycho-), the nervous system (Neuro-), and the immune system (Immuno-), which in turn affect every other system in the body. In themselves placebos may be nothing, but by modulating psycho-neuro-immunologic interactions, they release a powerful healing force that can effectively cure real diseases.

Although placebos are effective in helping some children sometimes, they do not always cure everyone. Another active ingredient in addition to the PNIM is sometimes necessary to kill the bacteria or restore balance. The difficulty in scientific research is in figuring out how much better the proposed therapy is than a placebo.

"Well," Toni said, *"as good as placebos may be, I want something better for my family. What kinds of studies prove a treatment is better than a placebo?"*

Comparative Studies

"American children who are not immunized have the same rate of polio as those who are immunized."

"Children who are breast-fed are smarter than those who are fed formula."

These kinds of claims compare one group of patients to another. Such studies look very scientific and they are a big improvement over case studies, but they can also be misleading. What would you think of the first example if I told you that the rate of polio was nearly zero for both groups? The risk of polio for any child in the United States is nearly zero because of the success of the polio vaccine. What about breast-feeding and IQ? IQ scores are generally a few points higher among infants who are breast-fed than formula-fed infants. However, American babies who are breast-fed generally also have parents with higher IQs and better education. Is the higher IQ due to the breast milk, the intelligence genes of their parents, or the environment?

Sometimes this type of comparative study is the best that can be done in the situation. A scientist can't tell 100 new mothers: "Now you 50 are going to breast-feed, and you other 50 are going to feed your baby nothing but formula. Then we'll measure your babies and see how smart they are." You just can't make people have certain health habits (what they eat or how much they exercise or whether or not they smoke or meditate or pray). It's likely that the people who practice one healthy habit are healthier in other ways too. All of this makes it challenging to figure out exactly what's making children healthy (or sick). Many factors must be taken into account before you can conclude that the one thing you're interested in is the thing that matters.

"Well," Toni continued, "what do you think about using remedies that have been proven to work in adults?"

Children Are Not Sophisticated Rodents or Small Adults

"Rats who are fed only 80 percent of their needed calories live twice as long."

"Essential oil kills bacteria in test tubes."

"Cold medicines reduce symptoms in adults."

What's good for other animals isn't necessarily what's best for human beings. Calves thrive on cow's milk while human infants do best with their own mother's milk. What happens in isolation in a test tube bears little resemblance to what happens inside a complex human being. Therapies that work in adults don't necessarily work in babies and young children. Cold medicines do reduce adults' symptoms, but are useless in infants. Unless a therapy has been proven useful in humans, I don't recommend it. If it has been proven useful in adults, it's worth testing in children, and may be worth trying if it is safe.

"Aren't there any studies I can really trust?" Toni pleaded.

The Gold Standard: The Randomized, Controlled, Double-Blind Clinical Trial

Although all studies have shortcomings, the gold standard of scientific evidence is the randomized, controlled, double-blind clinical trial. Randomization means that each child in the study has a 50–50 chance of getting the study treatment or a comparison. The treatment decision is determined by the luck of the draw. That way the group of children who receive the experimental treatment are nearly certain to be similar to the group of children who receive the comparison treatment. This is important because without randomization, certain families would be more likely to believe in and choose one treatment over another. The differences between the families who choose different treatments might very well outweigh the differences in the therapies themselves.

Good studies have a control or comparison group. If you give aspirin to children with fever, the fever will go down. But the fever would go down eventually anyway. A comparison group is needed to see if the fever goes down faster with aspirin than without it. (Yes, aspirin is an effective fever fighter.) The strongest evidence for an effective therapy is one that has been tested in children (not adults, animals, or test tubes), randomizing them into active treatment and placebo groups and assessing the results blinded to treatment assignment.

Double-blinding is when neither researcher nor parent nor child knows whether the child received the study treatment or the comparison treatment until after the study is over. If the child and family don't know what treatment is being given, but the researcher does know, that's *single-blinding*. In a study of the effectiveness of acupuncture, the patient and parent may not know whether the needles are being placed in real points, but the acupuncturist must know whether the needle is going in a real point or not. If the family knew the child was getting the "better" study treatment, their expectation about its benefit would modify the psycho-neuro-immunologic response. This would give the new treatment an unfair advantage or a bias toward appearing to be more effective. Keeping everybody in the dark until the study is over helps minimize the bias that might arise if everyone knew which treatment the child was getting.

"Well that randomized, placebo-controlled double-blind trial sounds just about perfect—that is, if it was easier to pronounce!"

It does sound perfect, but remember, if something sounds too good to be true, it probably is. Randomized controlled trials aren't perfect either. Because they are so costly and difficult to perform, they are usually limited to just one patient group (in terms of age, race, gender, geographic area, or stage of illness), which may not be all that similar to your particular child. Also, the folks who enroll in studies are a pretty special breed. How desperate

do you have to be to agree to be randomized and perhaps endure extra doctor visits, blood draws, questions, phone calls, and so on, especially when there's a chance you aren't going to get the "new, improved" treatment? The people who actually complete the study are only a subset of those who start, and your family might be more like those who stop participating than those who stick it out to the bitter end. And that's just the logistics of the study itself. We haven't even begun to talk about who funds or publishes these studies (which determines which therapies and conditions get studied and heard about) or about the possibilities of the patients somehow learning what treatment they receive.

Summary

"Whew! Let me see if I've got this straight," Toni said. *"The lowest level of evidence is the case report and the case series. Stories about eventual success with a certain doctor or therapy are really just the same thing. Studies of success in adult patients may not apply to children. Studies in children are good if there is a comparison group to make sure the therapy is better than a placebo. And the strongest evidence is from a randomized, double-blind, controlled trial in children. But even those aren't perfect and have to be taken with a grain of salt, right?"*

Exactly. If the only evidence for a therapy is based on tradition, a case series or case report, I consider it an interesting possibility, but unproved. If the evidence is from a comparative trial in adults or children, I'll tell you it's worth trying, but not recommended until better studies are done in children. The therapies I'm most likely to recommend in this book are those that have been tested with two or more controlled clinical trials in children. I also like to see them used for at least a few years so we know something about the safety and long-term effects of the therapy.

Do people combine faith with evidence? Of course we do. If you generally believe in herbal medicine, you may just hear one neighbor's story about how *Echinacea* conquered her daughter's cold before you try it for your own child. On the other hand, if you're skeptical about stimulant medications, you may doubt your doctor even when she cites randomized controlled trials showing they offer short-term benefits for kids with attention deficit hyperactivity disorder (ADHD). In all cases, you will end up judging my recommendations by your own faith, values, and experiences. We all have our own biases. Be open-minded and remember the levels of evidence when you hear or read about possible therapies for your child.

Ask yourself:

- Have clinical studies in children demonstrated that the therapy is more effective than a placebo treatment?
- How will the therapy affect my family's lifestyle and pocketbook?
- What are the risks and alternatives?
- Who is recommending the therapy, what are their qualifications, and do they have a vested interest in my choice?

"New truths commonly begin as heresies, but all too often end as superstitions."
—T. Huxley, 1885

3
ACNE

Gerald Taylor came to see me last spring about his acne. He was an industrious 17-year-old who was active in the Drama Club at school, and he worked part-time at a fast food restaurant to earn money for college. Gerald was upset about his skin; the senior prom was coming up in two months, and he wanted his skin to be cleared by then. I asked him what he'd tried so far. "Oh, the usual stuff," he replied. "I tried some cream from the drug store, but it didn't work and it just made my face red. My mom says I should stop eating french fries, but I like them and I get them for free at work. My friend, Marco, got this prescription from his doctor for some kind of antibiotic, and his face looks real good. I'd really like something natural, but if that doesn't work, I guess I'm ready for those antibiotics." Before writing a prescription to clear up Gerald's face, I wanted to clear up some of his ideas about acne, where it comes from, what causes it, and what he needed to do to make it better.

If you suffer from acne, like Gerald, you are not alone. At some point in their teens, 70–90 percent of people are affected by acne. Many worry that their acne is the first thing other people see about them. About 10 percent of adults continue to deal with acne into their thirties and forties. Most people treat their acne on their own with changes in diet, hygiene, and over-the-counter creams, ointments, and other remedies; only about a third ask their doctor for a prescription medicine.

Beauty may be only skin deep, but most of the acne action occurs deep beneath the skin surface in the tiny hair shafts of the face, the chest, and the back. The principal players in the acne story are *sebum, hormones, bacteria,* and the *immune system.* Sebum is a complex oily lubricant made in glands lining hair shafts. A year or two before other signs of adolescence are visible, increasing levels of *hormones* (such as testosterone and progesterone) stimulate sebum production. *Stress* can also stimulate sebum production. Excessive or blocked sebum forms a plug below the skin surface. Dead skin cells mix with sebum and worsen plugging. Pores plugged with sebum and dead cells create the perfect breeding grounds for *bacteria.* The bacteria break sebum down into irritating *fatty acids* and attract *white blood cells,* starting a cascade of inflammation—redness, swelling, heat, and pain.

As the plug of sebum, cells, bacteria, and fatty acids reach the skin surface, the complex appears as a *blackhead.* Contrary to popular belief, the black color does NOT come from dirt on the skin, but from pigment in the dead skin cells. Albinos (people who lack skin pigment) can get acne, but they do not get blackheads. Although blackheads are unsightly, they are eventually extruded without causing redness or irritation.

If a plug gets so big that it blocks its own path to the surface, it makes trouble. The expanding plug eventually bursts the walls of the hair shaft, spilling irritating fatty acids into the surrounding skin tissue. Immune cells rush to the site, releasing even more irritating chemicals. On the surface, this process initially appears as a whitehead (plugged hair shaft), eventually becoming a raised, red pimple, and then a pustule (a pimple that has come to a head), or, in the worst cases, a scarring nodule or cyst. It takes four to eight weeks for the plugs beneath the skin surface to become visible as acne. This is why *there are no overnight acne cures*; it takes several weeks for changes beneath the surface to become visible at the skin surface.

ACNE AGGRAVATORS

- More sebum production (higher androgen and progesterone hormones, stress, obesity)
- Higher hormone levels (puberty, males)
- Higher fatty acid levels (diet and stress)
- Blocking hair shafts with *cosmetics,* hair, oil in the air, or dirt
- Irritating the skin, which recruits more immune cells (picking and pushing)
- Aggravated immune system
- Higher bacteria levels
- Certain medications (iodine, isoniazid, lithium, phenytoin, prednisone)

Boys generally have more severe acne because they have more *testosterone* hormones. Many women experience acne flares just before menstrual periods when *progesterone* hormones are highest. Athletes who bulk up on *androgenic steroids* also tend to get worse acne, among many other more severe side effects. *Stress* from lack of sleep, major exams, family or romantic conflicts, a poor diet, sudden changes in schedule, or emotional tension also increases sebum production and free fatty acids, exacerbating acne. *Diet* is discussed in more detail later. Acne-causing medications include the seizure medicine, *phenytoin (Dilantin)*; a drug for manic-depression, *lithium*; an anti-tuberculosis drug, *isoniazid*; glucocorticoid medications like *prednisone*; and *iodine.* *Oily air* can block pores, contributing to acne. Fast food restaurants (filled with deep fat fryers), refineries, garages, and some chemical plants have high oil levels in the air. *Air-borne oil* settles on the skin and blocks pores.

Gerald's job may be contributing to his acne, because of the oily air in the restaurant. If he con-

tinued to work there, he would need to wash his face after every shift to help minimize blocked pores.

Common "treatments" can actually worsen acne. Despite the popular idea that *sun* dries up acne, increased *sweating* from summertime heat and humidity can also aggravate rather than improve acne.[1] Overzealous facial scrubbing, frequent rubbing, picking, and pinching increase the breakdown of sebum into irritating fatty acids and provoke the immune system. *Friction* from clothing, football pads, hair, headbands, and resting the face on the hand can also aggravate acne. *Oily cosmetic cover-ups* block pores, preventing the normal outflow of sebum and dead skin cells. Use water-based rather than oil-based cosmetics. Remove all make-up before bed. Read labels. Avoid skin lotions or cosmetics containing acne-aggravating ingredients.

COSMETIC INGREDIENTS TO AVOID

- Acetylated, PEG 15, PEG 75, or plain *lanolin*
- Butyl stearate
- Cocoa butter or coconut butter
- Glyceryl-3-diisostearate
- Hydrogenated vegetable oil
- Isopropyl anything
- Lauric acid or laureth 4
- Monostearate
- Myristyl myristate
- Oleyl alcohol
- Stearath 10

WHAT CAN YOU DO TO PREVENT OR TREAT ACNE?

You don't have to suffer with acne. There are numerous, proven effective treatments. Start with a healthy lifestyle, and then try safe, nontoxic, non-prescription remedies. If you need professional help, your physician or nurse practitioner can help with 95 percent of all acne. Refractory cases usually end up at the dermatologist's office. There are no overnight cures. Most treatments take *at least eight weeks to show an effect* because it takes that long for all the old lesions to heal and the new healthy skin to get to the surface. Let's consider the whole range of therapies to learn what works and what doesn't. We'll start with the foundation of all good health—healthy habits in a healthy habitat, that is, a healthy lifestyle. If you want to skip to my bottom-line recommendations, flip to the end of this chapter.

LIFESTYLE THERAPIES: NUTRITION, EXERCISE, MIND-BODY, ENVIRONMENT

Nutrition

Does a bad diet cause acne? There are an increasing number of studies confirming the common sense that what we eat affects our skin. Although a frequently cited study from 1969 showed that eating *chocolate* didn't make acne worse, the comparison group in that study was also eating a very fatty diet.[2] Furthermore, the study did NOT focus on patients who felt that chocolate made their acne worse, and did NOT control for fat or sugar intake. It was also conducted before modern research showed the many benefits of the Mediterranean diet and the low glycemic index diet.

People who eat a diet low in saturated *animal fat* (especially Mediterranean diets low in red meat, butter, and milk) and higher in olive oil have a lower risk of acne.[3] Furthermore, the Mediterranean diet helps prevent heart disease, cancer, and obesity. I recommend the Mediterranean diet to adolescents interested in lifelong nutritional health whether or not they have acne.

What about Sugar?

As early as 1937, researchers at Bellevue Psychiatric Hospital in New York observed that teenagers treated with insulin shock therapy (which dramatically drops blood sugar and was standard therapy at the time for schizophrenia) had a marked improvement in their acne.[4] Based on this observation, British physicians began to give insulin injections to improve acne in otherwise healthy teens.[5] When oral medications to reduce blood sugar became available in the 1950s, they were also tried for acne and reported to be effective.[6] However, patients who used these drastic approaches ran the risk of going into shock from low blood sugar.

Possible problems with sugar metabolism form the basis for selling *yeast supplements* to acne sufferers;[7] yeast contains high amounts of *chromium*, which plays an essential role in helping tissues metabolize sugar. However, studies have not proved that oral yeast or chromium supplements are effective in treating acne. I do not recommend chromium supplements as an acne treatment.

On the other hand, there is growing evidence that eating foods that rapidly raise blood sugar levels (high glycemic index foods like fruit juices, white bread, white rice, and sugar) increase the risk of acne, and that changing the diet to emphasize low glycemic index foods (low glycemic foods include beans, seeds, whole grains, vegetables, whole fruits) can reduce acne severity.[8] Why? Because rapidly increasing blood sugar triggers the body to release insulin and insulin-like growth factor, which can trigger the release of hormones that increase sebum production. Based on this emerging research, a growing number of dermatologists recommend avoiding high glycemic index foods and maintaining an even blood sugar to help reduce acne severity.[9] Eating a low glycemic index diet can improve both acne and weight.

Do Certain Foods Trigger Acne?

In some acne sufferers, a particular food can make symptoms worse. For example, higher milk intake, even skim milk, is associated with a greater risk of acne;[10] it's unclear whether the problem is the milk itself or the hormones or other constituents in the modern milk supply. For those who have observed that milk, cheese, meat, sweets, french fries, or some other food makes their acne worse, eliminating the suspected offender is worth a try.[11] Prepare to wait six to eight weeks before you see some benefit. It takes that long for just about any lifestyle treatment to start to make a difference.

Mind-Body

No doubt about it—*stress* makes acne worse. Acne often worsens just before exam periods. Acne itself is a major stress for most teenagers. In fact, it's usually rated among the top teen health concerns. When British teenagers suffering from severe acne were offered a choice of $1,000 or a cure for their acne, 87 percent chose the acne cure![12]

What Increases Stress?

Lots of things increase stress from the daily news to upcoming exams to breakups and fights with friends and family. Be sure to set aside adequate time for sleep, relaxation, and recreation to help you be more resilient in the face of stress. For some people, stress is best managed with vigorous exercise. For others, it means finding a friendly listening ear of a family member. Soothing music can also be a balm to frazzled nerves. Beware of relying on television to relieve stress; although it's tempting to turn on the tube, the abundance of bad news, conflict, drama, and self-esteem deflating advertising can actually make stress worse.

Acne sufferers should find a *relaxation technique* that works for them—listening to music, meditation, self-hypnosis, breathing exercises, yoga, tai chi, or biofeedback—and stick with it. *Relaxation techniques are most useful if practiced daily, not just when major stress occurs.* The number of smart phone apps, YouTube videos, and online recordings offered by reputable academic institutions has exploded in the past 20 years. Find one that works for you. Parents can help by offering emotional support. For severely stressed teens, *hypnotherapy* and cognitive behavioral therapy may be helpful in reducing stress and stopping behaviors that worsen acne such as picking at the sore spots.

Environment—A Healthy Habitat

Acne is *not* caused by *dirt*. Washing twice a day is plenty to prevent blocked pores, particularly if you use an antibacterial soap containing chlorhexidine. However, teenagers who work in fast food restaurants, like Gerald, or in a garage with lots of airborne grease, may find that washing after work helps reduce the oil on their skin.

Some people feel that acne improves during the summertime with increased exposure to the *sun*. However, there is a condition called *tropical acne*, in which acne is induced by the high temperatures and humidity in the tropics. I do not recommend sunbathing as a treatment for acne.

Although sunbathing is not an optimal treatment, other types of light therapy (phototherapy), including modern lasers, LED lights, and even smart phone apps, are sometimes used to treat acne and other skin problems. A 2009 review of 25 studies of light therapy concluded that some forms of light therapy offered short-term benefits for acne, but there are few studies comparing light to other acne treatments or following up patients over months and years to determine how long benefits last.[13]

BIOCHEMICAL THERAPIES: MEDICATIONS, HERBS, NUTRITIONAL SUPPLEMENTS

Although it's tempting to launch into medications, let's follow up on the earlier section on healthy diet by exploring the benefits of commonly used nutritional supplements. Remember that supplements are not a replacement for a healthy diet!

Nutritional Supplements: Vitamins and Minerals for Acne

Which vitamins and minerals help or hurt acne?

Helpful: None proved effective as megadoses in large randomized, controlled studies.

Uncertain helpfulness: Vitamin A, Vitamin B6, Selenium, Zinc, Omega-3 and -6 fatty acids, Lactoferrin

Possibly harmful: Excessive Vitamin B12, Iodine

Vitamin A

Vitamin A supplements were recommended historically, but there have been no rigorous studies showing they improve acne, and excessive vitamin A supplementation can cause severe headaches, dry skin, cracked lips, and blurred vision. I do not recommend vitamin A supplements in higher doses than are found in non-prescription multivitamins (10,000 International Units daily). *The best plan is to eat a diet rich in fruits and vegetables—at least seven servings daily.*

Vitamin B6

Some women who have acne flare-ups with their menstrual periods have reported benefits from taking supplemental *vitamin B6, pyridoxine*.[14] However, there are no comparison studies proving extra B6 helps improve acne. More research is needed on this supplement before I recommend it as an acne remedy.

Vitamin B12

Beware of *vitamin B12 supplements*. High doses can cause an acne-like rash.

Minerals

Iodine

Excessive *iodine* may worsen acne in some sensitive teenagers. Sushi wrapped in *seaweed* and *kelp* supplements are potential sources of excessive iodine. Regular iodized table salt is okay if used sparingly.

Selenium

A preliminary study from Sweden indicated that *selenium* supplementation (0.2 milligrams taken twice daily) could improve acne for people in areas that have low soil levels of selenium.[15]

This study lacked a comparison group and has not been duplicated in the United States where soil levels of selenium are higher. If you eat selenium-rich foods like garlic or you take a multivitamin supplement that includes selenium, additional selenium is unlikely to help your acne.

Zinc

Zinc is an important trace mineral that is vital for a healthy immune system and optimal wound healing. *Zinc* levels are lower than normal in many male acne sufferers. Adequate zinc inhibits sebum production and makes sloughed skin cells less sticky. Studies about the effectiveness of *zinc supplements* are conflicting; a Swedish study showed that extra zinc was more helpful than placebo in reducing acne;[16] other studies have not duplicated these results.[17] Excessive zinc intake can be harmful. If you aren't getting enough zinc in your diet, you may not be getting enough of several other nutrients either. Zinc is found in high-protein foods like nuts, beans, whole grains, dark chicken meat, beef, pork, and lamb.

I'm less in favor of supplements than of a healthy, balanced diet, especially during the teenage years of rapid growth and development.

Zinc is a common ingredient in many skin lotions, including some sunscreens. Because of its benefits on the immune system, it is an attractive remedy for decreasing skin inflammation. However, in a study in sunny Saudi Arabia, a lotion containing zinc was less effective than a lotion containing 2 percent tea in reducing acne.[18] Although it lacks proven effectiveness on its own, zinc may be combined with other topical treatments including benzoyl peroxide or antibiotic lotions.

Omega-3 Fatty Acids

The *omega-3 fatty acids* found in cold-water fish, fish oils, flax seeds, chia seeds, and elsewhere have become popular supplements to help decrease inflammation. A South Korean study of 45 participants with mild to moderate acne found that taking 2,000 mg daily of a supplement rich in two omega-3 fatty acids, EPA and DHA, plus 400 mg of borage oil for 10 weeks significantly improved their acne compared with a control group.[19] Larger studies are needed before routinely recommending this therapy, but many people take this amount of omega-3 and omega-6 fatty acid supplements for other health reasons, and they are generally considered safe in this amount.

Lactoferrin

Lactoferrin is a naturally occurring protein that binds iron and decreases inflammation. When a group of Korean teenagers were given 200 mg of lactoferrin supplements daily for 12 weeks, their sebum production and skin fatty acid levels fell and their acne improved compared with a similar group of teens who were not given lactoferrin supplements.[20] A study of lactoferrin supplements also showed substantial benefits on acne in German teens.[21] However, it's not clear whether teens on a standard American diet would have the same benefit from lactoferrin supplements.

Herbal Remedies for Acne

Green Tea (Camellia sinensis)

Green tea is widely consumed for flavor and health benefits. It scavenges free radicals, works as an antioxidant, decreases inflammation, and decreases sebum production. Bathing the skin in tepid or cool green tea has also been used to help with eczema, acne, warts, and other skin conditions. As mentioned in the section on zinc, a skin lotion containing green tea was more effective than a similar lotion containing zinc in improving acne. In another study, using a lotion containing 2 percent green tea daily for six weeks was associated with a significant decrease in acne severity.[22] Green tea is safe when consumed as a tea or when applied to the skin as long as common sense precautions are taken. Don't apply boiling hot tea to your face.

Resveratrol

Resveratrol is a natural compound found in grapes, blueberries, raspberries, and mulberries. It is produced when the plant is injured or attacked by bacteria or fungi. An intriguing Italian study asked acne sufferers to apply a resveratrol-containing gel to one side of their faces daily, and a similar gel without resveratrol to the other side; after two months, there was more than a 50 percent improvement on the resveratrol-treated side compared with the control side. This study bears repeating in a larger group of American teens.[23] In the meantime, it's a great idea to eat berries and grapes rich in resveratrol as part of a healthy diet.

Tea Tree Oil (Melaleuca alternifolia)

Tea tree oil has proven benefits for acne.[24] It works by killing bacteria. An Australian study showed that a 5 percent tea-tree oil gel was as effective as 5 percent benzoyl peroxide for acne.[25] Another study from Iran also showed that tea tree oil gel was three to five times more effective than a placebo for acne.[26] Do NOT use straight, undiluted tea tree oil because it can irritate the skin. Keep all essential oils out of reach of children because ingesting large amounts can be hazardous.

Other traditional herbal acne remedies include infusions of calendula, catnip, chamomile, comfrey, lavender, thyme, witch hazel, green tea, or yarrow root used as facial rinses. Folk remedies for acne include poultices of grated carrot, cucumber, and aloe vera and lemon juice rinses. Others have used aloe vera or colloidal oatmeal applied directly to the skin. An Ayurvedic remedy is a paste made with turmeric and sandalwood powder mixed with water. Herbal masks containing equal parts of green clay and corn flour, with one or two drops of the essential oils of chamomile, lavender, juniper, and patchouli, have also been recommended. Native American herbal cures include burdock root, echinacea, Oregon grape, and goldenseal. Until studies document the effectiveness of these herbal remedies, I do not recommend them, but because they are generally safe, I don't mind my patients trying them as long as they use common sense safety precautions.

Medications

The medications used to treat acne fall into four categories: those that prevent or destroy sebum plugs; potent relatives of vitamin A; products that affect hormone levels or inflammation; and antibiotics. Some medications combine these strategies.

ACNE MEDICATIONS

- Prescription and non-prescription sebum plug busters, available as creams, gels, ointments, or lotion
  - Benzoyl peroxide
  - Salicylic acid
  - Azelaic acid
- Vitamin A
  - Topical (cream or gel) retinoids including adapalene (Differin)
  - Oral isotretinoin (Accutane)
- Hormonal treatments (birth control pills)
- Antibiotics
  - Topical lotions (clindamycin, dapsone, doxycycline, erythromycin, minocycline, tetracycline)
  - Oral

Most pediatricians and dermatologists recommend a step-wise approach to using acne medications. For mild acne, start with a plug buster or topical retinoid (vitamin A–related product) or a topical antibiotic. For moderate acne (or mild acne that doesn't improve adequately with just one product), use a combination of two medications (like benzoyl peroxide plus a topical retinoid). For more severe acne or moderate acne that hasn't improved adequately with two medications, go for a triple combination or head for the heavy hitter (with the most potential for side effects), isotretinoin (the most potent vitamin A relative). Once symptoms are well controlled, consider backing off of one medication.

Non-Prescription and Prescription Plug Busters

Some of the most effective treatments for preventing and destroying sebum plugs are inexpensive and available without a prescription. Some of these plug busters also have antibacterial effects. Most of them cause some skin irritation, which is why the strongest concentrations require a prescription. Your face may actually look *worse* the first three days you try them; don't give up during those first few days.

Benzoyl peroxide is one of the most effective and widely used non-prescription acne treatments. Save money by buying generic brands. Benzoyl peroxide generates free oxygen radicals in the hair follicles, breaking down sebum plugs and killing acne-causing bacteria. Strengths range from 2.5 to 10 percent. Start with the lower strengths (2.5 or 5 percent) once a day at night until your skin gets used to it. Then advance to twice daily using the 5 percent strength. If you haven't noticed any improvement after six weeks and your skin tolerates it, you can advance to the 10 percent strength at night or to twice daily.

The *side effects of benzoyl peroxide* include irritation, stinging, drying, itching, redness, and peeling. Benzoyl peroxide can also inactivate tretinoin (topical vitamin A preparations), so the two should not be applied at the same time (you can apply one in the morning and the other at night). Non-prescription benzoyl peroxide is available in either alcohol-based or water-based formulas. Alcohol-based preparations are more drying and irritating than water-based forms. The prescription formulation of benzoyl peroxide comes as a gel rather than a cream; the gel penetrates the skin more deeply and is more effective, but it also causes more redness and irritation. Benzoyl peroxide is more irritating when applied to wet skin, so wait 20 minutes after washing and drying your face to apply it.

About 1 to 2 percent of adolescents who use benzoyl peroxide become *allergic* to it. Because it generates free radicals, benzoyl peroxide could theoretically increase the risk of developing *skin cancer* later in life. However, studies looking at users of benzoyl peroxide have not found any increase in cases of skin cancer. Benzoyl peroxide is a *bleach*; be careful around colored fabrics, especially silk shirts! It can also bleach skin; darkly pigmented teens may prefer non-prescription salicylic acid creams.

Salicylic acid breaks apart sebum plugs by "ungluing" dead skin cells. It has been used for over one hundred years to treat acne. Salicylic acid is as effective as benzoyl peroxide for most acne. It is the active ingredient in many over-the-counter acne remedies. Strengths vary from 0.5 to 2 percent. Start with the lower concentration; as your skin tolerates it, try higher strengths. As with other topically applied acne medicines, the main side effect is redness and irritation. This is usually mild and doesn't mean that you have to stop treatment; just cut back to a lower strength or once a day in the evening.

Azelaic acid (Azelex™, Azepur™, Azaclear™, AzClear™, and other brands) was originally derived from yeast and is also found in tiny amounts in wheat, rye, and barley where it serves as part of the plants' defense system. In human skin, it busts plugs, fights bacteria, and reduces inflammation and skin discoloration, but its expense and side effects (irritation, redness, and dryness) have made it less popular than benzoyl peroxide, retinoin, and topical antibiotics; it is mainly used as an alternative therapy for patients who can't use retinoids.

Tretinoin, Retinoids, Adapalene, or Topical Vitamin A Acid (Prescription Retin-A® Cream or Gel)

The topically applied form of vitamin A is actually a medication, and it's one of the most effective anti-acne preparations available. Retinoids break down sebum plugs, decrease inflammation, and remove the welcome mat for bacteria. Retinoids work synergistically with non-prescription plug busters (benzoyl peroxide or salicylic acid) and antibiotic lotions because they make the skin more receptive to the other medications. Do not apply retinoids at the same time as other acne medications; instead, alternate them: for example, tretinoin or adapalene in the morning and either benzoyl peroxide or an antibiotic lotion in the evening. Combinations work well for moderate acne.

Topical retinoids come in several strengths (0.025, 0.05, and 0.1 percent) and in both a cream and gel. All require a prescription. The gel formations and the higher concentrations are more potent and have more side effects (red skin and peeling); side effects can be reduced by starting with lower concentrations. Start by using it at night once every other day, gradually to every night as tolerated. Topical vitamin A also makes skin more sun sensitive, so *be sure to use a good sunscreen if you're using tretinoin*. Treatment usually lasts six to eight months. Tretinoin does not cause birth defects or any of the problems associated with overdoses of oral vitamin A supplements.

Adapalene (brand name Differin®) is less irritating than tretinoin (Retin-A®) and better at decreasing inflammation. *Tazarotene* (brand name Tazorac®) was originally developed to treat psoriasis, but it's also an effective acne remedy. However, it is more expensive and more irritating than other retinoids, so it is not prescribed as often. There are also combinations (adapalene plus benzoyl peroxide, brand name Epiduo®; and tretinoin plus clindamycin, brand name Ziana® or Veltin®). Before accepting a prescription for one of these, look into costs of the combination medication compared with separate prescriptions or over-the-counter medications. Smart shoppers can save hundreds of dollars a year by asking a few well-chosen questions and checking their insurance coverage.

Retin-A® is usually one of the first medications I prescribe for teenagers with acne who have already tried benzoyl peroxide or other non-prescription acne remedies. For Gerald, I wrote a prescription for Retin-A® (0.025% cream) to use at night after his evening wash with chlorhexidine soap. I suggested that he continue his morning treatment with non-prescription benzoyl peroxide, explaining that he may see some improvements by prom time with persistent use, but that he might have some redness and irritation for the first week or so. We took a picture in the office to compare to his later appearance because the improvement is sometimes so gradual that teenagers aren't aware of it until they compare before and after pictures.

Isotretinoin (Accutane), a form of vitamin A taken by mouth, is the most powerful prescription acne medication. Along with its power comes the potential for *serious side effects*. It is reserved for patients with severe, deep, widespread acne. Women should not use it unless they abstain from sex or use a reliable form of birth control because isotretinoin can cause serious birth defects if taken during pregnancy. It can also cause dry eyes, headaches, nosebleeds, and changes in blood lipids. It is available only with a prescription; if you are considering this treatment, please see a dermatologist.

Prescription Hormonal Treatments

Birth control pills containing estrogen suppress the body's production of acne-causing hormones. If a woman is old enough to suffer from moderate acne, she's old enough to use birth control pills to treat it. Most birth control pills contain a combination of estrogen and progesterone. Low-progesterone pills are the most effective for reducing acne, especially in women who have other signs of hormonal imbalance, such as facial hair. It takes several months for benefits to become apparent, and the course of therapy may last one to two years.

Estrogen hormones are *not* recommended for male acne sufferers.

Antibiotics

Antibiotics kill bacteria, decreasing fatty acids and inflammation. They do *not* clear up pimples that are already present. To prevent pimples before they reach the surface, antibiotics are applied to *all* acne-prone areas, not just those that have already broken out. Antibiotics can be applied topically (as a liquid) or taken by mouth (orally). Both topical and oral antibiotics take several weeks to have visible effects.

BACTERIA KILLERS

Antibacterial Soap: Chlorhexidine (*Hibiclens*)—no prescription necessary

Antibiotics (topical)—require prescription

Antibiotics (oral)—require prescription

Do NOT use abrasive cleansers and scrubs; they cause too much friction and irritation, drawing immune cells into the fray and worsening acne. Most soaps don't penetrate deeply enough or stay on the skin long enough to kill acne-causing bacteria. On the other hand, *chlorhexidine* (Hibiclens®) skin cleanser is widely used in hospitals because it does such a good job of killing skin bacteria. When used twice daily, it has *proven effective* as an acne-fighter in a randomized, placebo-controlled trial. It is as effective as 5 percent benzoyl peroxide and is less drying and irritating.[27] Washing more than twice daily confers no additional benefits.

Prescription antibiotics and combinations of antibiotics and zinc or antibiotics and benzoyl peroxide are also available. Topical antibiotics are effective, but there are individual differences in response. If the first one you try is too irritating or doesn't work, another may be more helpful. Antibiotic lotions must be applied to the entire acne-prone area, not just the pimples that have already broken out.

Remember, most of our modern antibiotics were originally found in soil yeast, bacteria, mold, and plants where they helped defend their host. Using small amounts over long periods of time tends to increase the risk that other bacteria will become resistant to them. In the future, you may see more research on using probiotics (healthy bacteria) on the skin to help prevent and treat acne.[28]

Antibiotic lotions are nearly as effective as taking antibiotics by mouth, but they cause fewer side effects. Antibiotic lotions are also less irritating than benzoyl peroxide and salicylic acid. Be careful with tetracycline creams; they glow under ultraviolet light. Your face may attract unwanted attention if you apply tetracycline to it before visiting a nightclub featuring "black" lights.

Oral antibiotics kill bacteria throughout the body, including the entire skin surface. Oral antibiotics are more effective than antibiotic lotions in treating severe, widespread acne on the trunk and back, but it takes about four months before the effects are really noticeable. I do not usually recommend oral antibiotics for my patients with acne for three reasons:

1. The increasing use of low-dose antibiotics contributes to microbial resistance. This means that bacteria have become resistant to common antibiotics, reducing their effectiveness in treating life-threatening infections. To limit the spread of resistance, I prefer to use oral antibiotics only for serious infections.
2. Side effects. Teenage girls who use oral antibiotics often get vaginal yeast infections. Other side effects include abdominal pain, nausea, gas, upset stomach, allergies, increased susceptibility to sunburn, and

changes in skin color. *Tetracycline should not be taken by pregnant women* because it turns developing teeth an ugly brown. Minocycline (Minocin™) can cause arthritis, hepatitis, and systemic lupus erythematosis (SLE).
3. Oral antibiotics kill normal intestinal bacteria. If you take an oral antibiotic every day, be sure to eat yogurt (with active cultures) or a probiotic supplement every day to replace the healthy bacteria in your system.

BIOMECHANICAL THERAPIES: MASSAGE, SURGERY

Massage

If massage is used to decrease stress, use water-based lotions rather than oils to avoid blocking pores. Some therapists add a few drops of *bergamot, chamomile, cedarwood, eucalyptus, juniper, lavender, lemongrass,* or *sandalwood* as essential oils to add to the basic massage lotion. Most essential oils have antibacterial properties, and including small amounts of them does not increase the risk of acne.

Surgery

Some dermatologists use a long-handled thin *extractor* to remove blackheads. The extractor has a round end that fits around blackheads. By pressing down on the round loop, the plug is forced up and out. Extraction is a less irritating technique than pinching or pushing. It does *not* prevent pimples, but can help reduce the number of cosmetically disturbing blackheads.

A variety of *plastic surgery* treatments are available to treat acne scars. With *cryosurgery* acne scars are sprayed with a freezing cold liquid; the frozen surface layers eventually peel away, leaving smoother skin. *Chemical peels* work the same way. *Dermabrasion,* in which the skin is scraped with a wire brush, is reserved for more severe scarring. For deep scars, the individual areas are cut out, then grafted or stitched. Collagen or silicon can be injected under the skin to push up broad and flat scarred areas. These procedures should only be undertaken under the supervision of a plastic surgeon and only when the acne is no longer active.

BIOENERGETIC THERAPIES: ACUPUNCTURE, HOMEOPATHY

Acupuncture

In 2013, Chinese researchers published a large summary of over 40 studies evaluating the effectiveness of acupuncture in treating acne;[29] these investigators concluded that acupuncture is safe and had modest benefits on acne, but that the quality of study designs was poor, and additional research is warranted.

Homeopathy

Homeopathic remedies for acne include *Antimonium crudum, Carbo animalis, Hepar sulphur, Kali bromatum*, and *Sulphur*. There are no scientific studies documenting the effectiveness of homeopathic remedies for treating acne. They are safe, but until there are more studies evaluating effectiveness, I do not recommend that you spend money on them.

WHAT I RECOMMEND FOR PREVENTING AND TREATING ACNE

*Remember: It will take 8–12 weeks to see the effect of almost any therapy.
Treatment must be persistent and you must be patient!*

1. *Lifestyle—nutrition.* Eat a healthy diet with at least seven servings of fruits and vegetables daily. Eat fewer foods that rapidly increase your blood sugar. Eat less red meat, milk, cream, butter, and other sources of animal fat. Eat regular meals to keep your blood sugar stable. Monitor your response to specific foods to determine whether you need to avoid specific foods.

2. *Lifestyle—environment.* Avoid prolonged exposure to oily environments such as fast-food restaurants. Avoid hot, humid environments. Minimize friction to acne-prone areas. Don't try to cover the acne with long hair. Avoid abrasive soaps and scrubs. Do *not* pick at your pimples. Wash acne-prone areas twice a day with chlorhexidine or a gentle, antibacterial soap.

3. *Lifestyle—mind-body.* Practice stress reduction techniques such as meditation, yoga, self-hypnosis, biofeedback; or progressive muscle relaxation or listen to soothing music daily to lower your stress levels. Get plenty of sleep. Avoid pulling all-nighters.

4. *Lifestyle—cosmetics.* If you use cosmetics, use only water-based kinds. Avoid the following ingredients: acetylated lanolin, PEG 15, PEG 75 or plain lanolin, butyl stearate, cocoa butter or coconut butter, glyceryl-3-diisostearate, hydrogenated vegetable oil, isopropyl anything, lauric acid or laureth 4, monostearate, myristyl myristate, oleyl alcohol, oils of avocado, mink or sesame, propylene glycol, red dyes, stearath 10.

5. *Biochemical—supplements.* Avoid excess or deficiency of any essential nutrient.

6. *Biochemical—topical medications.* If you want to try non-prescription treatments, consider:
- Herbal *tea-tree oil gel,* applied twice daily OR
- *Salicylic acid,* applied twice daily OR
- *Benzoyl peroxide,* applied twice daily; start with a once daily application of the 2.5 percent strength before bedtime. As tolerated, increase to twice daily; then increase to higher strengths.

See your health care professional if these measures haven't helped within two months. You may benefit from topical treatments with:

- *Vitamin A cream, retinoic acid, tretinoin,* or *adapalene.* These products can make skin sun-sensitive, so avoid sun exposure and use a good sunscreen—at least SPF 15.

- *Antibiotic lotions.* Do not use at the same time as other skin creams. If you're using another treatment such as benzoyl peroxide, use one in the morning and the other in the evening.

For women, consider talking with your doctor about starting birth control pills. If you take antibiotics daily, increase the amount of yogurt (with active cultures) that you eat or consider taking a probiotic supplement to reduce the side effects of the antibiotics.

For severe acne that does not get better even with oral antibiotics, it is time to consult a *dermatologist* for a thorough evaluation and consideration of even stronger medication (such as isotretinoin, *Accutane*). Remember, the stronger the medicine, the stronger the side effects.

4

ATTENTION DEFICIT HYPERACTIVITY DISORDER (ADHD)

Sandy Vincent called me in late September about her first-grader, Brian. She knew Brian was energetic, but his teacher had called to tell her she thought that Brian was hyperactive. The teacher said that Brian had a hard time sitting still and was usually out of his chair even before she'd finished calling attendance. He frequently shouted out answers impulsively, even if another child had been called on. The teacher thought he should take medications like several other students. Sandy wanted to ask me about other strategies. She didn't like the idea of Brian having to take medicine every day just to be able to go to school, and she wanted to know if there were any natural options to help him succeed in school. First, she asked, what exactly is ADHD and how common is it?

Hyperactivity is shorthand for the diagnosis *of attention deficit hyperactivity disorder* (ADHD). ADHD is characterized by impulsivity and at least one of two other behaviors (mental inattention or physical hyperactivity) that cause problems for the child. The classic image is that of an energetic boy like Dennis the Menace who talks a lot, interrupts others, acts as if driven by a motor, fidgets and squirms, has a messy room, acts impulsively, has trouble following rules, and often breaks or loses things; he is often admonished to sit still, pay attention, and clean up his room. The quiet girl who daydreams and is inattentive in class has a second classic type of ADHD (ADHD without hyperactivity). There are three official ADHD sub-types: (1) the primary problem is hyperactivity; (2) the primary problem is inattention; and (3) a mixed type.

What Is ADHD?

- At least six symptoms of severe *impulsiveness*:
  - Often acts before thinking
  - Has difficulty organizing work
  - Often blurts out answers or interrupts other's activities
  - Has difficulty waiting his turn in group games or activities
  - Shifts excessively from one activity to another
  - Needs a lot of supervision
  - Often engages in physically dangerous activities, e.g., runs into the street without looking

AND at least six symptoms of *short attention* and/or *hyperactivity*:

- *Short attention*
  - Shorter than average *attention* span; distractibility; inattention
  - Often fails to finish things he starts
  - Has difficulty sustaining attention to tasks or activities
  - Has difficulty following through on instructions
  - Is easily distracted
  - Does not seem to listen to what is being said to him
- *Hyperactivity*
  - Runs about or climbs things excessively
  - Difficulty sitting still, fidgets with hands or squirms in seat
  - Has difficulty staying seated
  - Moves about excessively during sleep

- Is always on the go or acts as if "driven by a motor"

To make an official diagnosis of ADHD, these symptoms must (1) *appear before the child is seven years old*; (2) *impair the child in at least two settings (school, social, sports, music, work, or home)*; and (3) *last at least six months without being explained by another condition*. However, the diagnosis usually is not made until there are problems in the classroom. Children don't have to have all of the above symptoms to be diagnosed as having ADHD. However, they do need to exhibit at least several of these characteristics *more than is normal for their age*.

To some degree, impulsiveness, short attention span, and high activity levels are normal. Normal toddlers are messy, impatient, distractible, and very active. But things that are normal for two-year-olds are not necessarily normal for ten-year-olds. Before diagnosing ADHD, one has to be sure that the child is not simply acting his age. This issue of comparing children to others of the same age is not trivial. If we look at all seven-year-olds on September 1, some turned seven in August, and some turned seven the previous September; so even though they are technically the same age, there are nearly 12 months of differences. In one study, 10 percent of kindergarten students born in August (youngest in class) were diagnosed with ADHD compared with 4.5 percent born the prior September. The youngest children even in higher grades are about twice as likely to receive stimulant medications as older children in the same grades.[1] If your child is one of the youngest, least mature in his class, he is at increased risk of being labeled as having ADHD.

One also needs to account for other causes of disruptive behavior (such as the acute stress of a death in the family, a sensory deficit, learning disability, anxiety, depression, cognitive delay, pervasive developmental disorder, petit mal seizures, sleep apnea, having an intense temperament, or even being gifted and bored). Certain illnesses, such as itchy skin rashes and pinworms, can also make children irritable, distracted, and squirmy. ADHD may also be the first sign of thyroid problems. These conditions must be addressed before diagnosing ADHD.

How Many Kids in the United States Are Diagnosed with ADHD?

ADHD is the most commonly diagnosed behavioral disorder in children in the United States. The number of American kids diagnosed as having ADHD has skyrocketed over the last 20 years at the same time the rate of emotional and mental health problems has *declined* in Great Britain.[2] From 2003 to 2007, there was more than a 20 percent increase in the diagnosis of ADHD in American children. ADHD has replaced asthma as the most common chronic disease of childhood in the United States, with 5.4 million children between 4 and 17 years old or 9.5 percent diagnosed with ADHD, compared with 9.1 percent of children diagnosed with asthma.[3] Among 3- to 17-year-old boys, 11 percent have been diagnosed with ADHD. ADHD is not just for children—2.5 percent of adults have been diagnosed with it, too.

As the number of ADHD diagnoses has risen, so has the use of stimulant medications to treat it. Prescriptions for hyperactivity medications increased more than six-fold between the 1970s and the 1990s and have continued to rise in the twenty-first century (Table 4.1).

ADHD is not just a problem for kids in grade school. Increasing numbers of *preschool* and high school kids are being diagnosed and treated with medications.[4] Furthermore, many children with problematic behaviors are given additional psychoactive medications in addition to stimulants. (See *Biochemical Therapies* later in the chapter.)

There has been a national uproar and a great deal of concern about these high rates

Table 4.1: Rising Number of American Children Treated with Stimulant Medications	
Year	Number of Children in the United States Treated with Stimulants
1970	150,000
1980	400,000
1990	900,000
2003	2,500,000
2007	2,700,000

of diagnosis and drug treatment. Are we labeling normal behavior as a medical problem (over-diagnosis)? Are parents failing to teach discipline? Are teachers failing? Are we creating a nation of addicts? Have we poisoned our children's brains through environmental exposures to BPA, pesticides, excessive screen time, or poor nutrition? Is everyone just stressed out and having unrealistic expectations for kids? What is going on?

Some people think we have just improved our ability to detect and treat children who had suffered in silence previously. Perhaps children who were previously labeled as slow, lazy, dreamers, or class clowns are now being diagnosed with ADHD. If children who were previously labeled negatively are now benefiting from appropriate diagnosis and treatment, perhaps the increases are a good thing.

Others believe that ADHD is being over-diagnosed or misdiagnosed. Perhaps we are labeling and drugging kids whose problems are really not that severe, who could benefit from better behavior management, increased exercise, better classroom design, and lower student-teacher ratios. Maybe it's just a matter of stressed-out, overworked teachers who need quiet, compliant, attentive kids to meet end-of-year exam standards. Maybe we're jumping to the ADHD diagnosis before we've adequately considered other physical, learning, emotional, and mental health challenges that interfere with behavior and academic performance. Or perhaps we've been too heavily influenced by pharmaceutical advertising that sends the message that drugs are the solution to every problem.

Many speculate that ADHD reflects a poor fit between children's temperaments and current cultural and social expectations. Rather than be confined to crowded classrooms year after year, children used to stay at home and learn from their parents by active example with fewer people and more time in nature. Although most children seem to have adapted to the demands of modern society, some simply don't have the disposition to sit still, pay attention, and wait their turn without one-to-one supervision. It doesn't take long before these children are labeled as troublemakers; the label often becomes a self-fulfilling prophecy.

Sandy was curious about what actually causes ADHD. Is it a genetic disease? Too much sugar? Or what?

POSSIBLE CAUSES OF HYPERACTIVE BEHAVIOR

- Physical, medical problems
- Genetic factors or congenital problems
- Learning problems and other mental health issues
- Poor nutrition
- Environmental toxins
- Excessive screen time
- Problems of poor sleep, low arousal, and low self-esteem

Physical problems such as impaired *hearing* and *vision* can cause problem behavior that is easily mistaken for hyperactivity. If your child can't hear the teacher, he can't very well pay attention to what she says. If your child can't see the blackboard, he's likely to pay more attention to what's on his neighbor's desk.

Another physical problem that can cause daytime hyperactivity is *poor sleeping* at night.[5] Among the many causes of sleep problems, one of the biggest culprits is partially blocked breathing, called *obstructive sleep apnea*. Sometimes breathing is temporarily blocked during a cold or sinus infection. In these cases, breathing, sleep, and behavior return to normal when the infection clears. Chronic blockages are usually due to large adenoids or tonsils. Nighttime breathing blockages result in loud snoring and frequent stops and starts in breathing, daytime sleepiness, and poor growth. Surgery can correct these obstructions, resulting in dramatic improvements in behavior and growth. Sometimes it's a simple case of allergies causing the obstruction; treating the allergy can improve sleep at night and behavior during the day.

Poor sleep is associated with problems overall with *arousal*, sleeping, and wakefulness. Hyperactive children actually seem to be less alert or vigilant than other children. It takes a certain amount of alertness to inhibit impulsive behaviors; when watchfulness is decreased, there is more impulsive behavior and less thought about the consequences of that behavior. Make sure your child is sleeping well. If he snores, be sure to ask your pediatrician about doing a sleep study to see if *obstructive sleep apnea syndrome* is contributing to his daytime behavior problems.

Pinworms are a common childhood *infection*. The worms live in the large intestine, and crawl out at night to lay their eggs on the skin of the buttocks. Pinworms lead to an itchy bottom at night and irritability and distractibility throughout the day. Pinworms can be easily eradicated with medication, dramatically improving a child's disposition and behavior.

Brain damage from any cause can result in problems in thinking, memory, and attention. Even minor damage from *head injury* can cause learning and behavior problems.

Brain damage due to *lead* poisoning has

long-term effects on language and concentration skills. Lead levels are often elevated in hyperactive children. The higher the lead level, the more hyperactive symptoms present. Thanks to stricter environmental regulations, the average blood lead level has dropped over the last 20 years, and lead poisoning accounts for very little of the ADHD seen nowadays. On the other hand, studies by the Environmental Working Group have documented that infants are now born with hundreds of other industrial chemicals and pollutants; of the nearly 300 chemicals detected in babies, 217 are toxic to the brain and nervous system. Persistent pollutants such as dioxins and dibenzofurans are related to a sharply increased risk in ADHD and learning disabilities.[6] Also, higher levels of organochlorine pesticides, DDT, and DDE have been linked to poorer cognitive development.[7,8] Higher levels of organophosphates are linked to higher risks of ADHD.[9] Eating organically raised fruits and vegetables is associated with an 80 percent reduction in the amount of organophosphate pesticides in children's systems. Whenever possible, grow your own organic fruits and vegetables or purchase them from a local farmer's market.

Cigarette smoke adversely affects many aspects of a child's health, including behavior. Maternal *smoking during pregnancy* subtly limits the oxygen supply to the baby's brain, and can contribute to hyperactivity. In studies of children whose mothers used alcohol, marijuana, and cigarettes during pregnancy, cigarettes had even more impact on intelligence and behavior than alcohol or drugs.[10] Maternal *smoking after pregnancy* also affects children's behavior; the more she smokes, the worse the child's behavior. Children of alcoholic fathers are also at increased risk of behavioral problems such as hyperactivity.[11] Exposure to smoke at home is a bigger risk factor for school failure than are recurrent ear infections.[12]

Increases in the amount of screen time may also be contributing to our ADHD epidemic. The average American child spends four hours daily watching television, and even more on computers and hand-held electronics. Increased television watching at two years old is associated with an increased risk of ADHD at school age.[13]

To summarize, many factors can contribute to ADHD symptoms including genetics, medical problems, toxins, tobacco, and TV.

DIAGNOSIS

The diagnosis of ADHD or hyperactivity is based on the child's symptoms. There are no blood tests, urine tests, X-rays, brain scans, or other laboratory tests that prove ADHD. Parents and teachers may be asked to complete questionnaires or daily diaries of the child's behavior to help confirm the diagnosis. The most widely used daily behavior diaries are the *Connors Parent and Teacher Rating Scales* and the *Vanderbilt Rating Scales*. Other questionnaires include the Child Behavior Checklist, the Yale Children's Inventory, the Behavior Rating Profile, and the ADD-H Comprehensive Teacher Rating Scale, ACTeRS. There is also a questionnaire specifically for adolescents who might have ADHD called the ADD/H Adolescent Self-Report Scale. If your child's symptoms and tests are *not* suggestive of any other physical, emotional, or learning problems, and they ARE consistent with ADHD, the diagnosis is made.

I gave Sandy copies of the Vanderbilt Parent and Teacher Rating Scales for her and Brian's teacher to fill out over the next two weeks. I asked them to complete them the same day of the week so we'd have a consistent picture. She said she'd do the home scale since her husband has also been diagnosed with ADHD and could never remember paperwork. She wondered whether ADHD ran in families.

Hyperactivity does *run in families*. Many parents whose children are diagnosed with hyperactivity report that they, too, have attention problems or learning disabilities; they also have an increased risk of other mental health diagnoses. Several different genes have been linked to ADHD, but there is not a genetic test to determine a diagnosis.

Sandy said she had dyslexia, and she wanted to make sure that Brian didn't have a specific learning disability. She scheduled an appointment for him to be tested by the school psychologist.

Between 30 and 70 percent of children who meet the criteria for ADHD also have learning disabilities, language problems, or slow learning. Some also have oppositional behavior, mood or anxiety issues, or tics. Learning problems such as dyslexia can also lead to distractibility. Language problems can be addressed by proper school evaluations. It's hard to pay attention if you can't understand what's being said or can't understand or remember the directions. Improved language skills can dramatically improve academic and social performance.

Without help, hyperactive children frequently become disruptive in a regular classroom, and are soon labeled as the "bad kids." Children whose behavior frequently meets with disapproval often end up with poor self-images. This is certainly true of hyperactive children who frequently suffer from *low self-esteem*. Low self-esteem exacerbates feelings of hopelessness and aggressiveness. Kids with ADHD are at increased risk for developing substance abuse problems, injuries, and poor social relationships. This means that focusing on positive goals such as developing a better memory, better organizational skills, more neatness, punctuality, safety, sociability, cooperation, persistence, and hope are keys to academic and social success. Healthy lifestyle approaches including nutrition, exercise, supportive, nontoxic environment, mind-body practices, and healthy communication skills are the backbone of treatment for hyperactivity.

STRENGTHS FOCUS

ADHD is more than a set of problems. It is also usually accompanied by significant strengths. Your child may show higher than average levels of creativity, imagination, sociability, adaptability, confidence, exuberance, spontaneity, and a desire to please others. She could be a natural comedian or entertainer.

OTHER CAUSES OF INATTENTION AND IMPULSIVITY

Children who are grieving for a significant loss, such as the loss of a parent through death or divorce, are often inattentive to their school environment, may act out their sadness through aggression, or act impulsively in dangerous situations out of despair and hopelessness. Anxiety and depression can also mimic attention deficit disorder. Before a child is diagnosed with attention deficit disorder, consider whether the child has a long-term behavioral disorder or a reaction to an upsetting situation.

Brian didn't seem to be suffering from any particular stresses this year and didn't appear to be anxious or depressed except about his school performance. While she was filling out the Vanderbilt scale and waiting for the results of the school psychologist's tests, Sandy wanted to do something to help Brian. She planned to start buying more organic fruits and vegetables and non-organic produce from the Clean 15 list prepared by the Environmental Working Group. She wondered if some of Bryan's problems weren't caused by his grandparents' frequent indulgence of his sweet tooth. Does sugar cause ADHD?

Sugar doesn't cause ADHD, but it isn't good for the teeth or the waistline. Many products that contain sugar also contain artificial colors, flavors, and preservatives that can contribute to hyperactive behavior. One the other hand, the kind of sugar found in milk and fruits is not problematic as long as children are getting it from whole foods.

The same thing may be true of *caffeine*. On a weight basis, children typically consume more caffeine than adults. For example, a twenty-five pound toddler who drinks a can of cola consumes about *twice* as much caffeine (for his size) as a 120-pound woman who drinks one strong cup of coffee a day. Caffeine *improves* attention in children and adults.[14] Hyperactive children tend to drink more caffeine-containing sodas than their unaffected peers.[15] Again, hyperactive kids may be unconsciously treating their symptoms with caffeine.

Dr. Ben Feingold, a San Francisco allergist, developed his theory about hyperactivity from his observation that hyperactivity increased as Americans increased their intake of *artificial colorings and flavorings*. He was also alarmed at the increasing use of children's aspirin in the 1960s and 1970s, knowing that its active ingredient, *salicylate,* could trigger asthma symptoms in some patients. He became convinced that subtle allergic symptoms to artificial colorings, flavorings, and salicylate were the basis for about 50 percent of cases of hyperactivity.

Based on this belief, he began treating hundreds of children with the Feingold diet, which excluded numerous foods. Enthusiasm spread with the hope that restricting children's diets would improve intolerable behavior.

The original Feingold diet avoids:

- Artificial colors and flavors, especially yellow dyes (tartrazine)
- Medicinal salicylate compounds such as aspirin, Pepto-Bismol, and oil of wintergreen
- Foods containing natural salicylates such as almonds, apples, apricots, berries, cherries, citrus, cloves, cucumbers, currants, grapes, raisins, nectarines, peaches, plums, prunes, strawberries, tea, and tomatoes

There has been a great deal of confusion about what items are actually excluded from Feingold's original diet. For example, artificial preservatives, stabilizers, and other chemical ingredients were *not* excluded from the original Feingold diet. However, many healthy fruits were excluded (see previous list).

Back in 1978, in the best study of the Feingold diet, three dozen hyperactive children and their families were placed on experimental diets for up to two months.[16] The researchers concluded that there may be some preschool children whose behavior is improved by the Feingold diet, but there was no benefit for school-age children.[17]

Since Feingold's original hypothesis, thinking about food–ADHD relationships has evolved. Several studies have evaluated the possibility that ADHD is due to *sensitivities to specific foods or food additives*. In a 1985 study, a "few foods diet" improved ADHD symptoms in 62 of 78 children, many of whom also reported improvements in other symptoms such as headaches, tummy pain, and seizures; 48 different foods and additives were implicated in these symptoms, most often artificial colors and flavors.[18] Results were repeated in a 1993 British study.[19] Artificial colors and preservatives were also blamed for significant increases in hyperactivity symptoms in a study of 273 three-year olds published in 2004 and a large group of eight- to nine-year-old children published a few years later.[20,21] In a 2011 study, 100 children with ADHD were randomized to be on their regular diet or a restricted diet that included only rice, meat, vegetables, and pears; after five weeks, 78 percent of children on the few foods diet had significant improvements on ADHD rating scales, a much higher percentage than the children eating their regular foods.[22] This means that a restrictive diet is as effective as stimulant medications for many children, and has led many schools, particularly in the UK, to limit or eliminate serving foods containing artificial flavors, colors, or preservatives. Products with artificial colors Yellow 5, Red 40, and others are required in some European countries to carry the label, "May have an adverse effect on activity and attention in children."

If you suspect your child has a food allergy or sensitivity, please seek help from a health care professional who is experienced in pediatric nutrition. This is particularly important if you want to conduct your own trial of a restricted or "few foods" diet. The diet may help, but excessive, lengthy restrictions can result in nutrient deficiencies and poor growth.

Sandy wasn't ready to restrict Bryan's diet that much—it would lead to too many fights about food. She wondered whether Bryan might just outgrow his hyperactivity.

Some children outgrow their hyperactivity, but many remain distractible and impulsive throughout adolescence and young adulthood. Some learn to compensate for their disability, developing different strategies for staying focused and attentive. When followed over four years, only 15 percent had a complete remission of symptoms, and these tended to be the kids who had fewer problems initially.[23] Even those without hyperactivity continue to have problems with distractibility and impulsiveness, impaired school work, and social interactions. Children who suffer from hyperactivity during school years remain more accident-prone and have higher rates of car crashes than other teens and young adults. They have much higher rates of serious injuries, hospital-izations, visits to emergency rooms and clinics, and a much higher overall cost of health care. They are about twice as likely as other teenagers to smoke cigarettes. Without ongoing treatment, they are also at much higher risk of developing substance abuse problems, unintended pregnancy, and antisocial behavior, resulting in rates of unemployment and imprisonment that are five to ten times higher than in the general population.

What Is the Best Way to Treat ADHD?

Now let's talk about what you really want to know: how to treat ADHD. Many parents are cautious about agreeing to put their child on a lengthy course of stimulant medications; in a survey of 381 Australian children with ADHD seen in a pediatric clinic, 69 percent of parents were giving stimulant medications, and 64 percent were giving or had recently given some kind of complementary or alternative medical therapy such as the Feingold diet, restricting sugar, avoiding allergens, or giving vitamins or herbs.[24] Let's tour the Therapeutic Mountain to find out what works, starting with Lifestyle Strategies. If you want to skip to my bottom-line recommendations, flip to the end of the chapter.

Lifestyle Strategies: Communication, Behavior Management, Nutrition, Exercise, Environment, Mind-Body

Reframe How You Communicate about ADHD

Take a strengths-based approach to communication. Recognize the child's strengths and where possible, reframe negative labels or challenges as positive opportunities or gifts (Table 4.2).

Table 4.2: Reframing Labels	
Negative Label	Reframed as a Positive
Hyper	Exuberant, vigorous
Distractible	Aware of details that others miss
Spacey	Rich inner life
Driven by a motor	Energetic
Off-task	Creative
Impulsive	Eager, enthusiastic, willing
Inattentive	Listening to a different drummer
Poor concentration	Flexibly aware of changes in the environment
Accident-prone	Fearless

Behavior Management: SMART Plans

Identify clear, specific rules with achievable, measurable behaviors and clear time-linked consequences. In behavioral pediatrics, this is referred to as a SMART plan: Specific, Measurable, Achievable, Relevant, and Timely. For example, "Start to get ready for bed by brushing your teeth at 8 pm" is more specific, measurable, and timely than "Go to bed soon." "Spend 15 minutes studying spelling words before 4 pm" is more timely than "You'll have to work harder on homework."

Frame rules in positive terms. For example, "Please play with your toys in your bedroom" is more positive than "Don't leave your toys in the kitchen."

Organizational Skills, Anticipating Barriers, Providing Feedback

Help your child learn organizational skills by breaking down complex tasks into simple steps. Then tackle one step at a time, with plenty of praise in between. Rather than asking a child to "set the table," start with just one part of the request: "Put a plate at each person's place at the kitchen table." When that task is done, give positive feedback and then the next step: "Thank you! That's so helpful. Now put out the forks." As the child's memory and capacity improve, you can increase the number of steps or the complexity of the request.

Anticipate that the child will test the rules. Testing rules and limits is how children establish a sense of cause and effect, trust, and reliability. This is normal and expected. For example, if you ask a child who dislikes peas to "eat his peas," he may well leave a few (or many) on the plate to find out exactly what you mean. Or he may bargain (What if I eat all the carrots and leave some peas?) or rationalize (I shouldn't have to eat peas since I had a salad) or compare (Suzy didn't eat all of *her* peas) or distract (look at Dad, while feeding peas to the dog) or sabotage (roughly reaching for something and spilling the plate on the floor). It may be helpful to practice or rehearse a few of these scenarios in the office in a playful way to help families anticipate how to handle these normal situations when they arise at home.

Give positive feedback frequently and negative feedback neutrally. This practice will help counter the pattern of criticism and sense of failure that are all too common among families confronted with ADHD. "Catch them being good" is a cornerstone of behavioral pediatrics. It is easy to pick on the faults, failures, and lapses, and I am not advising you to ignore problems, but instead to make sure there's a balance of at least three episodes of praise for every one of correction. Rehearse corrective language. "We all make mistakes. How do you imagine handling it next time Johnny forgets his homework?"

The benefits of nurturing communication strategies can be more profound and long-lasting than medications. For example, when the Tolson School in Tucson began applying this approach, they found a sharp drop in discipline problems, a dramatic decrease in the number of students requiring special education, a marked reduction in the number of students requiring medication for ADHD (down to less than 1 percent), and an improvement in test scores.[25]

Working with Schools

Children spend more time at school than anywhere other than home, so developing a strong relationship with the school is vital. Parents can help teachers and school administrators recognize the child's unique gifts and challenges. Please schedule regular meetings with your child's teachers to monitor progress and advocate for arrangements that help the child succeed, such as seating arrangements that put the child near the front of the classroom. Advocate for your child to receive the public services to which he or she is legally entitled. According to the 1999 addendum to the U.S. Individuals with Disability Education Act (IDEA), children and youth whose disabilities adversely affect their educational performance should receive special services or accommodations that address their problem (e.g., ADHD) and its effects. Section 504 of the U.S. Vocational Rehabilitation Act prohibits discrimination against any person with a disability. Under Section 504, students may receive services such as a smaller class size, tutoring, modification of homework assignments, help with organizing, and other assistance.

If the child has not received sufficient services or accommodation within six months of asking the teacher or principal, you can write or ask your pediatrician to write to the school district's director for special educational services. The letter should request an evaluation for specific learning disabilities and a functional assessment to determine how the disabilities are affecting the child's classroom performance. These evaluations are required to develop an Individual Educational Plan (IEP) or a 504 Accommodation Plan. Middle school and high school students diagnosed with ADHD are also entitled to these evaluations and, if appropriate, an IEP or accommodation plans. With an IEP, the child may qualify for extra help, special classes, extra time for tests or projects, an extra set of books for home study, permission to take notes on a computer keyboard rather than by hand, extra breaks in the day, fewer classes, and other accommodations. Support teachers and administrators who offer creative, effective strategies to promote children's strengths.

Encourage your child to try other activities that explore his or her interests, talents, and possible lifelong passions or vocations. When choosing activities, consider the adult-child ratio. Music, art, tutoring, and individual language lessons may offer more individual attention than soccer leagues. Look for consistency. A class that meets every Tuesday is easier to schedule and attend than a band that has inconsistent practice and performance schedules that require frequent changes in the family routine.

Nutrition

Even though it accounts for less than 5 percent of the body's total weight, the brain consumes 20 percent of its energy. Provide a steady supply of high-quality fuel with optimal amounts of essential nutrients: essential fatty acids (like omega-3s) for healthy brain cell membranes; proteins to make chemical messengers; vitamins and minerals necessary for metabolism; and a steady supply of complex carbohydrates to fuel the brain's endless activity.

Hydration is important, too. Dehydration can impair attention and mood. In a small study of first graders, drinking water before taking a test led to better attention and greater happiness.[26]

COMMON NUTRITIONAL APPROACHES TO ADHD

- Avoid sugar (limited evidence)
- Focus on low glycemic index diet to keep blood sugar stable (moderate evidence)
- Avoid artificial colors and preservatives (moderately strong evidence)
- Avoid food allergies (best evidence for those with allergic symptoms)
- Avoid deficiencies of omega-3 fatty acids, iron, magnesium, and zinc (good evidence)

Even though *sugar* doesn't cause ADHD, your child should not subsist on cookies, cake, and candy bars. Sugar is bad for your child's teeth, and it replaces many more healthy foods needed for a balanced diet. If sugar or specific snacks make your child's symptoms worse, temporarily eliminate that from his diet (without telling anyone else), and then ask his teachers or other adults how they think he's doing. If you remain convinced that sugar is the culprit, your child can live without it, but don't expect it to be easy. Rather than sugar alone, focus on the overall pattern and quality of foods.

Focus on foods with a *low glycemic index*. The glycemic index reflects how quickly the body absorbs the energy in foods and how quickly those foods cause a spike and crash in blood sugar levels. High glycemic foods rapidly increase blood sugar, followed within a few hours by a crash in blood sugar, leading to large fluctuations in the ability to concentrate. Raw sugar, fruit juices, plain bagels, plain toast, doughnuts, and candy have high glycemic indexes. Whole fruits, whole grain cereals and

breads, vegetables, and dairy products have low glycemic indexes. Healthy five- to seven-year-old children fed a low glycemic load breakfast did better on tests of memory and sustained attention, and showed less frustration than kids fed a high glycemic index breakfast;[27] similar findings were reported in a study of the effect of low and high glycemic breakfast in teens.[28] Low glycemic foods may help prevent obesity and diabetes as well as even out mood and attention swings; they also tend to be richer in fiber, minerals, and vitamins than high glycemic foods. I recommend eating a breakfast containing low glycemic index foods (whole grains, vegetables, whole fruits, seeds, nuts, and dairy).

Avoiding artificial dyes may be helpful for many hyperactive children.[29] On the basis of all existing evidence to date, I recommend avoiding processed foods and foods that contain a high burden of pesticides (see the Dirty Dozen list from the Environmental Working Group) and artificial colors, sweeteners, and preservatives.

If you suspect a *food allergy* behind your child's symptoms, consider getting help from a registered dietitian to try a "few foods" diet. This diet includes only lamb and turkey, rice and potato, bananas and pears, root vegetables, green vegetables, sunflower oil, milk-free margarine, and bottled water. It avoids the most common food allergens: cow milk, soy, corn, wheat, citrus, tomatoes, eggs, peanuts, and artificial preservatives and colors. If children improve after two weeks on this diet, one new food per week is gradually added. Allergic children who improve on this regimen may have better sleep and less irritability as well as improved attention. This diet is extremely difficult for most families to maintain, and only a few children need it if they receive healthy nonprocessed foods.

Eat foods rich in *omega-3 fatty acids* such as trout, salmon, and sardines. Humans require these essential fatty acids in the diet in order to make cell membranes for red blood cells and brain cells. This is why omega-3 fatty acids like DHA (docosahexaenoic acid) and EPA (eicosapentaenoic acid) are added to infant formula—to support early brain development. Although flaxseed, walnuts, and green leafy vegetables contain the omega-3 fatty acid *linolenic acid*, humans convert only 5 percent to 10 percent of linolenic acid to the useful EPA and DHA. Children with ADHD, dyslexia, and autism spectrum disorders tend to have lower levels of omega-3 fatty acids in their blood stream, but these levels can be increased by dietary changes.[30] Encourage your child to consume foods rich in omega-3 fatty acids, at least 500 mg daily.

Avoid *iron deficiency* by eating foods rich in iron. Foods rich in iron include lean, grass-fed meats, poultry, seafood, iron-fortified foods, beans and peas, nuts, and raisins. Iron can be low in toddlers consuming a lot of cow's milk, children with chronic diseases or inflammation, those with bleeding disorders, and those who take proton pump inhibitor medications that interfere with iron absorption. In our clinic, we've found that nearly two-thirds of children with ADHD tested low for iron using a ferritin blood screening test. Other research confirms that average ferritin levels (reflecting iron levels) are lower in children with ADHD than those without ADHD.[31] Furthermore, some stimulant medications are less effective in children with low iron stores. Make sure your child's iron level is optimal by providing healthy foods. Excessive iron can be toxic, so please do not give your child iron supplements without having him evaluated by a health professional.

Avoid magnesium deficiency. *Magnesium* is a cofactor in over 300 different enzyme reactions in the body. Rich dietary sources of magnesium include almonds and other nuts, spinach, soybeans and other beans, avocado, potatoes, bananas, brown rice, yogurt, and oatmeal. European studies have shown that children with ADHD have lower magnesium levels than children without ADHD.[32] This is a problem because so many children and teens consume less than recommended amounts of magnesium. Magnesium deficits are also linked to constipation and irritability, which are both common in children with ADHD. Excessive magnesium supplements can cause diarrhea and other side effects. Please focus on a healthy diet containing all essential vitamins, minerals, and other nutrients; consult a health professional if you are considering supplementation.

Avoid zinc deficiency by eating foods rich in *zinc*. Foods rich in zinc include oysters, lean red meat, poultry, seafood, beans, nuts, and whole grains as well as fortified foods. Zinc levels can be low in poorly nourished children, picky eaters, children with inflammatory bowel diseases like Crohn's disease and ulcerative colitis, and children with sickle cell disease. Low zinc levels are linked to inattentiveness, but not hyperactivity.[33] Supplementing children whose zinc levels are already normal does not improve intelligence or behavior.

Exercise

Exercise is great for hyperactive kids. Kids with ADHD are often a bit clumsy, struggling with eye-hand coordination. Go for coordinated, aerobic activities like running, soccer, and swimming. Aim for at least 60 minutes of moderate to vigorous exercise daily. Even indoor sports like table tennis (Ping-Pong) can help, but outdoor exercise in nature is even better. Dr. David Katz at Yale promotes the ABCs of exercise at school: Activity Bursts in the Classroom.

If your child is involved in team sports, let the coach know that extra directions may be needed. Keep the child close to the coach or team leader so his attention is less likely to wander. Martial arts training in small groups may be helpful in promoting discipline, coordination, and self-esteem. *Relaxing, mind-body focusing exercises* such as yoga, QiGong, and tai chi may have special benefits for the hyperactive child. A comparison study showed that hyperactive children who were taught simple yogic breathing exercises, slowly moving stretches, and relaxing postures had improved self-esteem.[34] Teaching a child how to relax through movement is a perfect therapy for hyperactive children, who are on the move and need to develop relaxation skills.

Impulsive, distracted people are prone to injuries. They often end up in the emergency room for stitches and casts. Make sure your youngster wears proper protective gear whenever he goes out to rollerblade, scooter, ride a bike, or engage in contact sports.

Sleep

Sleep deprivation impairs organizational skills, memory, diligence, self-discipline, and focus. Make sure your child has a well-defined bedtime routine. Stimulant medications can contribute to insomnia. Be sure the sleeping environment is quiet, cool, dark, and comfortable. Keep electronics out of the sleeping area. Consider a pre-bedtime massage with lavender aromatherapy, soothing mint, or a bedtime story. Improving sleep habits can significantly improve daytime behavior and performance.

Environmental Strategies

Several environmental strategies are helpful in promoting calm, deliberate behavior:

- Reduce exposure to lead and other neurotoxins
- Organize home environment
- Use music
- Limit screen time
- Encourage play outdoors in nature

Help prevent your child from developing *lead poisoning* by keeping your home free of paint chips and dust. Use a wet mop and a damp duster at least twice a month; vacuum and dust window ledges weekly; and have toddlers wash their hands several times daily (at least before all meals and bedtime). Do not store your child's juice in glazed pottery or pewter containers, because both may contain lead that can leach into your child's juice. Keep lead-containing objects, such as watch batteries, fishing weights, and old, soldered toy soldiers away from toddlers. If you or your spouse work in a high-lead occupation (foundries, firing ranges, etc.), change your clothes as soon as you get home so you don't spread lead dust from your clothes to your house. Removal of lead-based paint should always be done by professionals who are trained and have the proper equipment to get rid of the lead without increasing airborne lead levels.

Your health care practitioner can test your child for lead poisoning. Most pediatricians test at least once during infancy and toddlerhood if lead poisoning is a problem in your area. If levels are high, treatment can help your child excrete the lead.

As mentioned earlier, *pesticides* and other environmental *toxins* adversely affect brain development. Go organic when you garden whenever possible, and purchase organic foods. Avoid storing food or heating it in plastics.

If you have *guns* at home, please make sure they are locked with ammunition stored separately to avoid injuries from impulsive distractible children who are exploring and playing.

Help your child create an *organized environment* at home. Minimize clutter to minimize distractions. Calm colors and simple lines help reduce overstimulation.

Can *music therapy* help? An Israeli study compared normal and hyperactive boys' work performances while they listened to different kinds of music: fast-paced vs. slower tempo vs.

no music.[35] As expected, the hyperactive boys tended to make more mistakes than the normal boys. Their performance was even worse while listening to fast-paced music. While listening to calmer, slow-tempo music, however, they did nearly as well as the normal boys. These intriguing results bear repetition in a variety of classroom settings. Pending such studies, it makes sense to keep all environmental cues (musical and otherwise) as orderly as possible to help hyperactive children maintain a steady pace.

One of the most helpful things parents can do is to *turn off* the *television and encourage kids to go outside and play*. If you can, spend some time in one-to-one activities with your child. Do *not* allow the child to have a television in his room; it is far too distracting and is linked to obesity and violence as well as reduced attention spans.

Environmental and mind-body approaches overlap when it comes to optimizing the educational setting for children.

OPTIMAL EDUCATIONAL ENVIRONMENT FOR KIDS WITH ADHD

- Structured learning environment (order and predictability)
- Simple instructions, repeated frequently about work assignments
- Visual as well as verbal instructions
- Use of tape recorders, computer-assisted instruction, and other audio-visual equipment to reinforce assignments
- Frequent feedback, for example, daily checklists or report cards
- Tests given with extended time and in a quiet setting with few distractions
- One-to-one tutorials or very small groups
- Attention to other learning problems such as memory and language processing problems

Mind-Body

Mind-body or behavioral therapy is a valuable key to treatment for hyperactive children. Parents must learn to express disapproval of a child's unwanted *behavior* while still expressing love for the *child* himself. Take every opportunity to praise your child whenever he behaves as you would like, even if it is only for a few minutes. Over the long term of months and years, mindful compassionate communication can eventually rebuild self-esteem, but it is far easier to maintain and improve self-esteem from the time your child is young rather than waiting until it is already severely impaired.

MIND-BODY THERAPIES FOR ADHD

- Structured schedule
- Step-by-step instructions
- Positive reward for desired behavior
- Tutoring
- Biofeedback and relaxation training
- Meditation
- Professional counseling

First, review your child's *daily schedule*. Make sure there *is* a schedule. Everything should be as routine and predictable as possible—meal times, nap time and bed time, exercise time, story time, day care or school time, bath time, cleanup time, relaxing time, chore time, and so on. It may help for you and your child to make a poster or chart together, listing all the day's activities and the times and places they occur. Try to stick as much as possible to the schedule, even on weekends and vacation time. Organization and structure are very important for children with ADHD.

Break every activity down into smaller steps. Hyperactive children need specific, detailed, step-by-step instructions. You can't just say, "Get ready for dinner." Give them no more than three to four steps: (1) stop the game; (2) put it away; (3) wash your hands; and (4) come to the table. You may have to go through this exact same routine a thousand times before they get it. If you vary it, your child may not get it right, even though he's done the same thing a hundred times before.

Reward your child for getting things right. He should get at least three compliments for every correction! Many families find a chart and sticker system very helpful. Put all your child's daily activities on a chart hung on his bedroom door. He gets a regular star (or ordinary sticker) for doing each activity when you remind him and a gold star (or special sticker) for doing each activity without needing to be reminded. Some parents use vouchers instead of stickers. When he has earned a certain number of vouchers, he can trade them in for a special reward.

Tutoring can be extremely helpful. It provides one-to-one attention and addresses other learning disabilities better than a single teacher in a crowded classroom can. Tutoring also provides an opportunity for immediate feedback. Private tutoring can be expensive, but volunteer tutors are often available at community centers and literacy programs. Also, call your school to arrange an Individual Educational Assessment and Plan so your child gets all the school services to which he is entitled by law.

Biofeedback that focuses on reducing muscle tension in the forehead (EMG biofeedback) not only reduces muscle tension but also seems to result in more relaxed behavior. Muscle biofeedback also seems to improve language skills and helps children feel more in control of their behavior.[36] It can work as well as medication in improving behavior.[37] Biofeedback is especially helpful when used along with other behavioral therapies such as structured scheduling and rewards for relaxed behavior.[38] Biofeedback training takes at least six to eight weeks and possibly as long as six months with regular practice at home to achieve maximal effectiveness.

Brainwave (EEG) biofeedback became a popular treatment for ADHD back in the 1990s. Children with ADHD tend to show patterns of slower brain wave activity than normal children.[39] By learning to regulate their brain wave activity through biofeedback, children can improve their attention[40]—benefits that last beyond the time they are doing biofeedback.[41,42] This is an advantage over medications, which don't work when they aren't taken. On the other hand, biofeedback is time-consuming and expensive. EEG biofeedback typically requires two to three sessions weekly for at least eight weeks, and is often combined with tutoring and coaching for specific learning problems. A typical course of treatment involves 20 to 40 sessions; typical costs range from $75 to $200 per session. Most insurances won't pay for EEG biofeedback, but when it's offered by a psychologist, professional fees may be covered. Check with your insurance carrier to learn more about what's covered for your child.

Another new mind-body approach for ADHD aims at improving *working memory* (one type of executive function). Compared with normal children, those with ADHD typically have deficits in working memory. A program called CogMed Working Memory Training® has been tested in several studies; it generally shows improvements in working memory with training 45 minutes daily over five weeks, but the impact on other ADHD symptoms is less clear.[43]

Meditation comes in many forms. Moving meditation like yoga and tai chi or walking meditation may be easier for restless children than sitting meditation.[44] In general, whether it's sitting still or moving, meditation practice leads to better EEG patterns and better attention.[45] Regular meditation changes blood flow in the brain and increases the size of the brain areas dealing with attention, focus, planning, emotions, and mood regulation.[46] Students who practice meditation have fewer problems with absenteeism, behavior problems, and distractibility, and better cognitive function and grades.[47,48] Mindfulness meditation in particular has shown benefits on attention, emotions, behavior, and grades.[49]

Relaxation training can help a tense, restless child learn how to relax.[50] It can be done in groups or with an individual therapist. There are several types of relaxation training: *progressive muscle relaxation*, *deep breathing*, *autogenic training* (similar to self-hypnosis), and even *yoga* exercises. Relaxation training of any sort works best if the parents are involved and the child continues to practice regularly at home. Learning the relaxation skills can help improve the child's self-esteem as well.

Professional counseling by a psychologist or psychiatrist can be very helpful for a number of reasons. First, a mental health professional can help make an accurate diagnosis and can reassure you that your child is not suffering from another problem such as depression or anxiety, both of which are difficult to diagnose in young children. Second, a professional can provide more in-depth education about attention deficit disorder. Professionals can also provide additional behavioral, communication, and problem-solving strategies for dealing with the child's hyperactivity and self-esteem.[51] They can also help families deal with the stresses inherent in having a child labeled difficult, different, bad, or handicapped.

Last, but not least, seek the support of other parents and families with hyperactive children. You are not alone and you don't need to feel alone. National support groups, local support groups, and even Internet support groups are available. See the list of resources at the end of this chapter.

Neurophysiological retraining therapies, which have not been proved effective in comparison studies, include patterning, visual retraining, and vestibular stimulation. I do not recommend them.

BIOCHEMICAL THERAPIES: MEDICATIONS AND DIETARY SUPPLEMENTS

Medications

Stimulant medications have been the mainstay of medical treatment for hyperactive children for thirty years. On average, they help about 70 percent of kids who take them, at least for the first two years of regular use. However, the American Academy of Pediatrics recommends that medications never be used as the only treatment for hyperactivity. Lifestyle, especially behavioral, environmental, sleep, nutritional, and mind-body therapies, are integral to the success of a comprehensive treatment program.[52] Medications help with symptoms of inattention, impulsiveness, distractibility, memory, and hyperactivity, but they do not correct underlying medical problems or correct learning disabilities such as dyslexia. Because so many children do not respond to the first medication they try, a growing number of children are prescribed not only a sequence of medications, but also multiple medications simultaneously. There are no long-term studies evaluating the safety of many years of psychoactive medication used alone or in combinations.

Table 4.3 is a list of some different kinds of psychoactive medications used to treat children with ADHD.

It may seem paradoxical that the medications that are most effective in treating hyperactivity are stimulants, but by increasing arousal, stimulants decrease symptoms for many children with ADHD. Tranquilizers and sedatives are not helpful for hyperactive children, but are sometimes used for children who are intense or oppositional. Long-term safety data are lacking. I do not recommend sedative medications or atypical antipsychotic medications for children with primary ADHD.

In the largest study of ADHD treatment, the

Table 4.3: Four Primary Types of Generic (Trade Name) Psychoactive Medications

Stimulant Medications	Anti-Anxiety and Sleep Medicines	Antidepressant Medications	Antipsychotic Medications
Amphetamine	Alprazolam (Xanax™)	Amitriptyline (Elavil™)	Aripiprazole
Amphetamine-dextroamphetamine (Adderall™)	Buspirone (Buspar™)	Amoxapine (multiple)	Chlorpromazine (Thorazine™)
Atomoxetine (Strattera™)	Chloral hydrate	Bupropion (USAN™)	Fluphenazine (Prolixin™)
Caffeine	Clonazepam (Klonopin™)	Citalopram (Celexa™)	Haloperidol (Haldol™)
Dexmethylphenidate (Focalin™)	Clorazepate (Tranxene™)	Desipramine Doxepin (Deptran™)	Lithium
Dextroamphetamine (Dexedrine™)	Diazepam (Valium™)	Escitalopram (Lexapro™)	Loxapine (Loxapac™)
Lisdexamphetamine (Vyvanse™)	Lorazepam	Fluoxetine (Prozac™, Serafem™)	Pimozide (Orap™)
(Ativan™)	Midazolam (Versed™)	Olanzapine (Zyprexa™)	Prochlorperazine (Compazinel™)
Methylphenidate (Concerta™, Daytrana™, Metadate, Methylin™, Ritalin™, Ritalin SR™)	Oxazepam (Multiple™)	Fluvoxamine (Luvox™)	Risperidone (Risperdal™)
Pemoline (Cylert™)	Mirtazapine (Remeron™)	Imipramine (Tofranil™)	Thioridazine (Mellaril™)
Zaleplon (Sonata, Andante™)	Zolpidem (Ambien™)	Quetiapine (Seroquel™)	Ziprasidone (Geodon™, Zeldox™)
		Nefazodone (Serzone™)	
		Nortriptyline (Pamelor™)	
		Paroxetine (Paxil™)	
		Sertraline (Zoloft™)	
		Trazodone (Desyrel™)	
		Venlafaxine (Effexor™)	
Non-Stimulants			
Clonidine (Kapvay™)			
Guanfacine (Intuniv™)			

Multimodal Treatment of ADHD (MTA) study, 579 children were randomly assigned to medication, behavioral therapy, both or neither. After 14 months, those who received both medications and behavioral care were doing better than others, but by 24 months the benefits were waning, and by 36 months, there was no longer any significant benefit from medications.[53] However, those who had been assigned to the medication group had worse growth that did not catch up after they stopped taking medication.[54]

This brings up the concerning facts about side effects. About one-third of children who take stimulant medications have side effects. The most common side effects include decreased appetite, abdominal pain, poor sleep, higher blood pressure, tics, agitation, hallucinations, and psychological dulling. Parents report "he's just not himself" or "he's lost his spark" or "she's sad or teary." Teens with ADHD who participate in the arts may report feeling less creative when they take their medications, so they sometimes use short-term medications for math class, but allow them to wear off for art, music, drama, or dance.

Medications have several other problems that make them less attractive as front-line treatments than other treatments that promote healthy habits in a healthy habitat:

1. They don't work when they're not taken, and many families have a hard time giving the medication daily, especially if it causes nausea or other side effects.
2. Overreliance on medications can occur to the extent that families don't address the healthy lifestyle issues that will improve overall health.
3. They are costly: Three of the top five drugs prescribed for children are ADHD medications.
4. There are long-term effects: We are just learning from the MTA study that stimulants appear to stop working within a few years, but there is little research on the longer-term effects of taking them for 10 or 20 years.
5. There is the possibility of misuse, diversion, or abuse: As the number of prescriptions for stimulants has grown, so has the number of reports that these drugs are being diverted or sold to people who do not have ADHD. There has been a 76 percent increase in the number of calls to Poison Control Centers related to abuse of ADHD medications.
6. Stimulant medications may address attention, impulsivity, and hyperactivity, but they do not benefit aggressive behavior or poor social skills.

All of these problems have led many pediatricians and parents to be somewhat skeptical of stimulant medications.

The best way to tell if your child will benefit from a stimulant medication is *the double-blind, placebo-controlled crossover trial*.[55] This means that you, your child, your child's teacher, and your health care professional together conduct a brief (three-week) study. The physician will talk with the pharmacist who will make up three different preparations:

a. A placebo pill
b. A low dose of a stimulant medication
c. A higher dose of a stimulant medication

Your child will take each of the medicines for one week. Neither you, nor your child, nor your child's physician will know the order of the medications. During each week, you and the child's teacher, day care provider, or coach will keep careful track of your child's symptoms on a symptom diary (such as Connors or Vanderbilt Parent or Teacher Rating Scale).

At the end of the three weeks, you return to your doctor with the symptom diaries for each of the three treatment periods and your best guess about when your child was taking the real medication. After reviewing the records, the physician will contact the pharmacist to find out when the child was taking placebo, low dose, and higher dose of medication. This may sound like a complicated process, but it is the most objective way to determine if your child will really benefit from taking that medication. Most parents would rather go through a three-week trial and know for sure rather than having their child on months and years of medication without really knowing if that's what he needs. I do not prescribe stimulant medication to my patients until we have gone through this process. Not all pharmacies are geared up to do this kind of study. Your physician may need to contact a pharmacy at the closest university or hospital to prepare the trial medications; it's worth the effort.

If you decide to use stimulant medications for your child, remember they are controlled substances and have a potential for misuse, addiction, and misdirection. A lot of prescription stimulant medication is finding its way to the street these days, and an increasing number of Ritalin poisoning cases have been reported to Poison Control Centers.[56] Lock them up at home, and deliver them to the school nurse yourself if doses must be given at school.

Many parents are worried that giving their kids these powerful stimulant medications (formerly referred to as "speed") may lead them to become drug addicts. This is an understandable concern, especially since kids with ADHD have an increased risk of later developing alcohol and drug problems. However, a five-year follow-up of the large MTA study found neither an increased nor decreased risk of drug addiction among those children assigned to stimulants compared with those who had not received them.[57]

Several studies have shown that a non-stimulant medication, *pemoline,* can also help improve attention and learning. However, pemoline may cause serious liver problems. Children who take pemoline need regular tests to monitor liver function. Because of its side effects, it is never the first medication I choose when treating a child with ADHD.

Another non-stimulant medication, *atomoxetine,* is an antidepressant medication that is preferred for use among teens at high risk for drug abuse. It works about as well as stimulants, but benefits may take three to four weeks to become apparent.[58] Other medications used to treat hyperactive children include antidepressants and the blood pressure medicine, *clonidine*. Some children are given Ritalin during the day and clonidine in the evening. Clonidine can make kids very sedated and worsen

depression and lower blood pressure. I am hesitant to prescribe it, especially because clonidine overdoses reported to Poison Control Centers have skyrocketed; most prescriptions that resulted in overdoses were for ADHD.[59] *Antidepressant medications* such as desipramine can also result in serious side effects. Children have died as a result of cardiac side effects when taking antidepressant medication for hyperactivity. Atypical antipsychotic medications can also have serious side effects such as weight gain and diabetes. All of these prescription medications should be taken only under the close supervision of a qualified health care professional. They are not my first choice of treatments for ADHD.

Dietary Supplements and Herbal Remedies

Omega-3 fatty acid supplements like EPA and DHA have become the most popular supplements to prevent and treat ADHD. Back in the 1990s several studies suggested that supplemental *omega-6* fatty acids such as those found in evening primrose oil might be helpful for ADHD. Both omega-3 and omega-6 fatty acids are essential nutrients. More recent research suggests that a *combination* of the two is more effective than omega-6 supplements alone;[60] furthermore, omega-3 supplements alone appear to benefit children with ADHD.[61] In a 2010 study, 15 weeks of supplementation with 500 mg daily of EPA alone was associated with improvements in teacher-rated attention and oppositional behavior in children with ADHD.[62] More recent research supports the beneficial effects of omega-3 supplementation on working memory,[63] reading,[64] learning problems, and ADHD.[65] An analysis of multiple studies concluded that based on the evidence (as of 2011), "Omega-3's, particularly with high doses of EPA, were modestly effective in the treatment of ADHD"; modestly effective means about 40 percent as effective as stimulants, without adverse effects on appetite or growth.[66] The main side effect of omega-3 supplements is fishy breath and burping.

Remember that omega-3 supplements are most helpful for children who (a) have problems with reading, cognition, or behavior; and (b) have less than optimal levels of omega-3 fatty acids in their systems. They are not helpful for children who are already functioning well and consuming plenty of fish, seeds, and green, leafy vegetables.

Deficiencies of essential nutrients such as *iron, zinc, magnesium, pyridoxine* (*vitamin B6*), and *vitamin C* have been suspected of causing hyperactivity in some children.[67]

In children whose *iron* stores are low, reflected in low serum ferritin, supplements can optimize iron stores, improve cognition, and decrease ADHD symptoms.[68,69] Remember, excessive iron can be toxic, so please do not start supplements without an evaluation by a health professional.

Hyperactive children who have low *zinc* levels may be less likely to respond to stimulant medications than children with normal zinc levels.[70] Optimizing zinc levels through supplementation may decrease the need for stimulant medications.[71] As with iron, excessive zinc may be harmful, so please do not embark on long-term, high-dose zinc supplementation without an evaluation by a health professional.

A double-blind controlled study of hyperactive children given megadoses of *vitamins C and B* and *calcium* showed no benefits. If anything, children treated with megavitamins tended to have worse behavior, and over 40 percent developed signs of liver toxicity.

On the other hand, newer research has begun to test the effectiveness of nutrients that are often deficient in children with ADHD. In one study, a combination of omega-3 fatty acids, zinc, and magnesium for 12 weeks resulted in significant improvements in sleep,

attention, behavior, and emotional problems in 5- to 12-year-old children.[72] When I see a child with a poor nutrient intake or someone who has tested low for one or more essential nutrients, I often recommend a combination of omega-3 fatty acids and essential minerals for three months while we work on improving the diet.

Melatonin is a hormone produced by the pineal gland. Unlike most other hormones, it is sold over the counter without a prescription. Melatonin helps regulate normal sleep and wakefulness cycles and other biorhythms such as temperature fluctuations over the day. Two studies have shown that it can improve sleep for kids with ADHD (in doses of 0.5 to 6 milligrams nightly). Sleep is often a problem for children with ADHD. Melatonin supplements can help children (even those on stimulants) fall asleep faster and sleep more soundly.[73] Many parents find melatonin helpful for addressing just this one aspect of problematic behavior: sleep. Melatonin has few side effects even with long-term use, and it does not lead to hangovers or addiction.[74]

Supplements of the amino acid *carnitine,* found in meat and fish, had been thought to help improve attention and behavior for children with ADHD, but more recent placebo-controlled trials have not found an overall benefit with carnitine supplements.[75,76] Focus instead on a healthy diet.

Traditional herbal medicine has remarkably little to say about hyperactivity, probably because ADHD is largely a modern diagnosis. The German Commission E (which is roughly equivalent to the U.S. Food and Drug Administration) recommends *valerian* to treat restlessness, which has been interpreted to mean that it is approved for treating ADHD. However, sedative medications and herbs are generally *not* helpful for children with ADHD. On the other hand, sedative herbs like valerian, chamomile, hops, kava, and lemon balm may help children fall asleep. Although there is insufficient research to recommend these herbs alone or in combination with effective treatments for ADHD, I do not oppose using them occasionally to help a restless child fall asleep.

OTHER HERBS POPULARLY USED TO TREAT ADHD

- Coffee
- Ginkgo
- Ginseng
- Pycnogenol
- Blue-green algae, Spirulina

Supplementing your child with America's favorite herb, *coffee,* makes more sense than using just about any other herb to treat ADHD. *Caffeine* (found in coffee, tea, cola drinks, cocoa, and some other herbs) has been evaluated in numerous studies as an alternative treatment that is fairly safe, inexpensive, and readily available. However, caffeine is not as effective as prescription stimulants.[77–79] Low doses of caffeine may benefit children who are already taking stimulants, boosting the benefits without substantially increasing side effects.[80] Caffeine's well-known side effects include feeling jittery, nervous, anxious, and eventually feeling tired when the effects wear off. Caffeine should not be used as a replacement for a good night's sleep.

Once *ginkgo* became a popular remedy for adults concerned with memory problems, it wasn't long before it was marketed for children. It can cause bleeding problems in people who take anticoagulants or aspirin regularly, and it can interfere with some antidepressant medications. On the plus side, a pilot study from Italy indicated that ginkgo may help improve ADHD symptoms.[81] A German study using a standardized ginkgo extract (EGb 761) at a dose of 240 mg daily for three to five weeks showed a low rate of side effects and possible improvements

in ADHD symptoms, but there was no control group, so placebo effects may have accounted for the benefits. An Iranian study that did include a comparison group reported that ginkgo was less effective than stimulant medications. A Canadian product (AD-fX) that combines *ginseng* and ginkgo benefited patients with ADHD or dyslexia in one manufacturer-sponsored study.[82] Based on the limited research so far, I don't recommend ginkgo as a treatment for ADHD, but I do tolerate it if families wish to try it.

Pine bark extract contains antioxidant chemicals known as *pycnogenols* or oligomeric proanthocyanidins (OPCs). These compounds are also found in grape seed extract, apples, grapes, raspberries, and blackberries in lower concentrations, and are among the large family of flavonoids. They are potent antioxidants. Pycnogenols became widely marketed as ADHD remedies in the mid-1990s. There are numerous testimonials about their efficacy in treating ADHD.[83,84] They are very safe, but until there are studies evaluating their benefits compared with other treatments for ADHD, I do not routinely recommend them for this condition.

Another widely marketed natural product is *spirulina,* also known as blue-green algae, and affectionately called "pond scum" by those familiar with its source. Spirulina contains protein, a variety of B-complex vitamins, and several trace minerals. However, it can also contain pesticides, animal feces, bacteria, fungi and fertilizer, heavy metals, and other contaminants from the water in which it grows. There are several reports of patients becoming very ill after taking it. There are no studies showing that it is helpful for kids with ADHD. I do not recommend it.

A few herbalists recommend detoxifying teas such as *red clover, lemongrass, and milk thistle.* There are no scientific studies showing that these herbs help alleviate ADHD symptoms. A small Israeli study evaluated a proprietary blend of several herbs of the Nurture and Clarity brand and reported significant improvements in objective measures of attention.[85] Until more studies have evaluated the safety and effectiveness of herbal blends for ADHD, I do not recommend them.

BIOMECHANICAL THERAPIES: MASSAGE, CHIROPRACTIC

Massage

One study evaluated the benefits of daily massage among 28 boys with ADHD; they reported less fidgeting and greater happiness for those who received the massage.[86] Another study of 30 children with ADHD randomized half to receive massage for 20 minutes twice a week for a month; compared to those who served as controls, the massage group reported better moods and classroom behavior.[87] While these studies are promising, they are insufficient to convince insurance companies to reimburse for massage therapy as a treatment for ADHD. Longer-term studies and more objective outcome measures are needed before insurance companies are likely to pay for this sort of thing. Feel free to give your child back rubs or foot massage if these practices are consistent with your family's values.

Chiropractic and Osteopathy

Some chiropractors claim to improve ADHD symptoms by adjusting the bones and soft tissues in the neck and skull. There is little research evaluating these claims.[88] A small 2014 study suggested that compared with treatment as usual, *osteopathic* manipulative therapy could improve some test scores in children with ADHD, but this study bears repeating before osteopathic manipulation becomes a standard treatment for ADHD.[89]

BIOENERGETIC THERAPIES: ACUPUNCTURE, HOMEOPATHY

Acupuncture

Although there are case reports of dramatic improvement in ADHD with acupuncture treatments, acupuncture has only been evaluated in one randomized controlled trial for ADHD in China.[90] Although it is safe, acupuncture's effectiveness has not been established for ADHD treatment,[91] so I do not routinely recommend it.

Homeopathy

Homeopathy is another safe route that many parents have explored while looking for non-drug treatments of their child's ADHD. Again, a few cases seem to improve dramatically, but few studies have compared homeopathy with placebo remedies or with standard medications. Until there are at least two large studies comparing homeopathic treatments with other standard therapies, I do not recommend them routinely, but since homeopathy is safe, I don't discourage families who want to try it.

WHAT I RECOMMEND FOR ADHD

PREVENTING HYPERACTIVITY

1. *Lifestyle—environment.* Do not smoke, drink, or use recreational drugs during pregnancy. Do not smoke around your child.

Keep your child's environment free of lead. Get professional help in removing lead paint. Keep your dust levels down with weekly damp-mopping and dusting.

Do not put a television set in a child's bedroom, and limit all television and video games to less than two hours daily.

2. *Lifestyle—diet.* Eat whole foods rich in essential nutrients like fish, fruits, vegetables, beans, nuts, and whole grains. Avoid deficiencies of iron, zinc, magnesium, and omega-3 fatty acids. Avoid exposure to artificial colors, flavors, and preservatives. Reduce your child's exposure to pesticides by purchasing organic or the Clean 15.

3. *Lifestyle—exercise.* Encourage your child to turn off the electronics (an hour or less of screen time daily), and go outside and play in natural settings. Protect your child from injuries by providing proper protective equipment while bicycling, rollerblading, and engaging in contact sports. Always use seatbelts and child restraint devices when riding in a car.

Take your child to your health care professional to be evaluated for:

- Other health problems: hearing, vision, thyroid, sleep apnea, lead poisoning, pinworms, allergies, eczema, and other health problems that may interfere with his learning and behavior.

- Complete behavior rating scales such as the Connors or Vanderbilt parent and teacher scales. These will help you determine objectively how serious your child's symptoms are and how much he improves with various treatments.

Take your child to a clinical psychologist or educational specialist for:

- Testing for other emotional, mental, behavioral, or learning issues: intelligence testing, dyslexia, and testing for special learning problems; additional testing for emotional problems or stresses that may trigger problem behavior, additional advice about behavioral management; additional information about hyperactivity and support groups.

---

4. *Lifestyle—communication and behavioral management.* Frame your child's qualities in positive terms. Make SMART plans for behavior. Break complex tasks down into simple steps and praise progress frequently. Be persistent and consistent with expectations, rules, and communication. Avoid blame and harsh punishments. Use logical consequences to reinforce behavioral goals. Give your child clear structure. Make his routine consistent, orderly, low-key, and predictable. A wall chart of his daily activities may be very helpful. Have consistent meal time, bed time, chore time, study time, play time, and so on.

TREATING HYPERACTIVITY

1. *Lifestyle—as above for prevention.* Accept your child as he is. Avoid blaming him or yourself. Remember that he is not trying to misbehave. Be patient and persistent in reminding him of your expectations about his behavior. Reframe his problems as strengths and opportunities and focus more on goals and strategies than on diagnoses and labels. Set SMART goals. Reward him regularly for desired behavior. He should get at least three compliments for each correction. Give feedback immediately. Use time-out rather than physical punishment for discipline. Focus on building self-compassion, recognizing his common humanity with others, and treating himself and others with kindness.

Consider having him tutored in his most challenging subjects. Get as much academic help and help for any learning disabilities that the school provides.

Teach your child relaxation and mental-emotional self-regulation skills. These can be breathing exercises, progressive muscle relaxation, yoga, or other methods. Reward your child for his efforts to practice relaxation skills. Consider taking your child to a professional counselor for training in relaxation and self-regulatory skills or biofeedback. Support and encourage your child's practice at home.

Recognize that having a child with ADHD is stressful for most parents. Recognize your own needs for time-out and breaks. Seek support. Regularly talk with your child's teachers and other parents whose children have ADHD.

2. *Lifestyle—nutrition and supplements.* Ensure your child has adequate intake of essential nutrients like iron, magnesium, zinc, and omega-3 fatty acids. If you suspect a deficiency, see a health professional for guidance on appropriate supplements. Avoid foods containing artificial colors, flavors, and preservatives. If you suspect allergies or sensitivities, see a health professional for a thorough evaluation.

3. *Lifestyle—exercise.* Regular vigorous exercises and stretching, for at least 60 minutes daily, and relaxing exercise such as yoga or tai chi may be very helpful. Focus on sports with close interaction between child and coach (such as martial arts training or small teams).

4. *Lifestyle—sleep.* Make sure he gets at least 8–10 hours of sleep nightly.

5. *Lifestyle—environment.* Reduce excessive electronic or noise stimulation in your child's environment. Consider playing slow-tempo rather than fast-paced background music. Reduce the amount of time he spends in front of a television and do not allow one in his room. Reduce clutter. Create an organized schedule and orderly household. Spend more time in nature.

6. *Biochemical—herbal supplements.* If you are interested in milder, natural stimulants, consider giving your child extra caffeine in the form of coffee or tea. Remember that even natural stimulants can have side effects such as decreased appetite, feeling jittery, and insomnia. If sleep is a problem, consider melatonin or calming herbs.

7. *Biochemical—medications.* In conjunction with your health care professional and your child's teacher, do a double-blind crossover of medication vs. placebo to determine whether or not your child will benefit from medical treatment. If he does, make sure he gets it regularly and recheck his need for the same dose every year. Do not rely on sedative medications.

8. *Biomechanical*—Consider regular massage.

SUPPORT GROUPS

All Kinds of Minds, AKOM
  www.allkindsofminds.org

CHADD: Children and Adults with Attention
  Deficit Disorder
  http://www.chadd.org/

Learning Disabilities Association
  www.ldanatl.org

Mental Health America
  www.mentalhealthamerica.net

National Attention Deficit Disorder Association
  http://www.add.org/

National Center for Learning Disabilities
  www.ncld.org/

5

ALLERGIES

Every spring Yvonne sneezes and gets itchy, watery eyes and a runny nose.

When Cory eats strawberries, he breaks out in hives.

Aaron has chronic diarrhea, eczema, and asthma. He also has dark circles under his eyes and is not growing as well as his brothers.

When Anne wears inexpensive earrings, her earlobes swell painfully.

Twelve-year-old Joannie has spina bifida and severe reactions to latex balloons.

Fifteen-year-old Don says he's allergic to homework.

All but one of these are examples of allergies. Despite Don's complaints to the contrary, like most kids, he is *not* allergic to homework.

Allergies cause many different symptoms in different people at different ages and are triggered by different things. Symptoms can range from fussiness to fatal shock. Allergies are an overreaction of the immune system to a protein that is breathed, ingested, injected, touched, or innate. Allergies affect more than 50 million Americans, account for millions of days lost from school, and cost over $14 billion annually.[1] Food allergies affect about 4 to 8 percent of children, skin allergies affect 3 to 10 percent, and nasal allergies such as hay fever affect as many as 30 percent. The number of children and adolescents who suffer from various allergies is climbing, and no one is exactly sure why.

Allergies are confusing because the body

reacts to what it perceives as allergies in different ways; many allergic symptoms mimic other illnesses. On the other hand, many symptoms blamed on allergies are not true allergies. The vast majority of people who can't tolerate milk aren't allergic to it, but lack the enzyme needed to digest milk sugar. About 1 to 2 percent of people of Northern European ancestry have celiac disease, which is sensitivity to gluten (a component of wheat and other grains); this can cause upset stomach and diarrhea, problems with nerves and balance, and poor growth. Kids who have an upset stomach after eating a chili dog slathered with onions may have poor judgment, but probably don't have allergies to chili. Belching and passing gas after eating raw vegetables or beans are not due to allergies either. Chinese restaurant syndrome (flushing, sweating, headache, and facial pain or chest pain) is a direct chemical effect of the high levels of glutamate found in the flavor enhancer MSG, monosodium glutamate, not an allergy to bean sprouts. Red eyes after swimming are due to a chemical irritation from chlorine, not allergies.

COMMON ALLERGY SYMPTOMS

Skin: hives, eczema, swelling, itching

Nervous system: headache, fatigue, confusion, difficulty concentrating, decreased attention span, depression, insomnia, and other sleep problems

Eyes: red, itchy, watery eyes; dark circles under the eyes; swollen eyelids

Nose: watery, itchy, sneezing, congestion; horizontal wrinkle across tip of nose

Mouth: itchy roof of the mouth; swollen tongue, lips; sore, scratchy throat

Lungs: asthma, wheezing, coughing, feeling of tightness in the chest

Heart: rapid heart rate, irregular heartbeats

Muscles and joints: muscle pain, joint pain, arthritis

Intestines: colic, diarrhea, nausea, vomiting, constipation, abdominal pain, itchy anus, bloody diarrhea

Urinary system: sense of urgency or frequent need to urinate

Kids who suffer from allergies are truly miserable. Troublesome daytime symptoms seem worse at night, keeping restful sleep at bay. A tired, itchy, sneezing child is a kid whose apt to have a hard time paying attention in class and being well-behaved. As if that's not enough, the medications that are often used to control the symptoms (such as antihistamines) often make kids drowsy and dopey-feeling. All of this can interfere with school performance and contribute to underlying anxiety, depression, and low self-esteem.

WHAT CAUSES ALLERGIES?

Allergies are symptoms of an *oversensitive immune system*. They occur when the immune system decides that something that has come in contact with the body is dangerous and needs to be fought. Fights on one front may lead to symptoms in another system; for example, food allergies trigger breathing problems in about 10 percent of kids with asthma.

Allergies run in families. If one parent has allergies, children have a one in three chance of having allergies, too. When both parents have allergies, chances are more than 50:50 that their children will have allergies. If a sibling has allergies, a child is about seven times more likely to have allergies. Allergies are more common in boys than in girls, and more common in African than Caucasian Americans.[2]

A popular theory about the cause of allergies is called the "hygiene hypothesis." This theory says that humans evolved with animals, living close to the earth. As we've moved indoors and kept babies clean and separated from farm animals, worms, and dirt, their immune systems don't get the kind of workout that promotes a healthy immune system. There is some pretty good evidence for this theory since children who are raised on farms tend to have fewer allergies than kids raised in the city or suburbs; and kids raised with dogs tend to have fewer allergies than those raised without them. Children born vaginally have a lower risk of developing allergies than children born antiseptically via Caesarean section; also, children whose mothers took antibiotics during pregnancy have an increased risk of developing allergies.[3] Another intriguing recent finding is that children whose families hand-wash their dishes have a lower risk of allergies than children whose families rely on efficient, hygienic machine dishwashers.[4] Finally, babies whose moms consume probiotics (healthy bacteria and yeast) during pregnancy and who continue to take them while nursing have a lower risk of allergies and eczema than babies whose moms do not consume healthy bacteria.

Eating habits can also influence the development of allergies. We used to think that allergy-prone children should avoid eating allergenic foods like peanuts until after a year of age, but newer research indicates that earlier exposure, at least to peanuts, may be protective.[5] In a carefully conducted randomized controlled trial, parents whose babies were tested and found to be *not* allergic to peanuts were either told to completely avoid peanut products, or offer small peanut-containing snacks three times a week for four years; by the time they were five years old, 14 percent of those who had *avoided* peanuts had developed peanut allergies compared to less than 1 percent of those who had regularly eaten small amounts of peanuts in processed food (not whole peanuts, which can cause choking).[6] Many experts now recommend that infants be introduced to a variety of foods between 4 to 7 months (17 to 27 weeks) while continuing to obtain most nourishment from mom's milk.[7]

Living inside and using sunscreen help protect us from damaging UV rays, but they also limit our natural ability to use sunlight to make *vitamin D*. Vitamin D plays a critical role in regulating the immune system. Moms with lower vitamin D levels have babies with a higher risk of developing asthma, eczema, and allergies. And infants with higher vitamin D levels are less likely to develop food allergies.[8]

Certain *medications* can increase the risk of allergies. For example, taking acid-suppressing medications to treat reflux or heartburn antacids interferes with proper stomach acid breaking down the proteins in food; this can increase the risk of developing food allergies. If you want to reduce your child's risk of allergies, do not start him on acid-suppressing medications for simple reflux.[9]

Things that increase *inflammation* tend to increase the risk of allergies, too. For example, obesity increases inflammation and the risk of allergies and asthma.[10]

My child is allergic to pollen, cat dander, mold, and dust mites. How many different kinds of allergies are there?

HOW MANY TYPES OF ALLERGIES ARE THERE?

Even though a person could be allergic to hundreds of different things, there are just four main *types* of allergic reactions, sensibly called Types 1, 2, 3, and 4. Type 1 and Type 4 are most common, but we'll cover all the bases, just to be complete.

Type 1 allergic reactions are immediate; they are usually due to reactions of the immune molecule, immunoglobulin E (IgE). IgE molecules trigger the *mast cells* in the immune system to release *histamine*—the chemical that causes swelling, itching, and watering. Type 1 reactions cause watery eyes, sneezing, hives, itching, swelling, and in the most extreme cases, shock and death. People who are allergic to latex rubber (such as Joannie, who had had several operations in which the operating team used latex gloves) can become so sensitive that they develop symptoms if they even enter a room in which latex was used. Fortunately, severe Type 1 reactions are treatable with medication.

Just as we were about to close the clinic one warm summer evening, Bill Aylers rushed in the door carrying his pregnant wife, Judy. She was weak to the point of collapse, wheezing and dotted with hives. Bill told me that he and Judy had been visiting relatives in the country. As they got in the car, Judy was stung by a bee. She felt weak and started wheezing almost immediately. As he frantically drove down the road, he saw our clinic lights on and pulled in. Within minutes after a shot of adrenaline (epinephrine) and a dose of Benadryl (diphenhydramine), Judy roused, the hives faded, and the wheezing improved. Bill's quick action and our clinic's readiness for this kind of emergency saved his wife's (and unborn baby's) life.

This kind of immediate, life-threatening Type 1 allergic reaction is called *anaphylaxis*. In sensitive people, insect stings, certain medications, or even common foods, such as eggs, nuts, strawberries, or shellfish, can trigger anaphylaxis. Parents whose children have had an anaphylactic reaction to common foods such as eggs or wheat must be vigilant about the ingredients in prepared or processed foods such as breads, cakes, snacks, and candy because they may contain the fatal allergen. Type 1 allergic reactions to penicillin cause 400 deaths per year, and food allergies cause about 200 deaths in the United States each year—about twice the number of deaths due to allergic reactions to insect stings.

Most of the time, the IgE molecule causes Type 1 immediate reactions that are less serious, but still uncomfortable. The reactions most people get to ragweed pollen or cats happen within an hour or two of exposure—watery, itchy eyes, sneezing, wheezing, coughing, and hives. This less serious variety of Type I reaction is far more common than anaphylaxis. It is the main reason for the booming industry in antihistamines. Many food allergies are also caused by IgE molecules. These allergies trigger asthma symptoms and eczema for many children; in fact, among children who suffer from both severe, chronic eczema and asthma, eggs induce symptoms in 50 percent, and wheat causes reactions in 20 percent.[11]

In *type 2 allergic reactions* antibodies attack not just a molecule, but a whole cell. This is the kind of reaction that destroys red blood cells if the wrong type of blood is given during a *transfusion*. Type II reactions are uncommon nowadays because of careful cross-matching prior to blood transfusions.

Type 3 allergic reactions occur when antibodies bind to proteins in large clusters that settle in various tissues such as the kidneys and the joints. Wherever they settle, these clusters cause problems such as *allergic arthritis*. It usually takes several days after the initial clusters form to develop symptoms. This may be how allergies to certain foods and food additives trigger joint pain and arthritis.[12]

Type 4 allergic reactions are typified by the common responses to *poison ivy, perfumes,* and *nickel jewelry* (contact dermatitis). They usually take twelve to twenty-four hours after contact to become visible. Type 4 reactions cause intense itching and a blistering rash, but are rarely life threatening. Just about everyone is allergic to poison ivy. About 25 percent of children are allergic to other contact allergens such as nickel.

How Much Allergen Does It Take to Trigger Symptoms?

Some allergies don't depend on the dose of allergen—one bee sting or one bite of eggs may be enough to trigger a severe reaction. Other allergic responses do depend on the dose. A child who is allergic to cow milk may tolerate one-quarter cup of milk on his cereal in the morning, but experience severe bloody diarrhea if he drinks a quart of milk in a day. Some allergies (such as that to animal dander) are perennial (all year round), and some only occur during certain seasons (such as ragweed season in the fall or tree pollen allergies in the spring). Some allergies are lifelong, and others are outgrown; for example, about 6 percent of children less than three years old have food allergies, but this decreases to 2 to 3 percent by five years old and continues to decline as we age. The most commonly outgrown food allergies are to milk, egg, and soy; it is less common to outgrow allergies to shellfish, nuts, or peanuts. For example, among babies who had proven allergies to eggs at 12 months old, nearly half tested free of egg allergy a year later; among those who were initially free of egg allergy, those most likely to stay allergy-free were those had had eaten small amounts of egg regularly, thereby maintaining their tolerance to eggs.[13] Hay fever becomes more common in adults, peaking between 20 and 40 years old.

Why Do Some Kids Develop Allergies and Others Don't?

No one knows for sure exactly why some people develop allergies and others don't. Allergies run in families, but the number of people with allergies is rising much faster than our genetic heritage changes. Although being an only child of wealthy, older parents is good for preventing most health conditions, it is ac-tually associated with an increased risk of hay fever and grass allergies; when it comes to allergies, you're much better off being the youngest child in a large family or starting day care in the first six months of life.

Current theories about the increase in allergies focus on recent changes in the environment—from what we eat, to what we breathe, to how many people we live with, to what infections we get. There are concerns about the potential effects of processed foods, pesticides, hormones, genetically modified foods, and antibiotics given to livestock. Others believe that the problem is due to decreases in breast-feeding rates, increases in air pollution, decreases in exercise, or increases in urbanization (decreased exposure to pets and farm animals) or something else entirely. One of the weirder and most intriguing ideas about what's causing the increase in allergies is that we've been too successful in eliminating parasites from our intestinal tract. This theory says that our immune systems evolved over millions of years to combat these critters, and when we eliminate them, our bored immune systems started picking on other things.[14]

COMMON ALLERGY TRIGGERS

OUTDOORS

- Plants and their pollens such as poison ivy and poison oak, ragweed, grass
- Chemicals such as insecticides, pesticides, fertilizers, fumigants
- Air pollution such as sulfur dioxide and particulates
- Insect venom such as bee stings

INDOORS

- Animal dander, saliva, or urine
- Dust and dust mites

- Foods such as nuts and peanuts, cow's milk, soy, fish, eggs, wheat, oranges, strawberries, chocolate, tomatoes, corn
- Herbs such as feverfew, chamomile, yarrow, tansy
- Irritants such as wool, fabric finishes, dry cleaning solvents
- Chemicals such as perfumes, soaps, detergents, cosmetics, cleaning products, disinfectants, solvents, turpentine, paint, formaldehyde
- Food colorings, additives, preservatives, hormones, antibiotics
- Medications such as aspirin, antibiotics, morphine, X-ray contrast material

In addition to not feeling well, children who suffer from allergies frequently suffer from *other problems*. For example, they tend to *grow poorly*. So much of their energy is consumed with the allergy, there just isn't enough left over to grow. Chronic nasal allergies lead to nasal obstruction, mouth-breathing, and can necessitate orthodontic intervention. Children with hay fever–type allergies can have more frequent *ear infections, sinus infections,* and *delayed language development*.

When children eat foods to which they are allergic, the normal intestinal barrier is breached, promoting sensitivity to even more foods.[15]

COULD IT BE SOMETHING ELSE?

Yes. It is often difficult to diagnose allergies in babies because allergic symptoms can easily be attributed to other problems: colds, diarrhea, eczema, and sleeping problems. Sometimes it's hard to tell the difference between a cold and an allergy because both cause runny or stuffy noses and sneezing, but allergies do not cause fevers. Children who have

diarrhea and gas after drinking milk due to lactase deficiency have similar symptoms to children suffering from true milk allergies. If you are not sure whether your child's symptoms are due to allergies or something else, take him to be evaluated by a professional who is experienced in treating allergies.

DIAGNOSING ALLERGIES

The basis of all allergy diagnosis is the relationship of the child's symptoms to his exposure to allergens. Physical examination is also useful. Help your physician make a diagnosis by keeping a careful symptom diary, noting what kinds of symptoms your child develops, the timing, previous exposures, and so on. This will be the most helpful tool in figuring out what's causing allergic symptoms. Some triggers work immediately—the child develops hives within minutes of exposure to the trigger—whereas others are more delayed, with symptoms not appearing for 6 to 48 hours. Be alert for hidden exposures; one woman, for example, who was allergic to numerous foods and pollens and who made every effort to eliminate them, continued to have symptoms until she realized that her shampoo contained apricot—to which she was allergic. Changing shampoos dramatically improved her symptoms.

A variety of allergenic foods are hidden in common foods and ingredients.[16] Sometimes food manufacturers use the same equipment to make different products. One product may call for eggs, and even if the next recipe doesn't require eggs, tiny amounts of egg may remain on the equipment and contaminate the next product (Table 5.1). Milk protein also shows up in a lot of prepared foods (Table 5.2). Soy protein is called by several different names on ingredient lists. It's hard to avoid in processed foods (Table 5.3). All manufactured, processed foods must be suspect for highly allergic children. Kids may have severe reac-

Table 5.1: Ingredients and Foods That May Contain Eggs

Ingredients	Foods
Albumin, binder, coagulants	Baked goods, baking mixes, batters
Egg white, egg yolk, emulsifier	Béarnaise sauce, breakfast cereals
Globulin, lecithin, ovalbumin	Cake flours, cookies, custard, egg noodles
Powdered egg, vitellin	French toast, Hollandaise sauce, ice cream
Whole egg	Lemon curd, macaroni, malted drinks
	Mayonnaise, meringues, muffins
	Omelets, pancakes, waffles
	Puddings, sherbets, souffles, spaghetti, egg noodles, pasta
	Tartar sauce

Table 5.2 Ingredients and Foods Containing Milk Protein

Ingredients	Foods
Butter flavor, butterfat, buttermilk solids	Batter-fried foods, biscuits, bread
Caramel color, casein, caseinate	Breakfast cereals
Cheese, cream, curds, dried milk	Cakes, chocolate, cookies
Dry milk solids, high-protein flavor	Cream sauces or soups, custard
Lactalbumin, lactose, milk protein	Gravy and gravy mixes, ice cream
Milk solids, natural flavoring	Imitation sour cream, instant potatoes
Rennet casein, skim milk powder, solids	Margarine, muesli, muffins
Sour cream, whey, whey powder	Packaged soups, pies, puddings
Whey protein concentrate, yogurt	Sherbet, canned soups

tions the very first time they knowingly eat an allergenic food; or they might not have any problems the first few times and then develop a severe reaction.

Peanut allergies affect up to 8 percent of children and 2 percent of adults. This is a sharp, three-fold increase since the 1990s. A recent Australian study, for example, found that 10 percent of babies who were tested were allergic to one or more foods, and 3 percent were allergic to peanuts.[17] Peanut allergies can be quite severe and are seldom outgrown. Many products contain peanuts or peanut oil, a significant hidden hazard for those who are allergic to it. Be particularly careful of Chinese, African, and Thai cuisine, which often includes frying in peanut oil or using peanut butter in sauces. New research suggests that exposure to tiny amounts of peanuts especially early in life can help prevent peanut allergies and treat them if they develop. Do *not* try this on your own. Always seek help from a health professional when treating life-threatening serious allergies.

Table 5.3: Foods That May Contain Soy Protein

- Gum arabic
- Bulking agent
- Emulsifier
- Guar gum
- Hydrolyzed vegetable protein
- Protein or protein extender
- Soy protein, soy protein isolate, soy sauce, soybean oil
- Stabilizer, starch
- Textured vegetable protein, thickener
- Tofu
- Vegetable broth, vegetable gum, vegetable starch

How Do You Know If Your Child Is Allergic?

Keeping a careful symptom and trigger diary is the best method, but a variety of diagnostic *tests* have also been developed to test for allergies.

DIAGNOSTIC TESTS FOR ALLERGIES

- Symptom diary
- Skin prick, intradermal, scratch or patch tests
- Double-blind, placebo-controlled food challenges
- Blood tests: RAST, FICA, cytokines, CAP FEIA, levels of IgE, IgG, and histamine

Skin prick, intradermal, patch, or *skin scratch tests* are the most widely available and widely used allergy tests. The intradermal progressive dilution test is endorsed by mainstream medical groups as the most reliable type of test for food allergies (other than the randomized, controlled, double-blind food challenge described next). In a skin test, the child's skin is scratched and a liquid containing the suspected allergen is dropped on the scratched or pricked area. The test is considered positive if the skin develops hives at the scratch site. The prick or scratch tests have a fair number of false positive results. This means that your child could have a positive test and not really have an allergy. These tests can also have false negatives; this means that it looks like the substance is not a problem for your child when it actually is a trigger. If you strongly suspect a substance is allergic for your child, but the skin test is negative, ask the physician to repeat the test using *freshly* prepared extracts to avoid missing the diagnosis.

For diagnosing food allergies, the most common and reliable test is a *double-blind, placebo-controlled food challenge*. In this test, all potentially allergic foods are withheld for several days; then the child is given a capsule containing either a suspected allergen or an inert substance (placebo). Neither the child and the parents nor the physician knows whether the capsule contains the allergen or the placebo. After taking the capsule, the child is carefully watched for any type of reaction. A more extreme permutation of this test requires *hospitalization in a special low allergy unit* and going on a fast or severely restricted diet, gradually reintroducing substances and assessing the child's reactions as each thing is added—a food, a pollen, dust, and so on. This kind of evaluation is expensive, time-consuming, and stressful for the family and child, so it is not often done. In severe cases, it may be the best alternative.

A number of blood tests have been developed to try to spare kids the discomfort of having large areas of skin exposed to potential allergens in the various scratch, prick, and injection tests. Blood tests are also easier and less costly than doing a double-blind food challenge; on the other hand, sometimes you get what you pay for. Most of them are not very reliable. The *RAST* (radioallergosorbent test) is one of the most commonly used tests for allergies because it is convenient and pretty good for inhaled allergens such as pollen and dander, but it can be inaccurate for food allergies. The *FICA* (food immune complex assay) is another blood test. *Cytotoxic blood tests* for allergies are fairly controversial. In this test, the child's white blood cells are mixed with the suspected allergen; if the white cells react, the child is thought to be allergic to that substance. The problem is that white cells in a test tube don't necessarily react the same way they do in the body surrounded by other cells and molecules. The test is also very dependent on the skill of the person looking at the white blood cell reaction; there is a lot of variability in interpreting the test results. Another lab test is the *FEIA* (fluorescent enzyme immunoassay), which may eventually replace the gold standard for food allergies, the double-blind, placebo-controlled oral challenge test. Work is ongoing.

Kids who have allergies typically have higher than normal levels of several immunoglobulins (immune proteins) in their blood; these include IgE and IgG. They can also have higher than normal levels of certain immune cells such as basophils and other immune chemicals such as histamine. This just means that your child has a revved up immune system, which you probably already knew; it does *not* tell you what is triggering the allergy or what the best treatment is. There is no known scientific basis for *urine autoinjection, Vega* testing, or *electrodermal* diagnostic techniques despite the anecdotal testimonials you may hear from true believers or salespeople.

Remember, the best test for allergies is *careful observation*. Keep a diary of your child's symptoms and exposures. You may be able to figure it out yourself with some careful detective work. The son of one of my colleagues was

allergic to corn (among other things). She then noticed that he developed a severe diaper rash every time she gave him a common pain relieving medication. Finally she checked the label— sure enough, it contained corn syrup. When she switched to a product free of corn syrup, there were no more diaper rashes. Another child seemed to be allergic to a number of medications, but it turned out that all three contained red dye #40; it was the dye, not the active ingredients themselves that caused the problems. If you suspect an additive, and the label doesn't say what *all* the ingredients are, you may have to contact the manufacturer.

What's the Best Way to Prevent and Treat Allergies?

The best strategy is to nurture a strong, smart immune system. The second best strategy is to minimize exposure to the allergen. If your child is allergic to three things—cats, dust, and pollen, for example—and you markedly *reduce* her exposure to two of those things (say cats and dust), she will be much less likely to react to the third (pollen). If you can eliminate most of your child's allergy triggers by changing things you can control (dust and pets), you may also reduce her sensitivity to other triggers over which you have less control (pollen). Let's tour the Therapeutic Mountain to find out what treatments have proven effectiveness. If you want to skip to the bottom line, flip to the end of the chapter.

Lifestyle Therapies: Nutrition, Exercise, Environment, Mind-Body

Nutrition

Food allergies, remember, are much more common in children than in adults. The most common food triggers are cow milk, eggs, wheat, soy, fish, peanuts, and tree nuts (such as Brazil nuts, almonds, and hazelnuts), some of the most frequently eaten foods in childhood! Allergies to rice have been reported, but are rare.[18] Food allergies often run in families; for example, kids whose parents are allergic to peanuts are more likely to have peanut allergies than kids whose parents have no allergies.

By now you've probably noticed that I recommend *breast-feeding* as a preventive therapy for just about every condition. What about allergies? Yes, breast-feeding helps prevent allergies, especially for kids whose families are very prone to allergies and eczema and especially when parents are able to stop smoking. One study even found that breast-feeding for more than six months helped protect kids against developing allergies, eczema, and asthma for the next 17 years![19] Moms need to be extra careful about what they eat and drink while they're nursing because the foods they eat are often concentrated in breast milk; for example, the protein from cow's milk is concentrated in breast milk, so if a mom drinks cow's milk, her nursing child is exposed to cow milk protein. For breast-feeding moms who've battled allergies triggered by cow's milk, avoiding cow's milk during pregnancy and while breast-feeding protects the baby against exposure to cow milk proteins and reduces the risk of becoming allergic to cow's milk by about 60 percent.[20] If for some reason you need to feed formula, choose a hypoallergenic formula, preferably a hydrolyzed formula. Moms who are allergic to peanuts, fish, eggs, wheat, corn, or other common allergens should avoid eating them.[21]

You may have also recognized that I'm a big fan of *yogurt* and other cultured or fermented dairy drinks that contain probiotics or healthy bacteria. Women who are pregnant or nursing and children should consider boosting their intake of yogurt, kefir, and other foods rich in *Lactobacillus*, *Bifidobacterium*, and *Strep thermophilus*. These healthy bacteria suppress the growth of bad bacteria in the intestines and also seem to stimulate healthy immune responses. Eating yogurt regularly can decrease the production of IgE, the immune compound responsible for many allergic reactions.[22] There are very few contraindications or side effects. And you don't need a doctor's prescription. If your child is sensitive to cow milk and soy products, there are a growing number of probiotic products made with coconut, almond, and goat milk as well as fruit juices.

Children who are fed a variety of *solids before four months* of age are more susceptible to developing allergic skin rashes.[23] Stick to mother's milk in the first four months of life! Start introducing solids from four to seven months.

Omitting an allergenic food from the diet may be your only practical alternative if your child has a serious, Type 1 anaphylactic reaction to certain foods. Some clinicians recommend that a severely suffering patient go on a total fast for four to five days to clean all potential allergens out of the system before reintroducing possible offenders, one every few days. I do *not* recommend fasts for children under six years old; nor should they be undertaken in older children without the supervision of a dietitian or nutritionist. Eliminating major dietary staples such as milk or wheat without consulting a trained nutritionist runs the risk of developing serious nutritional deficiencies. If your child has had a severe allergic reaction to a food, please reintroduce it *only* in a professional health care setting to enable immediate emergency care if necessary.

Be aware of potential *cross-reactions between inhaled allergens and foods*. For example, many who suffer from *ragweed* allergies (worse in the fall) are also allergic to *melon* (cantaloupe, honeydew, or watermelon). Those who are allergic to *birch* tree pollen may react to *apples, carrots, cherries, pears, peaches,* or *potatoes.* Keep a diary of your child's exposures to foods and pollens and their allergic symptoms to help sort out these associations.

I do not recommend that most children eat much *red meat* because of its high fat and cholesterol content, the widespread use of hormones and antibiotics in the meat industry, and the tremendous drain on the ecosystem involved in producing meat compared with other sources of protein. However, lamb is one of the least allergic foods available, and you can include it as a source of protein in a "few foods" diet for your allergic child. Try to buy meats labeled as "organic" or free of hormones and antibiotics or use wild game to reduce your child's exposure to these chemicals.

Despite widespread recommendations to avoid *mucus-producing foods* such as dairy products, there have not been any scientific studies showing that dairy products produce more mucus than any other kind of food. True milk allergies typically cause diarrhea, upset stomach, anemia, chronic lung problems, and eczema.[24] On the other hand, *yogurt* may be a helpful preventive therapy for children suffering from allergies. Yogurt containing live cultures increases the level of gamma interferon, one of the body's own infection- and allergy-fighting chemicals. Start feeding your child yogurt several months before allergy season starts to build up gamma interferon levels. Better yet, eat fermented foods and yogurt all year round!

Remember, many children *outgrow their food allergies*. It's most common to outgrow allergies to cow's milk, eggs, and soy protein. Some food allergies are *not* outgrown. Kids who are allergic to peanuts, nuts, and fish tend to stay allergic to them, even if they're not exposed for years. Peanuts are especially hazardous because peanut oil is used in a lot of prepared foods and can trigger a life-threatening reaction in an unsuspecting youngster. If your child is allergic to peanuts, you *must* read food labels carefully, let the school or camp or coach know about the allergy, and be prepared with epinephrine (ask your pediatrician for a prescription for an Epi-pen) and an antihistamine. And if you decide to try to reintroduce an allergic food to see if your child still reacts, *please* do it in a physician's office, just in case your child has a severe reaction.[25]

Exercise

Vigorous exercise can make allergy symptoms worse. Not only does outdoor exercise expose children to pollens and air pollution, but some children actually get hives just from getting overheated. Some children have worse food allergy symptoms if they exercise immediately after eating.[26] For example, they might have no symptoms if they eat wheat (to which they are allergic) and didn't exercise, but they might have a severe reaction if they eat wheat and then exercise. One adult had repeated life-threatening allergic reactions when exercising after eating hazelnuts. No one knows exactly why this happens, but perhaps it's another reason for the conventional wisdom to rest for half an hour after eating before engaging in vigorous exercise. Some sensitive kids may have to wait 8 to 12 hours after eating certain foods to avoid having an allergic reaction to them;[27] I say, avoid the foods and get the exercise!

Environment

Be sure to keep your child's vitamin D level in the optimal range, at least 30 or higher. The easiest way to boost vitamin D in the spring, summer, and fall is to spend some time in sun. Don't get a sunburn, but do provide at least 15 to 30 minutes of sun exposure daily to help your child make his own vitamin D. During the winter months or cloudy days, supplement with vitamin D (*at least* 400 IU for infants and 600 IU for older children daily). Your pediatrician can check your child's vitamin D level with a blood test to make sure it's in the optimal range.

The primary treatment for allergies is environmental: Avoid the allergen! The three most common non-food allergens are: pollen, dust mites, and pet dander. The majority of people who are allergic to one thing are sensitive to other triggers as well. The bigger the burden of allergens, the more likely your child is to react to something. If you reduce your child's exposure to common allergens, it may reduce his reaction not only to those, but also to other allergens over which you have less control. Minimizing your child's exposure to house dust mites early in life can also help prevent a variety of later allergic reactions.[28]

If your child is allergic to pollens, take a few minutes each day to *check on the daily pollen count*. Radio and television stations usually have this information, or call your local weather service or check the Internet. Pollen counts are generally highest in the morning, so you may want to limit your child's outdoor activities until the afternoon and keep the windows closed in the morning.

Between 5 and 10 percent of Americans are allergic to *cats* or *dogs*. About twice as many children are allergic to cats as are allergic to dogs. Interestingly, children who live with cats as infants often develop tolerance to cats and are less likely to be allergic later on than children initially raised without them.[29] Children can also be allergic to *guinea pigs, mice, hamsters,* and *rats*. Even if the animal is removed from the home and you start an aggressive cleaning campaign, it can take four to six months for animal allergen levels to clear. Washing the cat makes for a miserable cat (and a pretty scratched up cat owner), and it needs to be done every five to six days to really reduce shedding of allergenic cat dander;[30] still it may be better than getting rid of a beloved pet. Same for dogs, though in my experience, they don't put up as much of a fuss at bath time as cats do.

TIPS FOR REDUCING ALLERGENS IN THE HOME

1. Thoroughly dust (using a damp cloth or mop) and vacuum the house every week. Change the vacuum's dust bag frequently. Keep your child out of the vacuumed room for an hour after you've finished to allow the dust to settle.
2. Avoid using chemical cleaners that leave a lingering smell. Cleaning with bleach helps eliminate molds from damp areas. Bleach (5 percent in water) eliminates other allergens, too. Or use a simple vinegar- or pine oil–based cleaner and avoid the chemicals.
3. Remove all dust catchers from your child's room. This means carpeting, stuffed animals, ruffled bedclothes, and draperies. Avoid wool blankets and synthetic pillows in your child's room. Wash all of your child's bedding in hot water (at least 130 degrees Fahrenheit) weekly to destroy dust mites. To avoid additional pollen, do not dry the bedding outdoors. Dry cleaning does not reduce allergens as much as cleaning in very hot water.[31]
4. Encase your child's mattress and pillow (favorite homes of the dust mite) in allergy-proof nylon or vinyl casings.
5. Install a High-Efficiency Particulate Arresting (HEPA) air filter in your child's room. Keep the windows closed between 5 and 10 a.m. when pollen counts are highest. Install an electrostatic air filter on your furnace or air conditioner.[32]

6. Add air-cleaning houseplants (such as philodendrons and spider plants) in your child's room. They help remove indoor air pollutants.

7. Do not allow furred or feathered pets in your child's room. Preferably, pets should be kept outside. If your pets are as much a loved and integral part of your family as mine are, you can keep them in the house if you bathe them weekly.

8. Keep the humidity in your house below 50 percent to discourage molds and dust mites.

9. Kill dust mites by spraying an acaricide (dust mite killer) or tannic acid on your carpeting, upholstery, and draperies at least twice yearly; thoroughly vacuum a day later to clean up the residue of dead dust mites.

10. DO NOT SMOKE and DO NOT ALLOW OTHERS TO SMOKE in your home.

Housecleaning is a key therapy to reduce allergy symptoms due to dust mites, dust, and danders. Unfortunately, I don't know of any health insurance companies that will pay to have someone come clean your home! The biggest exposure to dust mites is in bed, where children typically spend 8 to 10 hours a day. *Washing bedding,* even in cold water, reduces dust mites by a hundredfold, *but* it doesn't kill them all, and they re-accumulate within two weeks. If your child is sensitive to mites, you need to wash all his bedding weekly. Adding *eucalyptus oil* to the laundry detergent (three ounces of eucalyptus oil to one ounce of detergent for a single load of bedding in a top loading machine) kills significantly more mites than washing in detergent alone.[33] Keep the mattress enclosed with a finely woven polyester fabric to keep the mites in the bed away from

your child. House mites actually seem to prefer growing on synthetic pillows to down pillows; keep the feathers and down unless your child is specifically allergic to ducks or geese.[34]

High-energy particulate air filters (HEPA) will not eliminate dust mites, but they can help lower the amount of dander in the air. If you have a pet and your child is allergic to it, keep the pet out of the child's bedroom and run a HEPA air filter in the bedroom 24 hours a day; the filter will also help reduce the amount of mold in the air. If you make the wrenching decision to find another home for your dog or cat, recognize that it may take 6 to 12 months for levels of animal dander in your home to fall enough to make a dent in your child's allergy symptoms.

Keep the relative *humidity* below 50 percent in your home to discourage the growth of dust mites and molds. Molds thrive in damp basements, so if your basement gets wet every time it rains, you could be harboring some serious allergens. If your child's allergies or asthma flare up at school, talk with the school about the humidity there, as well. Allergies to molds can trigger sinus infections when molds are at their worst in damp weather.

Watch out for *volatile organic chemicals* (VOCs) that trigger allergy symptoms in many children; these chemicals are given off by vinyl flooring and many types of carpeting. Stick to hardwood floors, which don't accumulate molds and which can be damp mopped to remove dust and danders.

Environmental therapies can also help treat many allergic reactions. For example, to treat reactions to *insect stings*, immediately apply *ice*. Ice numbs the pain and reduces swelling. Remove the stinger by *scraping* it off; pulling it can release more venom into the skin. Children whose reaction consists of hives and swelling at the sting site are not at increased risk of a life-threatening reaction in the future.

HOME REMEDIES FOR POISON IVY AND OTHER ALLERGIC RASHES

1. Rub them with *ice*. Ice helps numb the area, reducing itch.
2. Make a paste of *baking soda* and *water*, and rub it in. Baking soda seems to be most helpful for blistering rashes, not plain old hives.
3. Some families have tried applying milk of magnesia (usually used for upset stomachs) when they were desperate and there was nothing else in the house. They reported that it was useful. I've recently learned that some of my colleagues actually recommend it as a soothing lotion for itchy rashes. It's readily available and free of side effects, but I don't think anyone has actually studied its effectiveness as an allergy remedy in scientific studies.
4. Other parents report that zinc oxide (the active ingredient in the strongest sun blocks and diaper rash preparations) is also an effective anti-itch treatment.

On the other hand, *rubbing ice* on the skin causes some people to break out in hives; watch what you're doing and stop if it makes things worse. The runny nose caused by going in and out of the cold is called "skier's nose."

Heat typically makes itching worse—whether the itching is due to eczema, chickenpox, or poison ivy. On the other hand, allergy sufferers can achieve four to six hours of relief from inhaling *hot steam*.[35] I don't know of any insurance companies that will pay for a gym membership so that your child (10 years or older) can sit in the steam room several times a week, but if you can afford a YMCA gym membership (and your child doesn't have any other medical problems that contraindicate heat), you might try it.

Running the *air conditioner* may also help reduce the pollen count in your home. For maximal benefits, keep the windows and doors closed (even when the temperature isn't high) and keep the pollen out! If you don't need the cold air, just run the fan; this will circulate air through the filter. Clean the filter at least twice yearly with a bleach or vinegar solution or use disposable filters.

Tepid water may be soothing for a child suffering from irritating allergies. Try putting a cup of plain, dry *oatmeal* in an old stocking and tossing it in the bathwater. Oatmeal is very soothing to irritated skin, no matter whether the underlying problem is allergies, eczema, or chickenpox.

Bathing and shampooing at night will help wash out all the pollens and other allergens that have clung to your child's skin over the day. Rinsing off in the evening ensures that your child doesn't have to deal with allergens when he's asleep.

Mind-Body

Several startling experiments have proven that the mind strongly affects allergic reactions.[36] In one experiment, hypnotized subjects were told that a leaf was poison ivy. After the leaf touched their skin, the subjects broke out in a typical poison ivy rash, even though the leaf was from a maple tree! In another experiment, subjects who were tested with varying strengths of allergen reacted differently depending on what mood they were in when the test was placed; they had much smaller reactions when they felt happy or lively than when they felt down or listless.

Fascinating evidence comes from unexpected quarters. Some patients suffering from multiple personality disorders display a unique allergic phenomenon: One of their personalities can be severely allergic to a food, such as oranges, while another personality has no reaction to that food at all. So if the person eats an orange in personality A, she may break out in hives; but if she eats oranges while she is personality B, she can eat as many oranges as she wants without any reaction.

This tells us that the connection between the mind and the immune system is far more complex than we currently understand. Clearly hypnotized subjects and patients with multiple personalities have only one body and one immune system, but their allergic reactions vary depending on their mental state. We don't need to fully understand the mechanism of this interaction to use it. Hypnosis and even conscious suggestion can markedly reduce symptoms of food and inhaled allergies in children.[37]

If your child suffers from chronic annoying allergies, consider taking him to a clinician who practices *hypnotherapy.* Hypnotherapy can help uncover the source of the problem, provide powerful suggestions to prevent reactions, and give you and your child useful images to help deal with symptoms such as being short of breath.[38] The image need not be elaborate. As you apply whatever treatment you use for your child's allergy, remind him that this is a healing treatment and that it will make him better. Hypnosis is not for everyone; in some studies, few people were able to affect their immune reactions to allergic triggers using hypnosis.[39] On the other hand, it's very safe and may be empowering, and I often recommend it if kids and families can make the commitment to regular practice. Children can also learn self-hypnosis, which can be helpful for common allergies like hay fever.[40]

Anne was able to cool off the itchy reaction to her earrings by imagining herself skiing down a snowy mountain on a clear, cold day with her cap off. She also decided to avoid jewelry containing nickel to reduce her risk of developing an allergic reaction.

BIOCHEMICAL THERAPIES: MEDICATIONS, HERBS, NUTRITIONAL SUPPLEMENTS

Medications

The overwhelming variety of allergy medications can be very confusing. There are several classes of effective medications that work different ways.

ALLERGY MEDICATIONS

- Antihistamines
- Leukotriene inhibitors
- Mast cell stabilizers
- Steroids
- Desensitization therapy
- Epinephrine
- Decongestants
- Saline
- Others

Antihistamines are the mainstay of medical allergy treatments. They work by blocking the histamine released from mast cells. There are many types, varieties, and brands of antihistamines. It's impossible to tell which antihistamine is most effective for your child unless you try different kinds. Most people start with the least expensive nonprescription remedy. If these don't help or have unacceptable side effects, see your physician about other (more expensive) prescription medications. More expensive antihistamines are not necessarily more effective.

ANTIHISTAMINES

- *Over-the-counter, OTC (non-prescription):* fexofenadine (*Allegra*); diphenhydramine (*Benadryl*); chlorpheniramine maleate (*Chlortrimeton*); loratadine (*Claritin*); brompheniramine (*Dimetane*);

clemastine fumarate (*Tavist*); cetirizine (*Zyrtec*); *Ocu-Hist* (an eye drop)
- *Prescription:* hydroxyzine (*Atarax*); cyproheptadine (*Periactin*); desloratidine (Clarinex); promethazine (*Phenergan*); levocetirizine (Xyzal); azelastine (*Optivar* or *Astelin*— eye drops/nasal spray); epinastine (*Elestat*—eye drops)

Many antihistamines are available without a prescription. Most of the older medicines require doses every 6 to 12 hours and have side effects such as a dry mouth, difficulty concentrating, and drowsiness; a few children also have problems with urination and constipation. Some children become irritable and hyperactive from the older antihistamines. You can *increase your child's tolerance* to the sedating effect of antihistamines by starting him on treatment before he goes to bed, so he can get used to it when it doesn't matter if he's sleepy. Then add low daytime doses and gradually increase the dose as his body gets used to it.

Because most people don't want to spend the entire hay fever season sleeping, they turn to *antihistamines* that are *less sedating,* such as *Claritin, Zyrtec,* or *Allegra.* These medications only require dosing once or twice daily. They have become less expensive since they became available over-the-counter without a prescription. They should *not* be taken at the same time your child is on erythromycin-type antibiotics or antifungal medications such as ketoconazole because of the risk of liver toxicity. Cetirizine (Zyrtec) doesn't seem to interact with as many other medications, but it is more sedating. Be sure to ask your pharmacist about potential interactions before starting any new medication. Choose the least sedating products if your child needs to drive or operate heavy machinery during peak allergy season.

Antihistamines do a good job with runny noses, but they don't help much with congestion. That's why many allergy preparations contain a *decongestant* as well as an antihistamine. Decongestants have additional side effects: increased blood pressure, irritability, insomnia, nervousness, increased heart rate, and decreased appetite.

Prescription steroid nasal sprays don't cause any sedation and may be more effective than antihistamines, particularly for kids with severe allergies and asthma; more and more physicians are turning to nasal steroid sprays instead of antihistamines for first-line treatment.

If your child's allergy symptoms are mostly related to the nose (runny nose, itching, sneezing), you may want to consider an antihistamine or steroid nasal spray. In comparison trials, using an antihistamine nasal spray twice a day was as effective as the non-sedating antihistamines that are taken by mouth. Also, because it's sprayed in the nose, it's less likely to cause systemic side effects and less likely to interact with other medications. It may sting for a minute, and some kids complain that it tastes bad, but it's less likely to interact with other medications.

When using an antihistamine, give it to your child *before* symptoms occur rather than after your child is miserable. Remember, antihistamines work by inhibiting the reaction to allergens. If your child's symptoms are worse when he goes to visit his cat-loving friend, giving him an antihistamine 30 to 60 minutes before he gets to the friend's house is more effective than delaying treatment till after he's coughing and sneezing.

Although antihistamines do help relieve allergic symptoms, they do not address the underlying sensitivity that causes allergies. Go ahead and use them if your child has mild or occasional symptoms and to help your child feel more comfortable while you take other measures to address the underlying problem.

Another class of medications affects the

production and attachment of leukotriene chemicals, which are inflammation-causing compounds produced throughout the body. These are preventive medications, best used prior to exposure. They include zafirlukast (*Accolate*), which is approved in children five years and older (mostly to prevent asthma); montelukast (*Singulair*), which is approved for children six months and older (again, mostly for preventing asthma, but also for allergies); and zileuton (*Zyflo*), which is approved for those 12 years and older; all of these medications can interact with other medications, and some can cause liver toxicity, sleep, or behavior changes.

Remember, a good source to look up common medications is MedLine Plus, a service of the U.S. National Library of Medicine and National Institutes of Health.

Mast cell stabilizers are medications that calm the mast cells that release histamine. These medications all *require a prescription.* They are also best given *before* symptoms start. Unlike antihistamines, they have very few side effects, and they can help children who have year-round or predictable allergies, such as to dust mites or ragweed. *Cromolyn* (when taken by mouth) is effective in preventing food allergy symptoms; when used as a spray, it helps control allergic asthma.[41]

MAST CELL STABILIZERS
(ALL REQUIRE PRESCRIPTION)

- Lodoxamide (*Alomide* eye drops), Olopatadine (*Patanol* eye drops)
- Cromolyn (*Nasalcrom*) nasal spray
- Nedocromil (*Tilade*) inhaler and eye drops

Medicines that are applied to just one area have fewer side effects than other allergy preparations that are taken by mouth. For example, *Alomide* and *Patanol* eye drops are just put in the eye, *Nasalcrom* is sprayed in the nose, and *Nedocromil* is inhaled into the lungs. They prevent the release of histamine from the mast cells where they're applied, but do not affect other tissues such as the brain or heart. They have specific rather than general effects. They are safe to use in combination with other more general treatments such as antihistamines or steroids.

Steroid medications are available in several forms to treat different kinds of allergic reactions. Steroid medications are very similar to the body's own anti-inflammatory messengers. They help reduce swelling, pain, and irritation.

STEROID PREPARATIONS FOR ALLERGIES

- *Creams or ointments* such as *Cortaid* (prescription and non-prescription hydrocortisone)
- *Nasal sprays* such as *Vancenase* and *Beconase* (beclomethasone), *Flonase* (fluticasone), *Nasalide* (flunisolide), *Nasonex* (mometasone), *Nasacort* (triamcinolone), *Rhinocort* (budesonide)—all prescription only
- *Metered dose inhalers* for allergic asthma such as *Aerobid, Azmacort, Beclovent, Decadron, Vanceril*—all prescription only
- *Oral steroids* for systemic symptoms such as Cortisone, *Decadron, Medrol, Pedi-pred, Prelone*, Prednisone—all prescription only

The mildest kind of *steroid creams* and *ointments* (0.5 percent and 1 percent hydrocortisone, Cortaid) are available without a prescription. They are helpful for allergic rashes such as poison ivy. Prescription strength steroids are available as *nasal sprays* to treat runny nose and congestion due to allergies. Some are effective even with once a day dosing; steroid nasal sprays are best used as a preventive therapy before symptoms start. It takes several days to a week for improvements to be noticeable. Steroids are available in *inhaled form* (metered dose inhalers, MDIs) to treat children with allergic asthma. For children with more severe, system-wide symptoms, *steroid* pills or liquid may be needed to get symptoms under control. When taken by mouth, steroids can have powerful side effects—suppressing the immune system, raising blood sugar, increasing blood pressure; they should be used for as little time as possible. When steroids are applied directly to the affected area (such as creams to rashes and sprays to the nose) they have few side effects and are safe even for young children.

Other Medications and Treatments

One example of an *antibody therapy* is the medication, Omalizumab (Xolair); this is actually an IgG antibody directed to block IgE. It is used to reduce moderate to severe allergies and allergic asthma when steroid medications are ineffective. It is approved in the United States for adolescents 12 years and older. This is not a first-, second-, or even third-line therapy. It is not a cure and requires ongoing use to work; it is expensive and can have serious side effects, so it should only be used with help from an experienced pediatric allergist. Some allergists combine this medication with oral immunotherapy to reduce the risk of side effects for a short period at the beginning of therapy; this is still a fairly experimental approach to care for children with serious life-threatening allergies.[42]

Allergy shots (*desensitization therapy*) are a series of injections of minute amounts of whatever is causing the allergy. By giving extremely small doses, the immune system is stimulated to produce the "good" kind of immune globulin, *IgG*. IgG blocks the allergy-causing immune globulin, *IgE*. Allergy shots are most effective for allergies to bee stings, pollen, dust mites, and animals; they are less effective, but can help with mold and food allergies. Desensitization therapy usually lasts three to five years, and may help protect against developing new allergies.[43] A few children suffer relapsing symptoms when the shots stop, but most (especially those getting shots for allergies to insect stings) continue to be protected.

In the last 20 years, allergists have figured out three new ways to desensitize patients without shots. The first is *oral immunotherapy* (OIT). The second is *sublingual* (under the tongue) *immunotherapy* (SLIT), and the third is *patch immunotherapy*. The first, OIT, started being tested for peanut and egg allergies, and has become a fairly widespread approach to treating moderate to severe food allergies; common side effects include sneezing and congestion, mild itching, mild hives, nausea, and diarrhea in up to 20 percent of children, but 50 to 60 percent develop tolerance to the allergen.[44] OIT can be combined with probiotic therapy for even higher (greater than 80 percent) effectiveness.[45]

A growing number of allergists offer sublingual desensitization immune-therapy (SLIT). SLIT involves giving drops of tiny doses of the allergen under the tongue; the doses are much smaller than those used for OIT, and SLIT has fewer side effects.[46] It can reduce peanut, grass, pollen, and latex allergies, but it may not be as effective as OIT or allergy shots.[47]

The formal name for patch therapy is *epicutaneous immunotherapy* (EPIT). The first studies about EPIT evaluated its effectiveness in allergic mice and dogs. Studies are underway to test its effectiveness and safety in children under the auspices of the U.S. National Institute for Allergy and Infectious Diseases. EPIT uses even smaller doses than SLIT, and since the protein is applied to the skin, local reactions may result in redness and hives, but are unlikely to cause swollen airways or wheezing.

Desensitization therapy can be dangerous because if the dose is advanced faster than IgG builds up, the child can have a severe allergic reaction. That's why allergy shots and other desensitization procedures should *only* be done with the supervision of a physician who has extensive experience and emergency equipment on hand.

Medic-Alert Bracelet

If your child ever has a major allergic reaction (anaphylaxis) to a food, bee sting, or anything else, please have him wear a *Medic-Alert bracelet* and ask your physician for a prescription for *epinephrine* for emergency use. Keep one at home and one with the child wherever he may come in contact with the allergen. Epinephrine can be life-saving if your child has an anaphylactic reaction.

Home Remedies: Saline Nose Drops

If you want to help keep nasal secretions loose, give your child simple *saline nose drops* (¼ teaspoon of salt in 8 ounces of water or buy premixed saline drops, *Na-Sal*). One or two drops on each side can be given as often as you like to help with congestion. Or you can use a Neti pot to wash pollen out of the nose before bed. In fact, some families find that frequent nose washes help rinse out the allergy-causing pollen and offer substantial relief from allergies and congestion.

Herbal Remedies

A variety of herbal remedies have been used historically to ease allergies.

HERBAL REMEDIES

- *Scientifically Proven Useful:* ephedra, but significant side effects—AVOID
- *Scientifically Unproven but Widely Used:* angelica, skullcap, coleus root, eyebright, goldenrod tea, goldenseal, licorice root, nettle leaves, plantain, green clay, and calendula
- *Herbs That May Cause Allergic Reactions*: chamomile, echinacea, others

Ephedra tea (Ma huang) has long been used by the Chinese as a treatment for allergies, asthma, hay fever, and the common cold. It is the original source of *ephedrine,* which has been chemically synthesized and is an ingredient in many cold and allergy medications. Ephedra is a decongestant, and it has some anti-inflammatory effects, but it should be used only in low doses (high doses can be fatal) and short periods of time (long-term use can lead to habituation and needing higher doses for the same effect). Ephedra has side effects such as high blood pressure and rapid heart rate, so it should not be used by children who have weak hearts or who already have high blood pressure. Due to the many reports of severe and even fatal reactions to ephedra, the federal government and many states have greatly restricted the availability and dosing recommendations for ephedra. Avoid it; safer alternatives are available.

In addition to having a beautiful ornamental flower, the *Angelica (Dong quai)* plant has been used by Chinese herbalists since ancient times to treat allergies, eczema, and hay fever. *Chinese skullcap tea* helps decrease inflammation and inhibits the immune system's allergic response. One of the chemical constituents of *coleus root* (forskolin) has demonstrated antihistamine effects. However, there are no studies that demonstrate the effectiveness of angelica root, Chinese skullcap, or coleus root in treating children with allergies. I do not typically recommend them.

Eyebright tinctures have long been used to treat the burning watery eyes and runny nose associated with hay fever–type allergies, but I do not advise putting herbal products in the eyes until they are regulated more like medicines than food. The anti-inflammatory properties of *goldenseal* have made it a favorite herbal remedy, but natural supplies of the wild herb have been overharvested and it is no longer commonly recommended. Scientific studies have not evaluated the effectiveness of any of these herbs in treating childhood allergies. I do not recommend them.

Goldenrod tea and tincture are controversial allergy remedies because goldenrod actually causes allergies when it is inhaled. There are no studies showing either marked benefits or severe risks of goldenrod tea in treating childhood allergies. I do not recommend it.

Licorice root has anti-allergy and anti-inflammatory effects because it blocks the breakdown of the cortisol, which is a naturally produced hormone that decreases inflammation. Large doses or chronic use of licorice can result in fluid retention, high blood pressure, headache, and significant potassium loss. However, licorice is safe for occasional use in modest doses either as a tea or poultice.

Nettles (*Urtica dioica*) cause stinging and burning when you brush up against them in a field or forest. They are also a traditional remedy for allergic reactions, particularly for runny nose and watery eyes. In a randomized controlled trial of 98 adult allergy sufferers, those given freeze-dried nettle were more likely to report dramatic or marked improvement in their symptoms over the next week than those patients given placebo pills; however, the differences were not significant statistically.[48]

Nettles are also commonly recommended as a *homeopathic* remedy for allergies. Again, there is insufficient scientific evidence that it is useful for children. It needs more study.

For those suffering from allergic skin rashes, *plantain* poultices are said to be soothing. Others recommend an application of a paste made of *green clay* and water to the affected area. Many people find *Calendula* creams and ointments soothing for allergic skin irritation. However, there are no scientific studies documenting the effectiveness of any of these remedies in treating allergic rashes. They are safe, but of unknown effectiveness scientifically, so I tolerate their use but don't actively recommend them.

Many herbs can actually cause allergic reactions. The most famous culprits are members of the daisy family—echinacea, chamomile, feverfew, and others. Life-threatening allergic reactions to these plants are rare.

Dietary Supplements

You may be surprised by the dietary supplements that can help with allergies.

DIETARY SUPPLEMENTS FOR ALLERGIES

- Probiotics
- Vitamin C
- Essential fatty acids—fish, flax, and evening primrose oil
- Other vitamins and minerals
- Red peppers (capsaicin)

Probiotics can be consumed in supplements as well as in naturally fermented foods like yogurt and kefir. There are many different kinds of probiotic products, and different types and dosages have been used in different studies. There are few studies comparing one brand to another, and there is not enough research for me to recommend a specific brand or dose for preventing or treating different kinds of allergies, but probiotics are safe for most children. If you decide to try probiotics, look for products that have been approved by an independent testing agency, such as ConsumerLab.com, and products that contain species that have some evidence of effectiveness in randomized, controlled trials.[49,50] These include *Bifidobacterium* and *Lactobacillus* species.

Large doses of *vitamin C* (2 grams per day) can help decrease histamine levels and alleviate allergy symptoms in adults with allergic asthma and hay fever.[51,52] One study suggested that vitamin C supplements (2 grams daily) helped improve exercise-induced asthma in teens,[53] and it is so safe and inexpensive, I often recommend it for children with allergies and asthma. If you'd like to try vitamin C, give about 250 mg twice daily for children between three and six years old and 500–1000 milligrams twice daily to older children. Vitamin C is excreted very rapidly, so look for extended release formulations. If your child develops diarrhea (an early symptom of vitamin C overdose), reduce the dose.

Other supplements, such as *vitamins A, B6, E, beta-carotene, selenium,* and *zinc* have been recommended as allergy remedies, but there is little research to support their use in treating children with allergies. Although it is always wise to avoid deficiencies and consume optimal amounts of essential nutrients in a healthy diet, I do not typically recommend these particular vitamins or minerals for allergy prevention or treatment.

Supplemental *essential fatty acids* (EFAs) are anti-inflammatory. Two main types of EFA are omega-3 fatty acids (found in fatty fish, fish oils, cod liver oil, and flax seed oil) and omega-6 fatty acids (found in evening primrose, black currant, and borage oils). Some allergic families have a blockage in fatty acid metabolism, which makes the immune system more prone to allergies and inflammation. For example,

children who are prone to allergies and eczema have lower than normal levels of EFA in their blood, and their moms seem to produce less omega three fatty acids in their breast milk.[54] By giving large doses of EFA, the immune system may become more stable. In scientific studies, doses of two capsules (1,000 milligrams) three times daily of evening primrose oil were helpful in treating children suffering from eczema (see Eczema chapter); it takes about eight weeks for any benefits to become noticeable.

A few years after the second edition of this book was published, I developed seasonal allergies to the abundant oak pollen in North Carolina. A month after starting fish oil supplements (1500 mg daily of the combined constituents, EPA+DHA), I was able to reduce my reliance on antihistamine medications. In subsequent years, I've continued fish oil and flax seed supplements year-round for their many health benefits, and found that I need far less antihistamine medication than I used to.

Unless they're allergic to fish, I think that all kids who are prone to allergies should eat fatty fish (such as salmon, sardines, or mackerel) at least twice a week, and try to include flax seed oil in their salad dressings. For children who have allergies and other conditions involving inflammation, I'm a fan of fish oil supplements: one to two grams daily for 12 years and older, one-half that much for 6- to 11-year-olds, and one-quarter that much for children ages three to five. Again, look for brands that have passed independent testing that are free of mercury, dioxins, and other contaminants.

Capsaican, the spicy, pungent molecule in *red peppers*, decreases airway sensitivity to irritants such as tobacco smoke. Spicy foods such as peppers, horseradish, and hot mustard have long been part of folk remedies for respiratory allergies. In animal studies, capsaicin decreases a variety of airway allergic responses.[55] There are no studies yet evaluating the effects of hot peppers in allergic children, but see if your child's symptoms improve following a spicy meal.

One theory holds that food allergies are due to the passage of small molecules of undigested food across a leaky gut wall into the bloodstream. Those who hold to this theory believe that *digestive enzymes* such as *papaya enzyme tablets* and *bromelain* may be helpful in preventing allergies. Others believe that allergic reactions are causes, not consequences, of leaky gut.[56] Although papaya is one of my favorite foods, it has not been scientifically evaluated for its effectiveness in treating childhood allergies.

Bee pollen supplements are another favorite allergy remedy that has not been scientifically studied in children. I do not recommend them.

BIOMECHANICAL THERAPIES

Spinal Manipulation

Osteopathic manipulation was the most commonly recommended treatment by the American psychic Edgar Cayce during his trance readings for children suffering from allergies. Cayce recommended relaxing adjustments of the upper neck and back areas and stimulating adjustments for the lower back. Although these recommendations are fascinating and the patients reported remarkable improvements, neither chiropractic nor osteopathic treatment for allergies has undergone scientific evaluation, and I do not routinely recommend them as mainstays for allergy treatment until more research confirms their value.

BIOENERGETIC THERAPIES: ACUPUNCTURE, HOMEOPATHY

Acupuncture

Acupuncture has been a mainstay of allergy treatment in China for hundreds of years.

Acupuncture is best when used as *preventive* therapy, *before* symptoms occur. Allergic patients who receive acupuncture three times a week for four weeks have gradual changes in their white blood cells that are associated with being less allergic.[57] A 2015 meta-analysis of 13 randomized controlled trials evaluating acupuncture for allergies concluded that acupuncture was a safe, effective treatment for allergic rhinitis (nasal allergy symptoms).[58] These results are impressive. If you are tired of the side effects of chronic allergy medication and are considering taking your child to an allergist for desensitization shots, you may want to try acupuncture. Please seek an acupuncturist who has extensive experience in dealing with children and their fears of needles. An experienced pediatric acupuncturist can make the treatment pleasant rather than fearful.

Prayer/Reiki/Therapeutic Touch

Although the power of *prayer* to help heal is acknowledged by just about every culture on earth, there are no studies specifically evaluating its benefits in treating allergies. Similarly, I use *Reiki and Therapeutic Touch* in my daily medical practice and would not hesitate to offer them to a patient suffering from allergies, but neither has been systematically studied for this problem. Scientifically, I cannot recommend you seek them out as allergy treatments.

Homeopathy

Homeopathy is more closely related to conventional allergy desensitization treatments than it is to any other medical therapy. Based on the principle of "like cures like," the primary homeopathic remedy for hay fever is *Ambrosia* (ragweed). Like all homeopathic remedies it is given in extremely minute doses. *Allium cepa* (spring onion) is a common homeopathic remedy for burning, watery eyes. *Apis* homeopathic remedy is an extract of crushed bees—rather like a crude form of the allergists' injections for bee sting allergies. *Euphrasia* (eyebright) is used for the allergic symptoms in which the eyes are primarily affected (redness, watering, burning, feeling rough or gritty). Euphrasia is also used in larger doses by herbalists to treat allergic symptoms. *Urtica urens* (stinging nettle) is the homeopathic remedy for treating hives. Homeopathic remedies for *hay fever–like allergies* (usually to inhaled allergens such as pollen) include *Arsenicum, Kali bic, Natrum mur, Nux vomica* (poison nut), *Pulsatilla* (windflower), *Sabadilla* (cevadilla seed), *Sulfur,* and *Wyethia* (poison weed). Many of these remedies are very poisonous undiluted. However, homeopathic remedies are *extremely* dilute and there is no danger of poisoning. Homeopathic remedies for skin allergies (contact dermatitis) include *Bryonia* (wild hops—said to be good for food allergies as well), *Rhus,* and *Sulfur.*

Europeans commonly rely on homeopathic remedies to treat allergies. Several European studies have demonstrated that homeopathic remedies help treat hay fever; however, some patients notice a brief worsening of their symptoms when they start homeopathy.[59] Despite the hundreds of studies done on homeopathy for allergies, it remains a controversial treatment in North America.[60,61] A 2013 comparison of homeopathic and conventional treatment of pediatric eczema concluded they had comparable effectiveness.[62] If you choose homeopathic remedies, please take your child to a homeopathic physician who has extensive experience in treating children with allergic disease. Do not rely on homeopathy alone for severe, life-threatening allergic reactions.

WHAT I RECOMMEND FOR ALLERGIES

PREVENTING ALLERGIES

1. *Lifestyle—environment.* Do not allow smoking around your child. Reduce exposure to pollution, allergens, and irritants. Do not smoke and do not allow others to smoke around your child. Use air filters, vacuum, and damp mop, and avoid chemical-containing cleaners. Use mattress and pillowcases to reduce exposure to allergens. Consider washing dishes by hand. If your child is not allergic to pets, consider having dogs as pets. If you don't live on a farm, at least consider raising a garden and get your child involved.

2. *Lifestyle—nutrition.* Breast-feed for at least the first year of life. Consume probiotic rich foods while pregnant and consider supplements. When your child starts drinking cow's milk, offer yogurt or kefir with live cultures. The best time to introduce new foods is between four and seven months old. Don't start beef until the child is at least six months old. Watch your child's weight; obese children have higher levels of inflammatory chemicals in their bloodstream and are more likely to have allergy problems. Avoid deficiencies of any vitamin or mineral, particularly vitamins A, C, D, and E, and the minerals zinc and selenium.

3. *Biofield—acupuncture.* Consider a series of three acupuncture treatments weekly for four weeks prior to allergy season to help prevent allergy symptoms.

TREATING ALLERGIES

1. *Biochemical—medication. For severe, life-threatening allergic reactions such as anaphylaxis from a bee sting or peanuts, seek emergency care immediately.* If your child has a severe reaction, ask for a prescription for an epinephrine (Epi-Pen) and learn how to administer it. Make sure your child has one available at home, school, and in between. Get your child a *Medic-alert* bracelet.

For symptomatic relief of minor symptoms, try non-prescription *antihistamines.* Follow package directions on doses and be aware of potential side effects.

For moderately severe respiratory allergies (affecting the eyes, nose, sinuses, and lungs), consider safe prescription medications such as cromolyn (*Intal®*), nedocromil (*Tilade®*), or steroid nasal sprays. Or ask your doctor about antihistamine nasal sprays or anti-leukotriene medications.

For moderate-severe allergies consider desensitization therapy. Ask for a referral to a pediatric allergist.

Consider saline nose drops or a neti pot to wash allergens out of the nose before bed.

Do not rely on herbal remedies alone to treat severe allergies.

2. *Lifestyle—nutrition.* If you suspect that your child has moderate or severe food allergies, seek the help of a health professional, nutritionist, or dietitian in developing an elimination or few foods diet. Consider making fatty fish such as salmon a regular part of your diet. Consider eating foods such as yogurt as part of a healthy diet. Avoid deficiencies of essential nutrients.

3. *Lifestyle—environment.* Wash thoroughly and immediately with soap after contact with poison ivy or other skin sensitizers. Use ice compresses to reduce swelling and itching. Try tepid baths with or without oatmeal to relieve itching; bathe and shampoo the child in the evening to wash out all the pollen before he goes to sleep. Keep pets away from kids who are allergic to them and consider running an air filter in the child's bedroom. Wash the child's bedding at least weekly. Cover the mattress with a hypoallergenic cover designed to separate your child from the dust mites that live in the mattress. If your child is allergic to pollen, check the weather report for pollen levels and keep him inside with the air conditioner running when the pollen count is high. Decorate your home with air-cleaning house plants. Keep the household humidity under 50 percent. Dust and damp mop weekly to keep dust levels down. Wash your child's bedding every one to two weeks to keep dust mite levels down. Do not smoke and do not allow others to smoke in your home.

4. *Lifestyle—mind-body.* For any type of allergy, consider taking your child for hypnotherapy or biofeedback therapy with a therapist trained in treating children.

5. *Biochemical—nutritional supplements*. For respiratory allergies, try Vitamin C supplements; start with 250 to 500 milligrams twice daily. Back off if diarrhea develops. Consider fish oil, flax seed oil, or evening primrose oil supplements—one to two grams daily. Consider probiotic supplements that have passed independent quality control testing.

6. *Bioenergetic—acupuncture*. If your child has moderate to severe allergies, consult an acupuncturist who is trained in treating children.

7. *Bioenergetic—homeopathy*. For hay fever symptoms, consider seeing a homeopathic physician who is experienced in treating children.

RESOURCES

American Academy of Allergy, Asthma and Immunology
http://www.aaaai.org/

Food Allergy and Anaphylaxis Network (also known as Food Allergy Network)
http://www.foodallergy.org

Food Allergy information from the National Institutes of Health
http://www.nlm.nih.gov/medlineplus/foodallergy.html

6
ANXIETY

One winter afternoon as they were making cookies and listening
to the radio, Holly's daughter, Rowan, asked if her mom knew about
Ron Weasley from the Harry Potter stories. Holly did. Rowan timidly
started talking about Ron and his fear of spiders; she said she felt like
that when she had to go to school—worried, shaky, and tense. She was
afraid that something terrible would happen to her mom if she left
her alone at home. Holly realized that Rowan's fears might be the real
reason she had been complaining about stomachaches every morning,
and saying she felt too ill to go to school. Holly wondered what she
might do to help ease Rowan's anxieties, while still helping her maintain
a healthy amount of caution and prudence. After all, there are real
dangers in the world; the radio had just aired breaking news about the
prison escapes of two convicted child predators.

Anxiety is one of the most commonly diagnosed mental health problems in America. Nearly one in ten children and youth suffer from a diagnosable anxiety disorder. Given the alarming increase in natural and manmade disasters, climate change, violence, and war, insecurity and anxiety are increasing around the world. If your child often feels anxious, worried, or insecure, *your family is not alone*.

TYPES OF ANXIETY

There are several different anxiety diagnoses. The most common types are general anxiety, post-traumatic stress disorder (PTSD), phobias, panic attacks, separation anxiety, and obsessive-compulsive disorder. (Yes, OCD is a type of anxiety.) Formal diagnoses are just the tip of the iceberg; with all the news (and color-coded warnings) about terrorism, a faltering economy, a collapsing infrastructure, bullying, climate change, and competition for scarce resources, many people feel worried or insecure at least part of the time.

It is normal for a nine-month-old to be afraid of a stranger. It is normal to have butterflies on the first day of school. But anxiety is more than just being scared of monsters under the bed. Disabling shyness deters children from making friends, attending school, or trying new sports or activities. Nearly 1 in 20 children suffer from significant school phobia. Anxiety is a chronic condition, so anxious youth typically continue to experience its interfering thoughts, emotions, and physical symptoms through adulthood.

Rowan's anxiety might be called school phobia, which, for many children, is related to fears about separation from parents. Holly was not surprised to hear about Rowan's fearful feelings, particularly since her uncle, who had just returned from the war, had been diagnosed with PTSD (post-traumatic stress disorder), and her cousin suffered from OCD (obsessive-compulsive disorder). Anxiety seemed to run in their family.

What Causes Anxiety?

Anxiety is partly environmental and highly genetic; that is, 65 to 74 percent of anxiety is explained by genetics.[1] The remainder is due to environmental exposures and experiences. Early adverse experiences (death of a parent, abuse, neglect, etc.) can make anxiety more likely to emerge and harder to overcome. Even the types of healthy bacteria living in our intestines can influence the development of anxiety.[2] Although certain genes and major stress can predispose someone to being anxious, the good news is that a strongly supportive environment can help children feel confident and flourish.[3]

Consequences of Anxiety

Because the brain and the rest of the body are so intricately connected, anxious emotions trigger a variety of distressing physical reactions and sensations—stomachaches, headaches, sweatiness, jitteriness, trembling, restlessness, dizziness, palpitations, dry mouth, lump in the throat, muscle tension, shortness of breath, chest pain, nausea, aches and pains, numbness or tingling in the fingers or toes, hot flushes, and cold sweats. Anxious people may tire easily or feel fatigued, sleep poorly, have trouble concentrating, or feel tense or irritable. Some have trouble concentrating or remembering things, while others cannot stop thinking about their worries. Anxiety's challenging symptoms can lead to many fruitless and costly medical tests.

The long-term consequences of anxiety can range from depression to sleep problems to avoiding social relationships and shunning professional help. Severe anxiety's impact on learning and memory can impair performance at school and work. An alarming number of anxious people suffer so severely that they try to dull their emotional pain by drinking alcohol, smoking, or using prescription or illicit drugs. The costs of anxiety to individuals, families, and society are huge, nearly as large as the impact of depression.[4] Reducing anxiety and worries is a huge benefit to the community as well as the individual.

Anxiety can also coexist with other health problems. The combined costs are enormous, and improving anxiety often improves other conditions too.

REFRAMING ANXIETY AS GOALS AND GIFTS

One way to reframe the diagnosis of anxiety is to focus on the *goals* of improving confidence, courage, and a sense of calm in the face of life's challenges. Building mental-emotional serenity, resilience, and inner security is like building strength. It takes practice, usually against some kind of resistance.

Surprisingly, at times a tendency toward being anxious might be considered a *gift*. Anxious people pay extra attention to information related to threat or harm. Anxiety creates a tendency to err on the side of interpreting ambiguous information as risky or threatening. Being cautious, prudent, methodical, protective, risk-averse, and a perfectionist may be very desirable characteristics in an accountant, airline pilot, or surgeon. On the other hand, worries that keep someone isolated, afraid to try new things, or in a state of panic or physical pain can make life miserable.

DIAGNOSING ANXIETY

There are no blood tests, microbiology cultures, or X-rays to confirm the diagnosis of anxiety. Instead, the diagnosis is based on behavioral criteria (that is, whether certain behaviors are more common in this person than most people) and whether anxiety hinders happiness in daily life. Medical evaluations are used to rule out other problems that can contribute to feeling anxious such as hyperthyroidism.

Standard questionnaires are used to determine whether a child meets diagnostic criteria and also evaluate how well a treatment is working. Popular anxiety questionnaires include: the Pediatric Symptom Checklist (PSC); the Screening for Children Anxiety-Related Disorders (SCARED); and the Patient Reported Outcome Measurement Information System (PROMIS) Anxiety Short Form. All are available online.

The formal diagnosis of *generalized anxiety disorder* includes six key components:

1. Excessive anxiety occurs more days than not for at least six months about more than one thing (generalized, not specific fears).
2. The worry is difficult to control.
3. The worry is associated with three or more symptoms such as restlessness, easy fatigability, difficulty concentrating, irritability, muscle tension, or disturbed sleep.
4. The anxiety is not a similar diagnosis like panic disorder, OCD, hypochondria, or PTSD.
5. The worry causes significant distress or impairment in daily life.
6. The worry is not due to a medical problem (like hyperthyroidism) or an herb (excessive coffee or other source of caffeine), medication (like some ADHD medications and appetite suppressants), illicit substance, or withdrawal from an addictive substance (like a benzodiazepine medication).

WHAT'S THE BEST WAY TO PREVENT AND TREAT ANXIETY?

Regardless of whether one meets criteria for an official diagnosis, a healthy lifestyle promotes being more calm, confident, and courageous and feeling more secure and serene. A healthy lifestyle (healthy habits in a healthy habitat) can dramatically decrease stress and anxiety without blunting the awareness of or capacity to affect the real challenges facing us. A healthy lifestyle means optimal nutrition, exercise, sleep, a healthy environment, managing stress, communicating skillfully, and living a life of meaning, purpose, and connection. Professional counseling (including therapies such as cognitive behavioral therapy and EMDR), massage, and acupuncture can also help. Medications and several herbal remedies can be considered, too. This chapter provides a quick and easy overview on both basic lifestyle and additional therapies that may be helpful. Let's tour the Therapeutic Mountain to find out what treatments have proven effectiveness. If you want to skip to the bottom line, flip to the end of the chapter.

LIFESTYLE THERAPIES: NUTRITION, EXERCISE, ENVIRONMENT, MIND-BODY

Nutrition

The brain requires a steady supply of healthy fuel to keep cool, calm, and collected. Low blood sugar (hypoglycemia) can make you feel jittery and nervous. Here's a quick list of helpful nutrition tips to help your child feel more calm and serene.

- Eat breakfast. Do not skip meals.
- Consider eating several small meals or snacks with proteins, healthy fats, and complex carbohydrates rather than one or two very large meals a day to keep blood sugar levels steady.
- Have a small snack of protein and carbohydrate before bed to help your child fall asleep (milk and graham crackers anyone?). A snack can boost brain levels of serotonin and melatonin that lead to sound sleep.
- Choose whole foods free from artificial flavors, colors, sweeteners, and chemical additives. Common ingredients in processed foods can cause reactions that can create a sense of unease in sensitive people. It

may be worth a short trial period (two to four weeks) of eliminating foods that commonly cause trouble. Talk with a registered dietitian or other health professional with expertise in nutrition before making radical changes to your child's diet in order to avoid unintentionally creating nutritional deficiencies.
- *Cut down on caffeine.* Caffeine can trigger rapid heartbeats and feelings of anxiety, time pressure, and panic attacks. If your child has been a cola drinker, switch to decaffeinated *green tea*; it contains *theanine,* which promotes a sense of calm. Or try herbal teas containing chamomile and lemon balm, which are both calming.
- *Eat more foods rich in omega-3 fatty acids.* Omega-3 fatty acids are important for healthy moods and response to stress.

Exercise

Exercise improves emotional well-being. Even children who are not suffering from diagnosable anxiety feel better when they participate in moderately vigorous activity *at least 60 minutes daily*. Youth who *stop* exercising due to an injury, illness, travel, or change in routine often feel more anxious or irritable. Resuming regular exercise improves anxiety. Exercise helps improve confidence and reduce worry in those with general anxiety disorder or panic attacks, patients undergoing cancer treatment, and those suffering from chronic fatigue.

Both aerobic conditioning and weight training can help ease anxiety. Exercise combined with breathing exercises and meditation, such as *tai chi, QiGong,* and *yoga*, have proven particularly helpful in promoting calm confidence.[5,6] If your child feels self-conscious or intimidated about competitive sports, consider yoga, tai chi,[7] QiGong, dancing, hiking, running, or brisk walking instead.

The other benefits of exercise include improved ability to concentrate, healthier moods, more manageable weight, and better sleep—a few less things to worry about! Speaking of sleep . . .

Sleep

Sleep problems are hallmarks of anxiety. Regrets from the day and worries about tomorrow keep churning through the mind, tightening muscles and keeping the sandman away. Then, worrying about not getting enough sleep can blossom into worrying about how the next day will go, what might go wrong, and snowballing into full-fledged panic. Worries can lead to sleeplessness that leads to even more worries.

Be sure your child's *medications* and *habits* are not part of the problem. Talk with your health professional about your child's medications. Steroids that are used for asthma and stimulants used to treat ADHD can lead to insomnia. Keep caffeine to a minimum, and don't let your child consume caffeine-containing beverages within four hours of bedtime. Exercise can contribute to alertness; do encourage exercise, but not within an hour of bedtime.

Create a calm, comforting *bedtime routine* and *sleeping environment* to make it easier to fall asleep. Keep your child's bedroom free of TVs, computers, cell phones, and other electronic devices. Turn off the TV and encourage her to start getting ready for bed. Listen to relaxing music or nature sounds. Consider practicing a stress management skill like autogenic training, meditation, or progressive muscle relaxation; extending goodwill and compassion to others; writing in a gratitude journal; or reading a soothing or inspiring book an hour before bedtime. Offer calming fragrances like lavender, roses, neroli, or chamomile as a few drops of essential oil on a washcloth or pillow-case, or incorporated into a hand or foot massage. Address your child's specific concerns by checking under the bed for monsters, or reassuring her that you've checked that the doors and windows are locked. Allow your child to use a favorite comforting object such as a stuffed animal or favorite pillow or blanket. Offer a soothing massage or back rub or read together.

Mind-Body

Stress triggers anxiety and fear. To promote a sense of security, confidence, and calm, help your child learn to anticipate and recognize threatening or stressful situations early, and learn to manage challenging situations, thoughts, and emotions skillfully. Stress is an inevitable part of life; stress management skills can also become routine. Healthy nutrition, fitness, sleep, promoting a positive environment, and avoiding unnecessary exposure to negative news and dramatic conflict can help reduce stress. Stress management practices are most effective when practiced regularly, not just saving them for a crisis. Table 6.1 shows many self-care techniques with proven effectiveness in reducing anxiety and promoting calm, confidence, and courage.

Finding and using a coach or therapist is an effective way to get started and improve existing skills. Psychologists and social workers can also teach additional skills and practices and go into greater depth with any of these strategies. Even without formal training, parents are the primary teachers helping children learn how to manage their feelings through their own examples and explicit teaching.

When your child starts to talk about feelings of worry or negativity, notice what she is doing, help her name it (feeling worried), notice how it feels in different parts of her body (butterflies in tummy, tight muscles, sweaty hands, etc.), and then gently encourage her to

Table 6.1: Mind-Body Strategies to Help Build Calm, Confidence, and Courage

Class	Examples
Body-focused	Slow, deep belly breathing; progressive muscle relaxation (sequentially tensing and relaxing muscle groups throughout the body)
Emotion-focused	Practice *positive anticipation by* picking a situation that might cause worry, and ask your child, "Wouldn't it be lovely in this situation if . . ." Imagine delightful or pleasant possibilities, and then invent strategies to make those dreams become a reality.
	Recall and celebrate successes to build confidence.
Mental-focused	Meditation (focused attention, mindfulness, or movement-based)
	Autogenic Training (see below)
	Guided Imagery/Self-hypnosis (recordings are available online)
	Biofeedback using heart rate, muscle tension, or finger temperature
	Different perspective—pretend you are at the Grand Canyon or the ocean; how big does your worry seem compared with the immensity of canyon, sea, and sky?
Spiritual-focused	Activate appreciation or gratitude through a gratitude journal; list three things you are grateful for before meals or bedtime; write thank-you notes; at dinner ask family members to tell stories about the best part of their day.
	Offer compassion to those who are suffering, and extend goodwill to others you know through simple acts of kindness. This builds self-compassion and kindness, which in turn help build courage.
	Generate generosity and cultivate happiness in others' accomplishments, like children who have overcome disabilities to participate in the Special Olympics to build security and trust.
	Find inspiration by reading stories or watching movies about people who have displayed compassion and courage.
	Prayer
	Participate in a spiritual or religious community.
	Laying on of hands or secular spiritual healing practices like Healing Touch,[8] Reiki, and Therapeutic Touch[9]
	Participate in the arts—involvement in music, painting, dance, or other arts can help build confidence in taking small risks, and allow a child to become caught up in an enjoyable activity that counteracts the tendency to worry.
Social-focused	Practice overcoming small challenges and celebrate success. If your child is shy, encourage her to practice introducing herself to dolls, action figures, pets, or relatives so she develops confidence in this skill.
	Reflect on ways others have been kind to you. This builds a sense of security and safety with other people.

switch to thinking of things for which she feels some genuine gratitude; help her notice that as her thoughts shift, her body feels different, and her mood improves. Soon she will learn for herself that building a sense of gratitude (or compassion, goodwill, generosity, inspiration, etc.) helps promote trust in the world and the people around us who have enriched our lives. It also lowers levels of stress hormones.

Holly asked Rowan to recall the scene in which Ron Weasley learned to overcome boggarts, creatures that assumed the form of one's greatest fear. Ron and his fellow students were instructed to turn the tables, imagining their worst fear (boggart) as something silly or laughable. When Ron imagined the giant boggart-spider slipping and sliding unsteadily on roller skates, it vanished. People who practice seeing the funny side of situations build their confidence and sense of resilience. Holly and Rowan started playing the silly boggart game about all kinds of things; it was fun and helped Rowan feel more secure.

Guided Imagery, Self-Hypnosis, Autogenic Training

Guided imagery and self-hypnosis can significantly decrease anxiety and distress. The brain responds to vivid images as if they were actually happening, decreasing anxiety and improving the sense of well-being, even in children undergoing painful procedures.[10] Over 60 publications have reported on the benefits of using practices such as imagining being in a safe, pleasant place or surrounded by powerful beings who deeply desire the well-being and safety for the child.[11] The Ohio State University Center for Integrative Health and Wellness offers over a dozen free online guided imagery recordings that can help improve sleep, increase a sense of safety and security, and ease worries before procedures and surgery (go.osu.edu/guidedimagerypractices). One of the most popular practices among my patients is autogenic training. It's also one of the most effective and easily practiced.[12]

Autogenic training is a self-hypnosis practice in which six phrases are slowly and gently repeated as the child relaxes:

1. My arms and hands are heavy and warm.
2. My feet and legs are heavy and warm.
3. My heartbeat is calm and regular.
4. My breathing is easy and free.
5. My belly is relaxed and soft.
6. My forehead is cool.

Repeating these phrases for 10 to 20 minutes leads to a profound sense of calm and relaxation, improved sleep, and reduced reactivity to stress.[13,14] You can repeat the phrases for your child, have her listen to a pre-made recording, or she can record her own voice repeating these six simple sentences.

You can also help your child *reframe* his fears. If he's worried about failing, help him reframe the question as "What will it take to succeed?" or "How will I celebrate when I complete this?" Being more specific can also be helpful: What will it take to succeed in math class? The next test? Or the questions on a certain topic? Once you've narrowed it down, you can address that specific concern. This helps counter the tendency to blow things up and catastrophize (e.g., if I don't know the state capitals, I'll fail social studies, and then fail this grade, and never get into college, and end up being homeless); by learning early to recognize the tendency to catastrophize, you can help your child reverse the thinking process. You can also help her see the signal in the anxiety. What is the need, goal, or value behind the fear? For example, if she fears losing you, this is a signal that she loves you very much and wants you to be safe and secure. Once you identify a goal, you can think together about strategies to achieve it. Simply noticing what happens in the moment, shifting perspective, and reframing worries as goals or values helps build confidence and problem-solving skills.

Biofeedback

Biofeedback strategies have proven to be very helpful in building confidence. They are so useful that biofeedback therapies are being used to help adults suffering from post-traumatic stress disorder. Child psychiatrists at Harvard call this strategy "Biofeedback-Assisted Relaxation Training" or BART. Biofeedback involves using a device to give the child visual or auditory feedback about his skin temperature, heart rate, muscle tension, brain waves, or skin moisture levels; with this information and their imagination, children can readily learn to regulate these autonomic activities, which gives them a great sense of power. By learning to create imagery associated with relaxation and inducing a sense of relaxation, children can learn to feel more calm and confident. An increasing number of low-cost biofeedback devices are available, and most insurance companies will cover training provided by a psychologist or other licensed health professional.

Journaling and Creative Writing

Encourage your child to keep a journal, especially writing before bed. If he writes down his worries, he doesn't have to keep thinking about them. They will be there tomorrow, so he can relax and sleep easily. He can address them in the morning when he's refreshed and has a new perspective. A fun variation on this is to imagine scary situations as scenes in a movie. He can write a creative ending or several alternative endings including the use of humor.

Courage-Building Communication

Promote calm, confidence, and courage by *communicating* kindly with yourself and your child. Communication practices become a habit. Just as mind-body practices help us find new and more skillful ways of communicating with ourselves, skillful communication with others can help us, too. Focus on your goal of helping yourself and your child feel more confident, clear, and calm in order to function more happily and effectively in the world (Table 6.2).

Extending goodwill toward yourself and others helps keep your child's mind off of worrying what other people think. This is a particularly useful practice before starting a conversation that worries you a little. Before you go in for that chat with the principal, spend a few minutes extending goodwill toward her. You will feel calmer and express yourself more clearly. Practice offering these thoughts to yourself, your child, and others:

- May I (you) live in safety and security.
- May I (you) be healthy, serene, comfortable, and resilient.
- May I (you) experience peace, clarity, and confidence.
- May I (you) feel free and loved.
- May I (you) live in beauty.
- May I (you) be comfortable, strong, and filled with vitality.
- May my (your) daily life unfold easily and calmly today.

Holly asked Rowan to help her send good wishes to the world as they waited for the bus in the morning. "May the bus driver be happy. May the mailman feel strong and fit while he walks. May the neighbors feel healthy and comfortable. May the garbage collectors enjoy a nice hot shower." Rowan giggled. The more they ex-

Table 6.2: Communication Dos and Don'ts	
Do	**Don't**
Wish yourself and others well.	Engage in the Blame Game. Avoid blaming yourself or others. Finding ways to improve and being responsible are not the same as blaming.
Focus on the positive—I spent 23 hours without worrying.	Dwell on the negative—I wasted an entire hour today worrying.
Use "I" statements: "I feel . . ."; "When this happens, I think . . ."; "I would like . . ."; "This is important to me."	Labeling others: "You are scary, mean, rude, a bully . . . ."
Focus on goals: "I want to be calm." "I want to have peace and understanding between us."	Focus on problems: "I'm worried." "We just don't get along."
Focus on what you learned. This builds strategic strengths.	Focus on how you failed. This is dis-empowering.
Talk with a family member or friend. Build a social network. Get involved in clubs, sports, volunteer efforts.	Isolate yourself.
Compliment others and cheer for their successes.	Engage in gossip and "ain't it awful" conversations.

tended goodwill (even to the trees, grass, animals, and birds!), the more confident Rowan felt.

Turn it around. Instead of putting the good news first and the bad news last in a sentence (e.g., it was a great day until the dog tracked mud in the house), save the good news until the end so you're left with the most inspiring perspective (e.g., although the dog tracked mud in the house, he was so happy to see me, it made me smile).

Help your child see her worries as a signal of an underlying goal or need. Helping someone else is a great way to build our own skills. For example, if your child is worried about an upcoming test, instead of saying, "Stop worrying. You'll do fine," consider commenting on how that worry shows how much they want to succeed in school (and please you), and ask what kind of help they'd like to prepare. If a child expresses fear about a parent's safety, the parent can thank them for wanting them to be safe,

and reassure the child that "I have done a good job of staying safe for many years, and I intend to stay around a long time to watch you grow up. You are a caring and fun person, and I enjoy watching you learn new things, just like I did when I was your age."

Use culturally meaningful practices. Guatemalans have a tradition of worry dolls; the dolls listen to your worries. Then, they do the worrying for you so you can go to sleep. Similarly, the Objibwa people (North American natives) have dream catchers, small hoops containing delicate webs that catch bad dreams, allowing only good dreams to reach young dreamers. What has your family done for generations to help build a sense of safety and connection?

Avoid engaging in the "ain't it awful" competitions in social settings. This means when someone else starts complaining about something or expressing fear about something, you encourage your child to refrain from joining in or upping the ante. Instead, encourage her to find something to compliment about someone else. Or listen for someone's pride in their accomplishment and cheer for their success.

Environment

Thoughtful attention to your environment can promote calm, confidence, security, and serenity. Changing the environment can be as simple as intentionally choosing calming, soothing, upbeat, uplifting, or inspiring *music.* Music is such a potent anti-anxiety remedy, it is used in hospitals, surgery centers, dental offices, and airports to promote greater calm and confidence.[15] Military marches strengthen courage.

Personal preference should generally guide your musical choices. Especially before bedtime, though, choose music that is calming, rather than something that makes your child want to get up and dance. Music designed to induce *binaural beats,** putting brain waves into relaxed delta or theta rhythms, can be even more helpful in reducing anxiety than music without these tones.[16] Nature sounds (waterfalls, ocean waves, crickets, bird song) can also be very soothing. Music is a safe, simple strategy to promote calm, confidence, and courage; protect your ears and keep your stress levels down by keeping the volume down.

Speaking of *nature,* spending time in natural settings such as local, state, and national parks is soothing, too. Becoming an active participant in protecting natural settings builds social bonds and a sense of empowerment.

Consider *aromatherapy.* Scents such as lavender, chamomile, neroli, vanilla, and ylang-ylang promote relaxation and calm.

Turn off the TV and other *electronic media.* Much of this media features fear and horror to hook us into watching more. The "if it

* The Monroe Institute in Virginia is a good source for many CDs featuring music with binaural beats.

bleeds, it leads" philosophy has affected the evening news and many dramatic shows. Post-traumatic stress disorder (PTSD) can occur, not just in people who directly experience a life-threatening event, but in those who watch such events repeatedly on TV. Consider what your child watches as carefully as what you feed her.

Holly decided to switch their radio from playing the news station to playing more relaxing and upbeat music, especially in the morning before Rowan had to head to school; she'd also get some relaxing music and nature sounds to play in Rowan's room before bedtime to help her ease into sleep. After school, she'd make a point of taking Rowan to the park to play amid the grass and trees and to help her feel more confident outside the house. She decided not to watch the news on TV when Rowan was awake. If they were someplace it was on, they would talk about what they saw to help Rowan put scary stories in perspective.

Professional Coaching or Counseling

Psychologists, social workers, teachers, pastors, and clinicians are trained to help provide *coaching and behavioral advice* for people suffering from anxiety. Psychotherapy (including group therapy) is as effective as medications in treating anxiety, with fewer side effects. The combination of cognitive behavioral therapy (CBT) plus medications is more potent than either therapy alone. However, therapy's powerful benefits can last for many months; because it has fewer side effects and benefits last longer, psychotherapy can be a more cost-effective treatment for anxiety than medications, particularly for children.

Another proven therapy for anxiety, particularly PTSD, is called *Eye Movement Desensitization and Reprocessing* (EMDR). Although most studies of EMDR have been done in adults, new studies show significant benefits for children, too.[17] EMDR requires professional training and guidance. Look for someone with extensive pediatric experience and solid credentials. Insurance often covers professional counseling or psychotherapy services. Check with your insurance carrier to make sure what is covered, what percentage of the visit is paid, and how many sessions are covered under your policy.

Also be aware that support groups, peer support, workshops, computerized on-line counseling, and Internet-based services are cost-effective ways to help overcome anxiety. A variety of effective approaches are available to meet individual needs for coaching and counseling.

BIOCHEMICAL THERAPIES: MEDICATIONS, DIETARY SUPPLEMENTS, AND HERBS

Medications

In adults, the combination of medications with CBT is more powerful in relieving anxiety than using either therapy alone, though medications alone can help. In children, CBT is considered the first-line treatment of conventional care for anxiety. Medications and CBT are not a replacement for an unhealthy lifestyle.

The medications used most often for pediatric anxiety are selective serotonin reuptake inhibitors (SSRIs), such as sertraline (*Zoloft®*) and fluoxetine (*Prozac®*) and selective serotonin/norepinephrine reuptake inhibitors (SNRIs) such as duloxetine (*Cymbalta®*). Medications work only as long as the drugs are taken daily. Side effects limit their desirability for long-term use. For example, prenatal use of SSRI medications may lead to irritability, feeding difficulties, and breathing problems in newborns. Other side effects include apathy, sleepiness, headache, vivid dreams, dizziness, nausea, vomiting, diarrhea, weight change, tremors, agitation, hostility, sweating, low blood pressure, liver or kidney problems, and sexual dysfunction. SSRI

medications can also trigger manic symptoms in patients with bipolar disorder and can increase thoughts of suicide in patients who are depressed. Suddenly stopping the use of SSRI medications can trigger withdrawal symptoms because these drugs are addictive and lead to dependence. Serious, life-threatening reactions (serotonin syndrome) have occurred in patients using SSRI medications and migraine medicines (triptans).

Talk with your physician or ask for a referral to a pediatric mental health professional to discuss the risks and benefits of using medications in the context of healthy choices in food, fitness, friends, and the environment as well as strategies to manage stress and build supportive relationships. Most prescription medications are reimbursed by insurance.

Dietary Supplements

Although a healthy diet is the best source of nutrients, selected supplements may serve as nutrition insurance for children who don't eat optimally.[18] Please talk with your nutritionally oriented clinician before embarking on long-term use of individual supplements. Some people benefit, but excesses can be toxic, and product quality varies.

Multivitamins/minerals (which include calcium, magnesium, and zinc) can help reduce anxiety, fatigue, and stress, even in healthy young adults who appear to be well fed.[19] I take a multivitamin daily, give one to my son, and routinely recommend them to my patients.

Children with shy-type anxiety should avoid taking high doses of the B vitamin, *niacin,* because it can cause flushing that is particularly troublesome for people struggling with social phobias. *Inositol*, formerly known as vitamin B8, can help people suffering from anxiety, obsessive-compulsive disorder (OCD), and even panic attacks. Given individual differences, inositol supplements may not help everyone who is feeling worried or anxious; however, it is extremely safe. Studies evaluating its benefits for patients with OCD, panic attacks, and anxiety have used doses of 12 to 18 grams daily for several weeks.[20]

Essential *minerals* are necessary for optimal mental health. Avoid deficiencies of iodine, iron, zinc, selenium, calcium, or magnesium; deficiencies are linked to stress, sleep problems, and anxiety. There is no evidence that high doses are helpful, and in fact, they may be toxic. If you are considering individual mineral supplements for long-term use for your child, check with your health professional.

Everyone knows that fish is brain food. Some studies suggest that *fish oil* supplements containing *omega-3 fatty acids* can help reduce anxiety as well as depression and other health conditions. In a study of medical students at Ohio State, omega-3 supplements (2.5 grams daily) taken for 12 weeks lowered anxiety levels; these students were not necessarily anxious (other than the normal stress of tests), so omega-3 fatty acids may be just part of a healthy diet for everyone.[21] If your child does not eat sardines, salmon, or mackerel twice weekly,* consider a supplement containing between 500 and 3000 milligrams of omega-3 fatty acids.

Holly decided she would start giving Rowan a multivitamin and mineral every day. Rowan hated fish, so Holly planned to give her a fruit-flavored fish oil supplement at breakfast, too.

Amino acids found in protein-rich foods are not just important for building muscle; they also promote calm, clear thinking because they provide the building blocks for neurotransmit-

* For more information about additional fish that are low in mercury and high in omega-3 fatty acids, see the Web site for the Natural Resources Defense Council (NRDC) and the Environmental Defense Fund (EDF).

ters.[22] Diets deficient in *lysine* (such as diets high in grains and low in protein) are associated with increased anxiety; in one study in a poor, marginally nourished population, fortifying wheat with lysine lowered anxiety and decreased stress hormone levels.[23] In adults, the combination of the amino acids *arginine and lysine* can help reduce anxiety and stress hormone levels.[24] Typical adult doses are three grams of each amino acid daily. Avoid giving more than three grams daily of arginine alone to minimize the risk of side effects such as diarrhea, belly pain, and bloating. Check with your clinician before starting arginine supplements; they can lower blood pressure. I do not recommend high-dose lysine-arginine supplements for anxious children until more research confirms their safety and effectiveness.

GABA (gamma-aminobutyric acid) is an amino acid found in green tea and other foods. Chemists have created a similar chemical *gabapentin* that is used as a medication to treat seizures, insomnia, and severe pain, not as a dietary supplement. The brain uses the GABA to promote calm and tranquility. Maintaining precise GABA levels is so important to the brain that it makes its own supply using *vitamin B6* and the amino acid *glutamate*. Some research suggests that GABA supplements can make it into the brain, boosting brain GABA levels, promoting relaxation, and easing anxiety.[25] The jury is still out on the role of GABA supplements in treating anxiety in children, but green tea is a safe part of a healthy diet.

Theanine is another amino acid found in green tea. Theanine helps protect brain cells from overstimulation. In adult volunteers given stressful tasks and students undergoing exam stress, theanine supplements enhanced tranquility and reduced physiologic measures of stress.[26,27] Offer your child decaffeinated *green tea* to get theanine without the caffeine.

D-cycloserine (DCS) is another amino acid

used to help battle anxiety.* Some studies in adults suggested that taking 50 to 100 milligrams of DCS just before psychotherapy sessions boosts the effects of the sessions in decreasing anxiety and phobias;[28] benefits were not observed when the supplements were taken four hours or more before therapy sessions. These results need to be repeated in children before DCS supplements become a routine part of most clinicians' treatment plans.

Acetylcysteine or *N-Acetylcysteine* (NAC) is widely used medically to help detoxify the liver after damage from acetaminophen overdoses. In the brain, NAC helps manage glutamate, an amino acid that tends to rev up the brain. NAC supplementation has proven useful in several studies of trichotillomania (compulsive hair-pulling, which is often associated with anxiety)[29] and obsessive compulsive disorder (a type of anxiety);[30] it has also been used successfully to help combat cocaine cravings and other addictions.[31] The doses used in studies range from 2000 to 4800 mg daily.

Tryptophan and its sister amino acid, *5-HTP,* are mainly used to treat premenstrual syndrome and depression, but some people also use them to promote calm, positive moods and better sleep.[32] Tryptophan and 5-HTP supplements can reduce stress and the tendency to panic or feel anxious.[33] Supplementation with 200 milligrams of 5-HTP reduced anxiety in adults prone to *panic disorder.*[34,35] In studies of adults with *obsessive compulsive disorder* (OCD) whose symptoms had not improved with medications alone, tryptophan supplements provided relief.[36,37] Side effects of 5-HTP and L-tryptophan include nausea, constipation, numbness, palpitations, aggression, and other symptoms. Doses higher than three grams daily can cause headaches. To minimize these

risks, start with low doses (25 milligrams daily) of 5-HTP; gradually increase the dose as tolerated with professional guidance. Typical doses of tryptophan are a bit higher—50 milligrams three times daily or 150 milligrams before bed on an empty stomach or with a few crackers (not with a protein rich meal). Tryptophan and 5-HTP can also boost the effects (and side effects) of SSRI medications; please do not give these supplements to your child without medical supervision if your child is taking sedative or antidepressant medications.

Probiotics are the healthy bacteria found in yogurt, kefir, and other fermented foods and supplements. Emerging research has found that probiotics are not just good for the gut; they are also good for the brain! In studies in adults, probiotic supplements improved anxiety.[38] It will not be long before research is conducted on probiotic supplements for children. Probiotic supplements are low risk and low cost. If you decide to try probiotics for your child, look for products containing *Lactobacillus* and *Bifidobacterium*.

Herbal Remedies for Anxiety

Herbal remedies and dietary supplements are not a replacement for a healthy lifestyle, but some herbs are helpful in temporarily calming anxiety. Calming herbs such as *chamomile, hops, lemon balm, passionflower,* and *valerian* can help promote sleep and relaxation and reduce stress. These herbs are generally recognized as safe; allergies are possible but uncommon. However, because they can cause sleepiness, don't give them to your child on the morning of exams or before driver's ed.

Brahmi, keenmind, or *Bacopa monnieri,* is a traditional remedy from India used to treat anxiety and improve overall mental functioning. Several adult studies support its benefits for calm and clarity;[39,40] adult doses in recent trials

---

* DCS is an old treatment for tuberculosis. Its use in treating anxiety is modern.

were 320–640 mg of a standardized extract, and no serious side effects were reported.[41] However, similar studies have not yet confirmed bacopa's safety and effectiveness in children.

While most people think of *ginkgo* as a memory enhancer, 240 to 480 milligrams daily of the European ginkgo product, EGb 761 or Tanakan®, can also help anxious adults feel more calm.[42,43] Ginkgo may cause bleeding problems, particularly in patients taking medications that affect clotting; talk with your clinician before offering ginkgo supplements to your child. I do not recommend it until more high-quality trials have been conducted in children and adolescents.

Kava kava (*Piper methysticum*) is a Polynesian herb that eases anxiety and stress-related insomnia;[44] however, its toxic effects on the liver (when manufactured using alcohol or organic solvents rather than water extraction) have curtailed its use.[45] Furthermore, it can intensify side effects from anxiety medications, and suddenly stopping after taking it regularly can lead to withdrawal symptoms. Until kava has undergone additional studies in pediatric populations, I do not recommend it as a treatment for childhood anxiety.

Lavender and *neroli* (*Citrus aurantium*) are used mainly as aromatherapy or as ingredients in bathwater or massage lotion to soothe agitation, restlessness, nervousness, or stress, and to promote relaxation and sleep; lavender is even used in some hospitals and hospice centers to decrease anxiety and improve a sense of security and well-being.

Holly started putting a few drops of lavender oil and neroli on the inside of Rowan's jacket and on her scarf to help her feel calmer. Rowan liked the fragrances so much, she asked to have a few drops on her pillow, too, so she could fall asleep with those nice smells.

Rhodiola rosea is used in Russia and northern Europe to reduce anxiety, improve the ability to cope with stress, and as a general tonic.[46] A study done at UCLA showed that 340 milligrams of rhodiola supplements daily helped adults diagnosed with general anxiety disorder.[47] Rhodiola is generally safe, but as with all herbal products, some people get an upset stomach and a few have allergic reactions. Until more pediatric research on rhodiola has been published, I do not recommend it for children.

St. John's wort is used primarily to treat depression, but some people also use it for anxiety. Results of clinical studies have been mixed; it appears to be more helpful for those with mild symptoms than those suffering from severe anxiety, obsessive-compulsive disorder, or panic attacks. Typical adults doses are 900 milligrams daily. Although it generally has fewer side effects than medications, St. John's wort can decrease the effectiveness of many other medications and herbs. If your child is taking any other medication or herbal remedy (even birth control pills), check with your health professional before starting St. John's wort. St. John's wort can also trigger manic symptoms in patients with bipolar disorder. Until research has addressed questions about its safety and effectiveness in children, I do not recommend it for pediatric anxiety.

BIOMECHANICAL THERAPIES: MASSAGE

Massage

Massage is a very ancient therapy to promote relaxation and restful sleep, reduce stress, and ease anxiety. Many studies support the benefits for children of receiving massage and to parents for giving massages, with or without aromatherapy.[48–51] Massage can also be combined with relaxing music to enhance its relaxing benefits. Therapeutic massage is safe, even for small infants. Massage therapists are licensed as health professionals in most states, with strict requirements for training and continuing education. Although it can be helpful, ongoing professional massage therapy can be costly. Consider getting some training so you can offer the benefits of regular massage to your child.

BIOENERGETIC THERAPIES: ACUPUNCTURE, MAGNETS, PRAYER, THERAPEUTIC TOUCH, HOMEOPATHY

Acupuncture

Although many children think of acupuncture as something scary, it has actually proven effective in decreasing anxiety.[52,53] Acupuncture has been used to reduce the fear of dental, medical, and surgical procedures, and ambulance rides. Acupuncture can help relieve anxiety and pain both for chronic sufferers and in those who are normally calm until confronted with the stress of a medical or surgical procedure.[54] Acupuncture has also proven useful in treating post-traumatic stress disorder and obsessive-compulsive disorder.[55,56]

Studies on over 2,000 adults suggest that acupuncture and acupressure can be effective remedies for patients suffering from anxiety-related insomnia.[57] This is a great benefit for anxious insomniacs.[58]

Acupuncture is generally safe, though minor bleeding and bruising are possible with any treatment involving needles. Acupuncture needles are regulated by the U.S. Food and Drug Administration, and most modern practitioners use disposable needles; this means that in the United States the risk of acquiring an infection from treatment is very small. No serious adverse events have been reported in several large studies; serious side effects are extremely rare. Acupuncture has gained so much credibility over the past 20 years that more than one-third of pediatric pain treatment programs in teaching hospitals have begun offering acupuncture.[59,60]

Be sure to ask about the acupuncturist's experience treating children, and to ease your child's anxiety, have the acupuncturist demonstrate on you before treating the child. When the child sees how calm you are, that you are not in pain and not worried, it will be easier for her to understand that acupuncture needles are not scary. Alternatively, ask the acupuncturist to start by using non-needle techniques. Acupuncturists are accustomed to patients who are worried about their first treatment, and they are prepared to be gentle, reassuring, and accommodating.

Finally, ask your insurance carrier whether and how many acupuncture treatments are covered and for what conditions before you start a series of treatments.

Magnets and Electromagnetic Field Therapies

Cranial electrotherapy stimulation (CES), or electro-sleep, was originally developed in the Soviet Union in the mid-twentieth century as a non-drug therapy to promote sleep and relieve anxiety. CES devices send very mild electrical currents to the brain through electrodes attached to the skin of the earlobes or just behind the ears; they do not require surgery and do not send the kind of strong current used in electroconvulsive shock therapy or transcranial magnetic stimulation. Controlled studies have shown that CES relieves anxiety and promotes sleep.[61,62] Dentists have used it to reduce anxiety in young patients undergoing dental procedures.[63] Although CES has not caught on as a mainstream approach to managing anxiety or stress in the United States, devices such as the Alpha-Stim™, the Fisher Wallace Stimulator™, and the Liss-Shealy CES™ are available by prescription from a health professional. They cost several hundred dollars and are not typically covered by insurance plans. The device is used 20–30 minutes daily for several weeks. No serious adverse effects have been reported from using the CES, but children who have implanted pacemakers, cardiac defibrillators, or electrical insulin pumps should check with their pediatrician before using any kind of electromagnetic therapy.

Prayer/Reiki/Therapeutic Touch

Prayer is a commonly used mind-body-spirit strategy that can lead to a calm, peaceful state, affirming trust and security in a higher power to provide protection and guidance. Religiosity and prayer appear to improve overall psychosocial functioning in high-risk children.[64]

Even those who are not religious may benefit from secular forms of laying-on-of-hands prayer, such as Reiki, Therapeutic Touch, and Healing Touch.[65] When these services are consistent with families' beliefs and are offered at no or minimal cost, they are safe to use to promote a sense of calm, confidence, and connection.

Homeopathy and Bach Flower Remedies

Some anxiety sufferers have reported dramatic benefits from homeopathic remedies. However, rigorous randomized controlled trials have not consistently found that homeopathic remedies alone (without the support of the homeopathic practitioner) are more effective than placebo pills for anxiety. The understanding and support of the homeopath may be more important than the remedies in achieving a sense of calm, trust, and security. Homeopathy is safe, but seldom covered by American insurers.

WHAT I RECOMMEND FOR ANXIETY

HEALTHY HABITS IN A HEALTHY HABITAT

1. *See your health professional* to make sure that there are no physical conditions or medications that are contributing to your child's anxious feelings. Ask to have her thyroid, iron stores, and vitamin D level checked. Your health professional can also help you find experienced psychotherapists, counselors, biofeedback trainers, massage therapists, and acupuncturists, and prescribe a CES device.

2. *Reframe your goals*. Focus on the positive outcomes of calm, confidence, and courage rather than worrying about worry.

3. *Exercise*. Make sure your child gets at least 40 to 60 minutes daily at least five days a week, preferably outdoors in a natural setting.

4. *Sleep*. Make sure your child gets at least 7½ to 10 hours a night. Use a routine; keep the room dark and quiet or use comforting music and aromas. Consider practicing together a stress management approach such as meditation, guided imagery, self-hypnosis, autogenic training, biofeedback, counting blessings, prayer, or

writing in a journal before bed. Minimize exposure to light, noise, and alarming or distressing news or dramas before bedtime. Keep TV screens and other electronic screens out of the bedroom. Consider offering a small snack containing a balanced protein and carbohydrate before bedtime.

5. *Eat well.* Keep your child's blood sugar steady by serving whole grains, beans, seeds, nuts, legumes, fish, fruits, and vegetables. Eat breakfast. Cut down on caffeine. Offer decaffeinated green tea.

6. *Supplement wisely.* Talk with your pediatrician about offering a multivitamin or B-complex vitamin with essential minerals to ensure your child's brain has the necessary nutrients to stay calm and confident. Watch out for high doses of B3 (niacin), which can cause flushing. Talk with your health professional about trying inositol and N-acetylcysteine supplements. Consider adding an omega-3 fatty acid supplement and extra L-tryptophan or 5-HTP.

7. *Herbal remedies.* Talk with your health professional about herbal remedies with a long history of safety in children, such as chamomile, hops, lemon balm, passionflower, as well as aromatherapy like lavender.

8. *Environment.* Increase what's helpful and decrease what's not. Help your child become a skillful selector of music—listening to calming, relaxing music or nature sounds before bed, and rousing marches before heading into a situation calling for courage. Consider using music with binaural beats. Spend more time in nature. Use calming, comforting fragrances such as lavender, chamomile, neroli, vanilla, and ylang-ylang. Free yourself from the tyranny of media news coverage. Make sure your child's environment is safe so you can offer honest reassurances.

9. *Practice effective emotional, mental, or spiritual stress management strategies that suit your family.* Help your child intentionally activate feelings of gratitude, affection, compassion, goodwill, generosity, courage, and positive anticipation. Learn and practice meditation, self-hypnosis, guided imagery, autogenic training, or progressive relaxation regularly. Keep a journal. Try a home biofeedback device or consider professional assistance to help your child learn to reframe anxious thoughts and learn to see the signals of her values, goals, and needs that lie behind her emotions. Catch catastrophizing thoughts before they snowball and avoid "ain't it awful" competitions. Pray, participate in a spiritual or religious community, and help your child to be inspired by stories of courage. Try prayer, laying on of hands, Therapeutic Touch, Reiki, or QiGong healing if they are consistent with your family's beliefs. Help your child express herself through art.

10. *Communicate kindly and encouragingly.* Wish yourself and others well. Focus on the positives. Turn self-criticism into strategic thinking. Reflect on others' kindness.

11. *Build community and seek fellowship.* Help your child join a club, team, choir, orchestra, band, or church. Encourage him to volunteer with you. Support him in caring for pets and plants, and playing with friends who are kind and nurturing.

12. *Offer your child a massage*—even just a foot rub, hand massage, or back rub regularly. A professional massage therapist can help you learn to feel more comfortable if massage is not part of your family tradition. You can combine massage with aromatherapy and/or relaxing music for even greater benefits.

13. *Acupuncture.* It's counterintuitive, but acupuncture can ease anxiety. I recommend it.

14. *CES.* Talk with your health professional and your insurance company about trying a cranial electrostimulation device.

15. *Medications.* Talk with your health professional about a short trial of medications in the context of healthy behaviors (exercise, nutrition, and stress management) and a healthy environment.

16. *Advocate* for changes in your community to promote a calmer, more secure, and less stressful society and a less toxic physical environment.

7
ASTHMA

Yolanda Jefferson awoke suddenly in the middle of the night, hearing her youngest son, 18-month-old Devante, cough again. Neither of them was getting any sleep. Yolanda was exhausted from the lack of sleep, caring for three children and two cats, getting the kids to day care while she worked, making meals, and trying to keep the house clean. Devante seemed to cough more than other kids when he caught a cold. His cough was worse at night; sometimes he coughed even when he didn't have a cold. Yolanda was beginning to wonder if he had asthma. His grandmother, who lived with them, continued to smoke even though she had asthma, and Devante's two older brothers had asthma, too. Yolanda wanted to know what she could do to ease his symptoms and if there was anything she could do to stop it from getting worse. His older brothers had both been hospitalized and required steroid inhalers to stay out of the emergency room.

If your child suffers from asthma, you are not alone. About 12–15 percent of American children suffer from asthma, making it the most common chronic condition of childhood. The 10 million U.S. children with asthma account for millions of missed days of school each year.

Over the last 30 years, asthma has become even more common and more severe, suggesting that the major reason is environmental and behavioral, not genetic. Between 1980 and 1995 pediatric asthma rates climbed over 70 percent and asthma deaths jumped a whopping 40 percent.

WHO IS MOST LIKELY TO DEVELOP ASTHMA?

Asthma is more commonly diagnosed in boys (16 percent) than girls (12 percent) and more often in African American (22 percent) than Hispanic (14 percent) or Caucasian (12 percent) children. This means that more than one in five African American children are diagnosed with asthma in the United States! It is more common among children living in poverty; children who have poor health for other reasons; children who were born prematurely; those living in crowded conditions; children who are obese; those who have had pneumonia or other lung infections; and those living with a smoker. Like allergies, asthma runs in families.

CAN CHILDREN OUTGROW ASTHMA?

Yes. Symptoms improve for many children when they reach adolescence. On the other hand, some children first experience asthma as teenagers.

CAN CHILDREN WITH ASTHMA LIVE NORMAL LIVES?

Yes. Children with asthma can do just as well in school, sports, and social situations as their peers without asthma. In the 2012 Olympics, 8 percent of athletes had asthma. They consistently outperformed their colleagues.

Asthma is one disease with several symptoms.

ASTHMA SYMPTOMS

- Dry cough (especially at night)
- Wheezing (high-pitched whistling sounds) while breathing out
- A feeling of tightness in the chest
- Having difficulty breathing; can't catch his breath

Sometimes children will have only one of these symptoms—usually the cough. Coughing and wheezing can be caused by other health problems, too.

OTHER CAUSES OF COUGH AND WHEEZING

- Viral infections
- Aspirating food or other objects into lungs
- Genetic lung diseases
- Airway abnormalities

Viral infections such as colds and bronchitis, aspiration (when food or a small object is inhaled into the airways), and genetic diseases such as cystic fibrosis are all marked by coughing and wheezing. Abnormal airways such as floppy tracheas or having a blood vessel mistakenly wrap around a breathing tube can lead to wheezy breathing, too. X-rays, blood tests, and lung function tests may be necessary to make the correct diagnosis. A child is called asthmatic only if his symptoms occur at least three times, several family members have asthma, or there is some other reason to suspect that symptoms will recur.

Devante had classic asthma symptoms. His diagnosis was confirmed by his physical examination and response to treatment. Yolanda wanted to know what was happening in his lungs to cause Devante's symptoms. She also wanted to know what might trigger his asthma so she could prevent flare-ups.

Three changes in the small airways (*bronchioles*) cause asthma symptoms.

LUNG CHANGES IN ASTHMA

- Inflammation and swelling of the walls of the small airways

- Increased mucus production, blocking the small airways
- Bronchospasm, muscle tightening around the airways

When the airways become inflamed during an infection or allergy, their walls become swollen and irritated, and this blocks airflow to the tiny air sacs (*alveoli*) at the end of the bronchioles. The irritation also triggers coughing. Normally, *mucus* helps clear the airways, but when there is too much mucus, it clogs air passages. Irritation also stimulates the airway muscles to tighten or constrict, further cutting off airflow. These changes lead to cough, wheezing (the sound the air makes as it tries to flow through narrowed tubes), and feelings of tightness and an inability to catch one's breath.

Very few children have asthma symptoms all the time; most just have symptoms triggered occasionally. The National Heart, Lung, and Blood Institute (NHLBI) of the National Institutes of Health (NIH) has published guidelines for diagnosing and treating asthma. It categorizes asthma in four levels of severity (Table 7.1)

The goals of asthma education and treatment are to restore normal function and minimize symptoms. The different degrees of asthma severity mean different kinds of treatments are needed.

Regardless of the severity of symptoms, every child with asthma needs an Asthma Ac-

Table 7.1: Asthma Severity for Children Five Years and Older	
Class	Examples
Mild, Intermittent (Step 1)	<ul><li>Daytime symptoms two or fewer times weekly</li><li>Nighttime symptoms two or fewer times monthly</li><li>No school absence; no need for emergency room</li><li>Needs steroid treatments less than twice yearly</li><li>Peak flow rates > 80% predicted</li></ul>
Mild, Persistent (Step 2)	<ul><li>Daytime symptoms more than twice weekly</li><li>Nighttime symptoms more than twice monthly</li><li>Needs steroid treatments two or more times yearly</li><li>Peak flow rates more than 80% predicted</li><li>Symptoms have minor impact on activities</li></ul>
Moderate, Persistent (Step 3)	<ul><li>Daily symptoms</li><li>Nighttime symptoms more than once a week</li><li>Peak flow rates between 60 and 80% of predicted</li><li>Symptoms moderately affect activities</li></ul>
Severe, Persistent (Step 4)	<ul><li>Symptoms every day and night unless treated</li><li>Peak flow less than 60% of predicted</li><li>Symptoms limit physical activities</li></ul>

tion Plan. This means that you and your child sit down with your child's physician and work out goals, strategies, and contingencies to maximize her health and minimize the chance that she will end up in the emergency room or hospital. The plan should be written and shared with school, sports, camp, and other places your child spends a significant amount of time, and reviewed regularly to keep it updated. Post it on the refrigerator so everyone at home (including babysitters) can easily refer to it.

Devante's big brothers had Step 3 asthma. They had action plans for their medications, but had not addressed common asthma triggers.

WHAT TRIGGERS ASTHMA AND HOW DO YOU KEEP TRACK?

Sometimes an asthma trigger works immediately—the child starts wheezing as soon as he walks outside on a pollen-filled spring day. Sometimes triggers have delayed effects; a child is exposed to a cat in the afternoon and does not have symptoms until he goes to bed. Keeping a diary of your child's asthma symptoms can help you figure out what triggers the flare-ups (Table 7.2).

Table 7.2: Common Asthma Triggers

- Airway irritants: cigarettes, air pollution, wood smoke, gas heat/stove
- Allergies, including food sensitivity
- Exercise
- Cold air
- Infections: colds, sinus infections, bronchitis
- Medications (aspirin, angiotensin converting enzyme (ACE) inhibitors)
- Stress
- Acid reflux

Cigarette smoke is the #1 preventable trigger of asthma. Being around smokers will almost certainly make your child's asthma worse. Maternal smoking during pregnancy affects a child's lung development and increases his risk of having asthma later. *Do NOT smoke and do not allow other people to smoke around your child!*

Air pollution also aggravates asthma. Today's energy-efficient houses trap indoor mold, dust mites, and animal danders in the house, aggravating asthma symptoms. Children who live near busy roads, factories, power stations, and other sources of air pollution are barraged by bad air. City emergency rooms log more asthma visits when pollution levels are higher. Rural residents have their own sources of pollution: wood smoke from stoves and fireplaces, dust, animal dander, and agricultural chemicals. Globally, the growing levels of sulfur dioxide, nitrogen dioxide, ozone, and particulate matter are contributing to escalating rates of asthma. It will take global policy changes to reverse this trend.

Ragweed and pollen provoke seasonal *allergies* in some children and asthma in others. Other asthma-inducing allergens are molds, grasses, dust, dust mites, fleas, and cat and dog dander. *Food allergies* (such as to peanuts, eggs, and wheat) can also trigger asthma symptoms in some sensitive children. Cow's milk is a well-known but rare trigger for asthma symptoms; children who are truly allergic to milk almost always have other symptoms as well such as hives, diarrhea, or eczema. *Food additives,* such as yellow dyes, sodium benzoate, and sulfites, also trigger asthma in some children. If you think your child might be allergic to foods or food additives, please have him tested by an allergist before you make radical changes in his diet. Many children outgrow their food allergies, though no one knows why.

Exercise, especially in cold, dry air, triggers symptoms in nearly 90 percent of asthmatics. Frequent, vigorous exercise in cold, dry air may actually induce chronic asthma. Many elite figure skaters and cross-country and downhill skiers have asthma. Fortunately, exercise-induced asthma can be effectively prevented and treated.

There is no reason for your child to avoid exercise just because he has been diagnosed with asthma.

Cold and sinus *infections* are among the most common triggers for asthma. Cold viruses trigger wheezing in 80 to 85 percent of asthmatic children. They can even provoke wheezing and coughing in people who don't have asthma. If your child has frequent bouts of asthma, is troubled by nighttime coughing, or has had a cold that lasts more than ten days in a row, talk with your health care professional about the possibility of a sinus infection. Treating the infection can dramatically improve asthma symptoms.

The most common *medication* causing asthma flare-ups is aspirin. Steer clear of aspirin and other non-steroidal anti-inflammatory drugs (NSAIDS) such as ibuprofen. Acetaminophen (e.g., Tylenol™) does *not* trigger asthma attacks. Prescription medications known as beta-blockers (such as propanolol), used to prevent migraine headaches and treat high blood pressure, can also trigger asthma symptoms. The high blood pressure medicines known as *ACE inhibitors* frequently cause coughing that may either trigger or mimic asthma. Some of the *contrast dyes* used in X-rays can also trigger asthma symptoms. Whenever your child sees a new doctor for any reason, be sure to mention that your child has asthma to minimize the chances of an inappropriate prescription.

Emotional *stress* can cause a tight chest and labored breathing in almost anyone. It's not surprising that children with asthma are more prone to having symptoms at stressful times such as the first day of school or during a divorce. Previously it was believed that dysfunctional parents contributed to childhood asthma, and children were sent away to boarding schools to get away from their "toxic" families. On the contrary, children's illnesses contribute to parental stress and dysfunction—the more serious the child's illness, the more disruptive for the family.

Does *dirt* make asthma worse? There's little doubt that air pollution, cat and dog hair, dust mites, and cockroaches can make asthma worse in sensitive children. But interestingly, kids growing up in the dusty countryside tend to have less asthma than urban kids. It might be that certain microbes and parasites actually offer some protection against asthma, allergies, and eczema. While I don't recommend that you intentionally expose your child to dirt or worms, there is emerging research that an extra bit of the healthy bacteria found in yogurt and kefir (probiotics) might help promote healthy immune responses.

Heartburn not only feels bad; it can trigger asthma symptoms. *Acid reflux* (also known as heartburn or gastro-esophageal reflux disease, GERD) triggers asthma symptoms for many people, particularly infants as well as those who are older or obese.

Asthma can be deadly. If your child has any of the following symptoms, take him for professional care.

SEEK IMMEDIATE PROFESSIONAL CARE IF YOUR CHILD IS:

- Having a hard time breathing
- Breathing much more rapidly than usual
- Making a grunting noise when he breathes out
- Getting tired or agitated with his shortness of breath
- Getting blue in the lips or fingertips
- Sucking in the spaces between his ribs with each breath

MAKE AN APPOINTMENT IF:

- Symptoms interfere with sleep or other activities.
- The child has a fever.
- The child has a poor appetite or is starting to become dehydrated.

DIAGNOSING AND MONITORING ASTHMA

Observations of coughing, wheezing, and breathing patterns have long been the mainstays of monitoring asthma symptoms.

MONITORING ASTHMA SYMPTOMS

- Daily diary: symptoms, triggers, treatments
- Peak flow meter
- Visits to health care professional four times yearly

Today you can measure your child's lung function at home using a peak flow meter. The cost of peak flow meters is usually covered by your insurance. A peak flow meter measures how forcefully your child can exhale (blow air out). A low reading on the peak flow meter is an early warning sign that symptoms will soon slide downhill. Children as young as six years old can learn to use a peak flow meter reliably. Those with persistent symptoms should measure their peak flow at least twice daily (morning and evening) and record it in an asthma diary along with symptoms and triggers. Such information will be very helpful to your child, you, and your health care provider in mapping out the best treatment plan for your child. See your doctor four times yearly to monitor your child's growth and response to therapy and to receive support and information about new treatment strategies.

PREVENTING AND TREATING ASTHMA

There are no cures for asthma yet, but many treatments can help control its symptoms. Optimal treatment is based on careful observation and relies on a combination of therapies. Let's tour the Therapeutic Mountain to find out what works; if you want to skip to my bottom-line recommendations, flip to the end of the chapter.

LIFESTYLE THERAPIES: NUTRITION, EXERCISE, ENVIRONMENT, MIND-BODY

Nutrition

Proper nutrition is a cornerstone of good health, including reducing the risk of asthma. Excessive eating leads to obesity, and obesity increases airway hypersensitivity; reducing obesity can reduce asthma symptoms. There are *no* studies suggesting that restricting your child's calories and making her anorexic are helpful.

Asian children who eat a typical Asian diet have a lower risk of asthma than Asian children who eat a more American type of diet. This has led to widespread recommendations that patients who suffer from inflammatory conditions like asthma eat an anti-inflammatory diet. This means eating more vegetables, fruits, beans and seeds, whole grains like brown rice and quinoa, fish and other foods rich in omega-3 fatty acids, and fiber; less butter, cheese, red meat, processed foods (foods sold in boxes with nutrition labels); and drinking water rather than artificially flavored, carbonated water (soda). The anti-inflammatory diet is pretty much the same as the Mediterranean diet, and there is growing evidence that this diet reduces the risk of childhood asthma.

Foods can trigger asthma as well as allergies and eczema. Avoiding food triggers can improve symptoms. Some physicians routinely advise asthmatics to follow an elimination diet (restricting major allergenic foods such as cow milk, wheat, corn, soy, peanuts, tree nuts, and shell fish), a minimal diet (allowing only a very restricted number of foods), vegan diets, or diets excluding certain foods. Although this may seem harsh, in a study of 504 pediatric asthma patients, 45 percent had evidence of sensitivity to one or more of these common foods, and those who were sensitive to foods were more likely to require steroid medications or hospitalization; they were also more likely to be allergic to pollen and dander.[1] In a study of adult asthmatics, 79 percent of those who had tried a restricted diet reported an improvement in their asthma symptoms.[2] In a Danish study, the symptoms of hospitalized asthmatic patients improved when they were placed on a hypoallergenic, elemental diet.[3] If you think foods might trigger your child's symptoms, please ask your health professional for a limited time trial of a few foods diet to evaluate the impact on your child's health.

Many parents believe that *milk* increases mucus and worsens asthma symptoms. A Swedish study of asthmatic adults showed an improvement on a diet low in *tryptophan,* an amino acid found in milk, cheese, turkey, and bananas.[4] On the other hand, in a randomized, controlled trial in asthmatic adults, milk did not cause coughing, wheezing, or impaired lung function.[5] Milk does *not* make asthma worse unless your child is allergic to milk. While restricted diets may be helpful in some asthma patients, they must be done only under the strict supervision of a nutritionist and for brief (two months or less) trials to avoid developing deficiencies of calcium, protein, iron, and other critical nutrients.

On the other hand, there are some interesting studies suggesting that certain foods may protect against asthma.

NUTRITIONAL THERAPIES FOR ASTHMA

DO:

- Breast-feed
- Give plenty of water
- Eat fish like sardines and salmon as part of a healthy Mediterranean diet
- Try onions and spicy foods
- Try coffee

DON'T:

- Eat sulfites, preservatives, yellow dyes

Start your child's life right by *breast-feeding*. Breast-feeding reduces wheezing in the first month of life, and the benefits in reducing the risk of asthma persist throughout the first 17 *years* of life. Make sure that your older child gets plenty of *water*. Water helps keep secretions thin and loose, preventing mucus in the lungs from getting dry, sticky, and difficult to clear.

Everyone knows that fish are brain food, but did you know that *fish* contain omega-3 fatty acids that can help stem inflammation throughout the body? In a 20-year follow-up study of over 4,000 American young adults, those who ate the most omega-3 fatty acid–containing fish had significantly lower risk of developing asthma than those who ate the least fish.[6] Fish are part of the healthy Mediterranean diet. Children whose moms ate more of a Mediterranean diet during pregnancy had a lower risk of asthma than children whose mothers consumed a more conventional American diet.[7]

Certain compounds in *onions* and *spicy foods* such as red peppers can reduce the release of histamine and other inflammatory chemicals responsible for asthma's symptoms.[8,9] Don't rely on pickled onions, which may contain irritating sulfites. Some parents believe that hot spices such as horseradish, mustard, and chili peppers are also helpful in preventing asthma symptoms, but just smelling these spices triggers coughing in some people. Since onion and hot spices are common cooking ingredients, you may want to experiment yourself: Note onion and spice intake and asthma symptoms in your child's asthma journal to see how much better (or worse) he is on the days he eats (or doesn't eat) these foods.

Coffee for asthma? In the 1800s it was the treatment of choice. Caffeine is chemically related to the asthma medication theophylline (see section on medications). There is no benefit to drinking more than three cups daily. A 2010 review concluded that caffeine improves airway function for up to four hours in people with asthma.[10] Don't try to substitute caffeine-containing colas; colas can actually trigger asthma symptoms in some children.[11] Coffee is relatively inexpensive and widely available. If you want to try it for your child, remember that less coffee would be needed for smaller children than for adults and that coffee can have side effects including anxiety, tremors, racing heart, and insomnia.

Avoid *margarine*. In a German study, frequent intake of margarine was related to an increased risk of asthma in young adults.[12] In an Italian study, eating fruits and vegetables was associated with a lower risk, but eating more bread and margarine was associated with an increased risk of wheezing and allergies.[13]

Yolanda decided that since she could control what she fed the boys, she would make major changes in their evening dinner. Instead of hot dogs one night a week, she'd switch to fish. And since one of the older boys had ADHD, she'd try giving him coffee for breakfast to see if that would help improve his breathing and his behavior. She would stop buying margarine, too.

Sulfite preservatives, benzoic acid, MSG, and *yellow dyes* can trigger asthma.[14,15] Sulfites (chemically related to the sulfur dioxide in smog, which clearly makes asthma worse) are commonly sprayed on fresh fruits and vegetables (beware of restaurant salad bars!) and added to certain snacks, orange drinks, beer, and wine. It may be that the real culprit for food-triggered coughs is either allergies to additives or sensitivity to the food or preservatives being eaten. If you are not sure if your child is sensitive to preservatives or dyes, read labels for all prepared food and wash all fresh fruits and vegetables thoroughly before serving them. Given all the uncertainties in today's food supply, I think the best option is to serve only foods that are organically grown, preferably those from your own garden or local farmer.

Exercise

Exercise, especially in cold, dry air triggers symptoms in nearly 90 percent of asthmatics. However, children with asthma should still be encouraged to exercise. The overall health benefits of cardiovascular fitness are well known. Asthma is easier to control in patients who are physically well conditioned, and exercise-induced asthmatic symptoms can be readily controlled. Many Olympic athletes have asthma and have set world records for athletic performance. Don't let your asthmatic youngster become a couch potato.

Yoga exercises, particularly yogic breathing exercises (pranayama), improve lung capacity and reduce the number of asthma attacks.[16] The physical exercises (asanas) alone are helpful, but they're even better when combined with breathing exercises as part of a comprehensive approach to healthy living.[17] These exercises emphasize slow, regular breaths in which the ratio of inhalation (breathing in) to exhalation (breathing out) is 1:2. For example, have your child breathe in for a count of 5 and breathe out to a count of 10. The benefits can be enhanced by breathing hot, moist air.[18] Regular yoga practice can even reduce the need for steroid medications and frequent inhalers.[19] Yoga has long-term benefits. Follow-up studies show improvements for at least two years when asthmatics keep practicing yoga; yoga helped them participate in other sports as well.[20] Yoga is straightforward, inexpensive, and free of side effects. I recommend yoga or tai chi for all of my asthmatic patients.[21] Classes are taught in nearly every community, and over a dozen books about yoga for children are available.

Another good breathing exercise for asthmatic children is called *pursed lips breathing*.[22] In this technique, children purse their lips and blow out as if blowing a kiss. Adults with severe chronic lung disease who practiced this technique were able to increase their blood oxygen levels during pursed lips breathing alone without any other therapy.

Other types of *breathing exercises* that are frequently suggested as complementary therapies for asthma combine aspects of physical training and mind-body interventions. Training sessions may include voice training, relaxation, postural changes, and breathing exercises. In a randomized, controlled trial, breathing exercises significantly improved lung function—long-term improvements that were comparable to the short-term benefits of inhaled beta-agonist medications.[23]

Another breathing technique was developed by Russian physician Konstantin Buteyko. He believed that deep breaths or "over-breathing" causes asthma, and that to reduce asthma symptoms, patients should be trained to "breathe less."[24] In one Australian study, teenagers and adults who received training in Buteyko's breathing techniques did show some improvement in their asthma symptoms over four months compared with a control group who received standard asthma education;[25] they needed fewer treatments with beta agonist medications and reported better quality of life.[26] Similarly, a Canadian study showed that asthmatic adults who practiced Buteyko breathing were able to cut back on the steroid medications;[27] additional studies are needed in children.

Swimming is a great exercise for youngsters with asthma.[28,29] Swimming improves overall fitness and self-esteem. Spa treatments (swimming in hot spring/mineral water) may be even better. In a study of adults with severe, steroid-dependent asthma, 69 percent given spa treatments reported significant improvement.[30] Similar studies have not been done in children, nor are insurance companies likely to race to reimburse for spa memberships. Remember that some kids are sensitive to the high chlorine levels in public pools, and that the most important thing is to keep your child exercising.

Yolanda decided to enroll all three boys in swimming classes at the local Y. That way they could exercise all winter long, regardless of the weather.

PREVENTION STRATEGIES FOR EXERCISE-INDUCED ASTHMA

- A fifteen- to thirty-minute warm-up period before vigorous exercise
- Breathing in through the nose—not the mouth—to warm and filter outside air before it hits the lungs
- Covering the nose and mouth with a loose-fitting scarf or bandanna when exercising outdoors on especially cold days

Sleep

Thanks to the benefits of modern lighting and electronic gadgets, many American children get insufficient sleep, and this is also true in children with asthma. Also, children with asthma who have allergies, too, may also have obstructions leading to snoring and intermittent breathing pauses while sleeping. This means that children with asthma have even more problems with breathing while asleep than non-asthmatic children.[31] Conversely, poor sleep is linked to worse asthma symptoms.[32] Please make sure your child has a healthy sleep routine, goes to bed early enough to get sufficient sleep, and if you notice loud snoring or starts and stops in breathing at night, have her evaluated to see if having her tonsils and adenoids out helps improve her sleep, breathing, and asthma.[33]

Environment

Avoiding environmental triggers helps prevent your child from having asthma flare-ups. If typical environmental triggers at home (such as tobacco smoke, furry pets, and gas ovens or stoves) were eliminated, there would be about a 39 percent reduction in the amount of asthma in the United States.[34]

REDUCING AIRBORNE ASTHMA TRIGGERS

- *No smoking* anywhere near the child
- Changing from gas to electric furnaces and stoves
- HEPA air filters; furnace and air-conditioner filters
- Houseplants: philodendrons, spider plants
- Windows closed 5 am to 10 am when pollen counts are high
- Eliminate cockroaches and dust mites
- Advocate to reduce air pollution: ride a bike or take a bus

Avoiding cigarette smoke reduces the risk of many heart and lung diseases, including childhood asthma. Second-hand cigarette smoke may be responsible for up to 50 percent of asthma episodes in children. Don't smoke and don't let anyone else smoke around your child.

Free-standing HEPA electronic *air filters* efficiently remove airborne asthma triggers inside the home, but filters alone don't remove dander and dust lying on the floor or furniture. Potting soil and common house plants (such as spider plants, philodendron, bamboo palm, and English ivy) also absorb some indoor air pollutants. Change or replace your *furnace filter* and air-conditioning filter regularly. Consider installing an electrostatic *air filter* on your furnace; such filters can effectively remove over 90 percent of airborne irritants and allergens. Keep your child indoors when air quality is bad. Pollen counts are usually highest between 5 and 10 am, so keep the windows closed during early morning hours. Advocate for public policies that improve air quality. Personally, consider carpooling, walking, using buses or bicycles, and using electric vehicles as ways of helping your child stay healthy throughout life.

About 50 percent of children who suffer from chronic asthma are sensitive to *dust* and microscopic *dust mites.* By reducing the levels of household (and car seat)[35] dust and dust mites, you can help your child reduce asthma and allergy symptoms. Eliminate dust accumulators such as old carpeting and dust ruffles. Dust mites love to live in mattresses and pillows. Keep them away from your child by encasing the mattress and pillow in a plastic or vinyl fitted sheet. Wash your child's sheets, pillowcases, and blankets weekly in hot water to kill dust mites. Engage in a weekly cleanup campaign to fight dust and mold in the child's room; damp mopping helps remove dust without stirring it up. Clean thoroughly after spraying any insecticides to remove all of the dead dust mite particles and insecticide residues.

Rid your home of *cockroaches*. About one-third of children are allergic to cockroaches, and those who are often exposed (such as urban apartment dwellers) have a much higher risk of asthma.[36]

If your child's symptoms are triggered by *furry pets* (like 50 percent of hospitalized asthmatic children who are sensitive to dog or cat dander), pets should not sleep in your child's bedroom. Pets that trigger symptoms should be either outdoors or kept in non-carpeted rooms that are easily cleaned and in which the child does not spend a lot of time. If your child is visiting a home with pets that trigger symptoms, pretreat the child with asthma medication prior to the visit.

Mold, a common asthma trigger, is a problem in humid climates and in households that cook a lot of pasta or rice because of all the water vapor in the air. Try to keep household humidity less than 50 percent with a dehumidifier, and frequently clean any visible mold or mildew with a 10 percent bleach solution. Promptly repair any leaks in the roof or plumbing to avoid a buildup of mold and mildew.

If you think housekeeping issues like managing smoke, dust mites, pet dander, and mold have little impact, the results of a study in Eastern North Carolina may help you change your mind. In this study, households of children with asthma were randomly selected for intensive home cleanup versus usual home care. After six months, those in the intensive cleanup group had a 58 percent reduction in symptoms, 76 percent decrease in the need for rescue medicine for asthma flares, and 33 percent fewer visits to the emergency room.[37] These results are a powerful testimonial to the benefits of optimizing the environment to reduce asthma symptoms.

Negative *ion generators* have *no* proven benefits for asthma.[38] I don't recommend them. While *mist* and steam from vaporizers are often soothing for children with colds, sinus infections, and bronchitis, they are not very helpful for children with asthma. If you use a vaporizer, please clean it at least weekly with a bleach solution to kill the mold and mildew that build up in the moist environment.

Infections are common triggers for asthma symptoms. This is particularly true for the common viruses that run rampant in the fall and winter when kids return to school. To minimize the risk, try to keep your child in less crowded classrooms, enforce frequent hand washing, make sure her vitamin D level is optimal, and protect her with the annual flu vaccine. Interestingly, there are some studies showing that a few patients with very severe, chronic asthma are actually suffering from a form of walking pneumonia (caused by *Mycoplasma* or *Chlamydia* bacteria); and that when they receive appropriate antibiotic treatment, their symptoms improve dramatically.[39] Sinus infections are also major asthma triggers. If your child has cold symptoms that are not getting better within 10 days, take her to the pediatrician to make sure she doesn't have a sinus infection or walking pneumonia.

Mind/Body

Stress is a major trigger for asthma symptoms. Depression and stress in the family combined with inadequate social support make asthma worse and increase the risk of emergency care and hospitalization.[40] Stress management can take a variety of forms and be helpful for both parents and children. Mind-body therapies for asthma include *hypnosis, guided imagery and relaxation training, autogenic training, meditation, biofeedback, music therapy,* and several relaxation therapy hybrids. These therapies can be very helpful, even with relatively brief training.

Hypnosis and guided imagery have proven useful in improving symptoms and lung function, enhancing parental confidence in managing their children's asthma, and reducing the amount of medication and number of physician visits for asthmatic children, even among pre-schoolers.[41] Hypnosis can help patients reduce allergic reactions, the need for steroid medications, and even hospitalization rates. Seek help from a physician, psychologist, or other health professional trained by the National Pediatric Hypnosis Training Institute.

Relaxation and *guided imagery* are extremely safe. A trained therapist should be able to get you started on a home program of your own, and your child can practice by listening to recordings of the therapist's voice, leading her through the relaxation exercises. It is also important for parents to remain calm and confident when their child has an asthma attack. Plan your strategy in advance. When parents stay calm and relaxed, children have an easier time relaxing, addressing the problem, and reducing symptoms. This means it can be very helpful for parents to be proactive in practicing relaxation strategies, too.

Like self-hypnosis, *autogenic training* has proven effective in reducing chronic asthma symptoms.[42] (Visit the Ohio State University Center for Integrative Health and Wellness Web site for a link to a free MP3 recording of autogenic training.) Autogenic training is one of the practices I most often teach my patients because it is so easy to learn and so helpful for reducing stress.

Meditation was developed as a spiritual practice, but it has become a popular strategy to manage stress and reduce the symptoms of chronic conditions such as asthma. For example, t*ranscendental meditation* has proven helpful for adult asthmatics who practice it regularly.[43] *Mindfulness meditation* training has also helped improve the quality of life for asthma patients.[44]

Biofeedback can also be an effective part of a mind-body approach to stress reduction for children with asthma. Several studies have reported success in children and adolescents.[45–47] If your child's symptoms are triggered by stress (or he's stressed from having unpredictable, poorly controlled symptoms), biofeedback is worth trying.[48] See a qualified pediatric health professional to see if biofeedback is right for you. Insurance companies increasingly cover this kind of care, but be sure to check your policy to find out if you're covered.

Another mind-body therapy for asthma in older children is "journaling." Asthmatics who were assigned to write about their stressful experiences three times a week for 20 minutes at a time had a marked reduction in asthma symptoms and had improved lung function over the next four months.[49] Interestingly, these journals were not discussed with a therapist, and no particular therapy was initiated on the basis of writing. Just getting those stressful thoughts out on paper helps reduce stress, which in turn can improve lung function. This is very simple and straightforward and I recommend you try it.

Seek *support* from other families who have asthmatic children. Sharing stories and strategies helps decrease isolation and make stress more manageable.

BIOCHEMICAL THERAPIES: MEDICATIONS, DIETARY SUPPLEMENTS, AND HERBAL REMEDIES

Medications

Modern medications are lifesaving for acute asthma attacks. Asthma can kill. Do not rely on non-prescription medications or home remedies for acute asthma episodes. They are not strong enough to help, and you may be giving yourself false assurance that you are doing something useful. Breathing is important! Healthy habits like nutrition, exercise, sleep, environmental optimization, and mind-body practices are great, but an integrative approach includes skillful use of the best of modern medicine, too.

MEDICATIONS (ALL ARE PRESCRIPTION ONLY)

- Prevention and treatment: beta-agonists (e.g., albuterol)
- Optional for mild, persistent symptoms: cromolyn or nedocromil
- Maintenance and treatment: steroids (inhaled sprays and oral forms)
- Preventive for moderate or severe: leukotriene inhibitors (e.g., montelukast)
- Other: adrenalin, terbutaline, theophylline, ipratropium, heliox
- Prevention: flu shots and allergy desensitization shots

Before we get to the different kinds of asthma medicines, let's talk about how inhaled medicines are used. Inhaled asthma medications are delivered directly to the lungs by either a nebulizer (a device that turns the medication into a fine mist) or a metered dose inhaler (MDI or "puffer"). MDIs are the most widespread and convenient way to take inhaled asthma medications. A canister containing the medication inserts in a plastic dispenser. Pressing the canister releases a premeasured amount of medication. The child inhales the medication, delivering it directly to the lungs. Some MDIs only deliver the medication when the child takes a breath, enhancing efficiency. Even infants can use MDIs *if they use a spacer and mask.*

HOW TO USE AN MDI

- Shake canister while taking deep breath in and out
- Hold can one inch from mouth (or preferably, use a spacer)
- Release medication at the beginning of the next breath in (inhalation)
- Breathe the medication mist in deeply
- Hold breath to count of 10; exhale
- Wait one minute; repeat

Although these directions sound simple, it takes a fair amount of coordination to get the timing right to dispense the medication just as the child breathes in. If the child has finished inhaling by the time the medicine is dispensed, the medicine just evaporates into the surrounding air rather than reaching the lungs. A *spacer* is a device that holds the medicine in a confined space after its release from an MDI. The medication remains suspended, waiting for the child to take the next breath. Some fold up so they can be carried in a purse or backpack; some make sounds to let the child know when the medication has been inhaled properly. Some spacers come with facemasks that cover the nose and mouth, so even infants can receive MDI medication without a nebulizer. Most spacers and masks are covered by insurance. I always write a prescription for a spacer at the same time I write the initial prescription for metered dose medications. And I always make sure the child and family can demonstrate proper MDI technique before they leave the office. It's one of the things we recheck at our quarterly visits. It's that important.

For mild, intermittent asthma, the only medicine your child may need is an inhaled, short-acting beta-agonist. *Beta-agonists* (such as albuterol, bitolterol, pirbuterol, and terbutaline) are the most commonly prescribed asthma medications. They quickly relax airway muscles and allow the air passages to expand. Beta-agonists effectively prevent exercise-induced asthma when used twenty to thirty minutes prior to vigorous exercise. You may have seen athletes using them before practices and competition. Beta agonists are also useful for the quick relief of flare-ups caused by an allergy or a cold. They start to work in about twenty minutes; benefits last for four to six hours.

Longer-acting (12-hour) beta-agonist medications, such as *salmeterol* (*Serevent*™), are preventive medications for children and teenagers who need medication every day. Using it regularly may decrease the need for steroid medications. It can also be combined with steroid medications for children with moderate or severe asthma.

The main side effects from beta-agonists are like the side effects from drinking too many cups of coffee: racing heart, high blood pressure, high blood sugar, reduced appetite, or feeling a little "hyper." If your child requires more than two doses in a day or more than three times a week, see your doctor about additional or alternative preventive therapies so you can cut down on the beta-agonist treatments. Overuse results in dependence, needing higher and higher doses to achieve the same benefit; this is a very dangerous situation that is best avoided before your child gets into serious trouble.

For children who have persistent asthma symptoms more than a day or two a week, the treatment of choice is an *inhaled corticosteroid*. Inhaled steroids are used both to prevent and treat persistent asthma. Inhaled steroids get the needed medicine right to the lungs.

EXAMPLES OF INHALED CORTICOSTEROIDS FOR ASTHMA

- Beclomethasone (Beclovent®, Vanceril®)
- Budesonide (Pulmicort®)
- Flunisolide (Aerobid®)
- Fluticasone (Flovent®)
- Triamcinolone (Azmacort®)

Corticosteroids for asthma are *not* the same as anabolic steroids that athletes use to bulk up. Corticosteroids decrease airway inflammation and swelling. Given when peak flow meter readings start to fall, these steroids can prevent asthma flare-ups, and are among the most cost-effective treatments for children with chronic asthma. Proper use can prevent emergency room visits and hospitalization. Inhaling these steroid medications in an emergency works even faster and better than taking them by mouth. Most of them only need to be taken once or twice daily to keep symptoms under control.

Also, when steroids are given by MDI, they go directly to the lungs; very little is absorbed into the bloodstream, resulting in fewer side effects than taking them by mouth. Inhaled steroids do *not* usually impair growth; however, some children who require very potent inhaled steroids several times daily may lose an inch or two in eventual height. If your child requires inhaled steroids, he needs to have his growth monitored three to four times yearly. You can reduce the impact on growth either by decreasing the daily dose or by giving a larger dose once a day as opposed to smaller doses twice daily. Other side effects include sore throat, hoarseness, and yeast infections in the mouth. You can prevent these by using a spacer (so more of the medicine goes to the lungs and less stays in the mouth) and gargling with plain water after "puffing" to rinse it out of the mouth. Even though many people are scared by steroids' potential side effects, children who have persistent asthma end up spending less time in the hospital and less time in the emergency room if they use inhaled steroids daily rather than relying solely on beta-agonist medication.

Children with *severe persistent* asthma may require *steroids by mouth* in addition to all of the other preventive medications. Because of their potential side effects (e.g., decreased growth, unstable blood sugar levels, cataracts, easy bruising, impaired immunity), oral steroids should be taken for as short a time as possible. Generally, three to five days of oral steroids are sufficient for most children with a severe asthma attack and may help prevent a hospitalization; taking steroids for less than a week does not increase the risk of serious infections or other side effects associated with long-term use of steroid medicines. For children who are coughing too hard to take medicine by mouth, a single steroid injection is as effective as three to five days of oral therapy. This is a good thing because most steroid syrups taste awful, and many parents would rather give their child a shot than struggle with them several times a day to take a vile-tasting medicine.

For children with *mild persistent* asthma symptoms, there are a couple of additional treatment options. Inhaled *cromolyn* (Intal®) and *nedocromil* (Tilade®) help *prevent* asthma symptoms. They block histamine release and prevent asthma symptoms. Cromolyn and nedocromil only work when taken *before* the lungs get irritated; they are useless in treating acute attacks. These are helpful for preventing symptoms in children who have symptoms three or more days a week, particularly for exercise-induced asthma. For maximal effectiveness, cromolyn and nedocromil must be taken several times daily for at least two to four weeks before improvements are apparent. Few side effects have been reported from either medication, but they are not of much use in an acute attack, and because of their inconvenient dosing requirements and lower potency than steroids, they have fallen out of favor.

Another class of asthma medications is used for children with moderate to severe persistent asthma because they help prevent inflammation and allergic-type reactions: *antileukotriene medications* (such as *montelukast*). These medications are taken to prevent symptoms. In general, they are quite safe and have fewer side effects than steroids; they can be combined with inhaled steroids for more powerful benefits.

Other Medicines for Asthma

In an emergency, one of the quickest treatments for children with an acute attack of severe asthma is a shot of *epinephrine* (Adrenaline). Adrenaline works well, but it wears off quickly. In the 1970s clinicians started using *terbutaline* instead of adrenaline because it lasts longer. Because adrenaline and terbutaline require injections, they were largely replaced in the 1980s by inhaled medications, but you may still find medics who use "epi" in an emergency.

In the 1970s and early 1980s, *theophylline* and its cousin, *aminophylline,* were the treatments of choice for children with asthma. Both are chemically related to caffeine; they expand airways and decrease inflammation. Blood levels need to be monitored closely to minimize side effects. They fell out of favor because of many reported side effects: hyperactivity, decreased attention span, decreased appetite, and increased risk of seizures. Low doses of theophylline may decrease the doses of steroids needed to control symptoms for patients with moderate or severe asthma. Aminophylline may also be useful if your child's symptoms are severe enough to land him in the hospital. If your child has mild to moderate asthma, it is unlikely that your physician will recommend these medications, but your grandmother may ask about them.

She may also ask about *ipratropium bromide* (Atrovent™), which is mostly used for adults and older children suffering from severe asthma who need emergency treatment. It is chemically related to the older, plant-derived drug, atropine. Some adults with COPD (chronic obstructive pulmonary disease) use it daily, but only for severe symptoms that are not well controlled with inhaled steroids. It works best when used with beta-agonist medications. One product (*Combivent*) combines ipratropium with albuterol; it is mostly used by elderly asthmatics and is less commonly used in pediatrics.

Preventive Medical Treatments

Children with asthma can have especially severe symptoms with *influenza*. If your child has asthma, please get the *influenza vaccine every fall*; it helps prevent asthma flare-ups. I get the flu vaccine myself to reduce my chances of getting influenza and passing it on to others. The vaccine is especially worthwhile for kids who require medications daily. If your child has moderate or severe asthma, I'd recommend having everyone in the household vaccinated against influenza to protect the child.

Allergic desensitization therapy is useful for children whose asthma symptoms are triggered by a few specific airborne allergens such as pollen. Talk with your physician or nurse practitioner about a referral to a pediatric allergist if allergies are a major trigger for your child's symptoms, and they cannot be controlled with environmental optimization alone.

In an emergency, for severe asthma being treated in the hospital, your physician may call on the old remedy *heliox* to treat your child's asthma. Heliox is a combination of helium and oxygen. The helium replaces the air's nitrogen; helium is lighter than nitrogen, and it's easier to breathe heliox than room air, so your child doesn't have to work as hard to get the oxygen into his system. Back when I worked in a burn and trauma unit, we did research showing that heliox helped kids suffering from narrowed airways after their breathing tubes were removed.[50] Since then, it has become a popular therapy in the intensive care unit for treating severe asthma, too.[51] It is quite safe as long as oxygen levels are monitored, which they are routinely in American intensive care units.

Dietary Supplements

Dietary supplements are not a replacement for medical therapy, but some may be useful adjuncts. The dietary supplements most often used for asthma include vitamin B6, vitamins C and D, magnesium, and fish or flax seed oils containing omega-3 fatty acids.

DIETARY SUPPLEMENTS FOR ASTHMA

- Vitamin B6 (pyridoxine)—no longer
- Vitamin C—maybe
- Vitamin D—make sure it's optimal
- Magnesium—make sure it's optimal
- Fish oils/flax seed oil containing omega-3 fatty acids—yes

Supplemental *Vitamin B6* (*pyridoxine*) was used in the past to help asthmatic patients taking theophylline because theophylline depletes B6 levels. Vitamin B6 supplements in doses of 50–200 milligrams daily over several weeks helped reduce the number of asthma attacks, the severity of symptoms, and the need for medications in asthmatic children.[52] However, it is unclear whether pyridoxine supplements are helpful for children taking beta-agonist or inhaled steroid medications. Additional research is needed before I return to routinely recommending B6 for asthmatic children.

Children who consume foods rich in *vitamin C* (such as citrus fruits, strawberries, and kiwi) have the least wheezing and asthma;[53] those who eat the least vitamin C have a five times greater risk of having twitchy lungs.[54] In ancient days, the relationship between scurvy (severe vitamin C deficiency) and asthma was well known. In modern times, three randomized controlled trials have shown that a dose of 500 milligrams to 2 grams of vitamin C reduces exercise-induced asthma symptoms by 50 percent.[55] Giving extra vitamin C (500–1000 mg daily for adults) also protects against ozone-induced asthma attacks;[56,57] and higher vitamin C intakes in children help protect against the asthma-inducing effects of air pollution. Furthermore, two studies have shown that vitamin C supplements of one to five grams reduced asthma attacks triggered by colds (viral infections) in adults by 50 to 90 percent.[58] Giving your child 500 to 1,000 milligrams (½ to 1 gram) per day is safe; I recommend sustained-release vitamin C supplements to my asthmatic patients. Excessive vitamin C can cause diarrhea.

In the last 20 years, much has been learned about the many benefits of *Vitamin D,* not just for bones, but also for the immune system. Sun exposure helps us convert cholesterol to vitamin D in our skin; as sun exposure has decreased, vitamin D levels have dropped. Children with low vitamin D levels have an increased risk of developing asthma.[59] And children with asthma who need steroid medications may benefit from supplemental vitamin D to protect their bones.[60] Furthermore, in children already diagnosed with asthma, supplementing with 500 to 2000 IU daily can reduce the risk of an asthma flare-up by 60 percent.[61] If your child has not been tested for vitamin D yet, please have her tested, and if her level is suboptimal (<30), talk with your physician about offering supplements to boost her into the optimal range.

Higher dietary intake and blood levels of *magnesium* are associated with fewer and less severe breathing problems such as asthma.[62] Magnesium (300 milligrams to 2 grams depending on the child's weight) is given by vein in many emergency rooms for children with acute asthma episodes.[63] In a randomized, controlled study, daily magnesium supplements (300 mg) significantly improved airway function and reduced reactivity in children.[64] Too much magnesium can cause diarrhea. Focus on feeding your child foods rich in magnesium and other nutrients. Talk with your health professional about supplementing with 100–300 milligrams daily of magnesium.

Omega-3 fatty acids, such as those found in *flax seed oil* and *fatty fish* reduce inflammation and the risk of asthma (Table 7.3).[65] Children who regularly eat fresh fish or fish oil have better lung function and a lower risk of asthma compared with those who avoid eating fish.[66]

MAGNESIUM-RICH FOODS	
Beans (pinto, northern, black, navy)	100 mg per one cup of cooked beans
Beans (garbanzo or lima)	80 mg per one cup of cooked beans
Cereal (All-Bran)	120 mg per one-half cup
Cereal (Raisin Bran)	80 mg per one cup
Lentils	70 mg per one cup cooked
Nuts (almonds, cashews)	80 mg per three tablespoons
Peanut butter	50 mg per two tablespoons
Rice, brown	85 mg per one cup of cooked rice
Spinach	75 mg per one-half cup cooked

Table 7.3: Omega-3 Fatty Acids in Fish	
Fish	Amount of Omega-3 Fatty Acid in Three-Ounce Serving
Cod	<200 milligrams
Fish Sticks	<200 milligrams
Flounder	200–500 milligrams
Grouper	200–500 milligrams
Halibut	200–500 milligrams
Herring, Atlantic	>1.5 grams
Mackerel, wild Pacific	>1.5 grams
Pollock	200–500 milligrams
Salmon, Atlantic	>1.5 grams
Salmon, wild	>1.5 grams
Salmon, canned	1–1.5 grams
Sardines	500–1,000 milligrams
Trout	500–1,000 milligrams
Tuna	1–1.5 grams

Do Omega-3 Supplements Help?

There are good reasons to think they will. Omega-3 fatty acids help decrease inflammation in general, which is the underlying problem in asthma. In one study, 29 children with chronic asthma were given daily fish oil supplements; over 10 months, their symptoms gradually improved, but they also underwent a major cleanup in their environment, which might have accounted for the improvements.[67] In another study, adults who took an *omega-3 fatty acid* (docosahexaenoic acid, DHA—the active ingredient in fish oil) supplement had fewer asthma symptoms than others who took placebo pills.[68] Maternal supplementation can help decrease asthma in her children.[69] It may take two to twelve months for supplemental omega-3 fatty acids to make a difference in asthma symptoms, and they work best in the

context of overall healthy nutrition, lifestyle, and medical management. Excessive intake of omega-3 supplements (more than six grams daily) can cause easy bruising. I recommend eating fish twice weekly, and for those who can't, supplementing with 1,000 to 2,000 mg of omega-3 fatty acids daily.

Canned tuna as well as fresh mushrooms, shrimp, dried fruits, guacamole, and salad bar ingredients may contain *sulfites,* which trigger asthma in some children. Beware of restaurant salad bars and be careful to wash your produce before consuming it at home.

Herbs

Although herbs have been used to treat asthma by traditional healers historically, there are few scientific studies evaluating their effectiveness compared with modern treatments. They are *not* as effective as medications, and they are not tightly regulated by the FDA; so you cannot be sure you are getting what you're paying for. Do not abandon your current medications in favor of herbs. Please discuss *all* treatments with your doctor before making major changes. Even if you're just using herbs as supplements, please tell your doctor and pharmacist so they can help you avoid adverse interactions.

Perhaps the most well-known herbal asthma remedy is *ephedra,* the original source of the old asthma medication ephedrine and the modern decongestant pseudoephedrine. The Chinese have used ephedra (*Ma huang*) for nearly 5,000 years to treat asthma. It is chemically related to epinephrine, and it has similar side effects: high blood pressure, rapid heartbeat, decreased appetite, and feeling overstimulated, eventually resulting in severe fatigue. Sadly, ephedra gained notoriety in the 1990s as people tried using it as a natural high and performance enhancer. The U.S. Food and Drug Administration received over 600 reports of adverse effects, including 22 deaths, related to ephedrine. This led to tighter state regulations governing the availability and strength of ephedra-containing products. I do not recommend it.

Other mainstays of Chinese herbal therapy are *Chinese skullcap* (Huang Qin) and *Ligusticum,* which have anti-inflammatory properties similar to cromolyn; and *Angelica (Dong Quai),* which relaxes bronchospasm and reduces reactivity to allergens. *Cordyceps sinensis* is a fungus that is a common ingredient in Chinese herbal asthma remedies; it affects immune reactions in test tubes, but there are no studies yet comparing it to standard asthma medicines in children. I do not recommend any of these herbs as a replacement for medical management or a healthy lifestyle.

Licorice root (Glycyrrhiza glabra radix) has been used in folk medicines around the world to treat coughs. Licorice root's active compounds block the breakdown of steroids, which means it offers both the benefits and side effects of steroids. As recently as the 1960s, licorice was used successfully by American physicians to treat Addison's disease (a serious steroid deficiency). On the other hand, long-term high-dose licorice intake can cause fluid retention, swelling in the feet and ankles, hypertension, headaches, blood chemistry imbalances, lethargy, and muscle weakness. No studies have yet evaluated the risks and benefits of including licorice root as a standard pediatric treatment regimen for asthma. Patients using licorice chronically should be closely monitored for steroid-like side effects.

Turmeric is the spice that gives Indian curries their yellow color and pungent flavor. It also has powerful anti-inflammatory effects. In a study of older Asian adults, those who regularly ate turmeric-rich curry had better lung function.[70] While waiting for pediatric research, consider including turmeric-rich foods in your child's diet. Additional research is needed on this supplement.

Like the asthma medication montelukast, ancient Asian herbal asthma remedies, *Shinpi-To* and *Saiboku-To,* inhibit the body's chemical pathway that leads to inflammation. Saiboku-To contains five herbs that slow steroid breakdown, possibly increasing the risk of side effects in patients who need to take steroid medications, but it reduces the need for anti-inflammatory medications in adult asthmatics taking it over several months.[71] Shinpi-To contains seven herbs, including ephedra, licorice, apricot, and magnolia. Until more research has evaluated their safety and effectiveness in children, I do not recommend them because of the high rate of contamination and adulteration in herbal products imported from Asia.

Many healing traditions rely on herbal remedies for asthma.

TRADITIONAL HERBAL REMEDIES

- China and Japan: Ephedra (*Ma Huang*): Chinese skullcap, Angelica (*Dong Quai*), Licorice root; Saiboku-To, Shinpi-To, Moku-boi-to, Sho-saiko-to, Sho-seiryu-to and others
- European and American: Coltsfoot, Ginkgo, Bee pollen
- Hawaii: *Sophora chrysophylla* (Mamane), *Piper methysticum* (kava kava), nightshade
- India: *Adhatoda vasica* (Malabar nut), *Coleus forskholii, Verbascum thapsus* (mullein), *Tylophora indica,* others
- Latin America: *Aloe vera, Galphimia glauca,* red onions, Siete jarabes; Agua maravilla, Jarabe maguey

Bee pollen is widely touted as a natural remedy for asthma, allergies, and eczema. There are no clinical trials evaluating the effectiveness of bee pollen in treating childhood asthma. Serious allergic reactions and even fatalities have been reported. This is *not* a safe adjunctive therapy for your child. Avoid it.

Ginkgo biloba, was used in Chinese medicine long before becoming one of the biggest selling herbal remedies in North America and Europe. Standardized Extract of Ginkgo biloba (EGb) is sold under several different brand names in Europe. Ginkgo's active ingredient, ginkgolide, decreases inflammation and is also a powerful *antioxidant.* Despite its long historical use and biochemical rationale, only a few studies have evaluated ginkgo's effectiveness as an asthma remedy, and they have had mixed results. Ginkgo can cause bleeding problems. It has not been extensively tested in children. I do not recommend it as a pediatric asthma remedy.

Coleus forskholii is an herb used in Indian Ayurvedic medicine to treat asthma. It contains chemicals that help relax lung muscles, but there are no pediatric studies evaluating its effectiveness. Another Ayurvedic herbal remedy, *Tylophora indica,* has proven beneficial in controlled, double-blind crossover studies of adults with allergies and asthma, but optimal dosages and long-term effects on children are unknown; I do not recommend it until additional research establishes its benefits for children. *Boswellia serrata* is another Ayurvedic asthma remedy; it blocks inflammation in animals, and it looked promising in a pilot study of asthmatic adults, but again, the lack of pediatric trials prevents me from recommending it. *Solanum xanthocarpum* and *S. trilobatum* (from the same plant family as potatoes, tomatoes, and eggplant) improved lung function in asthmatic adults, but these herbs were not as powerful as standard medications and have not been tested in children. Until additional research is conducted in children, I do not recommend these herbs, but I'm happy to refer interested patients to my Ayurvedic colleagues.

Other traditional remedies for asthma include: *coltsfoot, yerba santa, wild cherry bark, ginger root, peppermint, red clover, coleus root, comfrey, lobelia, marsh mallow root, nettle, parsley,* and *thyme.* Adding a strong solution of thyme tea to bathwater is believed to help all kinds of coughing illnesses, and it is very safe. Thyme is also added to home steam inhalation treatments to soothe irritated airways. *Slippery elm bark* and *wild cherry bark* are both helpful for scratchy throats from dry, hacking coughs. Additional research is needed to determine their role in the treatment of asthma.

Herbs are not necessarily safe just because they are natural. Herbal remedies are regulated more like food than like medicines, but most families who can tell when they're buying a ripe banana cannot determine the quality of an herbal product sold in a capsule. Up to one-third of Chinese patent medicines are spiked with medications such as steroids, resulting in significant side effects. During harvesting, sometimes the wrong plants, pesticides, herbicides, animal residues, and so on get into the material. During processing, products may become contaminated with heavy metals such as lead and arsenic; strengths and dosing are not standardized, and significant variation in purity and potency have been reported. This means that you might be paying for herbs you are not really getting; or you may be unknowingly giving your child an overdose, or another product, or a contaminated product. Until there is better research on their safety and effectiveness and better government regulation protecting consumers, I do *not* recommend herbal remedies for asthmatic children.

Biomechanical Therapies: Physical Therapy, Massage, Spinal Manipulation, Surgery

Physical Therapy

Physical therapy treatments (pounding on the chest to help clear mucus) do *not* benefit asthmatic children, though exercise programs can improve lung capacity.

Massage

Massage not only feels good, it helps reduce stress and anxiety, which may contribute to asthma. In a study of 32 asthmatic children, massage was compared with relaxation therapy; parents were instructed in either technique and provided the therapy to their child each night before bed for one month. The youngest children had the biggest gains with massage—less anxiety, lower stress hormone levels, better attitudes, and better lung function—but even the older children with more well-established asthma had an improvement in some lung functions.[72] Another study showed improved lung function in children who received massage from their parents five out of seven nights a week for five weeks.[73] Once parents are trained, they can provide the therapy; and if the therapy is provided by parents, the cost is low and the side effects, if any, are minimal. This is one therapy that I regularly recommend for families to assist their children with asthma.

Spinal Manipulation

Spinal manipulation or spinal adjustments are provided by chiropractors, naturopaths, and osteopaths to treat asthma. Randomized clinical trials have not established the benefit of chiropractic adjustments in treating asthma in children.[74] Until research establishes a positive cost-benefit ratio, I do not recommend chiropractic therapy for asthmatic children.

BIOENERGETIC THERAPIES: ACUPUNCTURE, THERAPEUTIC TOUCH, PRAYER, HOMEOPATHY

Acupuncture

Acupuncture has been part of the treatment for asthma in China for thousands of years, but data from controlled trials are mixed.[75] A Danish study involving over 100 preschool children suggested that 10 acupuncture treatments over three months helped reduce asthma symptoms and the need for steroid treatments.[76] An Egyptian study in school-age children confirmed benefits of 10 acupuncture treatments for preventing exercise-induced symptoms and decreasing the need for steroid medications.[77] Even in China, however, acupuncture is not usually used as the only remedy for people with asthma, but it is used in combination with herbs or Western medications. Additional studies are needed to determine the cost-effectiveness of acupuncture and optimal treatment regimens for asthmatic children. Given its safety, I'm happy for my patients to try it to see if it helps them, but I don't prescribe it routinely as an asthma treatment.

Therapeutic Touch

There have not been any studies of the effectiveness of Therapeutic Touch in treating asthma symptoms. However, in my own practice I have watched children's oxygen levels climb and their breathing relax during Therapeutic Touch treatments. I incorporate Therapeutic Touch into my therapy for children during acute asthma attacks and for children hospitalized in the Intensive Care Unit. A group of German asthmatics reported great improvement in their asthma symptoms and a decrease in their need for medications when they received treatments from a "hands on" healer;[78] it's not clear from this study exactly what kind of healing or how many sessions might be helpful for children with asthma. Hands-on healing, Therapeutic Touch, and Reiki are all very safe. Clearly, more research is needed to determine the optimal strategy for using these different techniques for children, but if your health professional can offer you a treatment at no additional charge, go for it!

Prayer

Many of my patients tell me that they feel more relaxed and peaceful when they pray; they also feel that prayer helps them manage their asthma better. I recommend prayer if it is consistent with your family's values and beliefs.

Homeopathy

Homeopathy shows some promise in treating asthma,[79] but the quality of the research studies is not very good, and treatment is highly individualized.[80] Common asthma remedies include *Arsenica, Antimonia, Chamomilla, Ipecac, Lobelia, Nux vomica,* and *Pulsatilla.* There are no published studies of homeopathy in preventing or treating asthmatic children. Homeopaths themselves advise parents not to treat symptoms of acute asthma with homeopathic remedies alone. If your child has allergies as well as asthma, homeopathy might be a reasonable route to explore, but only as an adjunct, not a replacement for asthma medications and lifestyle changes.

WHAT I RECOMMEND FOR ASTHMA

Work with your doctor to develop a WRITTEN ASTHMA PLAN. Help children monitor their symptoms; obtain and use a peak flow meter and keep a record of symptoms, triggers, treatments, and peak flow meter readings. Treat all symptoms promptly.

1. *Lifestyle—environment.* Do *not* smoke and do not allow others to smoke around your child. Avoid allergy triggers. Keep your house clean and free from cockroaches, mold, and dust. Thoroughly clean your child's room weekly. Wash your child's bedding in the hottest possible water twice a month to minimize dust mite exposures and encase his mattress and pillow in dust mite–proof covering. Consider using an electronic air filter and adding air-cleaning house plants such as philodendron and spider plants to your child's room. Do not let furry pets sleep in your child's room. Consider a HEPA air filter in your child's room or on your central furnace. Keep windows shut between 5 and 10 am when pollen counts are highest. If possible, use an electric rather than a gas stove. Work for clean air in your community. Try to avoid crowds where your child will catch respiratory viruses. Teach your child to wash his hands before meals, after using the restroom, and after contact with other people.

2. *Lifestyle—nutrition.* Start your child off right by breast-feeding. Eat a healthy Mediterranean diet rich in omega-3 fatty acids (fish), vitamin C (fruit), magnesium (beans and whole grains), and probiotics (yogurt), and make sure your child's vitamin D level is optimal. Consider regularly serving turmeric-rich curry. Avoid exposure to sulfites, yellow dyes, and other dietary triggers. Reduce the use of red meat, butter,

cheese, and processed foods containing sulfites, benzoic acid, MSG, and yellow dyes. Avoid sodas, but consider coffee for breakfast and lunch. Give plenty of fluids, especially when your child has a cold. Talk with your health professional about the pros and cons of a few foods or elimination diet.

3. *Lifestyle—exercise.* Keep your child moving. Asthma symptoms can be controlled even during vigorous exercise. Have your child try yoga (especially yogic breathing), pursed lips breathing exercises, or swimming. Have your child warm up before exercise; breathe through the nose, not the mouth; and cover nose and mouth with a scarf when exercising in very cold weather.

4. *Lifestyle—sleep.* Encourage a healthy bedtime routine. If you hear your child snoring or having stops and starts in his breathing, have him evaluated for obstructive sleep apnea; treating that can help reduce asthma and improve growth and behavior.

5. *Biochemical—medications.* Work with your physician or other health care professional to develop the best plan for using medications to prevent and treat asthma. The most widely used therapies include beta-agonists, inhaled steroids, and montelukast. Make sure your clinician demonstrates and your child understands how to use a metered dose inhaler (and a spacer). Make sure your child is up to date on vaccines, including flu vaccine (have everyone at home vaccinated against the flu every year). See your health professional to have potential sinus infections and walking pneumonia treated promptly with appropriate antibiotics. If your child has allergies, make sure they are treated with effective antihistamines, and consider desensitization therapy.

See Your Child's Health Care Professional If Your Child Is

- Having a hard time breathing

- Breathing much more rapidly than usual

- Making a grunting noise when breathing out

- Getting tired or agitated with his shortness of breath

- Getting blue in the lips or fingertips

- Sucking in the spaces between his ribs with each breath

- Having symptoms that interfere with sleep or other activities or has a high fever or poor appetite

- Or if you are having *any* concerns about your child's breathing or he is not improving with home therapies

---

6. *Biochemical—medications* (continued). Avoid giving your child aspirin or ibuprofen. Stick to acetaminophen for fever and mild pain management.

7. *Biochemical—nutritional supplements.* Consider supplemental magnesium, vitamins C and D and fish oil or flax seed oil.

8. *Lifestyle—mind-body.* Consider training in self-hypnosis, guided imagery, meditation, biofeedback, or autogenic training. Encourage your child to keep a journal to write about her day, her stress, and her worries. Get support. Connect with other adults who care for your child about your child's asthma. Teachers, day care workers, baby-sitters, and relatives need to know what symptoms to look for, what your care plan is, and how to reach your child's health care professional in case of an emergency.

9. *Biomechanical—massage.* Get a book, a video, or a massage therapist to teach you how to give a massage to your child at least five days a week.

10. *Bioenergetic. Acupuncture* is a safe preventative therapy, but more research is needed before it becomes routine.

Remind yourself and your child to have fun! Asthma can be serious, but with an appropriate treatment plan, your child can have a happy, normal childhood.

8
AUTISM

Julie Frazier brought Jeff, her 18-month-old, in for a second opinion about his chronic constipation and recurrent ear infections, and mentioned that his language had recently regressed. He's stopped saying words he knew, and wasn't making good eye contact when she spoke with him. She worried that he might have either hearing loss from the ear infections or autism. "I've heard so much about the rising rates of autism and something about the autism spectrum, but I really don't know what it is," she said.

WHAT IS THE AUTISM SPECTRUM OF DISORDERS (ASD)?

The fifth edition of the American Diagnostic and Statistical Manual of Mental Disorders (DSM-5) describes ASD as a group of neurodevelopmental disorders that includes Asperger disorder, autism, childhood disintegrative disorder, and pervasive developmental disorder not due to another condition. Symptoms generally appear by age three and include: poor social interaction; impaired communication; and restricted or repetitive behavior. Some children gradually develop these symptoms, and others who seem to have been developing suddenly lose language and developmental milestones, usually during the early toddler period.

Poor *social interaction* means a child with ASD tends to make less eye contact, smile, and look at others less often than peers, and have trouble understanding social cues such as pointing and taking turns. On formal testing, children with ASD generally do less well at recognizing facial expressions of emotion. Given these challenges, it is difficult for children with ASD to make and maintain friendships.

Communication difficulties may initially appear as delays in babbling, turn-taking in conversations, decreased responsiveness to hearing their name called, fewer words, and fewer word-gesture combinations. They may simply repeat another's words (echolalia). When someone points, they may look at the pointing hand rather than the object of attention. Difficulties with communication and social understanding can be frustrating, and some children with autism have especially severe tantrums and aggressiveness.

Repetitive behaviors include hand flapping, head rolling, or body rocking. They may also have compulsive behavior, insisting on arranging objects in straight lines. They may strongly resist change and insist on daily rituals and routines. They may develop strong, but limited interest in a single toy, game, movie, or program. Self-destructive repetitive behaviors include skin-picking, hand-biting, head-banging, and eye-poking.

Jeff had several symptoms of autism: He'd greatly reduced eye contact, developed communication difficulties, and had become fixated on fewer toys, resisting Julie's attempts to vary the routine and get him interested in new toys or games. He wanted to watch one Winnie-the-Pooh video repeatedly and became very upset if she turned it off in the middle. He had no self-harming behaviors.

WHAT OTHER SYMPTOMS ARE COMMON IN CHILDREN WITH ASD?

Over 90 percent of children with autism show *heightened sensitivity* to sounds, light, textures, tastes, or smells.

Most (60 to 80 percent) have *movement difficulties* such as poor muscle tone, toe walking, clumsiness, poor coordination, or poor balance.

Many (about 75 percent) are *extra picky eaters* who restrict themselves to very few foods; and many have *bowel problems* such as constipation, diarrhea, bloating, reflux, and pain.[1]

Although studies on internal ecosystem (microbes that normally live in our intestines, also known as the "microbiome") have expanded rapidly over the last 10 years, we are just beginning to understand that children with autism generally host a different variety of internal microbes and gut metabolic profiles than normally developing children.[2,3] What we don't know yet is why this happens or how best to address it.[4]

Jeff had also become a very picky eater, limiting his intake to soft white and yellow foods; he suffered from chronic constipation. He also looked like his stomach was bloated.

Increased *intestinal permeability* ("leaky gut") is much more common in children with ASD (37 percent) and their immediate family members (21 percent) than in control subjects (5 percent).[5] Leaky gut is associated with abdominal pain, constipation, diarrhea, and other GI symptoms. Of course, the intestines are supposed to "leak" nutrients into the bloodstream, but increased permeability is found in conditions like celiac disease (gluten sensitivity), diabetes, and inflammatory bowel disease, indicating that too much leakiness can activate the immune system in troublesome ways.

About 25 percent of children with ASD also have *seizures* or epilepsy-like patterns on EEG testing.[6]

A few (1–10 percent) autistic children show *unusual abilities* such as musical or mathematical genius.

Having a child with autism is *stressful* for families who need to provide extra attention to help the child reach his or her potential.

Julie had definitely been feeling stressed as she tried to deal with Jeff's peculiar behavior and limited interests in activities and food. She wanted to know what caused autism and why rates were increasing. Jeff's father was fifty and Julie was forty when she became pregnant with

Jeff; she'd developed pregnancy-related diabetes. Could that have something to do with it? She felt incredibly guilty. She also wanted to know whether she should stop his immunization series to avoid any other setbacks.

What Causes Autism?

Just as sore throats are all characterized by throat pain but can have many different causes (such as strep infections, overuse, allergies, viruses, etc.), the symptoms of ASD may be caused by different combinations of risk factors. Genetics plays a role, and so does the environment. Much more research is needed to sort out all the different factors contributing to the different types of ASD, but these factors appear to include some combination of: genes, environment (including maternal health during pregnancy as well as diet and toxic exposures), cellular metabolism, inflammation and immune problems, oxidative stress, brain and body health (especially the GI system), and sensory, processing, and communication difficulties.

Although *genes* are certainly involved in setting the stage for autism, a single genetic mutation has not been identified. This means there are no blood tests that can be done before or after birth to screen for autism.

Children with autism have an increased risk of food and skin *allergies*, and children with early allergies and eczema have an increased risk of being diagnosed with autism.[7] It is not clear whether allergies cause autism or whether the same factors that increase the risk of allergies also increase the risk of autism. Nevertheless, it appears that the *immune system* is involved in ASD.

About 5 percent of children with autism (compared with fewer than 1 percent of non-autistic children) also have dysfunctions with the *cell's* energy factories (mitochondrial *metabolism*) producing insufficient energy.[8] Mitochondrial function can be disrupted (and the risk of autism increased) by environmental toxins such as bisphenol A[9], mercury, and lead.[10]

Environmental toxins linked to autism appear to act during the first eight weeks following conception, before most women know they are pregnant. Numerous toxins have been linked to an increased risk of autism, so it is unlikely that we will discover just one factor responsible for it.

Emerging data suggests that *pre-pregnancy maternal health* is important in preventing autism. The risk of having a child with autism is higher among children born to *older parents*, and among babies who are born *prematurely* or at *low birth weight*. Autism is also more common among children who have Down syndrome or Fragile X syndrome. Children whose mothers were *severely obese* before pregnancy have a significantly increased risk of developmental problems including autism and ADHD[11]. Maternal conditions such as *diabetes* and *hypertension* also increase the risk of autism in offspring.[12,13] By increasing hormones like leptin and insulin, blood sugar levels, and inflammation, a mother's *high-fat diet* can also interfere with optimal brain development.[14]

Health during *pregnancy* is important, too. Low maternal *vitamin D* levels and exposure to high levels of *particulate air pollution* are associated with an increased risk of autism.[15,16] Taking *folic acid supplements* before and early in pregnancy is associated with a lower risk of autism.[17] Women who develop pregnancy complications such as *pre-eclampsia* also have a markedly increased risk of having a child diagnosed with autism.[18] Maternal use of certain medications, including valproate (often used to treat seizures)[19] and *selective serotonin reuptake inhibitors* (SSRIs, which are widely used to treat depression, anxiety, and premenstrual dysphoric disorder), are also linked to an increased risk of autism.[20] It is wise for mothers to adopt a healthy lifestyle even before conceiving a child[21] and to take good care of themselves, including optimal prenatal care, cautious use of medications, and early, continuing use of prenatal vitamins to reduce the risk of having a child with autism.

Early infant *health practices* are also linked to the risk of autism. For example, there is a strong correlation between the use of acetaminophen (Tylenol™) and the risk of autism. Some scientists speculate that the higher risk of ASD in boys is due to the use of acetaminophen to help ease the pain of circumcision.[22] This speculation has not been proven and is not a reason to ignore or fail to treat infants' pain.

Thousands of research studies have addressed the question of whether *vaccines* cause autism. They don't.[23,24] Please don't let the fear of autism prevent you from protecting your child against serious childhood diseases like measles, mumps, and rubella.

I reassured Julie that immunizations were an important part of protecting Jeff, and that she needn't feel guilty about "causing" his autism. He was lucky to be here and have her as a mom, and the best thing now is to focus on the present and how to prepare him to have a successful future. Jeff is one of many children with autism, and there are an increasing number of resources to help.

How Many Children Have Autism?

Prior to the 1980s, autism was an uncommon diagnosis. But by 2014, 1 in 68 American children had been diagnosed with ASD, a 30 percent increase from just two years before. More than 3.5 million Americans have been diagnosed with autism. Rates are about five times higher in boys (1/42) than in girls (1/189). Some people think that the higher numbers of ASD diagnoses are due to increased awareness, screening, detection, and diagnostic labeling, but most people think the increase is due to a combination of a real increase in rates and an increased rate of diagnosis.

How Do the Brains of Children with Autism Differ from Other Children?

This is an area of ongoing research. Soon after birth, children who are eventually diagnosed with autism seem to have rapid brain growth. There are unusual connections between brain cells (which may account for their increased risk of seizures)—fewer connections in areas of the brain associated with executive and associative functions (the frontal cortex) and more connections in other areas. The brain and nervous system also show increased inflammation in autism, which could reflect problems with the immune system due to an infection soon after conception.

When Should I Suspect Autism?

By 24 months of age, most children who are eventually diagnosed with autism have developed early signs. These include:

- No babbling, gesturing, or pointing by 12 months
- No single words by 16 months
- No two-word phrases (aside from copying or echolalia) by 24 months
- Loss of language or social skills

The burden is not on you alone to make the diagnosis. Your pediatrician or family doctor will probably screen your child for autism at both the 18-month and 24-month checkup. Most physicians use a screening tool such as the Modified Checklist for Autism in Toddlers (M-CHAT).

What Unique Strengths Do Children with ASD Exhibit?

When we see a child with ASD, we tend to focus on problems, but many children with ASD show a range of strengths. By focusing on the strengths and building on abilities, we may find more energy for achieving our long-term goals for the child's success. Some of these strengths may include:

- Heightened sensory sensitivity
- Ability to recognize patterns
- Ability to solve puzzles
- Content with his own company; doesn't require constant external stimulation
- Fascination with machines and computers
- Ability to focus intently for long periods of time
- Ability to hear and see things in very concrete terms
- Loves repetition, not easily bored

Julie started to look for Jeff's strengths and realized that his fascination with puzzles and how things worked may help him have a future career in computer engineering like his dad. She also remarked that the Sheldon Cooper character from the TV series, Big Bang Theory, *had many ASD qualities, and he was a successful physics professor with many friends.*

COULD IT BE SOMETHING ELSE?

Yes. Many genetic disorders can lead to delays in development and communication. Sensory problems such as deafness can delay language development. Lead poisoning can cause developmental delays and abdominal pain. Anxiety can also cause behavioral differences. If your child is adopted, early life experiences such as abuse and neglect can cause behaviors that look very similar to autism. Do not leap to conclusions. Take a breath and meet with a health professional who has experience caring for children with developmental differences.

CAN ASD BE CURED?

Autism used to be viewed as a stable neurodevelopmental disorder. That view has changed. Autism is now seen as a dynamic system of metabolic, mitochondrial, immune, inflammatory, and behavioral abnormalities involving the whole body.[25] Some children with ASD have temporary or long periods of remarkable improvements in behavior. For example, some children have fewer ASD-related symptoms when they have a fever or are treated with corticosteroid medications for another problem, suggesting that inflammation may play a role in ASD symptoms.[26,27]

Research is ongoing into the best ways to treat and improve ASD symptoms by untangling and addressing all these different factors in a truly holistic approach. With careful attention to a healthy lifestyle, good health care, and ongoing behavioral management, some children outgrow or learn to overcome some of their behavioral and social challenges, just as older adults who have had a stroke can learn to use a paralyzed limb with intensive physical therapy. Since we do not know how to predict which children will overcome their symptoms or to what extent they will adapt or normalize, it is our obligation as caregivers to provide every child the best chance of flourishing while additional research points the way to ever-more-helpful therapies and preventive strategies.

DIAGNOSING AUTISM

The diagnosis of autism or ASD depends on careful observation, not on a single test of blood, urine, or other fluids, or on X-rays or other laboratory tests. Your physician may order tests to make sure symptoms aren't due to other medical problems.

Diagnosis is based on structured, consistent observations using a scale or instrument such as

- Autism Diagnostic Interview Scale–Revised (ADI-R)
- Autism Diagnostic Observation Scale (ADOS)
- Childhood Autism Rating Scale (CARS)

If your pediatrician or family physician suspects autism, you may be referred to a hearing specialist to check hearing and make sure that deafness isn't responsible for communication difficulties. You may also seek a referral to a pediatric neuropsychologist who can provide specialized assessment of the child's behavior and cognitive skills and suggest individualized accommodations for school and other learning situations. You may also ask for a referral to a geneticist for testing to make sure your child doesn't have a genetic problem such as Fragile X that can cause developmental disabilities.

We ordered a hearing test for Jeff and made a referral to a pediatric neuropsychologist at the Children's Hospital to obtain a thorough assessment of his strengths and challenges to help us develop a plan to help him flourish.

WHAT'S THE BEST WAY TO PREVENT AND TREAT AUTISM?

The best strategy is to nurture healthy habits in a healthy habitat. This means excellent nutrition, behavioral therapy, an individualized educational plan, a solid balance of exercise and sleep, avoidance of environmental toxins, and diligent attention to emotional and cognitive self-regulatory strategies. Other therapies, including medications, supplements, and professional care may be helpful, but they do not replace the basics. Let's tour the Therapeutic Mountain, starting with lifestyle therapies, to find out what works for children with autism. If you want to skip to my bottom-line recommendations, flip to the end of the chapter.

LIFESTYLE THERAPIES: NUTRITION, EXERCISE, ENVIRONMENT, MIND-BODY

Nutrition

Although the brain comprises less than 3 percent of the body's weight, it consumes 20 percent of its metabolic output. It is particularly important to provide high-quality fuel and avoid deficiencies of essential nutrients for children with ASD. It is important to start with breast-feeding, followed by whole foods, and avoiding processed foods, particularly those with artificial colors, flavors, and preservatives. In general, I recommend a whole foods, plant-based diet, including organic foods whenever possible to decrease exposure to pesticides. One of the most common strategies parents try to help their children with autism is to change the diet.

What about Gluten-Free, Casein-Free Diet?

The gluten-free, casein-free (GFCF) diet excludes all gluten (wheat, barley, and rye products) and casein (products derived from cow milk) products. This means that many common items—macaroni and cheese, cheese pizza, milk, pasta, bread, and cereal—are excluded. Many parents and several studies report success with this diet, despite the difficulty of implementing it in a family eating a standard American diet.[28,29] It is unclear whether the benefits of the diet are due to the absence of protein antigens from gluten or casein; the elimination of many processed and "junk" foods containing artificial flavors, colors, sweeteners, and preservatives; or the replacement of these foods with more nutritious whole foods such as fruits, vegetables, beans, seeds, non-gluten whole grains, fish, and meat (a Paleo-type diet).

What about a Ketogenic Diet?

The ketogenic diet is a high-fat, low-carbohydrate diet sometimes used to help children with very specific types of refractory seizures. A small Greek study showed that a subset of children with ASD, who also had blood and urine tests indicating an underlying metabolic disorder, had some symptomatic improvement after starting a ketogenic diet.[30]

Because these diets are so different from what many families eat and can run the risk of deficiencies if implemented incorrectly, I recommend that if you're having trouble getting your child to eat a well-balanced healthy diet, or if you decide to try a special diet for three months, you seek consultation from a pediatric registered dietitian who can guide you to healthy choices and share tips for successful implementation.

Julie was struggling with Jeff's food preferences. She wanted to try the GFCF diet for eight weeks, and she wanted help incorporating more high-fiber foods to address his constipation. I referred her to the pediatric dietitian in the Children's Hospital GI clinic who had extensive experience providing in-depth counseling and support.

Exercise and Sleep

Vigorous *exercise* for 30–60 minutes at least five days per week is good general health advice for all children. Because children with autism often have problems with strength and coordination, exercises building these skills can be helpful. Exercising just before schoolwork may help decrease problem behaviors and promote better engagement.[31] Mind-body exercises like yoga and dance can help improve behavior while building strength, stamina, and coordination.[32] Why not combine the best of both worlds and engage in mind-body exercise before school?[33]

Sleep is a common problem for children with autism. It is important to use common sense measures such as a dark, cool, quiet room; regular bedtime and bedtime routine; skillful use of slow, relaxing music; consideration of use of before-bed bath and/or back rub; consideration of relaxing aromatherapies like lavender; and possibly even melatonin supplements (usually 1 to 5 milligrams, but can be up to 10).[34]

Environment

Children with autism are generally very sensitive to small changes in the environment—visual cues, patterns and locations of furniture, auditory stimuli, aromas, tastes, and even tactile elements such as textures and shapes.

Music. Children with autism usually have communication problems, and they can benefit from the individual attention involved in music therapy to develop nonverbal communication and social skills. A review of 10 studies of music therapy for children with autism concluded that music therapy can help improve social interaction, verbal and nonverbal communication, and social-emotional reciprocity.[35] I recommend music therapy as part of a holistic approach to helping children with ASD.

Sensory integration therapies (SIT). Sensory integration therapies are usually offered by occupational therapists as part of an overall approach to helping children manage daily activities like eating, dressing, and bathing. Small studies have shown improvements in social behavior, communication development, and aggression, but other studies have had mixed results.[36] In other words, the scientific jury is still out on the effectiveness of SIT, but the therapies are safe, and many parents and pediatricians find them helpful.

Hyperbaric oxygen therapy (HBOT). We all need oxygen to survive. Hyperbaric oxygen therapy is an established therapy for decompression sickness (coming up too fast from the depths while scuba diving), carbon monoxide poisoning, and serious, non-healing wounds. There have been a few anecdotes about remarkable benefits of HBOT for autism, but no randomized, controlled trials comparing HBOT to sham therapy.[37] Given its expense and potential side effects (like high-pressure trauma to the ear drums and sinuses), I do not recommend it.

Transcranial magnetic stimulation (TMS). TMS sends rapidly alternating strong magnetic fields into specific areas of the brain. It is an established, effective therapy for patients with depression, certain kinds of epilepsy, and Parkinson's disease. There have been eight small trials in children with ASD suggesting improvements in social relatedness, social anxiety, and repetitive movements.[38] TMS is about as safe for children with ASD as it is for other conditions. Currently TMS is not widely available outside research settings.

Reducing exposures to environmental toxins. Pesticides, heavy metals, air pollution from traffic, phthalates (softeners in plastic), and other toxins are associated with an increased risk of autism.[39–41] To the extent possible, reduce your child's exposure to organophosphate pesticides in your home and grow your own or purchase organically grown foods.

Julie was already concerned about the impact of pesticides on bees and butterflies, and now she decided she'd buy organic as often as possible. In addition, she decided to plant a small garden at home to provide fresh, homemade vegetables. It would also give her another excuse to spend time outdoors in nature with Jeff.

Mind-Body, Behavioral, and Communication Strategies

The primary treatment for children with autism is behavioral therapy. Behavioral therapies are time intensive, and best results are ob-

tained when they are started early and applied consistently.[42] Some behavioral interventions have proven feasible and helpful even in very young children (7 to 24 months old) at risk for autism before a formal diagnosis has been made.[43,44]

Applied Behavior Analysis (ABA) was introduced in the 1960s and has gained widespread support as one of the best ways to promote flourishing for children with autism. ABA is a behavioral therapy that includes positive reinforcement of desired behavior and breaking down tasks into simple steps with frequent rewards and corrections to help a child master socially meaningful behaviors.[45] Formal ABA interventions are often provided in 1:1 settings, which means they are expensive. A slightly less expensive model pairs one therapist with three children for 15 to 25 hours per week; this approach has also had promising results.[46] Ask your insurance company if behavior training is covered.

Parent-implemented training is an extension of ABA in which the parents learn to function as behavioral therapists to promote social and communication skills in their young children. Preliminary research suggests that training parents to provide early intervention can help improve children's language and reduce the severity of other symptoms.[47,48] Training parents may increase stress, but it may also increase a sense of empowerment and self-efficacy in doing something to help the child.

For older children (6 to 21 years old), *social skills groups* can improve social competence for youth with autism.[49] I recommend them. Check with your insurance company to find out if this kind of support is covered under your plan.

Neurofeedback is a specific kind of biofeedback therapy that helps people learn to modulate their brain's electrical activity. A review of four randomized controlled trials concluded that neurofeedback promoted sustained attention, communication, sociability, and flexibility for children with ASD and ADHD.[50] Neurofeedback is safe but expensive outside research settings.

Parenting a child with ASD is stressful. A review of ten studies of *mindfulness* training interventions for parents of children with ASD concluded that mindfulness-based training helped decrease parents' stress and improve their well-being, which in turn helped them more effectively improve their children's behavior.[51] I strongly recommend behavioral therapy and social skills groups for children and youth with autism. Consider neurofeedback if it is financially feasible and an experienced, licensed practitioner is available.

Julie started ABA therapy, which was covered by their insurance. The therapist helped her meet other parents of toddlers with autism. The parents enjoyed the support of talking with each other and sharing resources. Julie also started looking for mindfulness training opportunities she could complete online given her busy schedule.

Animal-Assisted Therapy

Many people have experienced the benefits that therapy dogs offer during a stressful hospital stay. Numerous anecdotes support the comfort and powerful supportive nonverbal connections offered by dogs, horses (equine therapy or hippotherapy), and dolphins for people with mental health and behavioral challenges.[52] For example, in a study comparing 10 children with autism who received weekly equine therapy for nine weeks to other children who didn't, parents reported significant improvements in the horse-connected children's physical, emotional, and social functioning.[53] In another study, six children who participated in weekly horse therapy for 12 weeks showed improvements in posture, adaptive behavior, self-care, communication, and social interactions.[54] An Australian study showed improvements in

social skills and behavior when animal-assisted programs were offered in the classroom.[55] Altogether, over a dozen studies support the use of animal-assisted therapies to help children with autism improve social interactions and communication while decreasing stress.[56] I recommend it. Urge your insurance company to begin covering this therapy.

Aquatic or Hydrotherapy

Just as many people find dogs and horses comforting, others find comfort being buoyed by water. Preliminary studies are beginning to suggest that some kinds of hydrotherapy or aquatic therapy can be helpful in improving social interactions for children with autism. Until more research is done, I can't recommend a specific type, dose, or frequency of therapy, but I support parents who wish to try this approach to help their children.[57]

BIOCHEMICAL THERAPIES: MEDICATIONS, HERBS, NUTRITIONAL SUPPLEMENTS

Medications

There are no medications that cure autism. However, because so many children with autism also have other challenges with their health and behavior (such as ADHD, anxiety, and insomnia), most take one or more prescription medications. There is little rigorous research to support most of the medications that are prescribed for children with ASD and other symptoms.[58]

The two most commonly used medications to treat aggression, severe, damaging temper tantrums, and deliberate self-injury are *risperidone* (Risperdal®) and *aripiprazole* (Abilify®). For risperidone, a review of 22 studies found an overall improvement in problem behaviors, but often significant weight gain; other side effects include sedation, drooling, and tremor.[59] A review of two randomized trials concluded that aripiprazole can help treat some symptoms of autism, primarily irritability and hyperactivity.[60] Most experts recommend limiting their use to children whose aggressive behavior is so severe that they risk injuring themselves or someone else.[61] Talk with your physician, but if you decide to start one of these therapies, your child's behavioral response and weight must be carefully monitored.

In the 1990s, *secretin* became a popular therapy for autism, but at least seven randomized controlled trials evaluating it found that it did not significantly improve language, behavior, cognitive, or social skills.[62] I do not recommend it.

Chelation Therapy

Chelation therapy uses an intravenous infusion of a compound (usually EDTA which is used to treat lead toxicity) to remove heavy metals and other toxins. While some studies suggest that children with autism have higher levels of aluminum, arsenic, cadmium, lead, mercury, thallium, tin, and tungsten (and lower levels of iron, calcium, zinc, and other essential nutrients) than comparison children, there are no studies showing that chelation improves autism symptoms.[63–65] Chelation is expensive and can have severe side effects.[66] Unless your child has clear evidence of heavy metal toxicity (lead poisoning), I suggest avoiding chelation therapy for ASD.

Dietary Supplements

Dietary supplements such as vitamins and minerals are the most commonly used therapies for autism. Every child is different. There is no one dietary supplement that will meet every child's needs. However, given the high prevalence of food sensitivities, it is wise for families to be aware that many supplements contain artificial flavors, sweeteners, preservatives, gluten, casein, and other compounds, and to be careful readers of labels to avoid giving products that might make the child worse.

The primary role of vitamins and minerals in the brain is to act as cofactors in the production of neurotransmitters, the chemical messengers that help brain cells communicate with each other. Some essential nutrients also function as antioxidants, helping reduce the damage from oxidative stress.

Many American children eating a standard American diet do not consume recommended amounts of these essential nutrients, and this problem is even worse among children like Jeff who are picky eaters. In a study comparing vitamin and mineral levels in the blood of ASD children to typical children, the children with autism on average had lower levels of vitamins A, B, C, D, and E as well as magnesium, selenium, and zinc.[67] In a pilot study of 20 children with autism, in which half were offered a multivitamin, multimineral supplement, mothers reported a significant improvement in sleep and stomach problems with the supplements, but no overall difference in language, sociability, or behavior.[68] In a subsequent study, researchers randomized 141 children and adults with autism to a placebo or a multivitamin/mineral supplement; they found significant improvements with supplementation for many metabolic and oxidative stress biomarkers and on several behavioral measures: the Parental Global Impressions-Revised score and scores for hyperactivity, tantrums, and receptive language. These benefits were supported in a similar study that tracked the impact of supplements containing multiple essential nutrients.[69] Because multivitamin/mineral combinations are generally safe, particularly when started at low doses and gradually increased as tolerated, I recommend a three-month trial of a multivitamin/multimineral supplement for children with ASD.

DIETARY SUPPLEMENTS FOR AUTISM

- Multivitamin/multimineral combination
- B vitamins: B6 (pyridoxine), folate (B9), and B12
- Vitamin C
- Vitamin D
- Probiotics
- Coenzyme Q10
- N-acetylcysteine (NAC)
- Omega-3 fatty acids

Some children with autism have an increased risk of oxidative stress and problems metabolizing or using essential nutrients (converting them to their active form), and may benefit from additional supplementation to meet unique metabolic needs. Others are picky eaters and may miss essential nutrients unless they receive supplements.

B vitamins. B vitamins are essential in making many brain messengers (neurotransmitters). A chemical called homocysteine can build up without an adequate supply of B vitamins, especially B6, folate (B9), and B12. Some children with ASD have elevated homocysteine levels, suggesting they might benefit from these B vitamins.[70] Furthermore, although some children with ASD have above average levels of vitamin B6, they may have less of the enzyme needed to transform it into its active form.[71] A few studies have shown improvements in social interaction, stereotyped restrictive behavior, and delayed functioning with vitamin B6 and B6-magnesium supplements.[72]

Folate is a B vitamin found in leafy green vegetables. It is an essential cofactor necessary to make several neurotransmitters. Methylene tetrahydrofolate reductase (MTHFR) is an enzyme that is necessary for the body to convert one amino acid (homocysteine) to another (methionine) and several other metabolic reactions. Methionine is the key ingredient in s-adenosylmethionine (SAM-E), which plays a key role in mental health and behavior as well as energy production in mitochondria and gene expression. There are at least 24 subtle variations (genetic polymorphisms) in the gene that makes MTHFR, and these variations are linked to a variety of diseases ranging from dementia to heart disease to colon cancer. People with some of these MTHFR variations require extra folate or a different form of folate (activated or methylated folate) for their folate-dependent metabolic pathways to work well. Inadequate folate early in pregnancy increases the risk of ASD as well as other serious neurologic disorders.[73] Methylated *vitamin B12* is thought to help reduce oxidative stress (common in autism) by increasing the active form of glutathione (a potent antioxidant); a few limited studies have not demonstrated that methylated B12 supplements improve ASD symptoms,[74] but these supplements are safe. Additional research is underway to determine whether some children with autism have MTHFR variations and whether they might benefit from supplementation with folate, B6, and vitamin B12.

Vitamin C is sometimes recommended as an antioxidant. Although a small pilot study from the early 1990s reported improved symptoms in children taking large doses of vitamin C (equivalent to an adult dose of eight grams daily), no controlled trials have followed. Few physicians recommend vitamin C alone as a supplement for children with autism, but it may be included in a multivitamin/mineral mixture. Vitamin C is generally safe; excess (more than one to two grams daily) can cause diarrhea.

Vitamin D. About two-thirds of American children have suboptimal levels of *vitamin D.*[75] Vitamin D insufficiency is even more common among children with autism.[76] Furthermore, lower levels are linked to worse autism symptoms, and in preliminary studies, restoring vitamin D through daily supplements of vitamin D3 helped improve behavior.[77,78] Because vitamin D plays such an important role in both brain and immune function, I believe we should try to ensure every child has an optimal vitamin D level.[79] Talk with your physician about getting your child's vitamin D level checked (unless it's the summertime and the child has a suntan—I haven't found low levels in kids with tan lines). If the level of 25-hydroxy vitamin D is less than 30 ng/mL, use supplemental sun or vitamin D3 supplements to raise the level above 30.

Other essential nutrients like the *minerals iron, magnesium*, and *zinc* are also necessary for the brain to make its chemical messengers. There are no studies showing that megadoses help improve autism symptoms, but it makes sense to avoid or correct deficiencies.

Julie decided to start Jeff on a multivitamin/ multimineral supplement. When we tested him, his vitamin D level was also low, so we added vitamin D3 supplements to optimize his immune function and help prevent another round of colds and ear infections.

Probiotics. There is growing evidence that the community of bacteria that live in the intestines (microbiome) of children with ASD differs from the microbiome of typically developing children. It is not clear how long after birth this difference develops or whether it is a cause or consequence of ASD. (Does picky eating change the microbiome or do microbial differences drive appetite differences?)[80] While current research is insufficient to recommend a specific dose or type of probiotic for children with ASD, eating fermented foods such as yogurt, kefir, and kimchi is part of a healthy diet that supports a healthy gut microbiome.

Coenzyme Q10 (CoQ10) is chemically related to vitamin K. It plays an essential role in mitochondrial function. Since 5 percent of children with autism have mitochondrial problems, some people have advocated for coenzyme Q10

supplements. These supplements are typically used for children with muscular dystrophy and adults with congestive heart failure, but there are no studies to inform us about the optimal dose or duration of CoQ10 for children with autism. CoQ10 is safe, but costly.

N-acetylcysteine (NAC) is a precursor on the road to making glutathione, one of the body's most important antioxidants and detoxifiers. NAC is generally used medically to help detoxify the liver after acetaminophen overdose and in combination with nitroglycerin to treat angina; patients with cystic fibrosis use an inhaled form of NAC to help break up and loosen sticky airway mucus. Researchers at Stanford randomized children with autism into two groups, one receiving a placebo and the other receiving NAC, increasing from 900 milligrams daily to 2.7 grams daily by 8 weeks. After 12 weeks, the supplemented group was less irritable and had no major side effects.[81] In another study, children who were already taking risperidone for severe behavioral problems were randomized to placebo or 1,200 milligrams daily of NAC; after eight weeks, the NAC group had much lower scores for irritability.[82] Side effects included constipation, increased appetite, fatigue, drowsiness, and nervousness. In a third study, children 4 to 12 years old who were taking risperidone received in addition either placebo or NAC (600–900 milligrams daily); again there were improvements with NAC in irritability and hyperactivity.[83] NAC supplements are relatively inexpensive, and all brands tested passed ConsumerLab.com's quality testing. You may notice a sulfur smell when you open a bottle; this is normal. Higher doses (1,200 milligrams daily) may cause upset stomach, nausea, constipation, or diarrhea. Given its benefits in several randomized trials, I recommend that if irritability is one of your child's major symptoms, you consider an 8- to 12-week trial of NAC after discussing it with your child's physician.

Omega-3 fatty acids are increasingly used to

reduce inflammation and have been used as dietary supplements for children with ASD. This is in part because children with autism have below average blood levels of omega-3 fatty acids, and correcting deficiencies is an attractive treatment approach. In a Canadian study, children two to five years old were randomized to either placebo or omega-3 fatty acid supplements (1.5 grams daily); at the end of six months, there was no significant improvement in the omega-3 fatty acid group compared to the placebo group.[84] A 2011 review of several older studies showed no significant benefits, but noted that the studies had been small and that larger studies with longer follow-up were needed.[85] In a small Japanese randomized controlled trial published after the review study, 16 weeks of essential fatty acid supplementation was associated with improvements in social withdrawal and communication.[86] Omega-3 fatty acids are a healthy part of a good diet, and are useful for depression and ADHD-related learning problems; and I often recommend them as supplements for children with autism.

Herbs

Because autism is a relatively modern diagnosis, there are few traditional herbal remedies for autism. I do not recommend any specific herbs as autism treatments.

BIOMECHANICAL THERAPIES: MASSAGE

Children with autism may be particularly sensitive to sensory stimulation from touch. In a study of daily massage provided by parents plus weekly massage by staff members over five months, preschool children with autism showed improvements in language, behavior, sensory abnormalities, and parent-child interactions as well as decreased parenting stress.[87] QiGong massage is part of traditional Chinese

medical treatment; in a randomized study of QiGong massage in preschool children, the therapy was associated with improvements in touch sensitivity, self-regulation, and parenting stress.[88] A 2011 review of six studies on massage for autism concluded that massage was safe and showed a number of significant benefits, though more rigorous trials are desirable.[89] Given its safety and effectiveness in several studies, I recommend parent-provided massage on a regular basis for children with autism. Ask your insurance carrier about coverage for professional massage therapy, so you can get some tips from a pro and learn new skills to help your child.

BIOENERGETIC THERAPIES: ACUPUNCTURE, HOMEOPATHY

Acupuncture

As with traditional herbal remedies, there is no traditional Chinese medical diagnosis for autism. Over thirty studies have reported improvements in comprehension, cognition, motor skills, independence, and social communication with acupuncture treatment,[90] but most researchers point to flaws in the study designs.[91] More rigorous research is needed before I routinely recommend acupuncture for communication and relationship symptoms for ASD, but acupuncture is safe. I have no objections to a trial of acupuncture therapy, particularly to treat other conditions for which acupuncture has stronger evidence of effectiveness (pain, anxiety, insomnia).

Homeopathy

Homeopathy is another safe therapy with little research to test its effectiveness in helping children with autism. Again, I do not recommend it, but I readily tolerate it.

WHAT I RECOMMEND FOR AUTISM

1. *Lifestyle.* Healthy habits in a healthy habitat—good general health habits. This means nourishing nutrition; a balance of exercise (optimally 60 minutes daily of moderate to vigorous activity) and restorative sleep; stress management and emotional self-regulation; support from family and friends; and avoiding environmental toxins.

2. *ABA therapy* and *parent-implemented training; social skills building groups.*

3. *Music therapy* to promote nonverbal communication and to soothe anxiety and irritability.

4. *Animal-assisted therapy* to promote nonverbal communication and social skills

5. *Massage* provided regularly by parents; encourage your insurer to cover services from a licensed massage therapist

6. *Neurofeedback therapy* if you can afford it or convince your insurer to pay for treatment from an experienced licensed therapist.

7. *Dietary supplements* to prevent and treat deficiencies and metabolic differences.

8. *Melatonin* supplements to help with sleep.

9. If you choose a special diet, like a *gluten-free/casein-free diet* or a ketogenic diet, please work with a pediatric *registered dietitian* to avoid deficiencies.

10. Consider *mindfulness-based training* for the child's parents and other caregivers to master more skillful responses to stress and provide role models of effective emotional and mental self-regulatory skills.

9
BURNS

Brian West brought his two-year-old daughter, Cora, into clinic to be treated for a scald burn first thing one morning. Heavy-hearted, he told me he'd had a cup of coffee and a bran muffin sitting on the sink, so he could eat breakfast while getting ready for work. Before he knew it, Cora came in and pulled his cup off the sink, splashing hot coffee on her left shoulder, arm, and chest. The coffee had been sitting out for a few minutes, so he didn't think it was very hot, but Cora had cried immediately and loudly. He'd put cold water on the burn, which was pink and had a few small blisters. He'd also given her some Tylenol®. Brian's wife was out of town on a business trip and he wasn't sure what else to do. He'd called his mother, who said to put butter on it. Brian wanted to know if there was anything else he could put on the burn to prevent complications. Because Cora was screaming, he also wanted something stronger to help with her pain.

Each year over 23,000 American children are hospitalized and nearly 1,500 die from burn injuries. Burns can be caused by heat, chemicals, electricity, or radiation. The most common burns in children are sunburns and scald burns. The most deadly burns are from fires and electricity.

It may take a day or two after a burn before you can really tell how deep it is. Have you ever spent the day at the beach and returned home thinking you had escaped sunburn only to awaken the next day to red, painful skin? Many burns appear to be less serious initially than they are. It sometimes takes forty-eight hours

Table 9.1: Three Degrees of Burns		
Degree of Burn	How Deep	Appearance
First-degree	Superficial	pink, no blisters; dry; painful
Second-degree	Partial thickness	pink or red with blisters; moist; very painful
Third-degree	Full thickness	red, white, or charred; not painful in center

after the burn before even a physician can accurately determine how deep it is (Table 9.1).

Sunburns damage the skin in the short term and the long term. They increase the later risk of skin cancer and premature aging of the skin. Sunburn is typically a *first-degree burn*. The skin is pink or red, dry, and painful. Even a soft sheet or breeze can cause agony, but the pain resolves in a day or two. Regardless of how you treat it, it heals in three or four days without scarring; within a week the burned skin becomes itchy and peels. Severe sunburn can damage deeper layers of skin. *Second-degree sunburns* are signaled by the development of blisters within twelve hours after exposure.

Scald burns are the most common *second-degree* or *partial thickness* burns, accounting for more pediatric hospitalizations than any other type of burn. The peak age for scald burns is between six months and two years of age—when toddlers need constant, close supervision. Scalds typically occur when a toddler pulls a container of boiling water or a cup of hot coffee off the counter, splashing it onto her face and upper body. Also, dipping braids into boiling water to "set" them can unintentionally scald the scalp, face, or back. Second-degree burns should be evaluated by a health professional if they occur on the face or hands or if they are larger than a silver dollar.

Even tap water can cause burns if the hot water heater thermostat is set too high. Tap water at 130°F takes less than thirty seconds to cause a second-degree burn. Prior to 1978, 80 percent of homes had hot water heaters set at these dangerous levels! In 1983 Washington State began to require that new hot water heaters at the factory be preset to 120°F. As a result of this simple law, the number of hospital admissions for tap water scald burns was cut in half within five years.

Scalds can also cause deep burns. Nearly five hundred Americans die each year from scald burns; 20 percent of these deaths are in children under five years old.

Chemicals, such as lye and corrosive acids, and even vinegar and other household products can cause burns, too; these burns can range from superficial to quite deep depending on the strength of the acid, the length of time it is in contact with skin, and the skin thickness.

Flame burns, electrical burns, and contact with hot stoves, hot wood stoves, irons, or space heaters are the most common causes of deep *third-degree* or *full thickness* burns. Flame burns from fires are often accompanied by smoke inhalation, which can be even deadlier than the burn itself. *All third-degree burns and any burns accompanied by shock or smoke inhalation should be evaluated by a health care professional immediately.*

Many small, mild burns such as sunburns can be handled at home. Deeper or more extensive burns should be evaluated by a professional. Your child may need more fluids than she can take at home by mouth and may need stronger pain medications, antibiotic creams, physical therapy, or even surgery to minimize disfiguring scars.

Cora's burn looked like a typical scald burn with a mixture of first- and second-degree burns. It

TAKE YOUR CHILD TO YOUR
HEALTH CARE PROFESSIONAL IF:

- Any burn (other than a mild sunburn) covers more than 5 percent of her skin.
- Your child is less than two years old with any kind of burn.
- The burn contains blisters that cover more than a palm-size area of skin (1 percent of total skin surface).
- A blistering burn occurs on the hands, feet, face, or genitals.
- Any burn encircles the child's arm or leg.
- Any electrical burn.
- Any possible smoke inhalation, regardless of the size of the burn.
- The burn is deep or contains white, charred areas, or non-painful areas.
- The burn looks infected.
- Your child is in severe pain or refuses to eat or drink.
- You are concerned about the appearance of the burn.
- Your child has swallowed a caustic chemical (such as lye or Drano).

- Don't count on clouds—up to 80 percent of burning rays can get through.
- Wear protective clothing, such as a hat with a broad brim.
- Use sunscreen with an SPF of 30 or higher.
- Eat a healthy diet rich in antioxidant fruits, vegetables, nuts, and fish.
- Don't count on miracle treatments that sound too good to be true.

covered about 5 percent of her total skin area. It was bright pink, and a few spots had blistered and peeled already, leaving raw, tender skin exposed.

What Is the Best Way to Prevent Burns?

The best way to prevent burns is to use common sense, create a safe home environment, and supervise your child closely.

PREVENTING SUNBURN

- Minimize sun exposure between 10 am and 2 pm

Keep your child out of the sun between 10 am and 2 pm. Be especially cautious around water or snow because they can reflect burning rays back and intensify the trouble. Don't count on clouds to reduce sun exposure; they let about 80 percent of burning rays through to the skin. When your child is outdoors, protect her face with a brimmed hat. Use sunscreen generously. The higher the SPF (sun protection factor), the more protection your child has against the sun's burning ultraviolet rays. Choose a sunscreen rated at SPF 30 or higher that protects against UVA and UVB radiation. Zinc oxide (the white cream worn by lifeguards) is the most effective sunscreen. If you're going to apply insect repellent at the same time, put the insect repellent on *first* and *then* apply sunscreen; applying bug spray after the sunscreen dramatically reduces the effectiveness of the sunscreen.[1]

Serve meals rich in antioxidants.[2] Two small studies have tested this theory by providing supplemental vitamin C (found in fruits and vegetables), vitamin E (found in nuts, seeds, and leafy greens), proanthocyanidins (antioxidants found in grapes and berries), carotenoids (found in carrots and leafy greens), and selenium (highest amounts found in Brazil nuts) before administering ultraviolet light to test areas of skin. Both studies found that the skin was protected in those who consumed the antioxidants.[3,4] In an Italian study, after controlling for sun exposure, a diet rich in fish and vegetables was protective against development of sun-related melanoma.[5] You can't just eat a salad before being out in the harsh mid-day sun to expect a benefit. It can take at least 10 weeks for the levels of carotenoids and other protective antioxidants to build up to protective levels.[6]

Consuming alcohol, on the other hand, increases the risk of sunburns and decreases the protective effective of a healthy diet.[7] That's one more reason not to let your teenagers drink alcohol.

Don't count on miracle treatments. For a while there was an appealing rumor circulating that you could prevent sunburns by eating dark chocolate. Sadly, a randomized controlled trial testing this theory showed no benefit to eating chocolate.[8] You may enjoy it for other reasons, but don't count on eating chocolate or cocoa butter to prevent your child's sunburn.

You can also take steps to reduce burn hazards in your home.

REDUCING BURN HAZARDS IN THE HOME

- Lower hot water heater temperature to 120°F.
- Have smoke detectors and fire extinguishers on every floor.
- Cover electrical cords and outlets.
- Be careful with space heaters, radiators, wood stoves, and fireplaces.
- Practice stove and microwave safety.
- Ban baby walkers.
- Keep vaporizers and steam machines out of children's reach.
- Keep matches and lighters out of children's reach.
- Prohibit use of gasoline and fireworks.

Make sure your *hot water heater* is set no higher than 120°F. Your dishwasher and washing machine can still do an excellent job at these temperatures. Many dishwashers now include heat boosters to increase the water temperature in the unit without overheating the rest of the water in the house. By turning down your hot water heater, you'll not only reduce the chances of a scald burn, you'll save money on your utility bill, too.

Install a *smoke detector and fire extinguisher* on every floor of your home. A smoke detector is the single most inexpensive investment you can make to protect your family from fire injuries. Most states require landlords to have working smoke detectors in every apartment. Change the batteries regularly; I change mine every fall, the same day I turn the clocks back. I also keep two fire extinguishers on the main floor of my home—one in the kitchen and one in the living room next to the fireplace. I've never had to use them, but I review the directions and check the expiration dates every fall when I turn the clocks back and replace the batteries in smoke detectors.

Toddlers chew everything they can get their hands on. The word *no* just doesn't make sense to an exploring fifteen-month-old. She isn't being bad; she's simply trying to figure out the world. You can't expect toddlers to behave safely; you have to protect them from hazards in the environment. Keep appliances close to outlets, so the cords don't trail across the room. Install *outlet covers*, so your curious toddler isn't tempted to explore electrical outlets with fingers, paper clips, or hair clips.

Space heaters get very hot. Clothes or draperies near the space heater can easily catch fire. Turn off your space heaters when you go to bed. Keep all clothing and fabric well away from them whenever they are on. *Radiators* have caused burns in infants who rolled off of beds onto the radiator and toddlers who tried to pull themselves to a stand by holding onto the radiator. Put a barrier between toddlers and wood stoves, radiators, and fireplaces.

Keep the handles of pots and pans turned away from the edge of the *stovetop* to prevent curious toddlers from pulling hot pans onto themselves. Whenever possible, use the back burners of the stove to put more distance between your child and the burner. Before giving your child anything from the *microwave*, test the temperature first yourself. Do not heat your child's formula in the microwave. If you heat any liquids in the microwave, shake the container thoroughly to distribute the heat throughout the contents, and test the temperature on yourself before giving it to your child.

Infant walkers do *not* help children learn to walk any faster. Walkers were the culprits behind many of the scalded children I saw in the Harborview Hospital Burn Unit. The children "walked" over to the stove or table, and quicker than the parent could stop them, pulled pots of boiling water, coffee, or spaghetti down onto themselves, resulting in severe burns. If you already have a walker, remove its wheels. Better yet, when friends and relatives want to get you something for the baby, ask for a high chair, a baby gate, a car seat, or an infant swing instead.

Vaporizers are a comfort for many children struggling with coughs and colds, but steam vaporizers and steam machines can burn curious toddlers. If you have a hot air or steam vaporizer or steam machine, make sure it is safely out of reach of children.

Matches and lighters should always be kept out of children's reach. I have seen several tragic house fires in which families lost everything they owned and children were scarred for life because a youngster got hold of his parent's cigarette lighter. Children can unintentionally ignite their clothes while playing with matches or lighters, resulting in severe burns to the face and chest. This is another good reason for parents to stop smoking.

Gasoline and kids just don't mix. Some of the worst burns I saw in the Burn Unit were in boys between 8 and 15 years old who were burning leaves and threw gasoline (or lighter fluid) on the fire to get it going. Gasoline spilled on their pants, caught fire, and resulted in se-

vere, disfiguring burns. Don't let a child use gasoline or lighter fluid until she is at least old enough to pay to put it in the family car.

Fireworks also cause disproportionate damage among school-age boys. I have seen fingers, hands, feet, and other body parts blown off or burned from fireworks. Do *not* let your children use fireworks without your direct supervision.

Brian took the day off work to stay home with Cora. He took advantage of his day at home to do a careful inventory of burn and fire safety in his house. He drew up a checklist to show his wife and decided they would review it every year on the anniversary of Cora's burn as a way to remind themselves of the importance of prevention. They also decided to make a fire escape plan and practice that every year as well.

No matter how safety-conscious their parents are, some children still get burned. What treatments work for burned children?

Treatments for Burns

Children who have suffered different kinds of burns (mild to deep) require different kinds of therapy. Children who suffer severe burns go through several predictable stages, and they need different therapies at different times. For the first day or two, they are likely to be in shock, sleepy and withdrawn from the world. The burned area and other areas may become swollen, and the child requires much more fluid than usual. After a day or two, as the child begins to mobilize reserves to begin healing, the heart rate and blood pressure rise and the child may run a fever. As a burn heals, it gets very itchy. Finally, deep burns tend to leave ugly scars without appropriate therapy.

Let's look at the effectiveness of different kinds of therapies for burns. If you want to skip to my bottom-line recommendations, flip to the end of the chapter.

LIFESTYLE THERAPIES: NUTRITION, EXERCISE, ENVIRONMENT, MIND-BODY

Nutrition

One of the biggest challenges in caring for severely burned children is getting them to eat. Severe burns are a major stress, resulting in metabolic rates (and calorie needs) 50 to 100 percent higher than normal. Children feel miserable, are often nauseated, and just aren't hungry. However, hospitalized burned children heal much faster and have fewer infections if they are fed from the day they are admitted to the hospital, even if they aren't hungry.

High-protein diets help maintain muscle strength, overall weight, immune function, and blood levels of essential amino acids in severely burned children.[9] Some burn centers include specific amino acid supplements in their feeding regimens for severe burns. If your child is severely burned, consult a professional nutritionist at a burn center. If you are interested in dietary supplements, see the section to come.

Extra fluids are very important, especially in the first few days after a burn. Our bodies are mostly water, and our skin helps prevent our bodily fluids from evaporating. Burns destroy this barrier between body fluids and the environment. Children whose burns cover more than 10 percent of their skin often lose so much fluid they need intravenous fluid to replace their losses.

Brian stopped at the grocery store on the way home to pick up extra orange juice and some of Cora's favorite foods (milk, ice cream, and peanut butter) to encourage her to eat. After she ate, the first thing she'd do was take a nice nap.

Exercise and Sleep

Initially, children suffering from burns need to rest, so their body's energy can go into heal-ing the burn. For the first few days, it may help to keep the burned area *elevated*. Just as in a sprained ankle, the injured area is likely to swell. Swelling is not only uncomfortable, but it may block blood flow and slow the healing process. Elevating an injured extremity means keeping it above the level of the heart.

After the initial healing phase, it is important for children with healing burns to keep *moving their joints,* even though it may be very painful. Burns that extend across joints can cause scars that eventually contract and limit movement. Physical therapists and massage therapists can help a child keep motivated and keep moving, preventing long-term disabilities. Exercise programs improve self-esteem and speed returns to work and school. Slow stretching such as yoga or tai chi can be very helpful in maintaining limber joints.

Much healing occurs during sleep. Allow your child as much as she wants while she is recovering from a burn.

Environment

When a child is burned, the first and most effective therapies are environmental. *Remove the child from the source of the heat.* Usually in scald burns, the hot water or coffee has already run off or evaporated, but if the hot liquid or grease is still on the child's clothes, remove the clothes. If your child's clothes have caught on fire, smother the flames.

Second, *put cool water on the burned area.* Don't use ice. Ice may cause even more damage to the already injured skin. Your child doesn't need to get frostbite on top of the burn. Cool water helps numb the pain. If the burn covers more than 10 percent of her skin, do not run cold water over her; she will get chilled and be even more uncomfortable. *Large burns need immediate professional treatment.*

Third, *cover the burn with a clean, dry dressing.* You can use bandages or a clean handkerchief, sheet, or towel. Covering the burn reduces pain and the risk of infection.

Pressure dressings similar to Ace wraps are used to prevent scars from becoming heaped up and disfiguring. For maximal effectiveness, they should be measured, applied, and monitored by a physical therapist.

If you are planning to take your child to a burn center for evaluation and treatment, just wash the burn with cool water, cover it with a clean bandage, handkerchief, or sheet, and bring the child right away. For anything other than a very minor burn, a health care professional should be consulted regarding appropriate burn dressings and dressing changes.

Mind/Body

Hypnosis helps both adults and children to manage the pain of burns and the pain of dressing changes. At regional trauma centers, the staff psychologist who provides hypnotherapy and guided imagery is a key member of the burn team. Hypnotherapy and virtual reality therapy can help decrease pain and improve sleep and appetite.[10] In addition to easing pain, hypnosis also reduces the anxiety and distress associated with dressing changes. Effective hypnotherapy can markedly reduce the need for pain medications, improve wound outcomes, and reduce the cost of care for burn patients.[11]

Distraction with cartoons, stories, songs, and interactive games on portable electronic devices can also help burned children deal with the pain of their injuries and dressing changes.[12] Parents can leave recordings of their child's favorite stories and songs to play when the parent can't be there. Computer and video games are favorites among the children on the burn ward. You can also use these simple distraction techniques at home to help your child manage the pain of a burn or other injuries.

At the opposite end of the spectrum from distraction is having the child *fully engaged* in treatment. Some children scream and fight dressing changes despite maximal pain medication and attempts at distraction. Often these children feel wildly out of control. When they are offered the opportunity to help change their dressings, they do much better.[13] Giving children a sense of predictability and control decreases their anxiety and pain, making the whole experience much easier for everyone. Most children benefit from a sense of structure and predictability. If your child is hospitalized, work with the child-life team to create a poster showing what happens each hour of the day and pictures of all the professionals involved in her care.

BIOCHEMICAL THERAPIES: MEDICATIONS, DIETARY SUPPLEMENTS, AND HERBS

Medications

Burns are very painful. Your child will heal much faster if the pain is effectively managed so she can sleep, eat, and drink. No matter how much pain medication your child requires during the early phases of treatment, she will not become a drug addict from her burn treatment. In all the years I worked at the Harborview Burn Center, I never saw a child whose pain was effectively managed become a drug addict. People do crave drugs if they are not given enough medication frequently enough. Far more damage is done by withholding pain medications or under-treating pain than by giving the large doses necessary to make her as comfortable as possible.

BURN TREATMENT MEDICATIONS

- Pain medications (prescription and nonprescription)
- Tetanus immunization
- Antibiotic and non-antibiotic ointments
- Other salves

Effective *non-prescription analgesic medications* for mild burns include aspirin, acetaminophen (Tylenol®), ibuprofen (Advil®), and naproxen (Aleve®). These medications are generally safe and effective for mild to moderate pain. If the burn is deep or extensive, your child may need stronger *prescription pain medications*. The most commonly prescribed pain medication is codeine or a combination of codeine and acetaminophen (Tylenol III®). If your child is hospitalized, other pain medications such as morphine may be used.

Burns are highly prone to *tetanus* infection. If your child is burned and has not finished her initial series of immunizations (usually given at two, four, and six months of age), she needs a tetanus shot and a dose of tetanus immune globulin to prevent tetanus. If she has had all of her required immunizations but it has been more than five years since her last tetanus shot, she needs a booster. Nobody likes getting shots, but the risk of tetanus is far worse.

Burn wounds are extremely susceptible to infections. Petrolatum gel (Vaseline®) is a non-antibiotic ointment used to cover wounds and prevent the dressing from sticking to the wound. It is less expensive and in one study proved as effective as a commonly used antibiotic ointment in promoting wound healing while minimizing allergic reactions and infections.[14]

Antibiotic ointments help reduce the risk of infection. The most commonly used antibiotic cream is a prescription sulfa drug, silver sulfadiazine (Silvadene®). If your child is allergic to sulfa drugs, let your physician know. Antibiotic ointments are applied once or twice daily following thorough wound cleaning. The process of removing dressings, washing the burn, and replacing the antibiotic ointment can be very painful. Hospitals and burn centers routinely give pain medications before changing dressings. Oral antibiotics are not helpful and are not necessary unless your child has another known infection such as pneumonia.

The most interesting burn salve I have run across is Preparation H®. Believe it or not, many people put the famous hemorrhoid medication on minor burns to "soothe inflamed tissues." Even more amazing, several investigators have actually studied the active ingredient, live yeast cell derivative (LYCD), and found that skin grafts did heal faster with it than with a placebo ointment![15,16] The ointment stings a bit, and I'd like to see more studies in children before I routinely recommend it for *minor* burns.

Cora was up to date on all her immunizations. We gave her some ibuprofen in the office, waited 45 minutes, and while distracting her with videos on an iPad, carefully cleaned her burns. The nurse applied antibiotic ointment and covered it with a large dressing. Brian took the day off from work to stay home with Cora and watch cartoon videos to distract her from the pain. I gave them a prescription for Tylenol III® to help keep her comfortable and asked him to return the next day for another assessment and dressing change.

Herbs

A number of herbal remedies have been traditionally used to treat minor burns.

HERBAL REMEDIES FOR BURNS

- Aloe vera—yes, for minor burns
- Calendula and arnica—insufficient evidence
- *Avoid* raw herbs, undiluted essential oils, topical garlic

Aloe vera gel first gained popularity during the 1930s with reports of its success as a treatment for radiation burns.[17] It is now a common ingredient in many nonprescription burn remedies. Aloe grows well on windowsills with minimal care, and I grow it in my own kitchen to use on the inevitable occasional minor burns.

I recommend that you use the gel from fresh, homegrown leaves. In one study, aloe vera was more effective than an antibiotic cream in reducing pain and promoting wound healing.[18] I recommend it to my patients as a remedy for minor burns such as sunburns.

Calendula, another popular skin soother, is available in several brands of skin cream sold in health food stores. Calendula seems to have anti-inflammatory properties, but there are *no* studies evaluating its effectiveness in treating burns in children. *Arnica* is an ingredient in many herbal salves, creams, ointments, and oils. However, there are no studies on using it in children's burns. *Gotu kola* (*Centella asiatica*) is a native herb of Eastern Asia and the Pacific Islands used for numerous skin problems and wounds including burns, skin grafts, and scars. Gotu kola extracts stimulate wound healing by promoting collagen (connective tissue) formation. Gotu kola extracts can also cause irritation and allergic reactions in some people, and there are no studies of its use for pediatric burns. I do *not* recommend gotu kola for burns.

Geranium oil is a traditional remedy for cuts, bruises, and burns. Oil of geranium has shown some antibacterial, antiviral, and antifungal properties in the test tube and in experimental mice.[19,20] *Tea tree oil* is another potent antiseptic, but it can be very irritating and cause allergic reactions, especially when used undiluted or straight; and if accidentally ingested, it can cause serious neurotoxicity. Directly applying undiluted essential oils to the skin can cause irritation, and I do not recommend that you apply any essential oil directly to burned skin.

Folk remedies for burns include marsh mallow root, burdock leaf, plantain, comfrey, mullein leaves, turmeric,[21] and slippery elm bark powder.[22] Many of these herbs are added to herbal ointments as skin salve for minor skin irritations or made into poultices. Despite their popularity, they have not been evaluated in scientific studies in the treatment of burns in children. Because of their potential for contamination with bacteria and fungal toxins, I do *not* recommend raw herbal products as burn treatments unless you have grown, picked, and cleaned the herbs yourself.

Honey was cited in an Egyptian papyrus from 2,000 B.C. as a treatment for burns; it continues to be used today in many parts of the world as a burn salve. A review of over 25 studies evaluating the benefits of using honey as a burn salve concluded that it helps heal partial thickness (first- and second-degree) burns more quickly than conventional treatment without honey, and helped heal infected wounds more quickly than antiseptics and gauze.[23] Like all natural products, honey varies in its purity and potency. Honey appears to be a safe home remedy for sunburn and other mild burns.

Garlic can also lead to irritation and burns when applied directly to a child's tender skin.[24] Do *not* use garlic poultices on burns.

Dietary Supplements

Vitamin and mineral supplements along with certain amino acids and omega-3 fatty acids are often given to children hospitalized with serious burns. The most commonly supplemented vitamins and minerals are the B vitamins, vitamin C and E, and zinc.[25] Supplementation with trace elements (copper, selenium, and zinc) significantly reduce both hospital stays and infectious complications in burn patients.[26,27] Burn patients also tend to run low on *magnesium*, and many patients hospitalized for severe burns receive magnesium monitoring and supplementation if needed.[28] Iron supplements are avoided because extra iron seems to feed the bacteria that thrive in burn wounds. The studies on antioxidants as burn treatments have had mixed results.[29] The amino acid most often supplemented for burn patients is *glutamine*.

POSSIBLE NUTRITIONAL SUPPLEMENTS FOR BURNS

- Vitamin A (orally)
- Vitamin C (orally)
- Vitamin E (orally or as a salve)
- Zinc, selenium, magnesium, copper (orally)
- Glutamine (orally or in intravenous fluids)
- Honey (salve)
- *No* iron (orally), milk, butter, shortening, or lard (salves)

Vitamin A is essential for proper wound healing. Supplemental vitamin A (approximately 10,000 International Units [IU] daily) seems to decrease the risk of diarrhea that frequently complicates burns.[30] There are no studies evaluating the effectiveness of higher doses of vitamin A. Vitamin A overdoses can cause serious side effects. I do *not* recommend vitamin A supplements higher than the doses typically contained in multivitamins for burned children. However, your child can eat as many carrots and sweet potatoes (natural sources of vitamin A) as she wants.

Vitamin C is also necessary for healing. Many seriously ill and injured children can tolerate large doses of vitamin C. High doses of vitamin C given immediately after a burn help decrease swelling and fluid requirements and speed recovery.[31] You can try between 60 milligrams and 250 milligrams given three to four times daily for a burned child, depending on her size. If your child develops diarrhea (an early side effect of vitamin C overdose), reduce the dose. Your child can eat as many oranges, strawberries, and other vitamin C–rich foods as she wants.

Vitamin E has been recommended both as a nutritional supplement and as a salve for minor burns. In a study of seventeen patients with extensive burns, vitamin E supplements led to higher counts of immune T-cells, possibly improving resistance to infection.[32] Vitamin E *may* be useful, but until more studies are done in humans, vitamin E supplements and salves must be regarded as experimental, not proven therapy.

Children with major burns lose a lot of minerals, particularly *zinc and copper.*[33] Zinc is important for immune function and healing. Zinc and copper normally exist in balance, so their replacement needs to be balanced as well. The amount of replacement minerals needed depends on the size of the child and the severity of the burn (up to 50 milligrams daily for an adult with a burn requiring hospitalization, smaller doses for children and less severe burns).[34] For most small burns, foods rich in zinc or a multivitamin/multimineral are sufficient.

Glutamine is the amino acid most often given to supplement the diet of hospitalized burn patients. An analysis of several studies concluded that glutamine supplements could help reduce the risk of severe infections and death in burn patients.[35] It is not clear that glutamine helps with minor burns in pediatric patients, but it is important to avoid protein deficiencies while healing from any kind of wound.

Do *not* put *milk, butter,* or *lard* on a burn. Though they may initially feel soothing, they provide a great environment for bacteria to grow and multiply.

BIOMECHANICAL THERAPIES: MASSAGE, PHYSICAL THERAPY, SURGERY

Massage

Exposing the burn to air or touching it can be excruciating. I do *not* recommend massage directly on burns while they are healing. Some children find massage to other areas comforting, and research studies suggest that massaging the non-burned parts of the body helps patients relax, feel less pain, and have lower levels of stress hormones.[36] Massage is also very helpful *after* the burn has healed; it decreases itching, tightness, and pain; improves circulation; and promotes comfort and flexibility for patients with scars resulting from burns.[37] I recommend massage for burn patients; I think it is one of the most underused, yet safest therapies around.

Physical Therapy

Physical therapy can be helpful in maintaining flexible joints for children whose burns cross over a joint like a shoulder, finger, knee, or ankle. Special pressure wraps can also help reduce bulky scars from burns, too. Physical therapy is generally unnecessary for minor burns like those from sunburn.

Surgery

If a burn blisters, leave it alone unless the blister is larger than a quarter or has already broken. A small, intact blister helps protect the burn from further injury. Once the blister breaks and the fluid drains out, removing the dead skin on the top of the blister minimizes the chances of infection.

If your child's burn is very deep, *skin grafts* may be necessary to minimize scarring and maximize function. It takes a few days for the burn to declare itself; that is, to be able to tell how severe the burn really is. Young children can sometimes heal burns that adults can't heal on their own. Most burn surgeons wait at least ten days before doing any skin graft surgery. The child remains in the hospital before surgery to receive fluid therapy, dressings, pain medications, and other necessary treatments. Hospitalization is also required after the surgery to make sure that the skin graft "takes" and the child continues to receive optimal therapy.

BIOENERGETIC THERAPIES: ACUPUNCTURE, THERAPEUTIC TOUCH, HOMEOPATHY

Acupuncture

Acupuncture is becoming more and more widely used to treat pain. One comparison study in hospitalized adult burn patients indicated that acupuncture significantly improved their pain.[38] Some researchers are even evaluating acupuncture with electrical stimulation to see if it helps heal burn wounds more quickly.[39] Few studies have specifically evaluated acupuncture for treating children with burns; but based on its safety profile, if it is offered in your pain treatment center or burn unit, I would not object to trying it.

Therapeutic Touch

Therapeutic Touch and Reiki can help ease the pain of burns as well as diminish anxiety in hospitalized burn patients.[40] More studies are needed in children to determine the optimal length and frequency of treatments, but in my clinical experience, I have found it very soothing when offered daily for 5 to 10 minutes to hospitalized children.

I did Therapeutic Touch with Cora in the office and showed Brian what I was doing. He decided to practice it at home with Cora, while she was relaxing and watching cartoons.

Homeopathy

Commonly used remedies for minor burns include Arnica, Calendula, Causticum, Hypericum, and Urtica urens. Only one scientific study evaluated homeopathy's effectiveness for treating burns, and it found that the remedy was no more effective than a placebo.[41] There is insufficient evidence for me to recommend homeopathic remedies for burn therapy.

WHAT I RECOMMEND FOR BURNS

PREVENTING BURNS

1. *Lifestyle—environment.* Take sensible precautions regarding sun exposure: Avoid the hours between 10 am and 2 pm; don't count on the clouds; wear a hat; use sunscreen that is at least SPF30 and provides protection against both UVA and UVB radiation.

Make a burn safety inventory for your house:

   a. Hot water heater temperature 120 degrees F

   b. Smoke detectors and fire extinguishers on every floor

   c. Electrical cords and outlets covered

   d. Space heater, radiator, and steam vaporizer safety

   e. Stove top and microwave safety

   f. No wheeled walkers

   g. Matches and lighters out of reach

   h. Gasoline, fireworks out of reach

   i. Escape plan

2. *Lifestyle—diet.* Offer your child meals rich in antioxidant fruits and vegetables and fish rich in omega-3 fatty acids.

TREATING BURNS

1. *Lifestyle—environmental.* Get the child away from the heat; smother flames; remove clothing that has had scalding food or liquid spilled on it. Run cool water over the burn. Cover it with a clean, dry dressing.

2. *Biochemical—medications.* Give your child an analgesic medication such as acetaminophen. (Use an antibiotic cream, Vaseline®, aloe, honey, or ointment on the burn.)

3. *Lifestyle—mind-body.* Distraction may help make dressing changes more bearable; consider stories, songs, and videos. Other children do better if they can participate in their dressing changes. Consider guided imagery or hypnosis.

Take Your Child to a Health Care Professional If

- Any burn covers more than 5 percent of her skin.

- Any electrical burn occurs.

- Your child is less than two years old with any burn.

- The burn contains blisters.

- Any blistering burn occurs on the hands, feet, face, or genitals.

- Any burn encircles your child's arm or leg.

- There is any chance of smoke inhalation.

- The burn contains white, charred areas, or non-painful areas.

- Your child is in severe pain.

- Your child refuses to eat or drink.

- You are concerned about the appearance of the burn.

- Your child has swallowed a caustic chemical.

4. *Lifestyle—exercise and sleep.* As the burn heals, keep your child moving to decrease scarring and limited joint movement, and consider *massage* and *physical therapy*, too. Keep the burn elevated to reduce swelling. Encourage extra sleep.

5. *Biochemical—nutritional supplements.* Consider giving a multivitamin supplement. Consider additional vitamin C-60 to 250 milligrams three to four times daily. If your child is hospitalized, ask your health care professional about supplements with glutamine and zinc.

6. *Lifestyle—nutrition.* Encourage your child to drink plenty of fluids and eat a balanced, high-protein diet.

7. *Biochemical—herbs.* For minor burns, use pure aloe vera gel or pure honey as a salve.

8. *Bioenergetic—Therapeutic Touch.* Learn how to use Therapeutic Touch to help your child relax and heal faster from burns and other injuries.

10
COLDS

Janet Black brought her 2-year-old son, Jamal, to see me because he was congested and had a fever. His temperature had gone up to 103° the night before, and at the time of the office visit, it was 101.8°. He wasn't hungry, but he was drinking juice and playing just as he usually did. It seemed like Jamal had one cold after another since he started attending his new day care center. Janet, who smoked about a pack a day of cigarettes, had suffered from sinus infections all winter; she wanted to make sure Jamal wasn't getting one, too. Jamal's physical exam showed that he didn't have a sinus infection, bronchitis, or pneumonia. He was simply suffering from the common cold. She asked me why kids get so many colds, what she could do to help Jamal get better faster, and what she could do to prevent another bout with the cold bug.

WHAT CAUSES COLDS?

Colds are caused by a variety of *viruses* in susceptible children. There are over 200 different kinds of cold-causing viruses. Because so many viruses cause colds, there will probably never be a single vaccine to prevent all colds. This means the mainstays of preventing colds are (a) preventing the spread of viruses and (b) building a strong immune system to resist infections.

We prevent the spread of viruses by washing our hands frequently; keeping sick people home and out of school or day care until they are less contagious; and keeping the items we touch (doorknobs, faucets, computers, remote controls) clean.

We keep our immune systems strong

through healthy nutrition, vitamin D levels, a balance of exercise and sleep, a positive attitude, and supportive friends and family.[1] When study subjects are intentionally exposed to a cold virus, only about half to two-thirds develop symptoms. The others remain well, despite the same exposure. Although some do not get sick, cultures of their noses indicated that they are infected. That is, the virus takes hold, but symptoms do not develop. Those who remain healthy have a more effective immune response to the cold virus.

A child's immune system is less experienced than an adult's immune system. This is one reason children have more colds than adults. The average infant has 10 to 11 colds annually, preschoolers have eight colds a year, and school-age children and adults have about four colds a year. When they do get sick, younger children have cold symptoms longer than older children—an average of nine to ten days for infants less than a year old and six to seven days for toddlers. Colds last less than two weeks in 90 percent of children; if your child's symptoms last longer, have her checked for other conditions like allergies and sinus infections. Unlike allergies, colds often lead to low-grade fevers.

Jamal's symptoms were classic for the common cold.

Although colds are rarely serious or life threatening, they pack a powerful punch in terms of the *cost* they generate for physician visits, medicines, and absence from normal activities. Colds are the leading reason for doctor visits. Nearly a billion dollars a year are spent in the United States on cold medicines, most of which are ineffective in relieving children's symptoms. Each year colds cause children to miss 26 million school days and result in adults missing 23 million workdays. For children under three years old, colds can lead to sinus infections (5 to 10 percent) and ear infections (30 percent). Pretty powerful effects from tiny viruses!

WHAT ARE RISK FACTORS FOR GETTING A COLD?

At any given moment, 20 to 25 percent of children less than five years old have colds. Children whose *mothers smoke* get 60 percent more colds than children of nonsmokers. Passive smoking (when the child inhales the fumes of someone else's cigarette) increases the number of infections and the number of symptoms. Children in *day care* experience 70 percent more colds than children cared for at home. This is because colds spread like wildfire among young children who have not yet learned to cover coughs and sneezes and who need constant reminders to wash their hands. *Low vitamin D levels* also increase the risk of common colds.

Jamal had many risk factors for colds—a mom who smoked cigarettes, being in day care, and having darkly pigmented skin, which put him at risk for low vitamin D.

COULD IT BE SOMETHING BESIDES A COLD?

There are times when your child's symptoms could be signaling something more seri-

---

TAKE YOUR CHILD TO THE DOCTOR IF YOUR CHILD HAS:

- A fever over 103°F or 39.4°C
- A sore throat or sore glands in the neck
- Ear pain or sinus pain
- A stiff neck or sore back
- Shortness of breath, wheezing, or trouble breathing
- Cold symptoms for longer than ten days
- Is too sick to drink or behave normally

ous than a simple cold. A runny nose that lasts for weeks and months is more likely to be an *allergy* than a viral infection, especially if there is sneezing and itching along with the runny nose. A persistent or worsening cold and fever accompanied by facial pain may be signs of a *sinus* infection.

WHAT ARE THE BEST WAYS TO PREVENT AND TREAT THE COMMON COLD?

"The only way to treat a cold is with contempt."
—Sir William Osler, MD

Treated or not, most colds resolve on their own within ten days thanks to our strong immune systems. Let's tour the Therapeutic Mountain to find out the safest symptom relievers. If you want my bottom-line recommendations, skip to the end of the chapter.

LIFESTYLE THERAPIES: NUTRITION, ENVIRONMENT, EXERCISE, MIND-BODY

Nutrition

Most cultures have their own special dietary home remedies to treat colds. I grew up with *chicken soup* and *hot tea with honey and lemon.* Science has proved that chicken soup is helpful in thinning nasal secretions so that they can be more readily cleared.[2] Also, chicken soup blocks the immune reaction that leads to cold symptoms.[3] Nobody knows what the secret ingredient is: the chicken, the vegetables, or love. For best results, have your child sip the soup slowly rather than gulp it down. Other studies support the subjective benefits of sipping a hot beverage on symptoms such as runny nose, cough, and sneezing.[4] Some of the benefit may lie in inhaling the steaming broth. Effects only last for about half an hour, so it's better to have small amounts of soup through-

out the day than one big bowl at suppertime. Watch the temperature and avoid scalds.

Hot chili peppers, horseradish, mustard, garlic, galangal root, Tabasco sauce, salsa, wasabi (Japanese horseradish), and other spicy foods are commonly used in other cultures to help open nasal passages and "burn out" the infection or "warm up" the cold sufferer. Spicy foods thin nasal secretions, make noses run, and open the sinuses even when you're healthy. Such effects may benefit congested children, too. Although there are no scientific studies evaluating spicy foods in the treatment of colds, when eaten in moderation such foods are certainly safe and are probably worth a try.

When children are afflicted with colds, they tend to breathe through their mouths. This dries out the mouth. The low-grade fevers that often accompany colds also increase water loss. All of these symptoms mean that your child needs extra *fluids* when he has a cold. This is why physicians and grandmothers alike admonish parents to give their sick children *plenty of fluids.* How much is plenty? A good rule of thumb is that a child is getting plenty of fluids if he needs to urinate at least every two to three hours while he is awake.

What about *milk*? Some cultures consider milk and other dairy products "cold" foods and avoid them when someone has a cold. Others are concerned that milk makes mucus worse. There are no studies showing that milk makes children's cold symptoms worse, but if avoiding milk is part of your cultural tradition, it's safe to abstain for the 10 days of a cold in children two years and older. If your child is still nursing, please continue breast-feeding.

Many children lose their appetite temporarily while they're ill. There's no point in forcing a child with a cold to eat. Keep offering nutritious, hydrating foods that are easily digested such as broths, oatmeal, citrus fruits and juices, grapes, and pears, plums or peaches, lettuce, celery, and carrots.

Exercise and Sleep

Moderate exercise helps boot immunity and ward off colds. Several studies indicate that those at the highest risk of colds are couch potatoes and those who exercise intensively. Marathon running may increase the risk of developing a cold because that level of intense physical exercise is stressful, impairing immune function for several days after a long-distance running event.[5] Moderation is the key.

Make sure your child is resting well during cold and flu season so that his immune system is at full power to fight off attacks by viruses; insufficient sleep impairs immune function and increases the risk of getting sick.[6] Fatigue is one of the most common causes and symptoms of the common cold. Some kids continue to play normally. You don't need to force your child to go to bed when he has a cold, but do let him know that it's okay to take an extra nap. Sleeping frees some of the body's energy up to mend itself. It also reduces your child's exposure to other people he might infect.

Sleeping with an elevated head may help reduce your child's congestion. You may try an extra pillow or letting an infant sleep in his car seat.

Mind-Body

Psychological *stress* lowers resistance to colds. The more stress, the higher the risk of catching a cold. Among volunteers who were intentionally infected with cold viruses, those with the most stress were the most likely to develop symptoms; those with less stress had almost no symptoms despite having similar infections. One easily avoided stress is lack of sleep. Cuddling and reassuring your child that you love him are great antidotes to stress.

The power of a variety of placebos to improve cold symptoms also suggests that our minds have powerful effects on both getting colds and amount we suffer from them once infected. Given these facts, it is important to reinforce the idea that the child is strong and has the power and ability to overcome his cold symptoms. If you give your child the message that he is weak or susceptible to illness, you may be creating a self-fulfilling prophecy.

Environment

Tobacco smoke paralyzes the microscopic cilia that sweep cold viruses out of the nose and throat. Exposure to cigarette smoke increases the chance that your child will get a cold and will make it more difficult for him to fight it off. Please, do not smoke and do not allow others to smoke around your child.

Steam has been used by countless people to combat congestion. Studies done in adults have shown mixed effects on cold symptoms,[7] and none of them tell us about the effect of steam or mist on children's colds: whether repeated (as opposed to the one-time treatments tested in the studies) steam treatments are helpful or how much time a child must spend with the vaporizer or steam treatment to benefit from it. A cool mist vaporizer is less likely to cause unintentional burns. At this point the jury is still out. If your child benefits, keep it up; if not, discontinue treatment. If you decide to use a vaporizer, clean it daily. Mold and bacteria love the moist environment in a vaporizer and will quickly grow inside. Follow the manufacturer's instructions for regular cleaning.

Northern European immigrants may prefer the hot dry air of a *sauna* to prevent colds. Again, studies have had mixed results, and none show definitive improvements for children with colds.[8] Children can easily overheat and become dehydrated in a sauna, and I do not recommend sauna treatments for children under 12.

Eucalyptus, menthol, pennyroyal, pine, rosemary, wintergreen, and *tea tree* oils placed in the medicine cup of a hot air vaporizer or as part of a steam bath are believed to be helpful by many parents, but there are not enough studies evaluating their effectiveness for me to recommend them for children. They are safe (if you keep your toddler away from the hot vaporizer to avoid burns and away from the essential oils, which can be poisonous if swallowed), and worth a try. Warning: Do not take eucalyptus oil or tea tree oil internally or apply it to the skin; they are for inhalation only. Keep all essential oils in containers with childproof caps and out of reach of children.

Most kids under four years old can't blow their noses the way adults can. To help infants and toddlers clear their secretions, remove them with a *bulb syringe* or nasal aspirator. Squeeze the bulb syringe, then insert the small end in your child's nostril. Release the bulb and allow the suction to draw the mucus out. If the mucus is dry or thick, you can loosen it with a drop or two of water or saline drops. Wait a few minutes for the mucus to soften and then use the bulb syringe. Do one nostril at a time.

Wash your hands! Washing your hands and having your child wash his hands are the best ways to cut down the spread of colds and other infectious illnesses.

BIOCHEMICAL THERAPIES: MEDICATIONS, HERBS, NUTRITIONAL SUPPLEMENTS

Medications

The 1833 *Mother's Medical Guide* lists leeches as the preferred treatment for childhood chest colds. Leeches were to be applied to the chest until they dropped off or "until fainting takes place." Modern cold medications may be less gruesome, but most are no more effective for young children than leeches. Children get better over several days regardless of whether or not they take a cold medicine. Often parents try a medication in desperation,

the child's symptoms eventually improve, and the medication gets the credit.

COMMON COLD MEDICATIONS

- Antihistamines (e.g., *Benadryl®, Chlortrimeton®, Tavist®, Claritin®, Zyrtec®*)
- Decongestants—both taken by mouth and as nose sprays (e.g., phenylephrine, phenylpropanolamine, and pseudoephedrine)
- Cough Suppressants (e.g., *Dextromethorphan, DM,* codeine)
- Expectorants to loosen dry or thick phlegm or mucus (e.g., guaifenesin)
- Analgesics: (e.g., acetaminophen, aspirin, ibuprofen)
- Combinations
- Home remedies like honey
- Saltwater drops and washes (neti pots), xylitol
- Menthol lozenges and rubs
- Antibiotics

Cold medicines are commonly used, and though inexpensive, their cumulative costs add up. Hundreds of non-prescription cold medicines are available, costing desperate consumers nearly one billion dollars a year. However, none of these cough and cold medicines have proven safe and effective in children less than three years old and there is little evidence they help older children much either.[9] In 2007, the U.S. FDA recommended a voluntary withdrawal of over-the-counter pediatric cough and cold remedies for children two years and younger to reduce the risk of side effects from these products. Since 2007, there has been a dramatic decline in the phone calls to Poison Control Centers about pediatric cold medicines.[10]

Though they may be helpful in treating allergies, *antihistamines* have *not* proven any more helpful than cherry syrup in relieving children's runny noses.[11] Antihistamines are somewhat helpful for adult cold sufferers, probably because they make people sleepy. The active ingredient in Benadryl®, diphenhydramine, is so sedating, it is also the active ingredient in many over-the-counter sleeping pills. There are some children who have a paradoxical reaction to antihistamines; instead of becoming sleepy, they become awake, active, and irritable. Prescription antihistamines are also not effective in treating kids' colds. Unless your child's symptoms are keeping him awake and you want to try a dose of medication at bedtime, I do not recommend antihistamines.

Many varieties of *decongestants* are available to help unclog stuffy noses. The best known, pseudoephedrine (*Sudafed®*) and phenylpropanolamine decrease nasal congestion and sneezing in adults. Many adults also feel a burst of energy after taking decongestants because decongestants are related to caffeine. However, there are no studies documenting any decongestant's effectiveness in treating childhood cold symptoms. Up to 30 percent of people who take decongestants experience side effects, some of which can be serious: increased heart rate, increased blood pressure, dizziness, hallucinations, psychosis, decreased appetite, and abnormal heart rhythms. I do not recommend oral decongestants for children until they reach school age. Even then, they should only be used during the daytime because they may keep children awake at night.

Decongestant *nose drops* and nose sprays have fewer serious side effects and are effective in reducing congestion in adults. Use for more than a day or two may result in a side effect called *rhinitis medicamentosa*. This is "medicalese" for saying the spray eventually causes symptoms that closely mimic the cold itself. The nose becomes dependent on the spray for normal functioning. Within three or four days, the child's nose gets swollen and inflamed inside when the spray is stopped. To relieve the symptoms, more spray is used, but symptoms reappear when it is stopped—a vicious cycle. There are no studies showing that decongestant nose drops or sprays are any more effective than simple saline nose drops in young children. I do not recommend decongestant drops or sprays unless the child's congestion is so severe that it interferes with drinking or sleeping. Even then, decongestant drops or sprays should be used as little as possible and only for a day or two to avoid rhinitis medicamentosa.

Saline nose drops are a safe way to loosen sticky nasal secretions. You can make your own saline drops.

HOMEMADE SALINE (SALTWATER) NOSE DROPS

Dissolve:

- One-half teaspoon of salt in
- One cup of tepid water.

Put a drop or two of the saline mixture in one nostril. Wait a minute to give the saline a chance to soften and loosen thick or crusted mucus, and then remove it with a bulb syringe or nasal aspirator. Repeat on the other side. This process can be safely repeated as often as needed. Some doctors recommend adding *xylitol*, a kind of sugar that makes it very difficult for *Pneumococcus bacteria* to grow and develop into an ear infection. You can find xylitol in your health food store or over the Internet.

The nose can also be washed out using a neti pot filled with a saline solution. Neti pots have long been used in Ayurvedic medicine. They are small teapots (containing about four ounces of fluid), which are filled with tepid saline. The head is held over the sink, and the tip of the spout is placed in one nostril. The saline solution is poured into one nostril and flows out the other nostril. I usually pour for 10 seconds, take a break, and blow my nose, then switch to the other nostril, alternating nostrils until the pot is empty. It stings a bit at first and has a high "yuck" factor, but older children and teens may find significant relief from washing out sticky mucus. Be sure you are using very clean water and clean the pot thoroughly between uses to avoid contamination.

Neither of the two most commonly prescribed *cough suppressants*, codeine (prescription only) and dextromethorphan or DM (prescription and non-prescription), are more effective than honey or cherry syrup in suppressing children's coughs.[12] Save your money. Try a home remedy like one-half teaspoon honey instead of a cough syrup.[13] Do not use honey in children 12 months and younger.

Expectorants, such as guaifenesin, the active ingredient in *Robitussin,* supposedly loosen secretions so they are easier to cough out, but there is little evidence in children or adults that they actually do.[14] Expectorants are no more helpful than placebo cherry syrup in treating children suffering from coughs and colds.[15,16]

Analgesics such as aspirin, *Tylenol* (acetaminophen), *Motrin,* or *Advil* (ibuprofen) help relieve discomfort, but they do not cure the common cold. A child doesn't need medicine for a fever unless the fever is making him too uncomfortable to sleep or drink. If your child is acting okay, you don't even need to take his temperature. Do not wake your child to give him an analgesic. If he's sleeping, he doesn't need it. Fever may be one of the body's best defenses against infections. Nonprescription analgesic medications tend to suppress the immune system and may worsen cold symptoms over time.[17] *Aspirin* has been linked to the sometimes fatal illness, Reye's syndrome. Acetaminophen can cause liver damage. I do not recommend treating a child with analgesics unless he is clearly uncomfortable.

Most cold medications contain *combinations* of antihistamines, decongestants, expectorants, cough suppressants, and analgesics. If the individual ingredients aren't effective in reducing cold symptoms or hastening recovery, there's no reason to think combinations will be any more helpful. Combination cold medicines have proven *not* to be helpful in young children. Combining several different medications does increase the risk of side effects. Many children's cold syrups contain alcohol. Most contain sweeteners, artificial colors, artificial flavors, and preservatives as well as the active ingredients. The most common reason children die from cough and cold medicines involves excessive doses;[18] if, despite the lack of evidence of effectiveness, you decide to try a cough/cold remedy for your child, please do not exceed the package directions for doses.

Menthol, which is derived from mint and could be considered an herbal remedy, is an ingredient in just about every cough and cold lozenge. Menthol is cooling and soothing and reduces the sensation of being congested. Adult volunteers who were given menthol lozenges reported marked improvements in their ability to breathe even though objective measurements of airflow failed to show an effect.[19] Vicks VapoRub® was one of our family's go-to remedies for colds; most families rub it on the chest or under the nose. Be sure *not* to use Vicks VapoRub® on the face of infants to avoid side effects. Otherwise, this is an inexpensive, safe remedy that I keep in my medicine cabinet.

Despite their widespread use, *antibiotics* offer no benefit in treating the common cold. Antibiotics do nothing at all to the viruses that cause colds. Nor does taking them prevent a child from developing a more serious infection, such as pneumonia. Despite the clear research showing no benefit from antibiotics, many parents request them; and pediatricians, wanting to please the family, acquiesce. This leads to smarter bacteria that are better able to outwit common antibiotics, leading to the need for new and more powerful medicines. Please do not ask your pediatrician to prescribe antibiotics for your child's cold.

Among adults with colds, treatment with ipratropium bromide (*Atrovent*) nasal spray reduced runny nose symptoms about 20 to 50 percent compared with placebo spray.[20] There is insufficient evidence to recommend it for children; it also tends to make the mouth dry as cotton. I don't recommend it.

Although most cold medicines have *not* been demonstrated to be helpful in treating cold symptoms in infants and young children, many parents hopefully purchase them believing that at least they won't do any harm. Wrong. Even non-prescription cold medicines can have powerful and unpleasant side effects.

SIDE EFFECTS OF COMMON COLD MEDICATIONS

Antihistamines: drowsiness, irritability, dry mouth, fuzzy thinking, thirst

Decongestants: increased heart rate, increased blood pressure, decreased appetite, dizziness, abnormal heart rhythms, hallucinations, psychosis

Cough suppressants: sleepiness; possible addiction

Aspirin/acetaminophen: suppressed immune function; gastro-intestinal toxicity; Reye's syndrome

If medications aren't helpful and may be harmful, why do so many doctors recommend them?

Good question. Some cold medicines *are* useful for adults, and people assume they'll work for children, too. Two-thirds of parents surveyed in one study were convinced that their children *needed* medicine for their symptoms. Because of this strong parental demand, some doctors fear that if they don't give a prescription, their patients will be dissatisfied

and go elsewhere. Please don't pressure your doctor for prescription cold medicines. These products have not been proven to be better than placebos, but they are costly and do run the risk of substantial side effects. If you feel the need to give your child something, consider saline nose drops or neti pot rinses to wash out his sticky nasal mucus and one-half teaspoon of honey to soothe his cough. Some other dietary supplements and herbs may be useful under certain circumstances.

Dietary Supplements

Many dietary supplements are used to prevent and treat the common cold. Let's divide our discussion into two categories: (1) vitamins, minerals, and other non-herbal products; and (2) herbal remedies.

Vitamins, Minerals, and Other Non-Herbal Products

NUTRITIONAL SUPPLEMENTS FOR COLDS

- Vitamin A—no
- Vitamin C—yes
- Vitamin D—avoid deficiency with daily sunshine or supplements
- Vitamin E—no
- Zinc lozenges—maybe
- Beta glucan—maybe
- Brandy—no
- Bee pollen (propolis)—no
- Probiotics—yes

Vitamin A deficiency increases susceptibility to severe lung infections such as pneumonia. However, some studies have shown that vitamin A supplements can actually increase the risk of developing colds;[21,22] other studies have found no increased risk, but not much benefit either unless the children were malnourished before they became ill. I prefer that children get their vitamins naturally in the foods they eat. To make sure your child gets plenty of natural vitamin A, encourage her to eat vitamin A–rich foods: apricots, cantaloupe, carrots, sweet potatoes, spinach, cheddar cheese, eggs, and fortified milk. She does not need extra vitamin A supplements during cold season.

Vitamin C is widely used to prevent and treat colds. It is a good idea to eat a diet containing vitamin C–rich foods like fruits (e.g., citrus, strawberries, and kiwi)[23] and vegetables (e.g., red and green peppers) because even modest deficiencies increase the risk of developing a bad cold.[24] Those who eat the most vitamin C–rich foods have a 30 percent lower risk of developing colds than those who eat the fewest fruits and vegetables.[25] Vitamin C can reduce symptoms and the length of illness,[26] but the greatest benefits are for children, endurance athletes, and those operating under extreme stress (such as soldiers in the Arctic or the desert).[27] The doses many people take for colds (200 milligrams to several grams a day) are well beyond natural levels found in a single serving of fruits and vegetables. For example, a glass of orange juice contains 60–80 milligrams of vitamin C. Children need at least 200 milligrams per day to manage cold symptoms. I take sustained release vitamin C supplements (1,000–2,000 milligrams twice daily) when I feel a cold coming on, and I recommend it for my patients when they get colds. Your child needs to drink at least four or five glasses of orange juice throughout the day to achieve therapeutic levels of vitamin C.

Deficiencies of *vitamin D* are linked to an increased risk of serious respiratory infections like pneumonia. Many adults and children have lower than desired levels of vitamin D, and an analysis that included 11 studies showed that daily vitamin D supplements had a significant benefit in preventing colds whereas weekly or monthly supplements didn't.[28] My general rule of thumb is not to use supplements during the summer months when kids get their vitamin D from sunshine (and have the tan lines to prove it), but to recommend 600–2,000 IU daily (depending on size) during the late fall, winter, and early spring and for kids who avoid sun exposure.

Vitamin E is an antioxidant vitamin that protects against free radical damage. It is found in leafy green vegetables, seeds, and nuts. While it is healthy to eat vitamin E from foods, the benefits and risks of supplementation remain controversial. Until the risks and benefits are clarified, I do not recommend vitamin E supplementation as a strategy to prevent or treat the common cold in children.

Zinc lozenges may reduce the severity and length of colds in adults but they can cause nausea, upset stomach, mouth irritation, and an abnormal sense of taste. Most of the studies on zinc supplements in children have been done in developing countries with poorly nourished children; in these situations, zinc supplements may help correct a deficiency and reduce the risk of respiratory infections, but it isn't clear that children who are well nourished would benefit. Taking high doses of zinc over a long period of time can actually impair the immune system. Avoid putting any zinc products in the nose because it can lead to permanent problems with sense of smell. I do not recommend zinc supplements to treat the common cold in children, but I do recommend eating a healthy diet containing zinc-rich foods like beans, nuts, whole grains, seafood, and poultry.

Beta-glucans are natural soluble fiber compounds found in bran, baker's yeast, and some mushrooms. Some studies suggest that beta-glucan supplements can improve immune function and the number of days with cold symptoms in endurance athletes and people who are stressed.[29,30] One study evaluated beta-glucan supplements in children, and in this study, it was just one of several compounds in a follow-up formula given to three- to four-year-old children in China; they did report fewer and shorter colds and less need for antibiotics in children who received the supplement (which also contained omega-3 fatty acids and prebiotics). I'd like to see more data on the safety and effectiveness of beta-glucan supplements for children before I start routinely recommending them to prevent or treat colds, but eating a diet rich in natural soluble fibers is a healthy choice.

What about brandy?

Brandy has been used by parents for ages to soothe sick children, and it may well put a child to sleep. However, brandy (and all other alcoholic beverages) dilates the blood vessels in the nose, compounding nasal congestion. I do not recommend that you give brandy to a sick child.

Propolis, a bee product, kills bacteria in test tubes. However, there is no scientific evidence in human adults or children that it is helpful in warding off the common cold. Until more research is done evaluating its benefits and risks, I do not recommend it as a remedy for the common cold.

Probiotics are the healthy bacteria that normally live in and on human bodies. A review of 14 studies of probiotic supplements concluded that probiotics were better than placebo in reducing the risk of colds and school absences related to colds, and they had few side effects (such as occasional upset stomach).[31] Unlike many studies of cold medicines in adults, probiotics have even proved effective in preventing colds in children.[32] There are still not enough studies to recommend one specific probiotic product over another, and I recommend that you check with ConsumerLab.com to find the latest review comparing the quality and costs of different probiotic products.

Some over-the-counter products contain combinations of dietary supplements. In a British study of preschool children, a combination product containing 50 milligrams of Vitamin C and a probiotic containing multiple strains of *Lactobacillus* and *Bifidobacteria* significantly reduced the risk of developing a cold, school absences, and the length of symptoms in those who did get sick, compared with a placebo.[33]

Herbal Remedies

Just as there are a variety of cold medications, there is a similar array of herbal remedies. Most are no more effective than medications for treating children with colds. Although a few (such as ephedra) can have serious side effects, most have a very long track record of safety, even for children. This means that in general, I tolerate the use of herbal home remedies, but I don't recommend specific products or doses.

HERBS TRADITIONALLY USED TO TREAT COLD SYMPTOMS

Calming: *Chamomile*

Decongestant: *Ephedra* or *Ma huang, Eucalyptus, Pine oil*

Expectorant (to loosen dry or thick phlegm or mucus): *Angelica, Hyssop, Horehound, Lobelia*

Anti-inflammatory: *Angelica, Bromelain, Horehound, Hyssop, Licorice root, Slippery elm bark*

Immune-stimulating: *Astragalus root, Echinacea, Goldenseal*

Other: *Andrographis paniculata, Elderberry, Garlic, North American Ginseng, Pelargonium*

Chamomile tea is soothing and calming and may help your child get the rest he needs, but it has no effect on the actual cold infection itself.

Ephedra or *Ma huang* is the original source of the medicinal decongestants ephedrine and pseudoephedrine. There is great variability in the potency of different species and different areas of cultivation of ephedra plants.[34] Dozens of people have died from ephedra overdoses, and I do not recommend it for children.

Eucalyptus or *menthol* (from mint) oil added to a vaporizer or bath will add a soothing smell and may help your child feel less congested, but it doesn't actually affect airflow or the infection itself. These are clean, comforting fragrances. Do not apply essential oils directly to the skin or apply under the nose. They can be irritating and cause significant side effects.

Angelica was used by both American Indians and Russians as an expectorant. In animal studies (but not in human beings yet), angelica extracts have anti-inflammatory activity; there are no studies evaluating its use in treating the common cold in children.

Hyssop tea and *horehound* lozenges are safe sore throat soothers. *Licorice* root tea is soothing for sore throats accompanying colds and boosts the body's own virus-fighting chemical, interferon; however, chronic use can cause problems with blood pressure, swelling, and the balance of salts in the blood.

Tinctures combining *Echinacea* and *goldenseal* used to be widely available. These herbs seem to boost the body's own immune defenses rather than attacking cold viruses directly. You are less likely to find goldenseal nowadays since it practically went extinct with overharvesting.

Echinacea is a common North American flower. Numerous studies (mostly from Europe) have demonstrated that echinacea helps prevent and treat colds in adults,[35] but in children, it seems to be best at preventing colds.[36] A European study suggested that an echinacea combination product (containing beta glucan, vitamin C, and zinc) helped reduce the risk of colds in children.[37] However, no one knows what the right plant is (there are at least eight different species of echinacea), what the right product is, what the right dose is, how often you have to give it, or how long you have to give it. I may use it personally, but I can't recommend it professionally until more research has confirmed the best product and dose for children.

Used by several Native American tribes, *slippery elm bark* tea has demulcent or mucilaginous qualities that soothe inflamed throats and noses. I often brew a cup if I feel myself coming down with a scratchy throat. You can also find it as an active ingredient in some throat lozenges. There are no scientific studies evaluating its effects on children with colds.

Spices such as *ginger, cinnamon, cloves, allspice,* and *cardamom* may also help your child feel less congested. You can brew a small pinch of each of these herbs into two cups of water to make a very fragrant tea for your stuffed up child. *Ginger* root and *cayenne* are spicy hot. They may help combat the chills and fatigue of fever. No studies have evaluated the effectiveness of raw, cooked, or powdered ginger or ginger ale in easing cold symptoms in children. But it's easy enough to make a home remedy with ginger: Buy a ginger root at the grocery store. Cut up about one to two inches of the root and boil the pieces in a quart of water for 10–20 minutes. Strain out the bits, and let them cool and sweeten as desired. Slowly sip the ginger tea to feel warm inside.

Some mothers use old-fashioned cold remedies such as *onion, comfrey,* and *eucalyptus poultices* or hot flannel packs of oil and camphor. *Hot baths* with *essential oils* of *eucalyptus, citrus, thyme, rosemary,* and *tea tree* may be soothing and decongesting. Commercial bubble bath products are also available that contain these marvelous fragrances, and I often use them myself. There are no studies on the efficacy of poultices or baths in treating children with colds, but if such things are part of your family's healing tradition, they are certainly worth trying. If they irritate your child's skin, discontinue use and consult your doctor.

Andrographis paniculata leaves are a historical Southeast Asian remedy for respiratory infections and other ailments. Controlled trials in adults have found it effective in reducing the length and severity of cold symptoms.[38] Most of the studies showing benefits have used 200 milligrams daily of a Swedish preparation known as Kan Jang®.[39] Although the data are promising for this herbal remedy, I have little personal experience with it, and have not heard much about it from my patients either.

I have been using *elderberry* (*Sambucus nigra*) extracts since I read a study showing it could significantly reduce the length of time adults infected with influenza experienced symptoms;[40] doses used in that study were three tablespoons (15 mL) four times daily. These results supported earlier studies showing rapid resolution of influenza symptoms in rural adults.[41] While pediatric studies on using elderberry extracts to treat colds are pending, this is a safe (allergies are possible but rare) home remedy; I not only keep elderberry extracts in my medicine cabinet at home, but I also planted an elderberry shrub, which is thriving (it is bigger than the dwarf apple trees) with minimal care and serves as a haven and food source for birds.

Garlic is another popular home remedy for a variety of ills. Eating a lot of garlic surely helps keep other people at bay, reducing exposure to cold viruses. A study in Florida found that consuming an aged garlic extract (two to three grams daily) for 45 days was associated with enhanced immune function and fewer days of work missed due to colds.[42] Repeated studies have shown enormous variation in the active ingredients present in aged garlic products. There are not enough studies showing what kind of garlic products most effectively help prevent or treat colds in children. On the other hand, it is certainly safe to cook with garlic, and I try to eat extra garlic when I feel a cold coming on. Garlic poultices can cause severe skin irritation. If you try a poultice, do not leave it in place for more than 20 minutes.

North American *ginseng* has been mostly studied using a Canadian product called Cold-FX®. A review of five studies involving over 700 adults showed a significant, 25 percent reduction in the risk of developing the common cold when used preventively, and a reduction in the length of symptoms by six days.[43] Although there are no pediatric studies yet evaluating safety and effectiveness, this is another remedy I keep in my own medicine cabinet and give to my son. The strongest evidence suggests that it should be used preventively or at the very earliest sign of an infection; there are no known benefits to starting it after two to three days of symptoms.

Pelargonium sidoides is a South African herbal remedy used to treat colds, bronchitis, and sinus infections sold in the United States under the brand name Umcka™. Most of the studies on its effectiveness come from Europe, Russia, and Germany. A review of 10 of these studies concluded that there is some evidence that it helps relieve the common cold and sinusitis in adults and bronchitis in children and adults, but the authors were not excited about the quality of these studies.[44] Since I don't have much personal experience with this product, I will wait until additional research clarifies its benefits and risks for children with colds before I recommend it.

BIOMECHANICAL THERAPIES: MASSAGE, SPINAL MANIPULATION

Massage

One of the fondest memories of my childhood is receiving a Vicks VapoRub® massage when I had a cold. I have been unable to find much research evaluating the efficacy of this time-tested technique (although there are a few studies showing it helps treat toenail fungus!); but I'm sure that the combination of parental love, a warm bed, and Vicks has made colds more bearable for thousands of children, and I heartily recommend them. Other massage oils that help open up a clogged nose include *camphor, camphorated olive oil, eucalyptus, menthol,* and *pine. Tiger balm,* a fragrant balm found in many health food stores, contains a combination of camphor, menthol, cajeput, and clove oils. Allergic reactions are possible; avoid applying pure essential oils, which can be very irritating.

Folk remedies for colds include massaging the soles of the feet with *mustard powder.* If you try this, you must be careful to wash off the mustard afterward to avoid burns.

Spinal Manipulation

Osteopathic manipulation of the neck and upper back to aid the lymphatic drainage of the head and neck has been recommended, but has not been scientifically confirmed as effective, so it remains unproved. I do not recommend spinal adjustments as cold remedies.

BIOENERGETIC THERAPIES: ACUPUNCTURE, HOMEOPATHY

Acupuncture

There are no studies comparing acupuncture to any other treatment in children with colds. One series of adults treated with acupuncture for respiratory illnesses reported improved symptoms, but there was no untreated comparison group, so it's hard to tell how many would have improved without treatment. I do not generally recommend acupuncture for children with colds.

Homeopathy

Homeopathic remedies for the common cold include *Aconitum, Allium, Arsenicum, Belladonna, Bryonia, Euphrasia, Gelsemium, Kali bichromium, Nux vomica, Phosphorus, Pulsatilla,* and *Oscillococcinum.* However, there are no published studies showing that homeopathy benefits children with colds. Homeopathic remedies are generally safe.

WHAT ABOUT JANET AND JAMAL?

Janet had just enrolled Jamal in his new day care and didn't think she could transfer him any time soon. The first thing I advised her to do was to QUIT SMOKING. Quitting now would help reduce Jamal's frequent colds AND minimize her recurrent bouts with bronchitis and sinusitis. I had her choose a date she would quit, gave her the name of an internist who could prescribe a nicotine patch to battle her cravings, recommended a smart phone app developed at Yale to help her quit (Craving to Quit), and called her in a week to see how it was going. I recommended that until her quit date, she avoid smoking in Jamal's presence—especially indoors or in the car.

Second, I recommended that she make a big pot of chicken soup that she and Jamal could sip over the next few days. While she was at the grocery store, she decided to try the ready-made chamomile and slippery elm bark herbal teas and a probiotic supplement we discussed after reviewing the costs and quality of various options on ConsumerLab.com.

Third, we discussed the home vaporizer. Janet wanted to try adding eucalyptus oil to the medicine cup to see if that helped. She said that when she was a little girl, her mother always rubbed Mentholatum® on her chest, and that it made her feel much better. She decided to continue the family tradition.

Jamal's vitamin D level was low, so we put

him on a supplement during cold and flu season to boost him back to a healthy level. Jamal was back to normal within a week. Janet quit smoking, but she relapsed two weeks later. I reassured her that relapses are common and that one slip *did not mean she was doomed to failure. She has finally quit for good three tries later. She and Jamal enjoy making their cold remedy (chicken soup) together, and have had fewer colds every year for the last three years.*

WHAT I RECOMMEND FOR PREVENTING COLDS

1. *Lifestyle—environment.* Do not smoke and do not allow others to smoke around your child. If your child must be in day care, try to make sure he is in a setting with small groups of children or in small classes in separate rooms to minimize his exposure to cold viruses. *Wash your hands* and make sure your child washes his.

2. *Lifestyle—exercise.* Regular moderate exercise helps to keep the immune system in top shape. Avoid sleep deficits.

3. *Lifestyle—diet.* Eat vitamin C–rich foods, at least 5 to 11 servings of fruits and vegetables daily. Eat foods like yogurt, which contains healthy probiotics.

4. *Biochemical.* When prone to colds, consider using probiotics, echinacea, or American ginseng daily to prevent colds and flu as an adjunct to a healthy lifestyle. If you have a risk factor for low vitamin D (e.g., darkly pigmented skin, indoor lifestyle, winter, living in a northern latitude), consider getting your vitamin D level checked and take supplemental vitamin D to boost the level into the healthy range.

5. *Mind-body.* Support, encourage, and cuddle your child to ensure he feels supported, confident, and connected. A strong social support system builds a strong immune system.

See your health care professional if your child has:

- A high fever (over 103°F or 39.4°C)

- A sore throat, sore glands in the neck

- Ear pain

- A stiff neck or sore back

- Shortness of breath, wheezing, or trouble breathing

- Cold symptoms for longer than 7 to 10 days

- Too sick to drink

TREATING COLDS

1. Do not use antihistamines, cough syrups, decongestants, or expectorants in children less than five years old. Do not ask your physician for an antibiotic prescription for a cold. Do not give your child brandy or other kinds of alcohol for a cold. Older children may benefit from decongestants. Avoid decongestant nasal sprays for more than a day or two unless your child is having so much trouble breathing that he can't eat or sleep. Use analgesics only to treat discomfort. Antihistamines may put him to sleep or make him hyper. Treat the child, not the thermometer.

2. *Biochemical—home remedy.* Use saline nose drops and a bulb syringe in children too young to blow their noses. Consider honey to soothe irritated throats and dry coughs.

3. *Biochemical—nutritional supplements.* Give vitamin C, 200 milligrams to 2 grams, divided into several doses over the course of the day.

4. *Consider elderberry.*

5. *Lifestyle—nutrition.* Give plenty of fluids, chicken soup, or other hot (not scalding) fluids (sipped slowly through the day). Consider raw or baked garlic, ginger, or spicy foods if they are consistent with your family traditions.

6. *Lifestyle—exercise.* Ensure sufficient rest; encourage extra naps.

7. *Lifestyle—environment.* If you use a steam or cool mist vaporizer, with menthol or eucalyptus oils in the medicine cup, keep it clean to avoid mold, mildew, and bacterial buildup.

8. *Lifestyle—mind-body.* Give extra hugs and encouragement, and encourage positive thoughts.

9. *Biochemical—massage.* Consider offering gentle massages using Mentholatum®, Tiger Balm®, or Vicks VapoRub® if this is consistent with your family traditions.

11
COLIC

Ken Johnson called one evening about his baby girl, Nancy. He had just arrived home after a hard day at the office; his wife, Erica, was cooking dinner, and six-week-old Nancy was crying again. He had tried feeding her, rocking her, singing to her, carrying her, and bouncing her. Nothing seemed to work for more than a minute before she started crying again, and she looked as if her tummy hurt. He wondered how Erica had coped with this all day, but she said that Nancy was not that fussy until about the time he came home. He wondered if the baby didn't like him, if he did something to cause the crying, or if she had a medical problem. He called his mother for advice; she said it sounded like colic. He wanted to know what I thought, whether he should bring her in for an evaluation, and if there were any natural remedies that would help.

All babies cry. However, by the time she's six weeks old, the average baby (even without colic) cries less than two hours a day.[1] Colic is intense or excessive crying (more than three hours a day for three or more days a week) in babies who are between three weeks and three months old. It is *not* a sign of serious illness, bad temperament, misbehavior, or inadequate parenting, but it *is* distressing for parents. In fact, mothers of colicky babies have an increased risk of developing depression,[2] so it is worthwhile to learn effective ways to prevent and treat colic.

Most parents quickly learn the difference

between a hungry cry, a pain cry, an angry cry, and a bored cry. Colic is different. It sounds like a cry of pain—intense, high-pitched, and unpleasant.[3] Colic is usually worse in the evenings when parents are getting home, trying to fix dinner and unwind from the day. Colic usually peaks around the time the baby is six weeks old and is almost always over by three to four months. Sometimes the screaming and crying are accompanied by vigorous kicking, and the baby is difficult to console. While crying, babies with colic often pull their legs up, make tight fists, have swollen or distended tummies, appear to be in pain, and burp or pass gas.

About 10 to 20 percent of babies develop colic that is troubling enough that parents consult a health professional.[4] Colic is more common in first-born babies than in subsequent children and more common among babies whose mothers smoke cigarettes than in babies of nonsmokers. It is most common among babies whose parents are professionals and least common among babies whose parents are laborers. It may be that these babies really have different amounts of colic, or it could be that their parents just have different expectations about how much babies should cry. A Swedish study found that colic was less common when moms kept track of babies' behavior using daily diaries than when they were interviewed after the fact.[5] It may be that we tend to remember the crying as even worse than it seemed at the time.

Babies with colic may appear to be totally miserable, but they are generally very healthy from every other standpoint. They eat well and gain weight; they don't have fevers, diarrhea, or any other symptoms. However, colicky symptoms can be a signal that something else is wrong. Common crying contributors include: infections, constipation, injuries, reflux, allergies, and hernias.

WHEN TO TAKE YOUR BABY TO A HEALTH CARE PROFESSIONAL

- Poor feeding, diarrhea, blood in stools or urine, projectile vomiting, or weight loss; fussiness while eating
- Bruising
- Floppy muscle tone or lethargy
- Irritability
- Fever
- Bulging soft spot on the top of the head (fontanelle)
- Trouble breathing or spells of not breathing
- Bluish or mottled color to fingers, toes, or lips
- Unusually severe crying for more than three hours daily
- Initiation of colic before two weeks of age or persistence beyond three months of age

---

If your child has any of these signs, take her to the doctor for an evaluation. She may be suffering from another problem such as a bladder infection, ear infection, or other health problem.

WHAT CAUSES COLIC?

No one has come up with one right answer for what causes colic in all babies, but there are lots of theories.

THEORIES FOR CAUSES OF COLIC

- Developmental stage
- Emotions and family stress
- Differences in infant temperament and physiology
- Food intolerance

Developmental stage. Infants develop rapidly after birth. The coordination between swallowing, digestion, and peristalsis (waves of contraction that move food through the intestines) is still developing; some babies develop more quickly than others. The nervous system is also developing, and the baby may be plain old worn out or overstimulated by the time evening comes—at least until she has learned to pace her arousal and sleep cycles over the course of the day. Babies who have learned to soothe themselves may be able to tolerate small discomforts more readily than babies who have not yet learned to soothe themselves.

Emotions can also affect digestion and gas formation. Family tensions can trigger infant crying. Parents who feel they have plenty of help and support in caring for the baby and who feel they had a good childbirth experience are less likely to have colicky babies. Confident mothers are less likely to have colicky babies than mothers who are anxious about their parenting abilities. Families dealing with domestic violence have an increased risk of having a colicky baby, adding to existing tensions. On the other hand, having a colicky baby can make even a calm, happy mother lose confidence and feel anxious and depressed. (Sorry, dads, there just aren't very many studies about how fathers' moods and expectations affect infant colic.) Having a baby with colic doesn't mean you are a bad parent, but it can surely make you feel like a failure. If you feel yourself being pushed over the edge or having overwhelming feelings of failure, sadness, or anxiety, please see your health professional. Help is available.

Even very young babies have different *temperaments* and physical makeups. Some babies may just be naturally fussier and have different digestive dynamics than others. Babies who have difficult temperaments (for example, irregular sleep patterns, oversensitive, more squirmy, etc.) in the first two weeks of life cry and fuss more at six weeks than other babies. Colicky babies are more likely than non-colicky infants to grow into toddlers with sleep problems and frequent temper tantrums.[6] Babies diagnosed with colic are more likely later on to also be diagnosed with certain kinds of migraines.[7] These kids could benefit from learning self-soothing strategies early in life. Higher levels of *motilin* (a molecule that increases intestinal activity) are present in newborn babies who eventually develop colic than those who do not.

DIETARY RISK FACTORS: DO THEY OR DON'T THEY CAUSE COLIC?

- Formula vs. breast milk
- Soy or hydrolyzed formulas
- Frequency of feeding
- Cow milk sensitivity
- Diet of the nursing mother

There is no difference in the risk of developing colic between babies who are *breast-fed* and those who drink *formula*. Breast-fed babies' stomachs empty faster after a feeding, and they are hungry again sooner. Once their hunger is attended to, they usually stop crying.

For many other reasons, I recommend that you *breast-feed your baby at least 12 months*. Breast-fed babies who are fed *less often* (every three to four hours) tend to cry more than babies who are fed *more often* (every two hours). Feeding your baby every two hours may seem exhausting, but over the first few weeks of life it may result in less crying.

Babies whose parents *respond to their cries* more *quickly* and effectively fret much less than babies whose parents wait a bit longer to see if the baby will stop crying on its own. You can't spoil a young baby! In fact, you may make your life easier if you respond to the baby quickly and effectively in the first two to three months of life. And you may be helping your baby learn self-soothing skills that will be helpful for months and years to come. What's the best way to respond? Dr. Harvey Karp advocates that after you've ensured that the baby is not hungry, in pain, or needing a diaper change, you use the five S's: swaddling, side/stomach position (not for sleeping), shushing, swinging, and sucking.[8] Look for YouTube videos of "The Happiest Baby on the Block" to see what this means in more detail.

There are some babies whose colic improves when they stop drinking *cow's milk formulas* or their breast-feeding mothers abstain from drinking cow's milk.[9] Cow milk proteins are absorbed by mothers and concentrated in breast milk; in fact, the levels of some of these proteins are higher in the breast milk of mothers who drink cow milk than they are in cow milk itself.[10] Several scientific studies of formula-fed infants who seemed to be sensitive to cow's milk showed that colic disappeared when infants were fed a formula free of cow's milk protein; when the protein was reintroduced, the babies became colicky again.[11] In one study, babies fed soy milk had less colic than those on a modified cow's milk formula.[12] Small studies have also suggested that hydrolyzed milk formulas (like Nutramigen®) help reduce colic symptoms, when switching to soy-based formula hasn't done the trick.[13] On the other hand, *for most babies, what the mother drinks makes no difference at all*. If you decide to stop drinking cow's milk or soy milk, be prepared to wait two weeks to see an improvement in your baby's symptoms; it may take that long for all of the cow's milk proteins to be flushed out of your system. And make sure you're getting enough protein and calcium from other foods.

Some nursing mothers notice that when they eat certain foods, their baby has more colic. A New Zealand study showed that the only foods eaten by nursing mothers that were consistently associated with colic in their babies were fruit and chocolate.[14] Chocolate, as well as coffee, tea, and cola, contains caffeine. Some babies are sensitive to other foods in the nursing mother's diet, such as soy, corn, wheat, and eggs. These foods are also the most common food allergens in infants and young children. Other foods that find their way into breast milk and seem to distress some babies are cabbage, broccoli, onions, peppers, and beans. On the other hand, many babies like the taste of their mother's milk better if she's eaten garlic.

You can probably do well without some of these items in your diet, but it is important that the nursing mother's diet be well balanced, with plenty of green vegetables and other sources of calcium, vitamin D, and protein (such as calcium fortified orange juice, canned fish, or tofu). If you decide to omit or restrict your intake of major food groups, please check with your health care professional to make sure you are getting all of the nutrients you and your baby need.

Do *not* try to treat your child's colic by offering her a variety of juices or solids. Solids (such as rice cereal) have been proven *not* to be helpful in treating colic. Sorbitol-rich fruit juices, like apple and pear juice, may increase crying in babies who have a hard time digesting carbohydrates.[15] Hold off on juices and solids until your baby is at least four months old.

CAN COLIC BE PREVENTED?

Here are some things you can try to reduce your baby's chances of developing colic.

1. *Nutrition.* If you are *nursing* your baby, *feed every two hours* rather than every three to four hours. If you are feeding your baby formula, choose one that is iron-fortified to reduce the risk of colic and anemia. Do not introduce solids or juices to your baby's diet until she is at least four to six months old. Consider probiotic supplements (see section below).

2. *Behavior. Carry your baby* as much as possible. A Canadian study showed that parents who carried their babies four to five hours a day were rewarded by a 50 percent reduction in infant crying at six weeks of age compared with parents who carried their babies two to three hours a day.[16] The improvement was especially noticeable in the evening hours when colic tends to be worse. I recommend carrying your baby at least three hours daily because I believe it helps promote bonding between parents and infants and facilitates quicker responses to babies' needs.

3. *Respond to your young baby's cry quickly* (within ninety seconds). This reduces crying both in the short term and in the long term as your baby learns that you are ready, willing, and able to meet her needs. "Spoiling" the young baby (under four months old) actually improves his behavior. You don't make a two-month-old tougher, stronger, or more self-sufficient by expecting him to cry it out; you just get more crying. Use effective soothing techniques like Dr. Karp's Five S's: Swaddling, Side/Stomach position (not for sleep), Shushing, Swinging side to side, and Sucking on a pacifier or thumb.

4. *Don't smoke.* Smoking parents are more likely to have colicky babies, and their children are more likely to suffer from colds, ear infections, and other problems later on.

5. *Relax.* The more you and your spouse do to remain calm, gain the support and confidence you feel you need to be good parents, and nurture your own relationship, the less likely you are to have a baby who develops colic,

and the better you will be able to cope with colic or any other problem that comes along. Make sure you get enough *sleep* and support to cope with a crying baby. Many parents find it helpful to join a *parent support group* so they can hear how other parents deal with the same issues they face, and so they can feel less alone in learning how to be good parents.

Never shake a baby to stop her crying. You can do serious damage to the baby's brain, and you will not help her learn to soothe herself this way.

WHAT CAN YOU DO TO TREAT A BABY WITH COLIC?

All babies outgrow colic. Colic doesn't do any long-term or serious damage to babies unless the parents become very frustrated and lash out at the defenseless child. Remember that having a baby with colic doesn't mean that you are a bad parent or that you have a bad baby. Colic is definitely distressing to the baby and the baby's family, and it is worth trying safe remedies. The best remedies involve *lifestyle therapies*. If you don't feel you can cope, get help. Let's go around the Therapeutic Mountain to find what works in treating colic. If you want to skip to the bottom-line summary, flip to the end of the chapter.

LIFESTYLE THERAPIES: NUTRITION, EXERCISE, ENVIRONMENT, MIND-BODY

Nutrition

NUTRITIONAL APPROACHES TO COLIC

- Formula: cow milk–free (may be helpful)

- Rice cereal (not proven helpful)
- Fiber (not proven helpful)
- Frequent burping (not proven helpful)

Although there is no difference in the rate of colic between breast-fed and formula-fed infants, colic improves for some *formula-fed babies* if they are switched to a formula that does not contain cow milk or has hydrolyzed cow milk,[17] such as soy formula or Nutramigen.[18,19] One study found that switching to hydrolyzed formula was more effective than herbal tea or massage in managing infant colic.[20] A favorable response to eliminating cow milk does not necessarily mean that your baby is allergic to cow milk; she may tolerate it easily when she gets a little older. After a month or two on soy or hydrolyzed formula such as Nutramigen, many babies can return to their previous (less expensive) formula without difficulty.

For years, parents have believed that giving the baby a little *rice cereal* helps with crying. A study from Johns Hopkins disproved this common myth.[21] It turns out that most parents started the solids about the time their baby was getting over colic anyway, and the cereal got the credit for improvement. There's no scientific evidence that cereal cures colic.

Because babies with colic look as if they're in the same kind of pain as adults with irritable bowel syndrome (who are often helped by increased fiber), some people have tried treating colicky babies with *fiber*. However, a comparison study showed that adding fiber does not improve colic.[22] Another nice-sounding theory bites the dust!

Another old-fashioned, safe (and scientifically disproved) therapy is *frequent burping*. Colicky babies do not need extra burping.

Environment

Music has also proven useful in treating infant colic. In one study, psychologists trained parents in how to use music. When the colicky baby was quiet, calm, and not crying, the parents played a recording of the baby's favorite music (selected by the parents) and paid extra attention to the baby. When the infant started to cry, the parents turned off the music and withdrew attention briefly. This procedure was followed throughout the day. Within days, crying and colic had significantly decreased. When the parents returned to their old way of dealing with crying, the crying again increased.[23] Music therapy is definitely safe and worth trying, but it takes a fair amount of good observational skills and discipline for parents to carry it off.

A *warm pack* or *hot water bottle* on the baby's tummy is an old-fashioned colic remedy. Others recommend a warm bath as a way to relax an upset baby. There are also lots of recommendations in old medical texts about placing the baby on her tummy over your knees or over a rolled up towel to increase pressure on the abdomen. Some parents put the baby's tummy across their knees and then gently bounce the baby to help break up gas bubbles and relieve pressure. There are no scientific studies evaluating the effectiveness of these time-honored remedies. Because they are safe, you may want to try them. Just be sure that if you use a warm pack or hot water bottle, you test the temperature and make sure it is not too hot for a baby's sensitive skin.

Mind-Body: Behavioral Strategies

Behavioral strategies are the keys to effective colic management. Many parents have found it useful to get into a rhythm of responses to a colicky baby.

Behavioral Responses to Colic

1. Check to make sure that she isn't hungry. (Has she fed in the last two hours?)

2. Check to make sure nothing is causing pain (an open diaper pin or a hair wrapped around a finger, toe, etc.).
3. Check to see if the diaper needs changing.
4. Think about her level of stimulation and activity.
   - Is she bored?
   - Is she overstimulated? Too much playing, singing, rocking, or any other visual, auditory, tactile, or other stimulation can overwhelm a baby. Keep it calm, quiet, and organized.
5. Still no luck? After you have checked these things, you may find the following helpful:
   - Swaddling or wrapping her snugly and placing her in a dark, quiet room
   - Slow, rhythmic rocking (not shaking) or swinging side to side
   - Shushing (rhythmic, comforting hushing sound)
   - Going for a walk in a stroller
   - Side or tummy position for a few minutes (not to sleep)
   - Letting her suck her fist or fingers

Other options include activities that stimulate a sense of movement or white noise:

- Going for a ride in the car
- Putting her in her car seat and putting the car seat on top of the dryer while it is running on "fluff" (no heat). This provides vibration and white noise that soothes many babies. Please make sure someone stays with the baby to make sure the car seat doesn't vibrate right off of the dryer!
- Running the vacuum cleaner or a hair dryer next to the baby while she's in the car seat. This is simply a source of

"white noise" which some babies find strangely soothing.

As a last resort, you can let your baby cry himself to sleep. This is very hard for most parents to do. Most parents feel guilty and also worry that the neighbors will be annoyed at hearing the baby cry or worry that the neighbors will think they are bad parents. You may want to take turns taking responsibility for the baby for an hour or two so you can get a break. If you are a single parent, try to find another trustworthy adult to care for the baby for an hour so you can take the time to re-center yourself.

BIOCHEMICAL THERAPIES: MEDICATION, DIETARY SUPPLEMENTS, AND HERBS

Medications

Most medications are no more effective than placebo in treating newborn colic.

MEDICATIONS—*NONE* PROVED SAFE AND EFFECTIVE

- Simethicone: *Mylicon* (non-prescription) and cimetropium bromide
- Lactase (to reduce lactose in milk)
- Sedatives (prescription), antihistamines, alcohol
- Dicyclomine: *Bentyl*

One of the most commonly used medications to treat infant colic is *simethicone* (Mylicon®), which is used to treat gas. Another older medication is *cimetropium bromide*. Neither has proven consistently better than placebos or other therapies for infant colic.[24] I do *not* recommend simethicone or cimetropium for treating colic.

Because some babies seem to be sensitive

to lactose (milk sugar), some folks recommend using supplemental *lactase* (the enzyme that helps us digest lactose) for treating colic. Unfortunately, studies on lactase have had mixed results.[25,26] I do not routinely recommend it until additional research yields definitive data on benefits.

Sedatives, antihistamines, and motion-sickness medications, including *dicyclomine* (Bentyl) are outdated and potentially harmful colic remedies.[27] These treatments have potentially serious side effects, including Sudden Infant Death Syndrome. Like these old-fashioned medications, old-fashioned remedies such as giving the baby wine or other *alcoholic beverage* are dangerous and discouraged by most pediatricians. I do *not* recommend *any* medications as colic remedies.

Dietary Supplements and Herbal Remedies

Before diving into herbal remedies, let's take a look at another supplement, *probiotics.*

Probiotics are the healthy bacteria that normally live in our intestines in the trillions. In 2010, Italian researchers reported that in a randomized, controlled trial, *Lactobacillus reuteri* probiotic supplements significantly improved colic in breast-fed babies.[28] These results were confirmed in a Polish study in 2013 and a Canadian study in 2015.[29,30] The probiotic supplements proved more useful than the old standard medication, simethicone, in another study.[31] Based on the success in using this specific probiotic (*L. reuteri*) in *treating* colic, researchers started evaluating whether giving the same probiotic to newborns could *prevent* it.

Indeed, a large Italian study published in 2014 showed that starting probiotic supplements in newborns reduced crying, spitting up, and constipation, and reduced the cost of health care, including doctor and emergency room visits and parents' lost time from work.[32]

Again, the results were confirmed in another study, showing clinical benefits and overall cost savings from using *L. reuteri* supplements to prevent colic.[33]

The dose used in most of these studies was 100 million colony-forming units.* Another study examined the benefits of prebiotics (galacto-oligosaccharides the healthy probiotics consume) added to formula and found that the babies fed the prebiotic-enriched formula had more *Bifidobacteria* (healthy bacteria) in their stool and less colic than babies who received regular formula.[34] Based on this research, I recommend probiotic supplements to prevent and treat infant colic; some brands that contain this specific *L. reuteri* probiotic included BioGaia® and Gerber Soothe®.

HERBAL REMEDIES FOR COLIC

- Effective: Combination of lemon balm, chamomile, fennel, licorice, and vervain; combination of chamomile, fennel, and lemon balm; fennel
- Other traditional remedies: anise, caraway, catnip, chamomile, cumin, dill, ginger, mint

A 1993 study documented that three to four ounces per day of an herbal tea (containing *chamomile, fennel, vervain, licorice,* and *lemon balm*) was significantly more effective than a placebo (a tea with simple sugar and flavoring but no herbs) in eliminating infant colic.[35] A 2005 study showed that another herbal tea containing chamomile, fennel, and lemon balm helped ease colic.[36] A Russian study showed that fennel oil alone offered significant symp-

---

* This sounds like a lot until you realize that human intestines normally contain trillions of bacteria. As one million seconds equals 12 days, a trillion seconds equals 31,688 years. So, 100 million probiotics is a tiny fraction of the total amount of gut bacteria, but it's enough to improve colic.

tom relief in colicky babies.[37] Chamomile is a traditional tummy-settling tea in many parts of the world. Allergies are possible. A California baby developed botulism from contaminated homegrown chamomile tea. Despite this one report, herbal teas are one of the safest remedies around; and I regularly recommend them to parents who have a colicky baby, as long as they keep the quantity under six ounces a day. If the baby starts to vomit, becomes constipated, loses her appetite, or has any other concerning symptoms, stop the herbs.

Other traditional herbal recommendations include anise, catnip, peppermint leaf, fennel, caraway seed, chamomile, and ginger root. The old remedy, *gripe water,* was made from dill. Newer products contain additional herbs. Dill, fennel, and caraway are all from the same family of plants that help ease intestinal spasms. Although they have not undergone extensive, rigorous scientific study, they appear to be safe, and I often recommend them.

WARNING: Do not give your baby more than six ounces per day of herbal tea. Filling up on tea means there is less room for the milk your baby needs to grow. Start slowly, say one ounce at first, and watch the baby for several hours for any side effects or reactions before trying more. Be sure you obtain your herbs from a trusted source and that you know exactly what you are giving your baby. The side effects of contaminated herbs can be deadly. A three-month-old baby girl whose parents thought they were giving her Chinese star anise for her colic ended up taking her to the emergency room when the product turned out to be Japanese star anise.[38]

BIOMECHANICAL THERAPIES: MASSAGE, MANIPULATION, SURGERY

In general, parents of babies with colic who receive manipulative therapies such as massage, chiropractic, osteopathic adjustments, and craniosacral therapy report improved symptoms. Scientifically, it is difficult to keep parents from knowing whether or not their children were actually treated,* and few studies offered another treatment for comparison; but parent-provided massage is certainly inexpensive, low-risk, and consistent with cultural practices, and I support it.

Massage

Many cultures around the world encourage mothers to provide massage for their colicky babies. The baby is held on her side, supported by the parent's arm, with the head somewhat down and the bottom elevated. The tummy is then massaged in a circular, clockwise fashion, starting with small circles at the belly button, and gradually increasing the size of the circles outward. You can do the tummy rub through the baby's clothes, or if he is naked, use warmed olive oil, almond oil, castor oil, or cocoa butter. Massage oils that include chamomile, lavender, and geranium are thought to be especially soothing, and there is some scientific evidence to support this.[39] The massage should be given about twenty to thirty minutes after a meal so the baby's stomach has had a little time to empty (and she doesn't spit up all over you!). You can extend the massage to include the baby's whole body. I think massage is terrific, helps bonding between parents and babies, and probably does not have a single bad side effect when it is done gently and lovingly. Give it a try!

Chiropractic and Osteopathic Manipulation

You may have heard that chiropractic manipulation helps babies with colic. One study

* When parents know their baby is getting a professional treatment, their expectations may bias them toward expecting a benefit, and these expectations can bias perceptions.

showed that six-week-old babies who were treated with three sessions of chiropractic manipulation cried less over the course of the next two weeks.[40] However, treatment started just about the time most colic was at its worst, and babies tended to improve over time with or without treatment. Since this study did not include a group of babies who were not treated (control group), we can't say that chiropractic was any more helpful than no treatment. Since then, research has improved.

A study from Denmark compared chiropractic adjustment to simethicone drops for colicky babies; the investigators concluded that two weeks of chiropractic treatment (typically three to five treatments total) was significantly better than two weeks of simethicone treatment, cutting crying time more than twice as much (two to three hours less daily in the chiropractic group compared with one hour less daily in the simethicone group).[41] A later British study actually kept some parents in the dark about whether or not their colicky baby had received a chiropractic adjustment; they found improvements in colic in the chiropractic group regardless of whether the parents knew the baby received chiropractic or not.[42]

A study of cranial osteopathic manipulation for colicky babies showed that those who received it improved more than those who didn't, but again, their parents knew they were receiving professional treatment, and their expectations may have influenced their perception of effectiveness.[43] Until additional rigorous research is conducted, I do not recommend professional manipulative therapies for infant colic.

BIOENERGETIC THERAPIES: ACUPUNCTURE, THERAPEUTIC TOUCH, HOMEOPATHY

Acupuncture

Small Swedish studies found that acupuncture had modest benefits for colicky infants and a greater impact on sleep.[44,45] However, another Scandinavian study failed to find any benefits of acupuncture therapy for colic.[46] While acupuncture is useful for many other conditions, I am not ready to recommend it as a treatment for infantile colic until additional research confirms the Swedish experience.

Therapeutic Touch

Therapeutic Touch has been reported to be helpful for crying babies and children. Although there are no studies on colic specifically, I know from my own clinical experience that TT relaxes and calms both the giver and receiver, and I use it in the clinic when confronted with a baby who is crying for any reason.

Homeopathy

The most common remedies are: *Chamomilla*, *Colocynth* (from the herb bitter cucumber), *Dioscorea* (wild yam), *Magnesia phosphorica*, *Nux vomica,* and *Pulsatilla*. Until scientific studies establish the effectiveness of these remedies in treating infant colic, I tolerate but do not recommend them.

WHAT I RECOMMEND FOR COLIC

PREVENTING COLIC

1. *Lifestyle—mind-body.* Respond to your baby's cries quickly using effective calming and soothing skills. Carry your baby in your arms as much as possible. Surround yourself with supportive family and friends. Take care of yourself, too. Learn and practice stress management techniques; you'll need them to handle toddler temper tantrums later on anyway!

2. *Lifestyle—nutrition.* If you are nursing your baby, nurse often (every two to three hours). If you are formula feeding, use iron-fortified formula.

3. *Lifestyle—environment.* Don't smoke and don't allow others to smoke around your baby. If you are dealing with domestic violence, seek help from your health professional, social worker, family services in your community, a shelter, a lawyer, or the police.

4. *Supplements.* Consider supplements with probiotics containing 100 million CFU (colony-forming units) of *L. reuteri.*

When to See Your Doctor

- If the colic persists or recurs or is causing you to consider drastic actions like shaking, slapping, or hitting your baby, see your doctor immediately.

- If you are considering changing formulas, talk with your clinician about a temporary change in formulas to one that does not contain cow's milk or one that uses hydrolyzed milk proteins; if you are nursing, consider altering your diet to omit suspect foods.

- If home remedies do not work, if your baby has other symptoms, or if she is over three months old and still has colic, see your health care professional. Severe, persistent crying may be a sign that your child has an infection (like an ear, bladder, or kidney infection) or other treatable condition. Other worrisome symptoms include black or bloody stools, vomiting, diarrhea, or fever in an infant less than two months old.

TREATING COLIC

1. *Lifestyle—mind-body.* Check to make sure she's not hungry, not in pain, not in a wet or dirty diaper, and not bored or overstimulated. Swaddle her; swing her gently from side to side while making a shushing sound; lay her down on her side or tummy and give her something to suck on. Try a ride in the car seat, five to ten minutes in a swing or rocker, or a vibrating seat. If you've done all of the above, and the baby is still crying, ask for help, put the baby on her back in her crib for ten minutes, close the door, and take time to get calm and regroup before you try again.

2. *Lifestyle—diet.* If you are feeding formula, consider switching to soy or hydrolyzed formula for six weeks. If you are breast-feeding, consider changing your own diet to avoid foods that upset the baby.

3. *Lifestyle—environment.* Try music therapy: When she's quiet, play her favorite music and praise her; when she starts crying, stop the music and ignore her for a minute or two.

4. *Supplements.* Consider supplements with probiotics containing 100 million CFU (colony-forming units) of *L. reuteri*. Consider offering an *herbal tea* containing chamomile, fennel, and lemon balm.

5. *Biomechanical—massage.* Consider a tummy massage with lavender aromatherapy.

12
CONSTIPATION

Belle Zagorski called about her four-week-old grandson, Walter, who was having a bowel movement every two days. She thought a breast-fed baby should poop every time he nursed. She was worried that Walter was constipated, although his stools were soft and he wasn't straining to have them. Belle's daughter wasn't worried, but Belle wanted my opinion about what was normal and whether she should be concerned.

Penny Brandenburg, a local nurse practitioner, called me to ask my advice about corn syrup. She had been taught that adding a little corn syrup to a bottle of formula was a good remedy for constipation; recently, she'd seen a child suffering from botulism related to eating contaminated raw honey. She knew that honey could harbor botulism spores, and she wondered if corn syrup, a similar sweetener, held the same hazard.

Eric Kay came in with his three-year-old daughter, Susan, because of her severe constipation. Susan had been doing fine until she started toilet training, but then she held her stools so long that they were hard and painful when they passed; so she was even more reluctant to sit on the potty. Eric wanted to know what to do.

There's a lot of variability in the normal frequency of bowel movements. Almost all babies (98.5 percent) have their first bowel movement within 24 hours of birth. Breast-fed babies can have a stool every time they nurse. Others have a stool every day or two. There's a wide range of normal. Among older children, more than 95 percent have bowel movements between three times a day and once every two days. Anything more than three times a day (diarrhea) or less often than every four days (constipation) calls for a professional evaluation.

Constipation simply means that your child is having stools less frequently than is normal for her, has hard or dry stools, or is having difficulty passing them. Ignoring constipation can result in complications: a tear or fissure in the delicate skin around the anus from passing large, hard stools; stool withholding; encopresis; headaches; and chronic stomachaches. *Encopresis* is the technical term for prolonged, severe constipation with stool buildup (impaction) in the rectum and occasional leaking of stool into the underwear (soiling).

Occasional constipation is a fairly frequent childhood complaint. The peak times it occurs are (a) when adding solid foods, formula, or cow's milk to the infant's diet; (b) during toilet training (especially if toddlers are rushed or pressured); and (c) when starting school (because of the change in routine and reluctance to use public toilets). Constipation becomes even more common during adulthood when active youth become more sedentary, and it practically becomes an epidemic among those over sixty-five as metabolism slows.

WHAT ARE THE COMMON CAUSES OF CONSTIPATION?

There are many contributing factors to common constipation:

- Insufficient dietary fiber
- Insufficient water or other fluids, especially in hot weather or with a fever
- Changes in diet or routine
- Ignoring the urge to defecate, postponing defecation (often during toilet training)
- Prolonged bed rest (being sedentary)
- Severe dieting (teens and adults)
- Medication side effects

The commonest cause of constipation is *insufficient dietary fiber and water*; the third most common cause is sedentary behavior. Americans and Western Europeans eat the least fiber and have the highest rates of constipation in the world. Fast foods and processed foods are typically low-fiber foods. If Americans doubled the amount of dietary fiber by eating 5 to 11 servings of fruits and vegetables daily, we could dramatically reduce problems with both constipation and colon cancer. How much fiber is enough? Some experts recommend a dose of "age in years" plus 5 to 10 grams daily. This means a three-year-old would need 8 to 13 grams of fiber daily. The Institute of Medicine recommends 20 grams daily for toddlers and young children. Excessive fiber provides no added benefit.

Constipation is more common in the summertime because children sweat more and may get dehydrated. *Dehydration* makes stools dry and harder to pass. Feverish children also lose more fluid than normal, increasing their risk of dehydration and constipation. Faced with dehydration, the body starts to draw on water wherever it can, including the bowels, resulting in drier, harder stool, and constipation. Another reason to give plenty of fluids to a feverish child!

Dietary changes are often accompanied by changes in stool frequency. Many babies who are switched from mother's milk to formula are constipated for a few days before they adjust to the new food. Changes in the regular routine of eating, sleeping, and playing, such as traveling, moving, or the arrival of a new baby in the house, frequently provoke temporary constipation. Bowel movements usually return to normal when the child becomes adjusted to the new routine or life returns to normal.

Cow's milk is linked to constipation in many children. This may be an early and only sign of allergy or sensitivity to cow's milk. In fact, allergies to cow's milk can be so severe that doctors suspect the problem requires surgery! Eliminating cow's milk (and dairy foods containing it) from the diet for two to four weeks improves chronic constipation in many children.

Some children are just so busy and preoccupied that they *postpone* going to the toilet until after the urge has passed. The body gradually reabsorbs more and more water from it until the stool becomes hard and dry. Dry, hard stool is more difficult to pass, and it may be uncomfortable or even cause tiny tears in the skin. A vicious cycle sets in, wherein the child postpones defecating, the stool becomes hard, defecation becomes painful, and the child delays further. It's best to avoid this cycle altogether by having a regular toilet time every day during which the child sits for five minutes, regardless of the outcome.

Bed rest slows down many bodily functions. Sedentary or wheelchair-bound children are more prone to constipation than their active counterparts. Even a little mild to moderate exercise keeps blood flowing to all the organs. Don't let your child spend hours glued to the television or computer screen. Have her get up and play.

Teenage girls who *diet* intensively frequently have problems with constipation. Interestingly, teenage girls are more prone to constipation during certain phases of their *menstrual cycle*. Due to hormonal fluctuations, they are most likely to suffer from constipation about sixteen to twenty-one days after the start of their periods.

Widely used *medications* such as antacids containing aluminum and prescription pain-killers containing codeine also slow down intestinal activity. Overuse of laxatives results in dependence on them; when the child stops taking the laxative (even natural ones such as senna and cascara), the bowels slow down.

Unusual causes of constipation in children include Hirschsprung's disease (in which part of the bowel does not have a proper nerve supply and doesn't move the stool along), cystic fibrosis, botulism, spinal cord tumor, and hypothyroidism. If you suspect any of these problems, see your doctor for a definitive diagnosis.

Constipated children often suffer from other symptoms such as irritability, tiredness, headaches, depression, and just feeling out of sorts. The ongoing pressure from large amounts of stool in the colon irritates the bladder; this makes children who suffer from chronic constipation prone to bedwetting and bladder infections, too.

WHEN TO SEEK PROFESSIONAL HELP FOR A CONSTIPATED CHILD:

- No stools for four or more days
- Swollen, distended, or painful abdomen
- Vomiting as well as constipation
- Pain when passing a stool
- Blood in the stool
- Dark black stools
- Constipation alternating with diarrhea
- Frequent, painful, or rushed urination

These symptoms may indicate a more serious problem such as an intestinal blockage, intestinal bleeding, impaction, irritable bowel syndrome, or a bladder infection.

HOW CAN YOU PREVENT CONSTIPATION?

You can take several easy steps to prevent constipation.

FIVE TIPS FOR PREVENTING CONSTIPATION:

- Begin by breast-feeding
- High-fiber diet (whole grains, fruits, and vegetables)
- Plenty of fluids (six to eight glasses per day)
- Exercise daily
- Regular routine

Breast-fed babies tend to have softer, more frequent stools than babies who are fed formula. Formula-fed babies are slightly more prone to constipation. Babies on soy formulas tend to have harder, less frequent stools than babies who are nursed or who are fed cow's milk–based formulas.

For older infants and children whose diet includes solids, the top remedy for preventing constipation is eating a diet high in natural *fiber* or roughage. Fiber increases the bulk of stools, draws water into the bowel, softens stools and makes them easier to pass, and stimulates contractions in the colon that move the stool along. Eating a diet that is high in fruits and vegetables, beans, and whole grains like steel cut oatmeal is the best way to prevent constipation.

Drinking plenty of *fluids* is also important for maintaining healthy bowels. Fiber draws water into the intestines, softening the stool. If there is not enough water in the system, the fiber can just clog up and make things worse. Water helps soften stools so they are easier to pass. Adding extra water substantially increases the benefits of extra fiber, especially in the summertime.

Regular *exercise* also helps keep things flowing. Children who are confined to bed for long periods (such as after surgery) have more trouble with constipation than children who are up and running around. On the other hand, if your child is normally active, adding a new, strenuous exercise routine won't necessarily improve her constipation.

Keeping a *regular schedule* is helpful in preventing constipation. Most adults have experienced the temporary constipation that accompanies a trip in which the normal daily routine is disrupted. Children who are being potty-trained benefit from having a regular schedule, including sleep, meals, and scheduled time on the toilet. It is best to schedule potty time after a meal to take advantage of the gastro-colic reflex. This is a physiologic reflex (on which most housebreaking routines are based for puppies): When the stomach is filled, a nerve reflex stimulates the bowels to move. This is why so many babies have a bowel movement right after they nurse. Children who are so constipated that they no longer feel the urge to go need a routine time in which they sit on the toilet no matter whether they feel they have to go or not. Just sitting on the toilet stimulates thoughts and reflexes that enhance regular bowel movements.

What Is the Best Way to Treat Constipation?

The best way to treat constipation is to prevent it by living a healthy lifestyle: high-fiber diet, fluids, exercise, and a regular routine. If the problem persists, there are additional proven remedies. Let's consider a variety of therapies to find out what works. If you want my bottom-line recommendations (so to speak), flip to the end of the chapter.

LIFESTYLE STRATEGIES: NUTRITION, EXERCISE, MIND-BODY

A healthy lifestyle is the cornerstone of therapy for constipation.

Nutrition

What are the top three remedies to prevent and treat constipation? Fiber, fiber, and more fiber (plus water)! Fiber also decreases blood cholesterol and stabilizes blood sugar levels. An Australian community education effort to increase the intake of whole grain bread led to a 58 percent increase in the sale of whole grain bread and a 49 percent decrease in laxative sales![1] It takes about a week for increased fiber intake to show results.

Non-digestible fiber is partially broken down by the bacteria that normally live in the intestines. (Fiber is called a prebiotic, and the bacteria are called probiotics.) The bacterial breakdown releases chemicals that help stimulate intestinal contractions. There are several different kinds of dietary fiber. You may want to combine different kinds to achieve the most balanced effect (Table 12.1).

Although fiber occurs in food, it is also sold as supplements. Commonly used fiber supplements include psyllium seed husks, which are sold as Metamucil™, and other over-the-counter medications.

Cellulose (the fiber in bran) is present in all plants. In humans it helps draw water into the intestine, making stools larger, softer, and easier to pass. Bran increases the size of the stool and decreases the time it takes to move through the intestines. *Psyllium seed (Plantago ovata)* is one of the most powerful laxatives from the plant kingdom. *Guar* and *pectin* are good for the bowel and help to lower cholesterol. *Gums,* such as xanthan gum, gum arabic, algin, and carrageenan, are used in many prepared foods to help stabilize them and give them substance.

Always avoid dehydration when taking fiber supplements. Fiber without fluid can lead to intestinal blockages. This is why many fiber supplement manufacturers recommend drinking a full glass of water with every serving of fiber. This is not a problem for those getting their fiber from whole foods like vegetables and fruits.

Start slowly and gradually increase your child's fiber intake. For kids who are used to a steady diet of white flour, fat, and sugar (such as doughnuts for breakfast and macaroni and cheese for dinner), suddenly switching to a high-fiber menu can spell gas, cramps, and diarrhea. Start by replacing doughnuts or white bread with whole grain bread or bran cereal. Then add one or two fresh fruits or vegetables a day. Then add high-fiber snacks such as Ry-Krisp crackers (seven grams of fiber per serving); Kavli 5 Grain, RyVita Crisp Bread, or Finn Crisp crackers (six grams of fiber per serving); or Wasa Fiber, Hearty, or Light Rye Crisp Bread (five grams per serving). If foods alone don't do the trick, consider adding wheat, oat, or rice bran to your child's cereal or sprinkling them on top of toast. Go slowly. Try adding one high-fiber food every few days. Do *not* give bran supplements to children under three years old without a doctor's advice. It could cause bowel

Table 12.1: Food Sources for Different Types of Fiber

Type of Fiber	Food Source
Cellulose and hemicellulose	Wheat, rice, or oat bran
Mucilages	Psyllium seeds, legumes, guar gum
Pectin	Apples, carrots
Others	Food thickeners and additives: gum arabic, xanthan gum, algin, carrageenan

blockages in young children who don't drink enough fluid.

HIGH-FIBER FOODS

- Bran: wheat, oat, or rice
- Bran cereals, puffed whole grain cereals, muesli, granola
- Bran muffins, bran cookies, bran crackers, whole grain pancakes
- Beans, peas, lentils, soybeans
- Dried fruit (prunes are best; figs, apricots, raisins, and dates are also good)
- Fresh fruit and vegetables, especially cabbage, carrots, sweet potatoes, jicama, corn, peas, broccoli, pears, apples, strawberries, papaya, and celery
- Seeds and nuts (for children over four who are unlikely to choke)
- Popcorn (for children over four who are unlikely to choke)

Prunes have a well-deserved reputation as laxative food. Other excellent dried fruits include figs and dates. You can make a yummy jam that will help keep your child regular. (See Right and Regular Jam.)

RIGHT AND REGULAR JAM

2 cups water
1¼ cup dried, chopped pitted dates
1¼ cup dried, chopped figs
1 tablespoon corn meal

Combine all ingredients in a glass, ceramic, or stainless steel pot. Bring to a boil. Simmer and stir until thickened. Cool to room temperature before serving.

Keep refrigerated. Use as desired on whole grain toast, pancakes, waffles, etc.

Most kids love fruit, and fresh fruits and vegetables are excellent sources of fiber. If your child doesn't like eating fruit, you can make a delicious fruit smoothie.

FRUIT SMOOTHIE FOR EXTRA FIBER

IN A BLENDER COMBINE:

½ banana
½ apple (with peel, but without seeds or core)
½ cup of yogurt with active cultures
½–1 teaspoon of wheat, oat, or rice bran (optional)
½ cup of pear, apple, or black cherry juice, plus ½ cup of water
¼–½ tsp of flax or chia seeds

You can add other fruits as you wish and adjust the amount of fruit and juice until you achieve the consistency your child prefers. This makes a delicious shake, and it contains all the fiber and fruit sugar your child needs to get going in the morning. Aim for a minimum of five servings of fresh fruits and vegetables daily.

Psyllium seeds are the source of several laxative medications. Rather than give your child a medication, look for psyllium seeds, *flax seeds,* or chia seeds. Try small amounts (¼ teaspoon) sprinkled on breakfast cereal, whole grain toast, or tossed in a smoothie. Alternatively, to make sure your child gets fluid along with the seeds, soak the seeds in a cup of water to soften them and allow them to start to swell; have your child swallow the whole thing—water and seeds. Allergies are possible. If your child develops a rash, hives, or any other allergy symptoms, stop the seeds. Flax seed is an ingredient in an increasing number of commercial cereals and breads. You can find flax seed in the natural food section of your grocery store. It tastes fine. You can add one-half teaspoon to your child's regular cereal or sprinkle it on toast.

Plain popcorn is a fun source of fiber. Don't wipe out the nutritional benefits by adding butter or salt. Stick to plain popcorn, flavored with a bit of your favorite herbs, Brewer's yeast, or Parmesan cheese.

As mentioned earlier, drinking enough *water is* also very important in maintaining regularity. Fiber draws water into the bowel to increase its bulk, stretch the colon, stimulate bowel action, and soften the stools. Pushing fluids beyond the body's basic needs doesn't offer any particular advantage. Offer your child fluids, but don't hound her into drinking more than she is comfortable with.

Some kinds of *water* may be better than others. For example, some brands of mineral water (like Gerolsteiner™ and Hepar™) and coconut water contain magnesium. Many Americans do not consume enough magnesium, which is important in keeping things moving in the GI system. In a study of French women, those assigned to drink a liter (quart) daily of magnesium-rich mineral water were much more likely to report improvements in constipation than those given similar water without the magnesium.[2] Although diarrhea has not been reported as a common side effect of drinking mineral water, it can occur in those who take too many magnesium supplements.

Fruit juice helps promote regular bowel movements but can contain a lot of empty calories. They are not as healthy as whole fruits. Fruit juice is so potent at loosening the bowels that it is often an unsuspected cause of chronic diarrhea. The most potent fruit juices are pear, apple, black cherry, prune, and syrup of figs. Fruit juices in moderation are safe even for children less than three years old, but too much fruit juice can lead to diarrhea, tooth decay, and obesity. Remember tooth brushing after fruit juice!

Many "sugar-free" candies and gums contain fructose and *sorbitol* as sweeteners. Both of these sugar substitutes are poorly absorbed by many children and adults. Because they are not well absorbed, they draw water into the intestines, softening the stool and making it easier to pass. They are also broken down by normal gut bacteria, creating gas and stimulating the bowel. Sorbitol is so good at moving things along it is used in an emergency medication to help speed the elimination of poisons. Children with sloppy stools may benefit from cutting back on sugarless foods, while constipated kids might benefit from having a few more "sugar-free" treats.

Like Penny Brandenburg, I was taught to add one to two tablespoons of Karo corn syrup to each bottle of formula for babies suffering from constipation. Since we now know that botulism spores have been found in corn syrup as well as honey, many practitioners strenuously recommended *avoiding* these products for infants under a year of age.

I advised Penny Brandenburg not to recommend Karo syrup but to stick with syrup made from simmering dried figs, dates, raisins, and prunes in plain water: one to two teaspoons with each meal. She was relieved to have an alternative remedy.

Many adults notice that their morning cup of *coffee* can really get things moving. Coffee stimulates the bowel, and it does so quickly. Even decaffeinated coffee can provoke increased bowel action in just a few minutes. I am *not* advocating coffee as a regular beverage for children; but for occasional use when a child is constipated due to travel or a disruption in schedule, a few ounces of coffee just might do the trick.

Cheese is frequently blamed for constipation, and many practitioners recommend avoiding cheese if a child is constipated. A different dairy product, yogurt, may actually be helpful in restoring healthy bowel bacteria and reducing constipation. Try including yogurt (with active cultures) to your child's regular diet. There is even emerging research that providing probiotic supplements can help improve intestinal "transit time"[3] and reduce constipation-related abdominal pain.[4]

Exercise

Bed rest slows down bowel activity. Mild to moderate exercise helps keep things moving. Exercise also helps reduce stress. Strenuous endurance sports such as long-distance running and cycling are associated with an increased risk of diarrhea, especially in girls. The best idea for maintaining balance and regularity is to keep moderately active—neither too much time in bed or on the couch nor marathon running. Make sure that your child has access to plenty of fluid during exercise to avoid dehydration.

Mind-Body

Stress upsets the stomach and aggravates all kinds of bowel problems. Some children react to stress with headaches, some have diarrhea, some have sleep problems, and some suffer from constipation. Anticipate the possible consequences of unavoidable stress (such as starting school, moving, death of a pet or family member) by increasing your child's intake of fiber-rich foods, water, and exercise during stressful periods.

To keep the gastrointestinal system functioning smoothly, try to keep meal times and toileting times low key. Make sure that your child's time on the toilet is pleasant; don't demand performance. Simply allow it to happen naturally. When your child does have the desired results, positive reinforcement such as praise and stickers can work wonders.

Much of childhood constipation is a result of the child ignoring the urge to defecate because she is preoccupied with something else. Ignoring the urge to go allows the colon time to reabsorb more water from the stool, making it dry and difficult to pass. Habitually suppressing the urge to defecate actually changes the bowel's normal dynamics and makes it more difficult to go once the child is willing. Remind your child *not* to postpone the urge to defecate.

Hypnosis and *biofeedback* have proven effective in treating chronic, severe constipation and encopresis and for patients with irritable bowel syndrome. They are unnecessary for the vast majority of children with short-term or mild constipation. Biofeedback helps retrain the defecation reflex, leading to improved control of bowel function that can last for months after the training is completed. It is most effective if used in the context of increased fiber, stool softeners, and behavioral techniques such as regular toilet-sitting times. Biofeedback therapy is reserved for children with severe constipation who have not benefited from other lifestyle therapies. New technologies have made it possible to develop biofeedback using interactive computer games and portable devices to help children re-learn how to use the muscles to poop.

BIOCHEMICAL THERAPIES: MEDICATIONS, HERBS, NUTRITIONAL SUPPLEMENTS

Medications

Though many medications are effective, do *not* use them until you've tried dietary changes first. A variety of non-prescription medications are available to treat constipation. Do *not* use any of them for more than a week without checking with your child's doctor. Do *not* use them if your child has severe abdominal pain, nausea, vomiting, blood in the stools, or cramping, because they could worsen a serious underlying condition. If your child has these symptoms, see your doctor.

The two principal natural types of fiber medications are based on *psyllium* (from psyllium seeds) and *methylcellulose* (plant fiber). There is also an effective synthetic fiber, *polycarbophil*.

Psyllium is the fiber in Metamucil™ and many other bulk laxatives. These medications come in a variety of forms: fruit-flavored, sweetened or unsweetened powder to mix with water or juice, or as wafers. Allergic reactions to psyllium are rare. Each teaspoon of Metamucil™ contains about 2.2 grams of fiber. This is the same amount of fiber found in one dark Finn crisp cracker (2.5 grams) and less than the amount in two Fiber Rich Bran Crackers (six grams). *Methylcellulose* is the fiber in Citrucel™. Each rounded teaspoon of Citrucel™ contains approximately two grams of fiber. *Read labels* to find out how much fiber each remedy (or food) contains.

NON-PRESCRIPTION MEDICATIONS TO TREAT CONSTIPATION

- *Fiber or bulk agents:*
  - Glucomannan (Konjac or elephant yam)
  - Psyllium (Metamucil™ and others)
  - Methylcellulose (Citrucel™)
  - Polycarbophil (Fiberall™ chewable tablets and others)
  - Wheat dextrin (Benefiber™)
- *Stool softeners*: Docusate (Colace™)
- *Osmotic laxatives*: Polyethylene glycol, PEG (Miralax™); Lactulose (Cephulac™)
- *Lubricants*: Mineral oil
- *Stimulants*:
  - Bisacodyl (Dulcolax™)
  - Cascara (Nature's Remedy Natural Vegetable Laxatives™)
  - Senna (Senokot™, Fletcher's Castoria™)
  - Castor oil

- *Magnesium salts* (Milk of magnesia, Epsom salts)
- *Others:* Maltsupex, Glycerin suppositories (Babylax™), Sorbitol syrup

Fiber works best if it is taken with lots of *fluid*. The liquid combines with the fiber, swelling up and creating the bulking effect that expands the colon and softens the stool. You may need to repeat doses two to three times daily for one to three days before you see results. Fiber-containing laxatives should *not* be used by children who have blocked bowels, because they can make the blockage worse. Fiber laxatives are too harsh for children under three years old.

Docusate (*Colace™*) is one of the most frequently used stool softeners in hospitals. Colace™ draws water into the bowel, softening the stool and making it easier to pass. Aspirin, ibuprofen, and other anti-inflammatory medicines may interfere with it. Colace™ is helpful when stools are hard and dry or when passage of a firm stool is painful, as in children suffering from anal fissures (tiny tears in the skin around the anus from passing hard stools). Colace™ is available as capsules, syrup, and as infant drops. It is one of the safest medicines around, though some children do not like the taste.

Polyethylene glycol, PEG (Miralax™) is FDA approved for use in adults for seven days but widely prescribed for children for intermittent or long-term use as an osmotic laxative.[5] This means it works by drawing water into the bowel and softening stool. It works best when taken with plenty of water. A 2014 analysis of multiple research studies concluded that PEG was more effective than magnesium-based laxatives, lactulose, or mineral oil for treating pediatric constipation. Side effects include diarrhea, abdominal pain, nausea, gas, headache, and bloating.[6]

Lactulose (*Cephulac™* and other brands)

is an artificial sugar that is not digested by humans. It is broken down by the gut bacteria, resulting in carbon dioxide and other compounds that stimulate bowel movements. Lactulose can cause gas, belching, cramping, nausea, and discomfort. It should not be used with other laxatives. It may take 24 to 48 hours to see an effect from lactulose since it has to travel all the way to the large intestine to start working.

Mineral oil has long been used to "grease the skids" of children suffering from severe, chronic constipation. The dose for children with mild constipation is one teaspoon to one tablespoon given at bedtime. For children with severe encopresis, higher doses can be used; check with your doctor for the proper amount for your child. It generally works in six to eight hours. Mineral oil tastes awful, but please do *not* hold your child's nose while getting him to swallow it. If he chokes on it, he could develop a nasty pneumonia, and you may end up with a gasping, choking child whose symptoms are far worse than simple constipation. Side effects include diarrhea, abdominal pain, watery stools, and distention.[7] Mineral oil is much more palatable if it is cold, emulsified, or mixed with frozen yogurt or ice cream, peanut butter, or chocolate syrup. Despite concerns that prolonged use of mineral oil may lead to loss of vitamins and minerals in the stool, a study that actually measured the levels of beta-carotene, vitamin A, and vitamin E showed that even prolonged (four months) use of mineral oil did not lower blood levels of these vitamins. If your child has severe, chronic constipation, mineral oil can be safely given for several months to get things back on track.

Bowel stimulants work directly on the intestines to stimulate the waves of contractions that move food and waste through the bowel. Stimulants generally produce results in 6 to 10 hours. They can work in as little as an hour if given as suppositories. Bowel stimulants can lead to dependence if they are used regularly.

They can be very powerful and cause cramping pain and diarrhea if too much is given. Several kinds are available. I do not recommend starting with these in children. Talk with your health professional.

Bisacodyl (*Dulcolax*™) is available as tablets or suppositories. Like fiber, it should *not* be used in children who may have blocked bowels or appendicitis, because it could aggravate these conditions. *Yellow phenolphthalein* is a bowel stimulant that used to be the active ingredient in several widely used non-prescription stool softeners. However, it was banned by the U.S. government in 1997 because large doses caused cancer in lab animals. Laxatives that used to contain this compound have been reformulated to rely on other bowel stimulants such as senna (ExLax™) and bisacodyl (Correctol™).

Cascara sagrada (e.g., *Nature's Remedy Natural Vegetable Laxative*™) is a natural bowel stimulant. Cascara is secreted in breast milk, so if you take it while you are nursing, your baby may get diarrhea. *Senna* (*Senokot*™, *Fletcher's Castoria*™, and others) is another potent natural bowel stimulant, which is safe even for nursing mothers. *Castor oil* is an old-fashioned natural bowel stimulant extracted from castor beans; several brands contain sweeteners and flavorings that may make castor oil more palatable. Frequent use of these natural bowel stimulants may result in dependence on them. Do *not* use them for more than two or three days in a row without professional consultation.

Magnesium salts are the active ingredient in *Milk of Magnesia*™ (magnesium hydroxide) and *Epsom salts*. Many families keep them on hand to treat indigestion, heartburn, and upset stomach. By drawing water into the intestines, magnesium salts soften the stool and increase bowel contractions. They generally work in one to three hours. The dose of Epsom salts for children is one to two teaspoons in a glass of water. Magnesium salts can impair the absorption of other prescription medicines. If your child is taking a prescription medicine, check with your doctor or pharmacist to make sure that Milk of Magnesia will not interfere with it. Ibuprofen and aspirin can interfere with magnesium's effect on the bowel. Do *not* go overboard on magnesium; too much can not only cause diarrhea, it can be fatal.

Maltsupex™ is a natural extract of barley malt, which helps draw water into the bowel, softening stools. Like fiber laxatives, Maltsupex™ should be taken with plenty of liquids. *Glycerin suppositories* (*Fleet Babylax*™) are among the safest remedies for constipated infants. They work within half an hour by stimulating the defecation reflex. *Sorbitol syrup* (70 percent solution) is a sweetener that has laxative effects; sorbitol is found naturally in fruits like apples, pears, peaches, and prunes. You may also find sorbitol in "sugarless" gums and other foods as a sweetener.

There are many varieties of nonprescription laxatives. The similarities in names can be confusing. Some products made by the same manufacturer contain different kinds of ingredients. *Read labels carefully*. To treat mild constipation start with fiber; consider Maltsupex™ or Milk of Magnesia™ because they have great safety records. Do not use bowel stimulants (synthetic or natural) without professional consultation; relying on bowel stimulants for more than a week could result in dependence.

Enemas

The vast majority of constipated children can be effectively treated without resorting to enemas. However, children who have severe stool buildup or impaction may need enema therapy to clean out the problem before starting on other treatments to restore normal bowel patterns. Enemas are *not* helpful for children who have a decrease in bowel movements because they have not eaten. Once impacted stool has been evacuated from the rectum, enemas do not have any advantage over oral laxatives in maintaining normal bowel function. I do *not* recommend them except for children whose severe constipation has not responded to other therapies.

Pediatric enemas contain a variety of ingredients. The most widely used are *Pediatric Fleet*™ products. Do not use phosphate enemas repeatedly except under the supervision of a health care professional, because they can result in abnormalities in the body's normal salt and water balance. Homemade enema solutions include milk and molasses, detergent, milk alone, coffee, and a variety of other substances. These solutions have *not* undergone rigorous safety evaluations and may cause severe side effects. Do *not* give your child coffee enemas, because they can also upset the body's delicate balance of salt and water if used repeatedly.

Many specialists in treating severe constipation recommend a combined approach: enemas initially to clean out impacted stool, followed by a high-fiber diet, daily mineral oil, a stool softener (such as Colace™) to help retrain the bowel, and counseling to address possible causes and consequences of this problem. Despite concerns that this regimen might deplete blood levels of vitamins and impair growth, with proper medical supervision these combination regimens are safe and effective.

Dietary Supplements and Herbs

Vitamin C overdoses can cause diarrhea. You can safely give vitamin C supplements to a child who is constipated, but the primary treatment for constipation should be diet, not supplements.

Magnesium supplements may help improve regularity, but excesses can be fatal. Ask your health professional for advice on dosing, and focus on offering magnesium-rich foods.

Avoid aluminum. Aluminum can make constipation worse. Aluminum is found in some antacids in the form of aluminum hydroxide. Read labels before you give your child an antacid for an upset stomach related to constipation. Also avoid cooking acidic foods (such as tomato sauce) in aluminum saucepans, because the acid tends to leach aluminum into the food.

HERBS TO TREAT CONSTIPATION

- Cascara sagrada (buckthorn)
- Senna
- Aloe
- Dong quai
- Others

Do not rely on herbs or medicines that are bowel stimulants until after you've tried dietary therapy, increased water, increased exercise, and regular time on the toilet. Do *not* use herbal laxatives for more than two days without consulting your health care professional. Chronic use can lead to dependence on the laxative for normal function and disturbances in blood chemistry.

Cascara and *senna* are powerful laxatives. Senna is one of the safest and most physiological of all laxatives. It generally works in eight to ten hours; bedtime doses should achieve results by breakfast. They are sometimes combined with fennel and peppermint, two traditional tummy soothers that may ease the cramps provoked by strong bowel stimulants. *Aloe juice,* prepared from the inner surface of aloe leaves, is a more potent cathartic (bowel stimulator) than *aloe gel,* which is typically used as a skin soother rather than a laxative. Aloe tends to cause more cramping than other herbal laxatives. Overuse of aloe and cascara has been linked with bowel cancer.[8] Treatments are not necessarily safer just because they're natural.

Dong quai (from the root of Angelica si-nensis), which is often used to treat menstrual cramps, is also a mild laxative. Other herbs that may help relieve the intestinal spasms that sometimes accompany constipation are chamomile, ginger, lemon balm, licorice root, mulberry, and wild yam. Herbs that may help soothe an irritated intestine are marsh mallow root and slippery elm bark tea. None have been tested as constipation remedies for children.

BIOMECHANICAL THERAPIES: MASSAGE, SURGERY

Massage

Despite its widespread use, massage did *not* prove helpful in one scientific study of its effectiveness in treating constipation.[9] More recent studies have had more promising results,[10,11] but do not suggest that massage should replace fiber, fluids, and increased activity in promoting normal bowel function. Massage is safe and can improve communication and attachment between parents and their children. As long as you are gentle, there is no harm in giving your child a tummy rub when she is constipated.

Surgery

Surgery is *not* needed to treat garden-variety constipation. It *is* necessary for illnesses such as Hirschsprung's disease, in which part of the bowel has an abnormal nerve supply. It is also lifesaving in cases of intestinal obstruction. If you suspect either of these problems, see your doctor.

BIOENERGETIC THERAPIES: ACUPUNCTURE, THERAPEUTIC TOUCH/REIKI, HOMEOPATHY

Acupuncture

Acupuncture is *not* a first-line treatment for pediatric constipation. Given its known effects on the entire GI tract, it would not be surprising if acupuncture was helpful for children with constipation resulting from ongoing treatment with narcotics for chronic pain.

Therapeutic Touch/Reiki

Although there are no scientific studies specifically evaluating the effectiveness of Therapeutic Touch/Reiki and other types of healing as therapies for constipation, I have used them for several hospitalized patients with excellent results. All patients who had complained of constipation who were treated with Reiki or Therapeutic Touch (usually for other medical problems as well as constipation) usually had a bowel movement within three hours of the treatment. On the other hand, all of these patients were also receiving numerous other therapies as well, so it is hard to give the credit entirely to "energy healing." However, because this type of healing work is so valued by patients and so safe, I continue to use it, pending the results of formal scientific studies.

Homeopathy

There are *no* scientific studies evaluating the effectiveness of homeopathic remedies in treating adults or children suffering from constipation. Commonly used remedies include *Alumina, Bryonia, Calcarea carb, Lycopodium,* and *Nux vomica* (especially if the child has vomiting as well as constipation). Although they haven't been formally tested, they are probably safe.

WHAT I RECOMMEND FOR CONSTIPATION

PREVENTING CONSTIPATION

1. *Lifestyle—nutrition.* Start your infant's life out right by breast-feeding. If you are feeding your baby formula, you may need to give him water supplements in the summer. Feed your toddler or older child a diet rich in whole grains, fresh fruits, vegetables, yogurt, and plenty of water. The goal in terms of number of grams of dietary fiber is: age in years + five = total grams of dietary fiber per day. Make sure your child gets plenty of fluids.

2. *Lifestyle—exercise.* Encourage your child to get 60 minutes of exercise or outdoor play.

3. *Lifestyle—mind-body.* Give positive rewards for positive results. Stick to a regular schedule for meal and potty time. Avoid embarrassing or punishing your child for "misses."

4. *Biochemical—medications.* Avoid codeine-containing pain relievers and antacids that contain aluminum.

When to Seek Professional Help for Constipation

- No stools for four or more days despite home remedies described below

- Swollen or distended belly

- Vomiting as well as constipation

- Pain with passing a stool or pain in the abdomen, especially on the lower right side

- Blood in the stool

- Dark black stools

- Constipation alternating with diarrhea

- Frequent, painful, or urgent urination

TREATING CONSTIPATION

1. *Lifestyle—nutrition.* For constipated *infants* try a little extra water, fruit juice (apple, pear, black cherry, or prune juice), or the juice from stewed figs, dates, raisins, or prunes. For constipated *toddlers* or *older children*, start by decreasing cow's milk and increasing fiber intake: high-fiber crackers and cookies, more fruits, vegetables, dried fruit, beans, popcorn, flax or psyllium seeds; add pear, apple, prune, or black cherry juice; consider sugarless gums or candies that contain sorbitol as treats; try a teaspoon of flax seeds on bran cereal or whole grain toast; add high-fiber crackers as treats; include probiotic-rich foods like yogurt. If you serve processed foods, look for labels that indicate three or more grams of fiber per serving. Make sure she's getting plenty of fluids. Try magnesium-rich mineral water or coconut water. Have her evaluated if you suspect cow's milk sensitivity is the problem.

2. *Lifestyle—exercise.* Encourage 60 minutes of exercise daily.

3. *Lifestyle—mind-body.* Give positive rewards (such as stickers) for positive results. Maintain a regular potty time schedule. Hypnosis and biofeedback can help in severe cases.

4. *Biochemical—herbs and supplements.* Try laxative herbs such as cascara or senna (Fletcher's Castoria or Senokot) given in divided doses over the course of the day. Do *not* use for more than two days without seeing your health care professional. Consider offering probiotics. Coffee can be offered occasionally.

5. *Biochemical—medications.* Try fiber-based medications (Metamucil), stool softeners (Colace), or mineral oil. For infants, try a glycerin suppository. If all else fails and your

child is severely blocked up, consider a Fleet's enema to clear out the impaction in the rectum initially.

Susan had impacted stool in her rectum. Eric began giving her two tablespoons of mineral oil at night (mixed with chocolate frozen yogurt) and in the morning (followed by pear juice). We gave her Colace to take twice daily as well to soften her stools. Her mom also began making fruit-bran-yogurt smoothies for breakfast and sending Susan to day care with a container of pear juice and bran crackers for snacks. Susan had a sticker chart that she filled in with a new sticker each time she had a bowel movement. Six months later, she was regular as a clock on her new high-fiber diet without any medications or herbal treatments. She never needed another enema or referral for biofeedback. Her parents were delighted and felt that the changes in Susan's diet had spilled over to theirs, benefiting the whole family.

13
COUGH

Nine-month-old Philip Koshi started coughing two days ago when he developed a cold. He sounded like a seal barking, and his cough was worse at night. His parents started to bring him to the emergency room at 2 am, but on the way, he improved so much they thought the doctors would think they were crazy, so they returned home. Philip seemed better this morning, but his parents wanted to make sure he didn't have another night like that one.

Terri Nguyen was a pale, tired-looking fourteen-year-old who had been coughing for several days. She had recently returned from a trip to visit her grandmother in Vietnam. The last two days she had noticed more phlegm and had developed a fever. This morning she'd coughed up some yellow-green mucus, flecked with blood. She wanted a shot of antibiotics so she could participate in a swim meet the next day.

Kris Masterson brought in his fourteen-month-old daughter, Emily, who had been coughing off and on for weeks, ever since she had bronchiolitis. Emily's lungs seemed to be extra sensitive; every cold she caught brought on more coughing spasms. Chris wanted to know if this was normal or if he should worry about cystic fibrosis.

Coughing is the body's natural method for clearing the airway. Anything that irritates or blocks the air passages stimulates a cough reflex. Generally the cough happens for a good reason. It is a mistake to suppress a cough unless you know for certain what is causing it.

TAKE YOUR CHILD TO A HEALTH CARE PROFESSIONAL IF SHE IS ...

- Persistently coughing for more than a week despite home therapies
- Coughing so hard she can't catch her breath, eat, or drink
- Coughing up blood
- Wheezing
- Breathing very fast
- Turning blue in lips or fingernails
- Lethargic, difficult to arouse
- Complaining of pain in the chest
- Complaining of headache or facial pain as well as coughing
- Has a fever over 103.9°F

Coughing has many causes. If you think your child's cough is due to asthma, skip to Chapter 7.

CAUSES OF COUGH

- Infections (see Chapter 10)
- Reflux or aspiration
- Allergies, smoke, irritants (see Chapter 5)
- Asthma (see Chapter 7)
- Cold air
- Reflex, nerves, habit
- Rare and serious illness—cystic fibrosis, heart failure, and others

Most coughs in children are due to colds and tend to get better over the course of a few days. However, some viruses such as pertussis, influenza, and respiratory syncytial virus (RSV) cause coughs that last for weeks or even months. Other cough-causing infections include croup, sinus infection, bronchitis, and pneumonia.

Coughs in infants may be due to a condition called *gastroesophageal reflux disease* (GERD). Reflux means that some of the baby's stomach (gastric) contents come back up the esophagus and a little is inhaled into the airway, causing cough and irritation.

Aspiration is when a piece of food goes down the wrong way and gets in the lungs.

Other cough causes are: allergies, cigarette smoke and other irritants, asthma, cold air, and even heart failure (when fluid from the heart backs up into the lungs). Sometimes tickling the ear canal (when we look inside to see if there is an ear infection) provokes the cough reflex. Most adults have experienced a nervous cough. Children can develop a chronic coughing habit following an illness in which their cough was rewarded by extra attention.

Dry coughs are usually due to irritants, allergies, a foreign body (such as a piece of food that got stuck down the wrong way), or asthma. *Rattle-like coughs* stem from phlegm in the back of the nose or throat and are usually due to colds or sinus infections. *Productive coughs* indicate that mucus is present in the airways and needs to be cleared. Most children under the age of eight swallow their mucus instead of spitting it out. The swallowed mucus usually passes through the intestines uneventfully. However, if a lot of mucus is swallowed, it can irritate the stomach, producing nausea and even vomiting.

Less common causes of cough include having a soft, floppy airway (tracheomalacia), an inflamed esophagus (eosinophilic esophagitis), abnormal blood vessels that circle the airway (vascular ring), and genetic problems like cystic fibrosis.

DIAGNOSING COUGHS

- Symptoms
- Physical examination
- Chest X-rays
- Other tests

Most of the time, the reason for the cough can be determined by its characteristics, other symptoms (such as fever and wheezing), and a physical examination. *Croup* coughs are dry, worse at night, and usually affect children between the ages of six months and three years during the fall and winter. Croup is caused by a virus. The cough sounds like a seal barking. It is relieved by mist, steam, or going out in the cool night air. Croup sounds terrible as the child struggles for breath and then coughs and coughs and coughs. The child's sudden improvement on the way to the emergency room is embarrassing for many parents who were sure their child was on the brink of death, but it is a completely normal response to the cool night air. If the child doesn't improve on the way to the doctor's office or hospital, medical treatments are effective. Croup usually lasts for three to four days and is worse on the second and third nights of the illness.

Philip Koshi's cough was due to croup. Emily Masterson had the typical, unfortunate, prolonged cough that follows a bout with respiratory syncytial virus (RSV). Though you might think that a cough from bacterial pneumonia would be worse, it usually resolves with a few days of antibiotics. Viruses, on the other hand, can damage the cells lining the air passages, leaving the child's lungs weakened for weeks or months. Kris was right to think about cystic fibrosis (CF). Infants with recurrent coughs or pneumonia (especially infants who are not growing well) should be tested for CF. Emily's tests turned out to be negative; she did not have CF. Emily eventually overcame her cough.

Sometimes an X-ray of the chest or sinuses is helpful. Less commonly, blood tests (to check for white blood cells fighting pneumonia) and skin tests (to check for tuberculosis [TB]) are needed. If the child is old enough to spit out the phlegm he coughs up, it can be checked for bacteria and TB.

Terri Nguyen's cough sounded like pneumonia because of her fever, cough, fatigue, and blood-flecked yellow-green sputum. We did an X-ray of her chest, took some of her sputum to the laboratory for a culture, and did a skin test for tuberculosis before starting her antibiotic treatment.

WHAT'S THE BEST WAY TO PREVENT COUGHS?

There are things you can do to decrease the likelihood that your child will develop a cough.

PREVENTING COUGHS

- Avoid cigarette smoke
- Immunize against pertussis (whooping cough), pneumococcus (pneumonia), and influenza
- Treat underlying illnesses
- Minimize exposure to sick children; use good hand hygiene; avoid fatigue
- Maintain healthy omega-3 fatty acid, vitamins A, C, and vitamin D levels

Avoid exposure to cigarette smoke. Smoke irritates the airways and makes even healthy kids cough. Do not smoke and do not allow others to smoke around your child, especially in enclosed spaces such as your home or car.

Make sure your child is *immunized against pertussis* (whooping cough). Despite its detractors, the highly effective pertussis vaccine has practically eliminated the old scourge of whooping cough. Pertussis can be fatal to infants and young children; those who survive often have lifelong lung damage.[1] The new acellular pertussis vaccine has about 90 percent fewer side effects than the old vaccine. Whooping cough epidemics have reemerged in countries where governments stopped pertussis immunization programs due to public fears about the vaccine. Immunization is very effective in preventing this potentially fatal illness during childhood. Protection wanes over the years. By twenty years of age, 90 percent of those who were fully immunized as children are again susceptible to pertussis. Nowadays, the biggest outbreaks of pertussis are among teenagers and adults. Most adults who get pertussis are only moderately ill and never even go to the doctor, but they can easily spread the condition to young unvaccinated children. This is why pertussis will probably never be completely eradicated and why childhood immunization programs must continue. There are now national campaigns to re-immunize adults to eliminate the reservoir of whooping cough.

I caught pertussis from one-month-old Dedra Shapiro (who hadn't yet received her first immunizations) during my pediatric residency training. Dedra was hospitalized with a severe cough. Soon after admission she stopped breathing. I immediately began mouth-to-mouth resuscitation. She did well, but an hour later her laboratory tests came back positive for pertussis. Two days later, I was coughing and had positive tests for pertussis. I took antibiotics for a week and continued to cough for six months. Even years later, my lungs were extra sensitive to infection. Like most young infants, Dedra caught pertussis from a parent who was sick with what he thought was simple bronchitis.

Vaccines are also available to prevent two other coughing diseases. The pneumococcal vaccine (*Prevnar*) is available for babies as young as two months old. It helps protect against the bacteria that is the most common cause of ear infections, sinus infections, and pneumonia. The *influenza* vaccine is available for babies six months and older. Many physicians, including me, believe it is a good thing for all eligible kids to get this vaccine to prevent illness in children and missing work for parents. I'm especially in favor of kids getting it if there's anyone at home who has a chronic illness, such as asthma, that could get really sick from influenza. The flu vaccine must be taken every year because different strains of the virus appear each year.

Treat any underlying allergies and asthma to reduce the risk of asthma attacks and the coughing that accompanies them.

As much as you can, minimize your child's exposure to other children who are *sick*. The more time spent with young children, the more exposure to infections. If your child is in day care, select a situation with fewer rather than more children, if possible.

Also remember that good hygiene (*handwashing*) and preventing fatigue through *adequate sleep* help reduce the spread of viruses and ensure optimal immune function. Knowing that we are loved and connected to others helps promote healthy immune function, too, so those hugs don't just feel nice, they actually help prevent symptoms from viral illnesses!

Good nutrition is also essential for optimal immune function and preventing viral illnesses that cause coughing. Maintaining healthy *vitamin A* and D levels is important for preventing viral illnesses. Low levels of vitamin A are associated with an increased risk of both pneumonia and diarrhea in children in developing countries who have poor nutrition; supplements help lower their risk of infection. On the other hand, Vitamin A supplements in healthy kids may actually increase the risk of pneumonia. Vitamin C may help prevent coughs due to colds in children but has no appreciable benefits for teens and healthy adults.[2] A low vitamin D level in pregnancy is a risk factor for infant coughing illnesses. Children with low *vi-*

tamin D levels have an increased risk of developing coughing-related illnesses, too.[3] Giving children vitamin D supplements to normalize their vitamin D levels helps decrease the risk of respiratory illnesses and coughing.[4] Consider having your child's vitamin D level checked and offer supplements to bring levels into the optimal range (above 30 ng/mL). *Omega-3 fatty acids* are essential fatty acids used by our brains and our immune system for healthy function. Babies whose mothers consumed omega-3 fatty acid supplements during pregnancy had fewer colds than babies whose mothers took placebo supplements.[5] Babies whose formula included supplemental long-chain fatty acids had fewer respiratory illnesses than babies fed regular formula.[6] Early supplementation with omega-3 fatty acids helps reduce the severity of coughing illnesses in children.[7]

What Is the Best Way to Treat a Cough?

Treatment should always aim at helping the child's body heal the underlying illness rather than just suppressing the cough. Coughs can indicate a serious problem that should be treated (such as asthma or pneumonia); nighttime coughs can also be very annoying and lead to sleep loss for the whole family. On the other hand, coughing may be the body's best defense and protection against pneumonia. Let's tour the Therapeutic Mountain to find out what works for most coughs; if you want to skip to my bottom-line recommendations, flip to the end of the chapter.

Lifestyle Therapies: Nutrition, Exercise, Environment, Mind-Body

Nutrition

Kris was frustrated with Emily's chronic cough. He'd read that milk made mucus worse and should be avoided when a child has a cough or cold. Millions of Americans avoid dairy products when they have a cough or cold. Does milk really make mucus worse?

An Australian study divided 169 adults into two groups: one was given a milk drink, and the other was given a soy drink disguised so as to be indistinguishable in appearance or taste from the cow's-milk drink. Before the study started, nearly half of the adults believed that milk made mucus worse. *Both* test drinks (cow's milk and soy) made some subjects' tongues and throats feel "coated" and made their saliva feel thicker.[8] So the culprit may not be the cow's milk itself so much as any beverage that is thicker than juice or water. Milk does *not* increase mucus production. For children with milk allergies, drinking milk may cause cough, but cough is seldom the only symptom of milk allergy.

Do give your child plenty of *fluids* when she has a cough. Fluids help soothe irritated airways and help clear the bacteria, viruses, or irritants that are causing the cough. Fluids also help keep phlegm loose so it is easier to cough out.

Feel like *fish*? Those who eat fish regularly (more than twice weekly) have a lower risk of having a nighttime cough.[9] I'm not aware of any studies testing how well fish fixes a cough once you have one, but fish contains fatty acids known to combat the kind of inflammation that triggers many coughs. I recommend fish as part of a healthy diet.

A favorite British remedy is *onion-honey* cough syrup. This is made by combining two to three cups of chopped onions with one-half cup of honey and cooking slowly for two or three hours over low heat. This combination is safe and can be given by teaspoon every one to two hours as needed; but when I tried it, I decided I'd rather have the cough than swallow the stuff.

Several studies suggest that *honey alone* is enough to help ease a cough;[10] in fact, it is as effective as the most widely used cough medicine, dextromethorphan.[11] Another tasty combination for dry, tickly coughs is *honey and lemon.* You can combine them as syrup or add a bit of each (to taste) to hot water or an herbal tea. It tastes much better than garlic or onion remedies, too. There are no studies comparing honey alone to honey with lemon or onions; let your cultural preferences be your guide. To avoid the risk of botulism, do *not* give honey remedies to children less than one year old.

Exercise

Despite her fatigue, Terri wanted to participate in a swim meet the following day.

When the body is fighting a serious infection, it needs all of its reserves to win the battle. I generally recommend that kids take it easy until their fever and cough have been gone for twenty-four hours before resuming their normal activities. Even then they should take it slow and be gentle with themselves. When your child is resting, prop her head on a pillow or have her lie on her side to minimize postnasal drip, the source of many a nighttime cough. Ambitious children may feel well enough to walk around the house when they're sick, but find that they tire and their cough returns when they resume more vigorous activities. Children tend to cough more with exercise, especially when exercising in cold, dry air (such as speed skating or sledding).

I advised Terri to sleep in and take naps rather than push herself to compete the following day.

If your child has a wet, productive cough, you can help secretions drain with a simple exercise. Have your child lie facedown on a bed and slide forward until the head and chest are hanging down off the bed. Your child can rest her head and arms on the floor or on a pillow on the floor. Being upside down helps the mucus drain out. In medicine, this is called *postural drainage* because it is the hanging-down posture that helps the lungs drain. It may also make your child feel light-headed, so don't do it for more than five to ten minutes at a time. After the child has been upside down for a minute or two, encourage her to cough to help bring out the mucus and phlegm. Studies of children with severe coughing due to cystic fibrosis have shown that this simple exercise results in a fivefold increase in the amount of mucus cleared compared with simply resting.[12]

Environment

Do not smoke and do not allow others to smoke around your child. Tobacco smoke, wood smoke, and air pollution aggravate coughing. Keep your child warm, away from cold drafts. If you keep the window open in your child's bedroom to provide fresh air, make sure she has plenty of covers to prevent her from feeling chilled.

Although mist or steam may be helpful in treating a child with a cough, living in a damp house with mold growing on the walls is not. Coughs are 80 to 90 percent more common among children living in homes parents describe as *damp or moldy* than among children in homes parents say are without damp or mold. Nasty mold and irritating dust mites prefer to live in damp environments. Remove mold thoroughly with a 10 percent bleach solution. To discourage mold and dust mites from living in your home, consider using a dehumidifier, vacuum regularly, and use an electrostatic air filter.

On the other hand, very *dry* air aggravates most coughs because it dries the secretions that normally soothe the air passages in the throat and lungs. *Mist, steam,* and *vaporizers* increase the moisture in the air passages and help loosen secretions. Despite the lack of controlled trials evaluating its effectiveness, mist is recommended by most health professionals because of its history of helping children with croup. One of my favorite home remedies for croupy kids is to run the hot water in the shower until the bathroom is very steamy. Then sit with the child on your lap in the steam and tell him stories or sing songs.

ENVIRONMENTAL THERAPIES

- Avoid smoke.
- Avoid damp, moldy environments.
- Try mist or steam.
- Suction nose with bulb syringe.

If you use a vaporizer, clean it regularly to prevent a buildup of mold or fungus in the unit. A seldom-cleaned print shop humidifier harboring fungi and bacteria was responsible for a serious coughing illness in sixteen out of twenty-eight workers in a shop.[13] Be careful with steam vaporizers to avoid accidental burns to curious toddlers.

If your child's cough is due to a runny nose (postnasal drip) and he is too young to blow his nose, you can try *suctioning* his nose *with a bulb syringe*. Suctioning may need to be repeated every few hours. If your child's secretions are too thick to be removed easily with a bulb syringe, place a few drops of salt-water solution (one-half teaspoon of salt in eight ounces of water) in each nostril before suctioning. Do one nostril at a time so your child doesn't feel as if he's drowning.

Mind-Body

As with any illness, it is helpful for the parent to remain calm and stay with the child to reassure him. Extra attention, such as reading stories, will help distract your child from whatever is ailing him and remind him that he is loved. Crying and being upset increase your child's need for oxygen. This is not a problem under normal circumstances, but if your child is having trouble breathing, crying may make matters worse.

Coughs often keep everyone in the house awake, especially as parents lie in bed wondering if their child is going to stop coughing or even stop breathing. If your child is having a hard time breathing, you and your spouse might want to take turns staying with him. If you know that someone is responsible, you can more easily relax and sleep yourself, so that you are better able to care for your child when your "shift" comes.

Some coughs are simply a kind of bad *habit,* which lingers on long after the initial cause for the cough is gone. In children with asthma, coughing can become a kind of conditioned response to any kind of stress. This kind of cough is dry, harsh, and *very* frequent; it usually stops when the child falls asleep. Children who have these coughs don't usually have any other behavioral problems or emotional problems, but they can often be helped with behavioral techniques, such as relaxation, self-hypnosis, and guided imagery.[14]

Hypnosis has also proven useful in treating the child with a habitual cough. An eleven-year-old boy who had such a severe, persistent cough that he missed a month of school was able to stop coughing with the help of a psychologist who taught him to use mental imagery and self-hypnosis.[15] The imagery involved characters from *Star Wars* who had a special medicine that would eliminate the cough. His cough quickly subsided, and he was able to return to school. When coughing recurred, he returned to his *Star Wars* companions, whose imaginary ministrations restored his health.

Another behavioral technique is to simply *count and record* the number of coughs per half hour. This is done several times over the course of the day. Sometimes the coughing will decrease with counting alone. You can also give your child stickers or other rewards for reducing the number of times she coughs per hour. As she achieves the goal, praise her, reward her, and set a new goal.[16]

BIOCHEMICAL THERAPIES: MEDICATIONS, HERBS AND OTHER DIETARY SUPPLEMENTS

Medications

There are three main types of cough medicines: cough suppressants, expectorants (to help loosen secretions), and throat soothers such as cough drops.

There is good news and bad news about *cough suppressants*. The good news is that you can save a lot of money. The bad news is that none of them are very helpful, which is why you can save your money. The two most commonly used cough medicines (*dextromethorphan,* the DM ingredient in many nonprescription cough medicines, and *codeine,* the most common prescription cough medicine) are no more effective than honey in reducing coughs in children under 12 years old. Cough medicines (aside from specific medicines for asthma) are *not* any better than placebos for young children, and they may have serious side effects such as irritability, fussiness, lethargy, and high blood pressure. Teenagers sometimes abuse dextromethorphan as a cheap way to get high. *The American Academy of Pediatrics Committee on Drugs says that cough medicines should not be used for children.*

Expectorants supposedly loosen secretions so they are easier to cough out. The most common one is *guaifenesin* (Robitussin®). Although guaifenesin helps thin secretions, it is no more helpful than placebo syrup in reducing the frequency or severity of coughs. I do *not* recommend it.

Cough drops work mostly by stimulating saliva, coating and soothing an irritated throat. Any hard candy will do the same thing; lemon and other citrus flavors are particularly potent saliva stimulators. Do *not* give cough drops or hard candy to children under four years old who might choke on them. Cough drops with menthol or eucalyptus oils also help a congested nose feel less stuffy. Sucking on a vitamin C lozenge also stimulates saliva and may help soothe an irritated throat. Hard candies and cough drops typically contain sweeteners that can damage teeth. Be sure your child brushes his teeth after using a cough drop.

The ineffectiveness of cough suppressants and expectorants does not mean that other medicines for specific types of cough are not helpful. *If your child is coughing because of croup, asthma, or allergies, use the treatments your health care professional has recommended and any others you have found to be helpful.*

For example, children suffering from severe croup that has not responded to home therapies can be helped by two prescription medications.

CROUP MEDICATIONS

- Racemic or L-epinephrine (epi)
- Steroids

Racemic or L-epinephrine is given by a mist machine (called a nebulizer) in a doctor's office or emergency room. Epinephrine helps shrink the swollen and inflamed blood vessels that line the airways, blocking airflow and causing coughs. Racemic epinephrine can be lifesaving. It works within minutes, providing rapid relief of breathing difficulties, but it only works for one to two hours, so it may need to be repeated several times. Rather than risk sending home a child who may need another treatment within an hour, many doctors automatically hospitalize a child who needs one racemic epinephrine treatment to be sure additional treatments are readily available if needed. Like caffeine (to which it is related), racemic epinephrine can cause a rapid heartbeat, shaky hands, agitation, and trembling. These side effects quickly disappear as the medicine wears off.

Steroids help reduce inflammation and swelling. Although they can have serious side effects when used over long periods of time (weeks or months), steroids are safe when used on a short-term basis (less than five days). The steroids that are used for children's illnesses are *not* the same kind of steroids athletes use to build muscles. Steroids given by mist machine (nebulizer) or with a facemask and spacer provide significant symptomatic improvement for several hours. Steroids are also effective when given by injection. Though a shot is more painful than a mist treatment, you can be sure your child is actually receiving the medication; some children do not cooperate with the mist machine, and the parent ends up getting most of the dose! Steroid syrups taken by mouth are also helpful, even in children suffering from severe, life-threatening croup. A short round of steroid treatment might very well spare you several days in the hospital.

We started Philip Koshi on a three-day course of a steroid syrup to help prevent more serious symptoms and a possible hospitalization. His parents were pleased to tell me at his next checkup that his croup symptoms rapidly improved, and there were no more midnight trips to the hospital.

Antibiotics have *no* role in the treatment of coughs due to viruses. However, if your child's cough is caused by bacterial pneumonia, bronchitis, or a sinus infection, antibiotics such as erythromycin, azithromycin, clarithromycin, and doxycycline can help clear the underlying cause. If you suspect that your child has one of these illnesses, have her evaluated by a health care professional.

Children whose lungs have been damaged by viral or bacterial infections early in life may need asthma-type medicines to manage their symptoms for several years to keep them from coughing and wheezing. If your infant or toddler has had a cough lasting for more than two weeks, see your doctor to see if prescription medicines might help.

Herbs

Different kinds of herbs are used for different kinds of cough, but *only* menthol (extracted from mint) has proven useful in the lab in terms of reducing irritant coughs. There are lots of historical anecdotes and rich cultural traditions supporting a variety of herbal remedies. There are infection-fighting herbs, expectorants, herbs that soothe an irritated throat, herbs to warm and stimulate a child who is fatigued and chilled, and herbs to sedate a child who has been kept awake by a cough. Herbal remedies are usually taken as tea. Doses vary by age, and are usually made up according to the "seat of the pants" rule. *There have not been any studies showing that any herbal remedies are useful treatments for coughing children.*

"SEAT OF THE PANTS" RULE FOR DOSES OF HERBAL TEA FOR COUGHS

- Children *under a year*: no more than one teaspoon three to four times daily
- Children *one to three years old*: up to one ounce four times daily
- Children *four to six years old*: up to two ounces four times daily
- Children *seven to twelve years old*: up to three ounces four times daily
- *Teenagers and adults*: three to four ounces every four to six hours

The heat from hot tea helps increase circulation to the throat, hastening healing to the whole area.

HERBAL COUGH REMEDIES

- *Infection-fighting herbs*: eucalyptus, garlic, hyssop, plantain, thyme
- *Herbs to stimulate the immune system*: borage, dandelion root, echinacea, garlic, marigold, nettles, and wild indigo
- *Herbs to help loosen mucus so that it is easier to cough up (expectorants)*: angelica root, anise seed, cowslip, elecampane, fennel, white horehound, hyssop, mullein, plantain, sage, senega snakeroot, thyme
- *Herbs to soothe a dry cough (demulcents)*: anise, comfrey, elecampane, horehound, licorice, lobelia, marsh mallow root, mullein, slippery elm bark, wild cherry bark
- *Herbs to stimulate and warm a weak and chilled child*: anise, cinnamon, cloves, fennel, ginger, ginseng, hyssop, sarsaparilla root, thyme
- *Relaxing herbs to help a coughing child sleep*: chamomile, catnip, lime flowers

Warning: Although coltsfoot has been used in Europe, Asia, and America for many years as an herbal remedy for coughs, coltsfoot flowers (the most commonly used part of the plant) are highly toxic to the liver and may cause cancer. Senega snakeroot can cause severe stomach upset. Overdoses of licorice tea can adversely affect the body's salt and water balance. No scientific studies document the effectiveness of any of these remedies in treating children with coughs.

POULTICES FOR COUGHS

- Mustard, garlic, or onion
- Turpentine or camphor
- Castor oil

An historical treatment for chest colds is the *mustard poultice*. Mustard poultices apparently increase circulation to your child's chest, creating a soothing sense of warmth. Do not let the mustard come in direct contact with your child's skin. It is irritating and could cause a burn. This same sense of heat and increased circulation can be achieved with a *garlic* or *onion poultice*. To avoid skin irritation, do not use a garlic poultice for more than twenty minutes at a time. Some herbalists recommend that the garlic poultice be placed over the soles of the feet to draw heat downward.

Other folk remedies placed on the feet to draw the circulation downward are *turpentine* and *camphor*. A North Carolina woman who was my patient during medical school swore that her homemade turpentine poultices were what were really curing her but asked me not to tell her doctor, because she didn't want him to be disappointed in his antibiotics!

Castor oil poultices are also used in many parts of the country and were recommend by the psychic Edgar Cayce as a remedy for many illnesses. Castor oil is very soothing, so it can be applied directly to the skin or as a poultice for the chest, abdomen, or back.

None of these folk remedies has undergone scientific evaluation. There is a huge potential of getting ripped off or getting dangerous products until the FDA starts regulating herbal products more carefully. If your child fails to improve in a day or two or becomes more ill, consult your health care professional to make sure there is not a serious and easily treated condition causing the cough.

Dietary Supplements

Vitamin A protects the mucous membranes of the nose, throat, and lungs, but it is *not* helpful against ordinary coughs. Vitamin A supplements do *not* reduce cough or pneumonia in well-nourished infants or children. Too much vitamin A may *increase* the risk of respiratory infections. If your child eats a healthy diet, he does *not* need supplemental vitamin A.

Vitamin C makes coughing and wheezing less likely, but there are no studies showing that giving children vitamin C helps improve symptoms once they have a cough. On the other hand, there are no contraindications to a diet rich in fruits and vegetables containing vitamin C.

BIOMECHANICAL THERAPIES: MASSAGE, SURGERY

Massage

Millions of parents and children around the world can attest to the healing power of a chest rub when a child has a cough or cold. The old standbys are Vicks VapoRub® and Mentholatum®. Alternatively, consider mixing vegetable oil (five teaspoons) with two to three drops of essential oils of eucalyptus, lavender, pine, or thyme for a pleasant-smelling rub. Place a bit of the oil in your palm, and rub your hands together until they are warm. Massage your child's chest, neck, and upper back with this mixture if this practice is consistent with your family and cultural values.

Surgery

Surgery is rarely necessary *unless* your child has gotten food (such as a peanut, raisin, or piece of hot dog) or a toy down the wrong pipe (aspirated it), and it has landed in the lungs. In this case, bronchoscopic surgery (in which a fiber-optic tube is placed down the airway so that the object can be seen and removed) might be necessary. *If your child aspirates (chokes on) something, take him to an emergency room immediately.*

BIOENERGETIC THERAPIES: ACUPUNCTURE, THERAPEUTIC TOUCH/ HEALING TOUCH, HOMEOPATHY

Acupuncture

In China, coughs have been treated with acupuncture for many years. In one series of patients with different kinds of cough, cupping (a variation of acupuncture treatment) for five to ten minutes was helpful.[17] Another study reported that moxibustion, another non-needle technique, helped children with chronic coughs.[18] However, these studies did not include a control (comparison) group of untreated patients, so it is impossible to tell how many would have improved anyway. I do *not* regularly recommend acupuncture as therapy for common coughs. On the other hand, I've seen a number of patients with cystic fibrosis whose breathing improved remarkably after acupuncture treatments; these patients now request acupuncture whenever they're admitted to the hospital with pneumonia because they *know* it makes them feel better.

Therapeutic Touch/Reiki/Healing Touch

Similarly, there are no studies evaluating the impact of any kind of hands-on healing technique in treating children's coughs. However, I've watched kids with asthma, pneumonia, and even on breathing machines (ventilators) have an easier time breathing when I treated them with these techniques. One young woman who had cystic fibrosis had such severe coughing following a bronchoscopy that the anesthesiologist was forced to put her back under anesthesia to help her breathe again; the woman's mother insisted that the anesthesiologist call me because she'd seen her daughter improve when I'd treated her. Within minutes of starting guided imagery and Therapeutic Touch, the young woman no longer required the powerful medicines to quell her coughing spasms and was able to breathe and talk easily. Was it the Therapeutic Touch, the guided imagery, or just good timing? More research is needed to answer this question for sure; in the meantime, I'm going to continue to offer these hands-on healing techniques because they seem helpful and calming and have no serious side effects.

Prayer

Similarly, despite the lack of randomized controlled clinical trials, I recommend prayer for folks who have values that support prayer. People who pray tend to be healthier and more confident and to experience less suffering. I've seen many children respond to the calm presence of people praying for them. Faith is a balm for the weary spirit and body.

Homeopathy

Commonly recommended homeopathic treatments for *dry, barky coughs* (such as croup) are Aconitum, Belladonna (use homeopathic doses only; more concentrated doses can be poisonous), Ipecac, Phosphorous, Rumex, and Spongia. For *productive (wet) coughs* with much mucus: Euphrasia (eyebright) and Natrum sulphuricum (sodium sulphate) are used. Pulsatilla is recommended for several types of childhood coughs. Homeopathic remedies are very safe, but should not be used as a substitute for proven medical therapies for serious coughing problems like tuberculosis, asthma, or cystic fibrosis.

WHAT I RECOMMEND FOR COUGHS

PREVENTING COUGHS

1. *Lifestyle—environment.* Avoid exposure to cigarette smoke and other irritants. Avoid exposure to sick, coughing children. Discourage cough-causing mold and dust mites by keeping the humidity under 50 percent and thoroughly cleaning your home weekly. Encourage your child to wash his hands before every meal, after playing with other children, and after using the restroom.

2. *Biochemical—medications.* Have your child immunized against pertussis (whooping cough), pneumococcus, and influenza. Treat underlying illnesses such as asthma and allergies. Maintain healthy vitamin A levels with a diet rich in carrots, sweet potatoes, and leafy green vegetables. Maintain healthy vitamin D levels with safe exposure to sunshine and vitamin D dietary supplements. If your child doesn't eat sardines or salmon at least twice weekly, consider an omega-3 fatty acid supplement.

3. *Lifestyle—diet.* Make sure your child eats a diet rich in fish and fruits that contain vitamins A and C as well as essential fatty acids such as omega-3 fatty acids, eicosapentaenoic acid (EPA), and docosahexaenoic acid (DHA).

4. *Lifestyle—sleep and social support.* Avoid fatigue by encouraging regular bedtime and adequate sleep. Let your child know through word and deed that he is valued, loved, protected, cherished, and connected with caring adults to promote healthy immune function.

Take Your Child to a Health Professional If She Is

- Coughing so hard that she can't catch her breath

- Coughing up blood

- Wheezing

- Breathing very fast

- Turning blue in lips or fingernails

- Lethargic

Or If She Has

- Pain in the chest for more than a day

- Headache or facial pain as well as coughing

- A fever over 103.9 degrees F

- A persistent cough that has not improved with home remedies

---

TREATING COUGHS

1. *Lifestyle—nutrition.* Encourage your child to drink plenty of fluids to keep the mucus and phlegm loose so it is easier to cough up.

2. *Lifestyle—exercise.* Encourage your child to rest on her side to reduce coughing from post-nasal drip. To help drain phlegm from the lungs of a child with a wet cough, have her lie with her head and chest hanging down off the edge of the bed for five to ten minutes. Then have her cough hard to clear the phlegm.

3. *Lifestyle—environment.* Try a humidifier, vaporizer, or steam treatment. Mist can be very helpful for croup. Consider adding essential oils of menthol (mint) if these are consistent with your family traditions or culture.

4. *Lifestyle—mind-body.* If you suspect that your child's cough has become a habit, consider hypnotherapy, or have him count his coughs, recording them in a cough diary.

5. *Biochemical—medications, herbs.* Honey can help soothe coughs, but don't offer it to children less than one year old. Hard candy, herbal cough syrups, cough drops, and hot tea may help soothe a throat irritated by coughing. Medications may be needed depending on the underlying cause (e.g., antibiotics for pneumonia). Please don't bother with over-the-counter cough medicines for children less than 12 years old. If your child has croup, he may benefit from epinephrine or steroid treatments.

6. *Biomechanical.* If your child's cough is due to aspiration, he may need bronchoscopy to remove whatever went down the wrong pipe.

7. *Biofield.* If consistent with your family's tradition or culture, you may try acupuncture, Therapeutic or Healing Touch, prayer, or homeopathy.

8. *Patience.* Coughs due to viral infections can last up to six weeks.

14
DIAPER
RASH

During a twelve-month checkup, Carmen Morales asked about a diaper rash her son, Marco, had developed. He had been on antibiotics for an ear infection for the past week and had developed a bright red, irritated rash that was worst on his scrotum and the creases in his thighs. Red patches and spots led up to his belly button. Carmen had been using cloth diapers, changing them five or six times a day. Carmen's mother thought Marco had a yeast infection. Carmen wanted to know if the antibiotics caused the rash; if she should avoid giving Marco bread and other foods containing yeast; whether she should switch to disposable diapers; and what was the best way to prevent and treat a diaper rash.

Very few babies escape infancy without at least one bout of diaper rash. Diaper rashes can be caused by several different things. All cases do better with good hygiene, yet individual management is guided by individual considerations.

WHAT CAUSES DIAPER RASH

- Irritation from contact with stool and urine

- Yeast (*Candida albicans*)
- Less commonly—allergy, eczema, seborrhea, psoriasis, infection

Irritation diaper rashes are pink or red; they look a bit like sunburn. They appear only in the diaper area and are usually more severe in the skin that directly touches the diaper than in skin creases and folds. Irritation rashes occur when the bacteria that are normally present in stool break down the chemicals present in urine, forming ammonia—an alkaline (high pH) irritant to babies' delicate skin. Prolonged exposure to wet, dirty diapers is the chief cause of irritation diaper rashes; many are also complicated by yeast infections.

Yeast diaper rashes are usually bright or dark red and are worst in warm, damp skin folds or creases. Typically, yeast diaper rashes extend out from the main rash in red spots. The yeast that causes these rashes, *Candida albicans,* is *not* the same as brewer's yeast, nutritional yeast, or baking yeast; it is passed from person to person, not from food or soil. Yeast infections are *not* aggravated by eating yeast. Food yeast and infectious yeast are completely different organisms. *Candida* also causes infant thrush (a white rash in the mouth) and vaginal yeast infections. *Candida* yeast can also be carried on mothers' nipples, re-infecting the baby every time he nurses. About 20 to 30 percent of women have *Candida* living in their skin and vagina, and babies who pick it up during the birth process often carry it as well; preliminary studies suggest that treating moms with an antifungal medication such as clotrimazole during pregnancy to reduce Candida helps reduce the risk of preterm delivery and the risk that her baby will develop thrush and *Candida*-caused diaper rash.[1] Applying probiotics topically in addition to the medication seems to be even more effective than medication alone.[2] In fact, some hospitals apply a probiotic (*Lactobacillus reuteri*) to premature babies' skin to help prevent infections.[3]

Candida yeast is usually kept in check by the normal bacteria that reside in babies' stool, intestines, and skin. When a child takes antibiotics, the normal bacteria are killed, making way for *Candida* to take over. Thus, children who take antibiotics risk developing thrush and *Candida* diaper rashes.

Less commonly, diaper rashes are caused by allergies to soaps, detergents, diaper wipe chemicals, the dyes, or other chemicals in highly absorbent disposable diapers, foods, medicines, or fabric softeners.[4] Eczema, seborrhea, or bacterial infections can also cause diaper rashes. Zinc deficiency, histiocytosis, Coxsackie virus, and Kawasaki disease are rare causes of diaper rash.[5] If your baby's diaper rash doesn't respond to simple measures in this chapter, see your health professional to figure it out and treat it together.

Bacterial infections due to *strep* or *staph* bacteria usually start at either the umbilical cord or the anus and spread from there. They tend to be painful and red, and the skin is tight looking. If you think your baby has a bacterial infection, see your health care professional.

WHAT INCREASES THE RISK OF DIAPER RASH?

Several things increase the odds of developing diaper rashes. Diaper rashes are more common among babies who drink formula than those who breast-feed. Rashes are more common in nine- to twelve-month-olds (after they've started eating solids but aren't yet toilet trained) than in newborns or toddlers. They are very common when babies have diarrhea. Diaper rashes are most common in babies who are kept in cloth diapers (especially if plastic covers are used) and least common in babies in super-absorbent disposable diapers. Babies who have more *Candida* yeast in their stool tend to have more frequent and severe rashes. Antibiotics such as amoxicillin suppress the body's normal protective bacteria, allowing yeast to multiply and precipitating diaper rashes. Diarrhea-inducing antibiotics (such as *Augmentin*™) are double jeopardy—reduced protective bacteria and increased dirty diapers.

DIAPER RASH RISKS

- Feeding formula instead of nursing
- Age: nine to twelve months old

- Diarrhea
- Cloth diapers with plastic covers
- Antibiotics, especially those that cause diarrhea

Carmen's mother was right. Marco had a typical yeast diaper rash, which may have been precipitated by the antibiotics he was taking for his ear infection. Carmen did not have to withhold yeast-containing foods.

TREATMENTS FOR DIAPER RASH

Irritant diaper rashes usually resolve within two to three days if treated promptly. *The best treatment for diaper rash is environmental: keeping the diaper changed regularly and keeping the bottom clean and dry.* Let's tour the Therapeutic Mountain to find out what works. If you want to skip to my bottom-line recommendations (so to speak), flip to the end of the chapter.

LIFESTYLE THERAPIES: NUTRITION, ENVIRONMENT

Nutrition

For babies over six months old, some mothers supplement their child's diet with yogurt (with active cultures) when he is taking antibiotics. The healthy bacteria in the yogurt help replace the yeast on the baby's skin, preventing diaper rashes. The yogurt must contain active cultures to have any chance of being helpful. Check the label; many commercial brands do *not* contain active cultures. Some mothers apply the acidophilus or yogurt directly to the baby's bottom. Others buy probiotic supplements at the health food store and make a cream by mixing the supplements with water or petroleum jelly, applying the mixture to the baby's bottom. I have not heard of any side effects from this treatment.

Many parents have heard that excessive sugar in the diet predisposes to yeast infections such as diaper rash. While it is true that diabetic adults are more prone to yeast infections, it is not clear that feeding a child sweet fruits changes his risk of having a diaper rash. Most nine- to twelve-month-old babies do not need and do not eat sugar, yet this is the peak age of diaper rashes. On the other hand, sugar intake *does* increase the risk of tooth decay.

Environment

The most effective treatment for diaper rashes is to keep the baby's bottom clean and dry. The best ways to accomplish this are by:

- Changing the diaper frequently (six to eight times daily)
- Avoiding plastic pants, which reduce air circulation to the diaper area
- Allowing the baby's bottom to be exposed to air

Change your baby's diaper immediately after he has a bowel movement or urinates to minimize the time his tender skin is in contact with urine or stool. If he already has a diaper rash, check his diaper every hour and change it immediately if it is wet or soiled. After cleaning his bottom with plain water or a hypoallergenic wipe, allow him to air dry before you put the next diaper on him. You can lay a diaper or towel under him so he doesn't wet his bed or the floor or wherever he is lying. The best times for air-drying are during naps and right after diaper changes. Some parents make sure their baby's bottom is really dry by using a hair dryer (set on low) to blow-dry the bottom. Be very careful; unintentionally, overzealous parents have severely burned their baby's bottom with hair dryers.

Be careful with the *diaper wipes*. Many of

them contain alcohol, which can sting broken or irritated skin. Some wipes also contain perfumes and other chemicals that could further irritate your baby's tender skin. Stick with hypoallergenic wipes or plain water and mild soap.

There is conflict between the values of health, convenience, and ecology over which is better—*cloth vs. disposable diapers*. There are fewer and less severe rashes in babies diapered exclusively in disposable diapers than in those using cloth diapers. The new super-absorbent diapers are even more effective in keeping babies dry and reducing diaper rash. Many parents are surprised to learn that disposable diapers take up less than 1 percent of space in landfills. The main thing is to keep the baby clean and dry with frequent changes rather than worrying about what the diaper is made of. If you use cloth diapers and wash them at home, use chlorine bleach to sterilize them and double-rinse them to make sure all of the potentially irritating detergent, bleach, and softener residues are removed before putting them in the dryer. Better yet, air-dry them in the sun.

You can add one-quarter cup of *vinegar* to the final rinse to make the diaper more acidic (discouraging the growth of yeast), but remember that vinegar does not kill germs as effectively as bleach.

Long-time home remedies for diaper rash include the use of *cornstarch or arrowroot powder* sprinkled on the baby at diaper changes. There are *no* scientific studies showing that cornstarch is any more effective than simply changing the diaper frequently and keeping the baby clean and dry. There's actually one sad case in which a baby died after inhaling a cloud of cornstarch. Avoid using cornstarch powder for babies' bottoms.

Do not use talcum powder. Talc creates a dust cloud when it is applied, and inhaling the powder has caused serious lung problems in some babies.

BIOCHEMICAL THERAPIES: MEDICATIONS, HERBS, DIETARY SUPPLEMENTS

Medications

Medications help heal irritant and yeast diaper rashes faster but are not a replacement for air, breast-feeding, and frequently changing the diaper to keep the baby clean and dry.

NON-PRESCRIPTION OINTMENTS FOR *IRRITANT* DIAPER RASH

- Barrier ointments: petrolatum (*A&D™, Vaseline™*), lanolin
- Drying agents: zinc oxide (*Desitin™, Dyprotex™*)
- Antibacterial: methylbenzethonium chloride (*Diaparene™*, A&D Medicated™)
- *Don't use*: clioquinol (*Vioform™*) or iodoquinol (*Vytone™*)

I recommend ointments rather than powders or creams because they are less likely to wash off when the baby urinates. Barrier ointments, such as A&D, help keep urine and stool away from the baby's skin. If the baby's diaper is changed regularly, a barrier ointment such as petrolatum may be all you need to prevent diaper rashes. Drying agents such as zinc oxide, the active ingredient in Desitin™, help dry up weepy, oozing skin. The active ingredient in Diaparene™ is mild for babies, but it kills the bacteria that break down urine into ammonia. Thoroughly clean and dry the baby's bottom before applying any medication.

Do *not* apply any products containing *clioquinol* or *iodoquinol* to your baby's skin. These two compounds have severe toxic effects on the brain and nervous system. Both the World Health Organization and the American Academy of Pediatrics recommend that they *not* be used to treat diaper rashes.

Antifungal medications can help cure diaper rashes caused by yeast.

NON-PRESCRIPTION CREAMS AND OINTMENTS FOR *YEAST* DIAPER RASH

- Nystatin: generic brands, *Mycostatin™, Nilstat™*
- Undecylenic acid: *Caldesene™, Desenex™*
- Fluconazole
- Miconazole: *Micatin™, Monistat™*
- Clotrimazole: *Lotrimin™*
- Gentian violet

I recommend *nystatin* ointment for yeast infections. Nystatin is a natural yeast killer derived from *Streptomyces* bacteria; it was discovered in New York State, which gave its name to the compound. It is safe, effective, and inexpensive; it has been used for over forty years to treat yeast infections of the mouth (thrush), vagina, and diaper area. Side effects are extremely rare, even if it is used recurrently or for long periods of time. It has one of the lowest rates of microbial resistance, and is one of the safest treatments available to treat Candida diaper rash.[6] Nystatin can be given by mouth to eradicate thrush and to kill the yeast hiding in the intestines, preventing reinfection and repeated diaper rashes.

Undecylenic acid (Caldesene™) is used to treat yeast diaper rash, prickly heat, and jock itch (which is also caused by a fungus). *Miconazole and clotrimazole* kill many common fungi and yeast, including *Candida*. They are very effective and generally safe, but allergic reactions and mild irritation, burning, and stinging have been reported. Resistance to mi-

conazole is rare. Symptoms tend to improve within two to three days, but it takes a full two weeks of treatment to completely eradicate the yeast.

Gentian violet is a very old remedy for yeast infections. (NOTE: It is *not* derived from the herb, gentian, which is a bitter tasting appetite stimulant.) It is very effective for both oral yeast (thrush) and diaper yeast infections. Because it is so messy, stains clothing a deep purple, and can cause skin and mouth sores, it is no longer widely used. Still, because it is so effective, some grandmothers request gentian violet when more modern treatments fail to do the job.

More powerful (and more expensive) antifungal medications such as amphotericin B (which was also derived from *Streptomyces* bacteria), itraconazole, ketoconazole, and fluconazole are available with a prescription. If non-prescription treatments have not worked within two to three weeks, take your child to be evaluated by a health care professional to make sure you are not dealing with something more complicated than a simple irritation or yeast. Your baby may benefit from a prescription medication.

Be cautious about what you put on your baby's skin. Infants' skin is very thin and can easily absorb medicines. For example, a Midwestern baby treated with a veterinary salve, Phillip's Corona Ointment, absorbed enough of the active ingredient (a hormone) through her skin to develop early puberty, with pubic hair and breast development.[7]

Herbs

Most herbal remedies are used based on tradition rather than modern scientific studies. Two herbs that have undergone study are aloe vera and calendula. In one study, three times daily use of a calendula-containing ointment was more effective than an aloe vera preparation in treating diaper rashes.[8] There are no studies comparing calendula to zinc-containing barrier ointments, but since calendula has a long track record of safety, I tolerate its use while we're awaiting more studies.

Many grandmothers, some herbalists, and even some physicians recommend *chamomile* tea as a wash for the diaper area.[9] Other herbal teas sometimes used to wash an irritated bottom include *chickweed, comfrey, elder flowers, lavender, marigold, marsh mallow root,* and *rosemary*. These herbs can also be mixed with almond oil (then strained out) to make a soothing bottom rub. Other herbal remedies for diaper rash are tinctures of *goldenseal* or *myrrh* (available in health food stores and some pharmacies). Ointments containing *comfrey* or *marsh mallow* (found in many health food stores and herbal catalogs) are also used to soothe irritated skin such as diaper rashes. This is another case where there are *no* scientific studies, but herbal remedies are generally inexpensive and have a low rate of side effects. There are rare children who are allergic to chamomile and other herbs. If any treatment makes your child worse, discontinue it and seek the help of a health care professional.

Another old-fashioned bottom wash is *witch hazel* (which is drying, as is rubbing alcohol). You can find witch hazel in your local pharmacy. *Apple cider vinegar* and *lemon juice* are also used as washes for the diaper area to create a more acid (lower pH) environment, which is less hospitable to yeast infection. Although lowering pH to reduce yeast infection is a good theory, there are no studies evaluating the effectiveness of these home remedies; they may cause stinging if applied to broken or irritated skin.

Garlic has antifungal properties in test tubes and animal studies. However, garlic is irritating, and putting a garlic poultice on too long can cause allergic reactions, irritation, and even minor burns. Do *not* put garlic on a baby's bottom. Berberine, the active ingredient in *goldenseal,* also has antifungal effects in test-tubes studies. These herbal remedies have not been formally evaluated for their safety or effectiveness in human infants suffering from diaper rashes.

Do not apply *egg white* to your baby's bottom as a treatment for diaper rash. Although this home remedy used to be common, it has caused severe egg allergies. Eggs are a fairly common allergen among infants under one year old; it is not surprising that applying a potential allergen to broken skin causes problems.

Dietary Supplements

Probiotics are increasingly used to help restore the balance of healthy bacteria that live in our mouth, intestines, and skin, and reduce the number of *Candida* yeast and harmful bacteria. Additional research is needed to determine the best brand and dose of probiotics to use for different health purposes. While we're waiting for the data, I encourage moms to eat a healthy diet that includes fermented foods like yogurt and kefir, and to offer them to babies who have started eating solids.

On average, babies with frequent diaper rashes have lower levels of zinc in their systems than infants with less frequent rashes.[10] This has led to trials of zinc supplements to reduce the risk of diaper rash. *Zinc* supplements (10 milligrams per day) may prevent yeast diaper rashes in some formula-fed babies, but they are *not* helpful in preventing diaper rashes in nursing babies. Breast-fed babies get all the zinc they need from mother's milk. If your baby drinks formula and has already had several yeast infections, you might want to give zinc a try to prevent future episodes. Overdoses of zinc can suppress the immune system. Do not give more than 10 milligrams daily and do not give it for more than two weeks without checking with your doctor.

Vitamin A is used to treat a number of skin conditions. Although it is widely used and recommended, vitamin A is no more effective than simple ointments such as petrolatum and zinc oxide.

Some parents simply break open a *vitamin E* capsule and rub the oil on their baby's irritated skin. This may be worth a try if you have vitamin E capsules on hand; there are *no* studies evaluating its effectiveness or safety.

Very rarely, nutritional deficiencies can cause diaper rashes. For example, one sign of biotin deficiency is a severe diaper rash. For babies suffering from biotin deficiency, supplemental biotin may clear up the rash. If you suspect biotin deficiency, check with your doctor. It's pretty rare.

BIOENERGETIC THERAPIES: HOMEOPATHY

No studies have evaluated acupuncture, Therapeutic Touch, Reiki, or prayer in treating diaper rash. *Homeopathic remedies* for diaper rash include creams, ointments, and sprays made from calendula and arnica. Extremely dilute homeopathic remedies given by mouth for yeast diaper rashes are *Arsenicum, Belladonna, Chamomile, Graphites, Hepar sulfur, Symphytum*, and the combination of *Hypericum* and *Calendula* known as Hypercal. There are *no* scientific studies showing that these remedies are any more effective than other creams or ointments. There are also no reported side effects.

WHAT I RECOMMEND FOR DIAPER RASH

PREVENTING DIAPER RASH

1. *Lifestyle—nutrition.* Breast-feed your baby for at least one year. Do not start solids until four to seven months old.

2. *Lifestyle—environment.* Change your baby's diapers at least six to eight times a day, more often if he has diarrhea. Keep the baby's bottom clean and dry. Consider using disposable diapers, especially the super-absorbent brands. Avoid brands that use chemical dyes or artificial scents. If you use cloth diapers, double-rinse them. Use vinegar in the last rinse to make them mildly acidic and discourage yeast from living there. Do not use plastic diaper covers. It is better to allow some air circulation.

3. *Biochemical—medications.* Consider using a barrier ointment containing petrolatum or zinc with each diaper change, especially when your child has diarrhea, to protect his skin from stool and urine. Consider prenatal treatment with topical medications and/or probiotics to reduce vaginal yeast levels.

> *Take Your Child to a Health Care Professional If*
>
> • The rash looks as if it may be infected (blisters or other symptoms).
>
> • The rash spreads outside the diaper area.
>
> • Your child has a fever, decreased appetite, or other symptoms.
>
> • The rash is not improving or is getting worse with home treatment.
>
> • You are concerned about the rash or any other symptoms.

TREATING DIAPER RASH (ABCS: AIR, BREAST-FEEDING AND BARRIER OINTMENTS, CLEANING THOROUGHLY)

1. *Lifestyle—environment.* Allow your baby to go for a period without any diapers on at all to allow the bottom to air dry completely. Change the diaper every two hours while the baby is awake, and every three to four hours overnight. Do not use commercial baby wipes containing alcohol, fragrances, or other chemicals, because these may further irritate your baby's skin. Use a clean washcloth and plain water or an unscented, chemical-free wipe. Do *not* use talcum powder or cornstarch; inhaling powders can cause lung problems.

2. *Lifestyle—nutrition.* If you are feeding your baby formula, consider supplementing his diet with 10 milligrams per day of zinc. Consider using yogurt with live cultures to restore the "good" bacteria to the diaper area. Try giving it by mouth (one-half to one cup daily), or apply directly to the baby's bottom. Alternatively, you can provide *Lactobacillus* supplements mixed in with baby's other food once he's eating solids.

3. *Biochemical—medications.* For irritant diaper rashes, use a barrier ointment containing petrolatum and zinc. For yeast diaper rashes, try non-prescription medications containing nystatin, clotrimazole, or miconazole. Consider a cream containing calendula if your child doesn't have an allergy to members of the daisy family.

15
DIARRHEA

The morning after her family picnic, Suzanne Cooper awoke with diarrhea. Her stools were so watery they ran out of her cloth diaper and down her leg. What a mess! No one else was sick, and she wasn't vomiting. Her parents wondered if it was something she ate or one of those viruses that made weekly rounds at her day care center. Her grandmother encouraged them to feed her flat soda. Their neighbor said to stop all milk products. Her aunt said to feed her whatever she wanted. Her father, Joe, had several questions for me:

- Did Suzanne need to come in to the office to be seen?
- How would he know if she was dehydrated?
- Besides food poisoning and the flu, why do children get diarrhea?
- How could he prevent Suzanne from getting it again?
- What was the best way to treat her now that she was sick?

Sooner or later all parents deal with diarrhea. The average child younger than five years old has two to three episodes a year. Diarrhea is simply when children have more stools than is normal for them, especially if the stools are runny or watery. The normal number of stools varies from child to child; it also varies with the child's diet. Breast-fed babies can have a bowel movement as often as after every meal.

Diarrhea is the body's way of getting rid of toxins, bacteria, and parasites. It usually lasts a day or two, but it may last for a week. Diarrhea is not life threatening unless the child becomes severely dehydrated or the diarrhea is a symptom of another problem.

This chapter is about *acute diarrhea* (diarrhea that lasts less than ten days). If your child has *chronic diarrhea* (diarrhea that has lasted weeks to months) and is not growing well or gaining weight, please take her to a health professional for a complete evaluation. Your child could have a problem absorbing nutrients (such as milk sugar or wheat gluten), inflammation in some part of the intestinal tract (such as Crohn's disease), a genetic problem affecting intestinal contractions, or a problem that requires surgery (such as Hirschsprung's disease).

Worldwide, diarrhea is the leading cause of childhood death; fatal cases almost always occur in malnourished children under five years old. There are many fewer deaths due to diarrhea in the United States (about 300 per year) than in most other parts of the world (450,000+ per year), but diarrhea causes plenty of suffering even here in the land of plenty. There are fewer cases of diarrhea since the rotavirus vaccine became routine. Almost all diarrhea deaths are due to dehydration. What are the most worrisome symptoms?

These may be signs that your child has something other than simple diarrhea or needs more therapy than you can provide at home.

WHEN TO CALL YOUR HEALTH PROFESSIONAL:

- The diarrhea lasts *longer than a week* or there are *more than ten* stools per day.
- Your child has a *fever higher than 101°F* for *more than one day*.
- She has pain in the *lower right* part of her abdomen.
- There is *blood* in the stools.
- Your child is less than six weeks old.
- She is losing weight.
- She is vomiting for more than one day.
- She is lethargic, disoriented, or confused or doesn't recognize you.
- She is not acting like herself.
- She has a high fever (over 103°F) at any time or has a seizure.
- She refuses to drink or isn't thirsty. [NOTE: It's normal for children to refuse solids when they're sick, but they should still be thirsty.]
- She looks dehydrated.

HOW DO YOU KNOW IF YOUR CHILD IS BECOMING DEHYDRATED?

- She voids (pees/urinates) less than four times per day or less than half of what is normal for her.
- She doesn't have tears when she cries.
- She loses weight.
- She has sunken eyes and a sunken fontanel (soft spot on baby's head).
- Her lips and tongue are dry; she has stringy saliva.
- Her hands and feet are much cooler than her arms and legs.

If your child has any of these signs, she needs more fluid. Children who are *vomiting* are more likely to become dehydrated than children who have diarrhea alone.

Suzanne did not have any signs of dehydration; she could be treated safely at home.

WHAT CAUSES DIARRHEA?

There are many causes for diarrhea. (See the box titled Common Causes of Diarrhea.) The leading cause of diarrhea in young children is a *viral infection*, most commonly *rotavirus*. Rotavirus is the number one cause of diarrhea worldwide among children who are between three and fifteen months old. It can be prevented now with the rotavirus vaccine.

COMMON CAUSES OF DIARRHEA

- Infections with viruses, bacteria, or parasites
- Reactions to food
  - Food poisoning (e.g., *Staphylococcus* bacteria)
  - Food intolerance, sensitivities, allergies
  - Excessive intake of fruit or "sugarless" gum
- Reactions to vitamins or medications
  - Excessive vitamin C or magnesium supplements
  - Excessive use of laxatives
  - Side effect of antibiotics, antacids, or other medications
- Other causes
  - Nervous diarrhea
  - Miscellaneous: teething, ear and bladder infections
  - Irritable bowel syndrome or inflammatory bowel diseases

Bacteria cause diarrhea in several ways. A few *bacteria*, such as *cholera*, attack the intestinal walls, while others, such as *Staphylococcus* (the cause of most food poisoning), indirectly cause symptoms by producing toxins that poison the gut. Other common diarrhea-causing bacteria include *Salmonella, Shigella,* and *E. coli.* Viruses and bacteria are usually passed from person to person, but they can also be transmitted by contaminated meat, poultry, and water; and they can be picked up at petting zoos when children touch animals and don't wash their hands. When you go camping and drink from a stream, you risk picking up the diarrhea-causing parasite *Giardia.* Giardia can also be passed from person to person, es-

pecially in day care centers with insufficient hand washing.

Suzanne could easily have caught a virus at her day care. Food poisoning was a less likely cause of her diarrhea because no one else was sick who ate the same foods.

Overindulgence in hot dogs and other greasy foods accounts for post–ball game diarrhea, but even healthy foods can cause diarrhea if consumed in excess. *Lactose* (milk sugar) intolerance is a common cause of cramping abdominal pain, bloating, and diarrhea in those who lack the enzyme necessary to digest it (lactase).

Cow milk, eggs, soy protein, and other *allergenic foods* cause diarrhea in fewer than 10 percent of children. Children who are sensitive to cow milk usually develop symptoms within the first six months of life. *Food protein enterocolitis syndrome* is a kind of allergy caused by IgG antibodies (unlike acute reactions to bee stings or peanut butter, which are usually caused by IgE antibodies).[1] The foods that are the most common triggers are cow's milk, fish, egg, soy, and corn. It takes an average of one to three hours for symptoms to occur, and the most common symptom is vomiting, not just diarrhea. This kind of food allergy is not readily detected with simple allergy skin prick tests.[2] If the child continues to eat the offending food, it can lead to dehydration, weight loss, and failure to gain weight normally. The good news is that many children outgrow their sensitivity by the time they're five years old.[3]

Undercooked hamburgers resulted in a deadly epidemic of *E. coli* diarrhea in 1993 due to a certain strain of *E. coli* bacteria. This strain (*E. coli* 0157:H7) carrying a certain toxin (Shiga toxin) can lead to a life-threatening condition called hemolytic uremic syndrome (HUS), which causes bloody diarrhea, anemia, and kidney failure. This type of *E. coli* is commonly carried by cattle that have no symptoms themselves, but pass the bacteria along in their waste. When their manure contaminates the meat, or is used as fertilizer, or runs off into agricultural fields growing other crops, the bacteria can contaminate whatever the manure contacts.

Drinking too much *pear or apple juice* accounts for astonishing amounts of toddler diarrhea; these fruit juices contain the stool-loosening sugars *sorbitol* and *fructose*. Chewing a lot of sorbitol-containing *"sugarless" gum* can also induce diarrhea.

Diarrhea is one of the early signs of a *vitamin C overdose*. Children or adults who suddenly start taking 500–1,000 milligrams of vitamin C several times a day to treat a cold may find they are having more frequent and looser stools. Children can usually tolerate higher doses of vitamin C when they are ill or if they gradually build up tolerance. The same thing happens with excessive intake of *magnesium* supplements.

Many *medications* cause diarrhea, either as a direct effect (e.g., chocolate-flavored Ex-lax, which children may mistake for candy) or as a side effect. The *antibiotics* prescribed for many ear infections kill the normal bacteria in the gut as well as the bacteria in the ear, resulting in diarrhea. Sugar-free medications, gums, and candy are sometimes sweetened with *sorbitol,* which can cause diarrhea. *Antacids* can also cause diarrhea by drawing water into the intestines.

Several other unrelated factors can also trigger diarrhea. Diarrhea due to *stress* is exemplified by the long lines to the bathroom right before a test. For reasons we do not fully understand, some children also get diarrhea when they are *cutting teeth* and when they get *ear infections. Bladder infections* cause sympathetic bowel irritation, occasionally resulting in diarrhea.

Unusual, serious *bowel diseases* and enzyme deficiencies can also cause diarrhea. Irritable bowel syndrome causes alternating diarrhea and constipation as well as belly pain; inflammatory bowel diseases such as ulcerative colitis and Crohn's disease can cause diarrhea as well as poor growth and a host of other problems, but these types of diarrhea are usually chronic (long-term) rather than acute (short-term).

The best treatment for your child's diarrhea depends on what is causing it. Regardless of the cause, give your child extra fluids to replace what she is losing with all those runny stools.

HOW TO PREVENT DIARRHEA

The most important things you can do to reduce the risk of diarrhea are to get your child immunized against rotavirus and use good hygiene in hand-washing and food preparation.

TO PREVENT YOUR CHILD FROM GETTING DIARRHEA

- Get the rotavirus vaccine.
- Use good hygiene.
  - Frequent handwashing, especially after diaper changes, when cooking poultry, and with toddlers who are toilet training
  - Thorough cleaning of cutting boards, utensils, and sink after handling raw meat
- Thoroughly cook all meat, poultry, and eggs.
- Use pasteurized products when possible.
- Breast-feed for at least six to twelve months.
- Offer yogurt and kefir to build up healthy bacteria.
- Make sure your drinking water is pure (boil or filter camping water).

Because rotavirus accounts for so many cases of early childhood diarrhea, the vaccine can cut your child's risk of developing diarrhea in half.

Because most diarrhea is caused by viruses that are passed from person to person, the second best prevention strategy is *frequent hand washing*. Toilet-training toddlers need reminders to wash their hands after using the potty and before eating. Frequent hand washing is absolutely essential in day care centers where there are children in diapers. Diarrhea outbreaks are more common in crowded day care settings than in those with fewer children. Hand washing is also important in the home before, during, and after food preparation. Much of the poultry in this country is contaminated with diarrhea-causing *Salmonella* bacteria. *Salmonella* is killed when the poultry is thoroughly cooked (eliminating the risk from the meat itself), but be sure to clean your cutting board, counter, knife, and hands before preparing other foods, especially raw salads. I recommend that you clean your cutting board and counter daily with a 10 percent bleach solution. Consider using a separate cutting board for meat and vegetables to avoid contamination.

Breast-feed your baby for at least the first six to twelve months. Formula-feeding markedly increases an infant's chances of getting diarrhea. Breast-feeding can also decrease the duration of diarrhea if your child does get sick. Nursing reduces the risk and severity of diarrhea from outbreaks of even very aggressive bacteria and parasites. Breast milk contains immune factors (immunoglobulin A), which help prevent bacteria from attacking the intestines and help reduce symptoms of diarrhea even when bacteria and parasites do get into the system.

Consider feeding *yogurt* to your child regularly to prevent diarrhea. A study at Johns Hopkins showed that regularly giving infants the "good bacteria" contained in yogurt markedly reduced their chances of getting rotavirus diarrhea. You can get these benefits only with yogurt containing live cultures or by giving probiotic supplements of *Bifidobacteria* or *Strep thermophiles*.

If your child requires antibiotics to treat a serious infection like pneumonia, please consider providing probiotic supplements to restore healthy bacteria and reduce the risk of antibiotic-associated diarrhea. Dozens of studies support using probiotic supplements to reduce the risk of severe antibiotic-associated diarrhea due to *Clostridium difficile*.[4]

Always *boil your water* or use a special filter when you go camping. When you travel abroad, use bottled, boiled, or filtered water if there is any question about its purity. Be sure that the "natural" fruit juices that you buy have been properly pasteurized, and wash all fresh fruit to make sure it's free of bacteria.

WHAT'S THE BEST WAY TO TREAT DIARRHEA?

The two principles of treatment are:

1. Address the underlying cause.
2. Give fluids to prevent dehydration.

If your child has diarrhea from drinking too much pear juice, the obvious treatment is to cut back on the juice. Even if the cause is not so obvious, you can still help prevent dehydration by making sure your child drinks plenty of fluids. Offer something to drink every fifteen to thirty minutes while she is awake. Over the course of a day the frequent fluids add up, preventing dehydration. Let's consider the whole range of therapies to find out what works best. If you want to skip to the bottom line, flip to the end of the chapter.

LIFESTYLE THERAPIES: NUTRITION, EXERCISE, ENVIRONMENT, MIND-BODY

Nutrition

Nutritional therapy is the mainstay of diarrhea treatment. Keep the child hydrated with electrolyte-balanced fluids and do not stop breast-feeding.

DO:

- Continue to breast-feed.
- Give electrolyte-balanced fluids.
- Continue full-strength milk or formula unless diarrhea is severe.
- Give yogurt, rice, lentils, potatoes.

DON'T:

- Give plain water alone without other fluids or foods.
- Give apple or pear juice.
- Demand antibiotics—often this makes things worse, not better, for diarrhea.

Breast milk is the best food for an infant with diarrhea. If your infant is dehydrated, breast-feed more often and consider adding other liquids that have a balanced solution of water, salt, and sugar.

The World Health Organization recommends that children with diarrhea be treated with *oral rehydration solution* (ORS). The most convenient, *electrolyte-balanced liquids* to replace fluid losses from diarrhea are premixed water-sugar-salt solutions such as Pedialyte™. Rehydration solutions are much better than half-strength Jell-O, soft drinks, or fruit juices, which have too much sugar (glucose) and not enough salt. Excessive fruit juice intake can actually cause diarrhea.

I asked Joe not to replace Suzanne's fluid losses with soda or apple juice, despite his mother-in-law's advice. These fluids just don't have the proper balance of salts and sugar that a child suffering from diarrhea needs.

You can make your own *rehydration solution* at home.[5]

HOMEMADE REHYDRATION SOLUTION

- 1 quart (1 liter) of clean water
- ½ teaspoon (2.5 grams) of salt *or* ¼ tsp salt + ¼ tsp baking soda *or* ¼ tsp salt and ¼ tsp salt substitute (with potassium)
- 4–8 teaspoons (20–0 grams) of sugar

Measure the ingredients with measuring cups and spoons and use pure water.

Very little salt is needed, and some salt can be replaced with baking soda (usually for children with more severe dehydration) or salt substitute (which supplies potassium).

In many parts of the world, salted *rice water* is used to treat dehydration in infants with diarrhea. In other parts of the world, *other cereal powders* (such as corn, millet, or sorghum) are mixed with water and salt to make the rehydration solution. An alternative recipe replaces the sugar with rice cereal.

HOMEMADE REHYDRATION SOLUTION:
ALTERNATIVE

- 1 quart of water
- ½ teaspoon of salt *or* ¼ tsp salt + ¼ tsp baking soda *or* ¼ tsp salt and ¼ tsp salt substitute (with potassium)
- 2 ounces (50–80 grams) or about one cup of rice cereal for babies

Joe decided that they would make their own rehydration solution at home. Because they had infant rice cereal on hand for the baby anyway, he decided to use the rice cereal-based solution. That way he could avoid giving straight sugar and save money. He wanted to know exactly how much rehydration solution he needed to give Suzanne. How much was enough?

A good rule of thumb is four ounces of extra fluid for each diarrhea stool. Give the fluid a little at a time. If your child is vomiting and you give her a large amount to drink at once, it could all be vomited back up. Giving a large amount at a time can also stimulate another bowel movement, making the diarrhea and dehydration worse. If your child is vomiting, try giving just a tablespoon (15 milliliters) or an ounce (30 milliliters) every fifteen to thirty minutes, and increasing the amount gradually as your child tolerates it. If your child is thirsty, give fluid more frequently.

Despite the neighbor's advice, you do not need to dilute milk, stop milk, or switch to a different formula, although it is fine to use a lactose-free formula or lactose-free milk.[6] Most children recover quite nicely from their diarrhea within a few days regardless of the type of milk they drink. When I was in training to be a pediatrician, I was taught that babies who had diarrhea should be fed clear liquids, advancing to quarter strength formula, then half strength formula, and finally full strength formula. Now we know that diluting milk or formula for children with diarrhea is unnecessary and may even be counterproductive. Children who are fed full strength milk regain weight faster than those who were fed diluted milk. Diluting formula will not help your child recover from the diarrhea any faster and deprives your child of needed calories.

There have been at least 29 studies involving a total of over 2,000 children with diarrhea to see if feeding a lactose-free diet is helpful. Lactose is the name of the sugar found in cow's milk. Children who are mildly or moderately ill do just as well on cow milk (or formulas based on cow milk) as those who are taken off of cow milk products.[7] That said, there *are* times when it *is helpful to stop milk and other lactose-containing foods*.

WHEN TO STOP MILK AND MILK PRODUCTS:

If your child:

- Has severe diarrhea requiring hospitalization
- Has diarrhea for more than two weeks
- Doesn't tolerate lactose (milk sugar) even when she's well
- Was malnourished even before she developed diarrhea

Physicians used to tell parents to stop all *solids* until the child was tolerating milk, and then restart solids slowly with *B*ananas, *R*ice cereal, *A*pple sauce, and *T*oast or *T*ea (the BRAT diet); we were also told to avoid fatty or fried foods. There are no scientific studies showing that this medical wisdom is right. Most children do just as well by continuing their usual diet as children who are put on special diets. The World Health Organization recommends that regular feeding should be continued during diarrhea episodes; the BRAT diet is unnecessary and possibly counterproductive. Another medical myth bites the dust.

Rice and *yogurt* are the most easily digested foods for children recovering from acute diarrhea. Rice, other cereals, potatoes, and other starchy vegetables are easy to digest and help reduce the duration of diarrhea. Children with persistent diarrhea who are fed yogurt do better than those who are fed plain cow's milk. The secret is the active acidophilus cultures. A French study compared children given jellied milk vs. those given yogurt every day for lunch; those given the yogurt (some of which was fortified with extra *Lactobacillus*) had *much* shorter spells of diarrhea—three days shorter, on average! Homemade yogurt contains active cultures and avoids the artificial colors, flavors, and sugars contained in many commercial brands. More and more studies suggest that giving acidophilus and other healthy bacteria (known collectively as probiotics) is a very effective way to cut down on diarrhea episodes and to reduce the severity and duration of diarrhea when it does occur.

Extra *fiber* may make the stools look more solid, but it doesn't prevent dehydration or shorten the course of illness. Some parents who are tired of dealing with the mess of sloppy diapers may want to consider fiber supplements. Fiber doesn't affect your child's illness, but it may reduce the mess of watery stools.

Plain *water* is okay as a supplement, but it should not be the only fluid you give. When your child has diarrhea, she is losing water and salt, so both need to be replaced. The intestines actually absorb water better if it contains a little sugar. Plain water does not contain any salt or sugar and is not absorbed as well as solutions that contain them. If a child gets too much water without replacing the salt lost because of the diarrhea, there can be blood cell damage, seizures, and even coma.

Fruit juices and flat soda are not as helpful as the rehydration recipes described previously. No one would treat diarrhea with prune juice, but many parents don't realize that as little as five to eight ounces of *pear* juice causes loose stools in many children. Apple juice and grape juice can also make diarrhea worse. White grape juice may be a little better and makes for far less colorful diapers. If you want to give fruit juice and avoid the mess, try white grape juice.

Joe realized that it was time to stop giving Suzanne soda. He put her back on milk, giving her one-half cup of rehydration solution to sip over 15 minutes every time she had a diarrhea stool. He went to the store to buy some yogurt with active cultures. He made a big batch of rice as a side dish for supper. He was concerned that Suzanne was starting to get a diaper rash from the diarrhea and wondered what he could do for that.

Environment

You can reduce the risk of diaper rash that often accompanies diarrhea by changing your child's diaper frequently and putting an ointment containing petrolatum or zinc oxide on the skin with each diaper change. Diaper rashes occur when the bacteria in the stool interact with urine, forming a compound that is very irritating to babies' tender skin. If you can keep the skin clean and dry, you can reduce the risk of rash. If possible, have your baby take naps with the diaper *under* her rather than around her, exposing her bottom to air and allowing it to dry out. Some parents also use a fan or a hairdryer set at a very low setting to help dry their babies' bottoms more thoroughly. Be sure you don't burn the skin!

Mind-Body Therapies

Although mind-body therapies are rarely used to treat short-term diarrhea problems in children, therapies such as hypnosis can be very helpful in treating older children and adolescents who suffer from nervous diarrhea and irritable bowel syndrome.

BIOCHEMICAL THERAPIES: MEDICATION, DIETARY SUPPLEMENTS, AND HERBAL REMEDIES

Medications

MEDICATIONS FOR DIARRHEA

Do use: bismuth subsalicylate (Pepto-Bismol™)

Don't use: paregoric, loperamide (Imodium™), diphenoxylate (Lomotil™)

Sometimes use: kaolin-pectin (Kao-Pectate™), antibiotics, immune globulin

Good old-fashioned *Pepto-Bismol* ™ (bismuth subsalicylate) given every four hours is helpful for sudden, watery diarrhea, even when the diarrhea is caused by a bacterial infection. Bismuth subsalicylate contains the active ingredient in aspirin (salicylate); to minimize the risk of developing the serious complication Reye's syndrome, do not give Pepto-Bismol™ or other aspirin-containing remedies when your child has influenza. Although their names are similar, influenza is not the same as intestinal flu. Influenza symptoms are usually high fever, exhaustion, weakness, headache, muscle aches, and coughing. Diarrhea is not a typical part of the influenza picture, but there's no guarantee that a child can't get influenza and a rotavirus infection at the same time. When in doubt, consult a health care professional.

Avoid the strong adult diarrhea medicines *paregoric, loperamide* (Imodium™), or *diphenoxylate* (Lomotil™). These medicines are from the same drug family as codeine and morphine. Although one of their main side effects is constipation, other side effects include sleepiness, nausea, and distended bellies. Some children have become comatose and died from taking loperamide. Keep these medicines out of reach of children.

Another commonly used medication is *kaolin-pectin* (Kao-Pectate™). Pectin is found in apple peels and is used to thicken jams and jellies. Pectin thickens stools and makes them less sloppy, but it does *not* decrease the overall amount of fluid lost or speed up the child's recovery from diarrhea. It gives the appearance that the child is better when she may still be battling the problem.

Joe had already given Suzanne some Kao-Pectate by the time he called me. Kao-Pectate is not actually harmful, but the improved consistency of the stools can be deceiving. I urged Joe to continue to offer Suzanne frequent sips of fluid to avoid dehydration even though her stools appeared to be less watery.

The vast majority of American children with diarrhea do not need and do not benefit from *antibiotics*. Many antibiotics actually cause diarrhea because they wipe out the "good" bacteria that normally live in the intestines as well as the "bad" bacteria that cause infections. Antibiotic treatment for diarrhea is only indicated for children who have diarrhea due to certain bacteria or parasites, such as cholera, shigella, amebiasis, or persistent Giardia. If your child is severely dehydrated or has bloody stools, a stool test is necessary to determine if she needs antibiotics. In some cases, antibiotics can actually make the picture much worse, resulting in more complications. For example, children with certain serious *E. coli* infections who were treated with antibiotics actually did worse than children with the same infection who battled the disease without antibiotics.

Dietary Supplements

Several dietary supplements are helpful for children with diarrhea. But some should be avoided.

DIETARY SUPPLEMENTS FOR CHILDREN WITH DIARRHEA

Do use: Zinc if you suspect zinc deficiency or the diarrhea is protracted; probiotics supplements such as *Lactobacillus* or *Bifidobacterium*

Don't use: Vitamin C, magnesium, copper

Maybe use: Vitamin A, glutamine

When children have prolonged diarrhea (more than seven to ten days), they can lose a lot of nutrients in their stools. One of the important minerals lost this way is *zinc*. Zinc is an essential nutrient, and zinc deficiency can also cause diarrhea. Dozens of studies from China, India, Bangladesh, Pakistan, and Peru have shown that zinc supplements help speed recovery from diarrhea.[8] Zinc deficiency is not common among children in developed countries, but one study from Switzerland also supported the short-term (10-day) use of low-dose (10–20 mg daily) zinc supplements for preschool children with acute diarrhea.[9] If you believe your child has a zinc deficiency, a reasonable dosage for children 6 to 60 months old is 10–20 milligrams of elemental zinc twice daily for one to two weeks according to the World Health Organization. If your child is younger than six months old, please consult your health care professional for a proper dose. Do not use zinc supplements for more than two weeks without seeing a health professional. Zinc supplements can have side effects, such as nausea and vomiting.

Recent research has highlighted the effectiveness of probiotic bacteria such as *Lactobacillus GG, L. acidophilus, L. bulgaricus, L. reuteri, Bifidobacterium bifidum, B. lactis, Saccharomyces boulardii,* and *Streptococcus thermophilus.*[10,11] These are the healthy bacteria that normally live in our intestines. They can be wiped out by antibiotic treatment and some infections. Giving supplemental probiotics to treat diarrhea episodes can cut the number of days of disease by two days. Giving probiotics to hospitalized children significantly reduces their risks of catching diarrhea in the hospital. Probiotics are also helpful in preventing traveler's diarrhea. I recommend yogurt and other probiotics to just about every child I see.[12] So do official bodies such as the European Society for Pediatric Gastroenterology.[13,14]

The data on *vitamin A* treatment for diarrhea have been conflicting. Initial reports from Thailand and Brazil indicated that supplemental vitamin A might reduce the risks of severe diarrhea. Subsequent research in India showed that it did not help. Excessive vitamin A can have serious side effects. Until research is done using vitamin A to treat American children with diarrhea, I do not recommend it unless the children have some complicating condition or malnutrition that increases their need for vitamin A.[15]

Glutamine is an essential amino acid that helps maintain a healthy bowel wall. A Turkish study suggested that glutamine supplements (300 milligrams per kilogram of the child's body weight daily) could help reduce the duration of diarrhea symptoms.[16] On the other hand, a study from Uganda found no benefit from glutamine supplements.[17] Until large trials are done with consistent results, I do not recommend glutamine supplements for well-nourished children with diarrhea in the United States.

Don't give supplemental vitamin C, magnesium, or copper to children suffering from acute diarrhea. You could make things worse.

Herbal Remedies

A number of herbal remedies are traditionally recommended for diarrhea. One of their advantages is that they are usually prepared as tea, so sipping these herbal remedies helps prevent dehydration on the basis of increased fluid intake alone. However, there are not enough randomized, controlled studies demonstrating that any herbal remedies safely reduce the length or severity of childhood diarrhea for me to routinely recommend them as specific treatments for diarrhea.

HERBAL REMEDIES FOR DIARRHEA

Maybe use: carob, barberry, chamomile, cinnamon, garlic, ginger, goldenseal, mint, Oregon grape, raspberry leaf, slippery elm bark

Don't use: Chinese herbal patent medicines

Carob powder is an old-time Mediterranean folk remedy for diarrhea, because it contains *tannin*. Carob reduces the growth of bacteria and binds some of the toxins produced by bacteria. In a few European and Turkish studies, carob-based products were significantly better than placebo in shortening the course of diarrhea.[18,19] Because carob is a biological substance, its potency varies; and it is not regulated by the FDA the same way that medicines are. Until larger studies are done using products available in the United States, I am not ready to recommend carob as a proven therapy for pediatric diarrhea.

Chamomile tea and *red raspberry leaf tea* are widely used in Europe as an aid to digestion, and *cinnamon* is commonly recommended for digestive disturbances. In a German study, a combination of chamomile plus apple pectin plus rehydration solution was more effective than rehydration solution alone in treating acute diarrhea in children.[20] You can also add a pinch of cinnamon to yogurt to help soothe the intestines of a child suffering from diarrhea. *Garlic* is believed to help the immune system fight whatever bacteria or virus is causing the diarrhea. You can flavor rice with a bit of garlic. *Ginger* (the basis for old-fashioned ginger ale) is recommended worldwide for treating upset stomachs. Ginger (one-quarter teaspoon chopped or grated), combined with the juice of half a lemon and one teaspoon of honey in a cup of hot water, makes a lovely tea for children with an upset stomach, though its effectiveness is much better established for nausea and vomiting than for diarrhea.

Goldenseal, barberry, and *Oregon grape* all contain the chemical berberine. Berberine kills many of the bacteria and parasites that cause diarrhea. Goldenseal was used by Native Americans to treat diarrhea. However, because of overharvesting, goldenseal is practically on the endangered species list, and most herbalists no longer recommend it.

Peppermint leaf tea and *catnip tea* are also recommended for all sorts of stomach pains, cramps, indigestion, and diarrhea. You can also add a leaf or two of mint to flavor yogurt and calm the stomach. Mint is readily grown (almost weedy) in the family garden, and you can brew up your own tea as a pleasant flavor component of a rehydration solution.

I do not recommend *Chinese patent medicines* to treat diarrhea. Chinese patent medicines are herbal mixtures formulated into pills. There are huge variations in the amount of active ingredients, and about one-third of the products tested have shown contamination with drugs, mercury, lead, or arsenic. They are not safe; they are not regulated by the FDA, and I do not recommend them.

BIOMECHANICAL THERAPIES

There are no studies documenting the effectiveness of any biomechanical therapies in treating ordinary childhood diarrhea.

BIOENERGETIC THERAPIES: ACUPUNCTURE, HOMEOPATHY

Acupuncture

Until more studies are done, acupuncture remains an unproved remedy for children with diarrhea.

Homeopathy

An analysis of three randomized controlled trials conducted in Nicaragua and Nepal concluded that homeopathic remedies could cut about one day off the length of a child's bout with diarrhea.[21] In a later study by the same group of researchers, homeopathic remedies were no better than placebo treatments.[22] Similar studies are needed in the United States. What follows are common recommendations for homeopathic remedies for diarrhea.

Arsenicum album for children who are vomiting, have tummy pain and severe, foul-smelling, burning watery diarrhea that is worse at night. Arsenicum is recommended for the child who is very anxious and restless. Children with this type of diarrhea also like to be cuddled under warm blankets. Arsenicum is the most commonly recommended homeopathic remedy for diarrhea due to food poisoning.

Calcarea carbonica (*Calcium carbonate*) is used for the diarrhea that accompanies *teething*. The child is afraid of the dark and doesn't like to be alone; there is a lot of sweating during sleep and the stools smell sour.

Chamomilla (German *chamomile*) is also used for children whose diarrhea is related to cutting teeth. The child is restless, whining, irritable, or clinging. He may ask for something and then reject it.

Podophyllum (*May apple*) is used for children who are very thirsty for cold water. The diarrhea is worse in the morning.

Ipecac (the very dilute, homeopathic form only!) is used for children who have intractable vomiting along with their diarrhea.

Pulsatilla is used for diarrhea from eating excessive amounts of rich foods.

Sulfur is used for children whose diarrhea is worse at night or very early in the morning, the diarrhea smells like rotten eggs, or there is redness around the child's anus.

I do not usually recommend homeopathic remedies, but I do not object to parents who want to try them because they are generally safe. Remember that the most important therapy is hydration.

WHAT I RECOMMEND FOR DIARRHEA

PREVENTING DIARRHEA

1. *Lifestyle—nutrition.* Breast-feed your child for at least the first six to twelve months of life. Avoid excessive intake of fruit juices and sugar-free candies and gum. *Biochemical*—Get your child immunized against rotavirus.

2. *Biochemical—nutritional supplements.* After weaning, consider daily doses of yogurt with live cultures or healthy bacteria or supplements of *Bifidobacteria* or *Strep thermophilus*. Whenever your child takes antibiotics, make sure they eat yogurt every day, and consider using a dietary supplement containing *Lactobacillus GG*, *Bifidobacterium bifidum,* and/or *Streptococcus thermophilus*. Avoid excessive doses of vitamin C or magnesium. Do not ask for antibiotics to treat viral infections.

3. *Lifestyle—environment.* Use good hygiene to prevent the spread of viruses and bacteria that cause diarrhea:

- Wash your hands.
- Wash food preparation surfaces using bleach.
- Thoroughly cook all meat products.
- If your child is in day care, choose one with just a few children in uncrowded settings to prevent infectious illnesses, including diarrhea.
- When traveling, drink bottled or filtered water. Do not drink out of the streams when you are camping unless you boil, filter, or treat the water.

Take Your Child to Your Health Care Professional If She

- Is less than six months old

- Has vomiting as well as diarrhea

- Isn't drinking well

- Looks dehydrated (no tears when crying, less urine/wet diapers, dry mouth, sunken eyes)

- Has a fever higher than 102° or is breathing faster than usual

- Has diarrhea for longer than a week

- Has blood in stools

- Looks sicker than you would expect with a simple virus—lethargic, irritable

- Has ingested a drug or known toxic substance

TREATING DIARRHEA

1. *Lifestyle—nutrition.* Give plenty of fluids. (Do not give apple, pear, or prune juice; avoid cola, boiled milk, and homemade soft drinks.)

Give a commercial rehydration solution or make your own:

- Four cups of water (one quart)
- One-half teaspoon of salt (*or* one-quarter teaspoon salt and one-quarter teaspoon baking soda)
- One to one-and-a-half cups (50–80 grams) of rice cereal for babies *or* two tablespoons of sugar

Offer one tablespoon to one ounce every fifteen to thirty minutes; increase as tolerated so that the child is drinking four to eight ounces of fluid for each diarrheal stool in the past hour.

If your child is breast-feeding, continue to breast-feed. Supplement with rehydration fluids. If your child did not have any problems with milk before, continue it or try lactose-free milk. You do not need to stop or dilute milk, formula, or solids unless the diarrhea is so severe that the child is hospitalized. If any foods seem to make the diarrhea worse, stop them for a day or two, then try again.

Do not force your child to eat; offer her solids if she's hungry and not vomiting. If your child is hungry for solids, try yogurt (with active cultures) or rice. You can flavor them with cinnamon, ginger, or mint leaves.

2. *Biochemical—medications*. Pepto-Bismol—every four hours (see label for dosing information); avoid this if your child has influenza.

3. *Biochemical—dietary supplements*. Consider dietary supplements with *Lactobacillus GG* or *Bifidobacterium bifidum* or *Streptococcus thermophilus*. Consider 10 days of 10–20 milligrams daily of zinc supplements.

4. *Bioenergetic—homeopathy*. Homeopathic remedies are best individualized to a child's particular symptoms; consult a homeopathic physician.

16
EAR
INFECTIONS

Sheryl Wu brought in her three-year-old, Nathan, because she was concerned that he might have another ear infection. He had a cold for a few days and then became more fussy and clingy. He had a fever when he was put to bed at 8 pm, and awoke at midnight, crying and holding his right ear; he was restless and irritable most of the night. He seemed better in the morning, but he'd already had three ear infections this year. Sheryl was concerned that Nathan had another ear infection, and she was frustrated that the infections kept coming back. She wanted to know why Nathan had so many ear infections, what she could do to prevent them, and the best ways to treat them. She didn't like the idea of giving him antibiotics all the time, but she didn't want Nathan to suffer, and she didn't want to jeopardize his hearing.

If your child suffers from ear infections, you are far from alone; most children develop at least one ear infection by the time they're three years old, with most occurring between 6 and 15 months of age. Ear infections account for more antibiotic use than any other childhood disease. The cost of treating ear infections runs about $3.5 billion a year in the United States, not counting parents' time away from work or extra day care. About 25 percent of children less than two years old have had six or more doctor visits for ear infections: That adds up to a lot of pain and suffering for the kids and a lot of time out of work for the parents.

What we usually think of as an ear infection is actually an infection of the *middle* ear (*otitis media*). The infection occurs between the eardrum and the inner ear, not in the ear canal. A middle ear infection that comes on suddenly is called *acute otitis media*. In acute otitis media, the middle ear is filled with pus. After the acute infection is over, the bacteria are gone, but some fluid persists for several weeks or months; 90 percent of kids are free of fluid within 90 days. The chronic or persistent presence of fluid is called *chronic otitis media* or *serous otitis*. Colds and allergies can also cause middle ear fluid; even food allergies can cause fluid in the ears. *Serous otitis* is usually painless, but it can decrease hearing (which can lead to delayed speech development) and predispose a child to another bout of acute otitis media. When children suffer from repeated ear infections, it is called *recurrent otitis media*. Confusing, isn't it?

Children with acute otitis media can experience a wide range of symptoms. Some children have *no symptoms* at all. These infections are usually discovered during a routine physical exam; the kids don't require treatment unless they have significant ongoing hearing loss or there's some other problem. Most children with acute ear infections have *pain* in the infected ear, especially when they lie down. Lying down increases the blood flow to the head, resulting in more pressure in the infected ear. This is why children with ear infections are often most fussy at night. Sick children tend to be more *whiny, clingy,* or *withdrawn* regardless of why they are ill. They also tend to *behave as if they are younger than they are*; this is called *regression* and is a normal response to stress. For example, a sick child may start sucking his thumb even though he stopped sucking it six months previously. Ear infections usually are accompanied by *fever* and a *decreased appetite*. Older children may also report that they *cannot hear as well* in the infected ear or that it feels full or stuffy. If the pressure in the middle ear builds too high, it *can rupture the eardrum*, allowing pus to drain into the ear canal and leading to scars on the eardrum. If this happens, you can see bloody, waxy, or milky white drainage from

the ear. If the fluid presses on the inner ear, it can affect the sense of balance, creating *clumsiness* and leading kids to complain of *dizziness*.

POSSIBLE SYMPTOMS OF MIDDLE EAR INFECTIONS

- None
- Pain in the ear, especially when the child lies down
- Pulling at ear
- Fever
- Irritability, whining, crying, sleeping less
- Clinging, withdrawn, wanting to be held
- Decreased activity, lethargy, sleeping more, decreased appetite
- Acting younger (regression)
- Decreased hearing
- Pus draining from ear
- Swollen lymph nodes in the neck
- Dizziness and clumsiness

DIAGNOSIS

It is easier to diagnose an ear infection in a four-year-old who can tell you that his ear hurts and sit still for an exam than in a four-month-old who is feverish, irritable, crying, and squirming. Ear pain is a pretty reliable sigh of an ear infection, but not always! Other causes of car pain include sudden pressure changes (such as riding a high-speed elevator or an airplane); headaches; infections in the sinuses, teeth, or throat; problems with the jaw joint (tempero-mandibular joint, [TMJ]); and objects (such as crayons, pebbles, or beads) stuck in the ear. You can't depend on the color of the outer ear to diagnose an ear infection, because the infection is actually deep inside the ear canal behind the eardrum. The only way to be sure your child has an ear infection is to check for fluid behind the eardrum or to see pus draining out from a ruptured eardrum.

DIAGNOSING EAR INFECTIONS

- Otoscopy
- Pneumatoscopy
- Tympanometry and acoustic reflectometry
- Tympanocentesis

A special device, an *otoscope,* is used to look into the ear canal and see the eardrum. Looking at the eardrum is called *otoscopy*. You can buy an otoscope to look yourself, but it is difficult to do without training and an unusually cooperative child. Medical students spend months and years learning to diagnose ear infections accurately. Exams can be tricky, especially when the child is squirming, there is wax in the canal, or the child is crying. When a child cries so hard that his face turns red, the eardrum usually turns red, too, and this redness can be mistaken for an infection. Redness doesn't predict the course of the child's illness.

Health care professionals use several additional techniques to diagnosis ear infections. *Pneumatoscopy* is the fancy name for seeing if the eardrum moves back and forth with mild air pressure. If there is pus or other fluid in the middle ear, the eardrum doesn't move very well. Accurate diagnosis depends on the combination of otoscopy and pneumatoscopy.

Before I did pneumatoscopy in Nathan's ears, I told him that I was looking for birds in his ear, and that he might feel their wings softly tickling him. He sat still during the exam, absorbed in the idea of birds in his ears.

Fancier tests such as *tympanometry* and *acoustic reflectometry,* in which sound waves are bounced off the eardrum to see whether or not it moves, are also available to help confirm the diagnosis, but they aren't necessary for most children. If your child has had persistent fluid (chronic otitis) for three months or more, he should have a *hearing test* to determine whether or not the fluid is affecting his hearing. Children with persistent fluid and hearing loss may benefit from therapy with antibiotics or surgery.

The gold standard for diagnosing an acute infection of the middle ear is obtaining pus from behind the eardrum. This procedure, *tympanocentesis,* is done by sticking a small needle through the eardrum and withdrawing middle ear fluid into a syringe. It was routine in the pre-antibiotic era because it helped relieve the pressure in the middle ear by drawing off pus. Nowadays, it is used in research studies, but is almost never necessary for routine diagnosis. I recommend tympanocentesis for only:

- Diagnosing ear infections in children less than one month old
- Treating children whose immune systems are weakened or suppressed
- Treating children whose infections have not responded to the usual antibiotics

Blood tests are not useful in diagnosing ear infections or in distinguishing simple ear infections from other serious illnesses such as meningitis.

Ear infections are caused by the interaction of three factors:

WHAT CAUSES EAR INFECTIONS

1. Blockage of the *eustachian tube,* which drains the middle ear
2. Bacteria buildup
3. The body's *white blood cells'* reaction to the bacteria (inflammation)

Under normal circumstances, the eustachian tube drains the middle ear into the back of the throat, equalizing pressure and draining bacteria and viruses out of the middle ear. Several things can block normal eustachian tube drainage.

EUSTACHIAN TUBE BLOCKERS

- Swelling (from an allergy or a cold)
- Swollen adenoids or tonsils
- An unfavorable position or weak muscles around the eustachian tube
- Bottle propping

When children have colds or allergies, their noses aren't the only things that swell and feel stuffy. The lining of the eustachian tube also swells and may block drainage. Swollen tonsils or adenoids near its outlet at the back of the throat can also block drainage from the eustachian tube. The tonsils and adenoids are large lymph node–like structures located near the eustachian tube opening. If they are larger than average or they swell up during an infection or allergy, they can obstruct eustachian tube drainage.

The eustachian tube is narrow in young infants and grows as the body grows. In infancy, the tube lies almost horizontally between the middle ear and the back of the throat, making it fairly easy for viruses and bacteria from the nose and throat to migrate to the ear. As children grow, the eustachian tube becomes longer and more vertical, and it is more difficult for bacteria to migrate from the back of the throat to the middle ear. This is why ear infections are much less common in older children and adults. Some children have less developed muscles around the eustachian tube, and it falls shut easily. Breast-fed babies have about half the risk of ear infections as formula fed babies. Breast-feeding takes a little more muscles than bottle-feeding, and helps develop the muscles around the eustachian tube as well as building a really strong immune system.

Propping a bottle with the baby on his back to feed him allows milk to flow back into the back of the throat, blocking the eustachian tube. Babies who are fed flat on their backs with their bottles propped up have more ear infections than infants who are held while they're fed.

Viruses and bacteria can both infect the middle ears. Viruses that cause ear infections include influenza virus (for which there is a vaccine) and respiratory syncytial virus. The most common bacteria involved are *Strep* species (including *pneumococcus*, for which there is a vaccine), *Hemophilus influenza,* and *Moraxella catarrhalis*.

The body patrols the middle ear with *immune molecules* called immunoglobin A (IgA). It takes some experience with bacteria for us to develop the specific kind of immunoglobin needed to manage those bacteria. As children grow older, their immune systems develop more experience and respond more quickly and efficiently. Getting immunized also helps the immune system respond more efficiently to bacteria. This is one reason why older immunized children tend to have fewer ear infections than infants.

When the eustachian tube is blocked, bacteria have no easy way out of the middle ear. They proceed to set up housekeeping, multiplying like mad. When the immune system gets wind of the situation, it rushes in to clean up the problem that should have been taken care of by an open eustachian tube. The white blood cells release chemicals to kill the bacteria, but these chemicals also cause swelling and irritation. With such pitched battle going on in the tiny space of the middle ear, it's no wonder the child is in pain.

WHICH CHILDREN ARE MOST LIKELY TO GET EAR INFECTIONS?

By the time they're six months old, nearly 40 percent of children will have had at least one ear infection, and 20 percent will have had two or more infections. There are two kinds of risk factors for ear infections: the kind you can do something about (modifiable), and the kind you can't do anything about (non-modifiable) (Table 16.1).

Several non-modifiable factors increase the risk of having ear infections. Because of the position of the eustachian tube and their inexperienced immune systems, younger children (under two years old) are at higher risk than older children are. No one knows why, but boys are more likely than girls to have ear infections. Caucasian, Native American, and Eskimo children are at higher risk than are African American or Hispanic children. Like many other characteristics, the tendency for recurrent ear infections runs in families. Ear infections are much more common during the cold and flu seasons of fall and winter. Children may be born with problems such as cleft palate that can also increase the risk of ear infections.

There are also several modifiable factors that influence the risk of ear infections. Children who are *exposed to lots of other children* are much more likely to get ear infections than

children who are spared exposure to other children's colds. Child care in large groups (10 or more children) increases the risk of getting an ear infection by about 50 percent compared with keeping a child at home or being in a small family day care setting. Nowadays, it is almost impossible to avoid day care of some kind. However, if your child has been getting a lot of ear infections, you may want to look into a setting with fewer children.

Infants who are fed *formula* have a higher risk of getting ear infections in the first six months of life compared with infants who are fed mother's milk. Exposure to cow milk before four months of age is associated with an increased risk of ear infections. *Breast-feeding* helps protect your child because breast milk contains immune factors that can help fight infections, and it spares your child exposure to potential allergens in formulas. Breast milk is protective even for children at very high risk of ear infections, such as those with cleft palates.

Infants who are fed while *flat on their back* are more likely to get ear infections early in life than children who are fed with their heads elevated. If you feed your child from a bottle, keep the child's head up to keep the milk from getting into the eustachian tube. Putting a child to bed at night with a bottle of milk in his mouth increases the risk of both ear infections and cavities.

Children whose *parents smoke* get about 50 percent more ear infections than children whose parents do not smoke. Smoking during pregnancy also increases the baby's risk of having recurrent ear infections. DO NOT SMOKE and do not allow other people to smoke around you when you're pregnant or with your child, especially in enclosed places such as cars and homes. Other air pollutants, including nitrogen dioxide, are also associated with an increased risk of ear infections;[1] support efforts to reduce air pollution.

Table 16.1: Risk Factors for Ear Infections	
Non-Modifiable	Modifiable
Age less than two	Exposure to lots of kids
Boys more than girls	Formula feeding
Race: Caucasian, Indian, Eskimo	Feeding flat on back or propping the bottle
Runs in families	Exposure to tobacco smoke
Season: winter and fall	No immunizations
Congenital problems: cleft palate	Early exposure to milk products
	Having allergies
	Pacifiers
	Low vitamin D levels

Sheryl's mother-in-law lived with her family and was a smoker. Sheryl didn't like the fact that her children were around tobacco smoke, but she didn't feel that she could change her mother-in-law's behavior. I wrote a prescription for Nathan saying that he should not be around tobacco smoke because it increased his risk of getting ear infections, colds, and asthma. Sheryl gave this prescription to her husband who gave it to his mother, who agreed to smoke only outside the house.

Pneumococcus bacteria are responsible for many childhood ear infections. Immunization with *pneumococcal vaccine* significantly reduces the risk of ear infections. The pneumococcal vaccine is available for infants as young as two months old. It prevents ear infections and other serious infections caused by these nasty bacteria. Not all ear infections are caused by pneumococcus, so the vaccine doesn't completely eliminate the risk of getting ear infections, but it does decrease the risk significantly,[2] and I think that's worth it.

Influenza vaccine is not just for senior citizens anymore! Influenza vaccine was traditionally given to seniors to prevent deadly winter infections. But studies have shown that lots of kids get influenza, too, and ear infections are a common complication of influenza in children. The vaccine prevents influenza and cuts the number of ear infections, too.[3] Infants as young as six months old are eligible to receive the vaccine. I recommend it annually since flu viruses change over time.

Pacifiers are another modifiable risk factor for recurrent ear infections, and they may interfere with breast-feeding. Although I used to think pacifiers were harmless, I now discourage their use except for soothing children who are in pain (such as newborns in the intensive care unit or babies getting an immunization or circumcision).

Children born with *low vitamin D* levels are at increased risk of developing recurrent ear infections.[4] Supplementing with vitamin D to restore healthy levels reduces that risk.[5]

A note about recurrent ear infections: For children who get an ear infection within one month of a previous infection, don't blame yourself or your treatment. Most of the time (75 percent) the recurrence is due to a new bacteria altogether. Recurrences are rarely due to giving the wrong antibiotic or an insufficient length of treatment. They are more likely to reflect problems in the environment, such as crowding or cigarette smoke, or problems in the child's anatomy or physiology, such as large adenoids or allergies.

What Is the Best Way to Treat an Ear Infection?

Even without any specific treatment, most children (70 to 80 percent) will get over ear infections on their own. Then why treat? There are two main reasons: Ear infections hurt, and they can sometimes lead to more serious problems ranging from a ruptured eardrum to hearing loss, to infections of the bones around the ear (*mastoiditis*).

About 30 percent of children under two years old with recurrent middle ear infections have some mild to moderate *hearing loss*. Hearing usually improves when the middle ear fluid finally dissipates. It takes an average of one month for the fluid to clear out even with optimal treatment.

Several long-term studies have examined the impact of middle ear infections and middle ear fluid on children's language development and school performance. In the 1980s doctors and psychologists in North Carolina followed a large group of children from birth until school-age, checking them frequently for ear infections and monitoring their development.[6] There was *no* relationship between the number of ear infections in the first three years of life and subsequent scores on any of several IQ tests or academic achievements in kindergarten. A *much* bigger factor in children's language development and school performance was parental responsiveness to the child and the creation of a stimulating home environment. The same researchers followed the children up to age 12 and did not find any relationship between early ear infections and later development.[7]

Because of the reasonable concern that persistent fluid in the middle ear would lead to hearing loss that would impede language development, many pediatricians in the 1980s and 1990s recommended inserting ear tubes to drain the middle ear fluid and restore hearing. However, meticulous research showed no long-term benefit of this surgery on child development.[8]

If you suspect that your child has a middle ear infection, you may try home treatments for 24 to 48 hours before consulting a health care professional *unless*:

- Your child is under 6 to 12 months old.
- The pain is severe.
- The fever is higher than 102°F.
- Your child seems sicker than usual, or you are concerned about his appearance or behavior.

In these cases, see your health care professional. Let's consider all the therapies to find out what works best to prevent and treat ear infections. If you want to skip to my bottom-line recommendations, flip to the end of the chapter.

LIFESTYLE THERAPIES: NUTRITION, EXERCISE, BEHAVIOR, ENVIRONMENT, MIND-BODY

Nutrition

Reduce your child's risk of developing ear infections by breast-feeding for at least 6 and preferably 12 months. Do not introduce cow's milk before four months of age. Mother's milk is also a good source of essential fatty acids. In one study infants fed formula supplemented with omega-3 essential fatty acids had fewer colds, ear infections, and less diarrhea than infants fed formula without these essential nutrients.[9] Moms, keep your omega-3 levels high by eating fish such as salmon or sardines weekly, and eating omega-3 rich walnuts, flax seeds, and chia seeds. Once the child is no longer nursing, consider adding omega-3 fatty acids to her diet.[10]

If your child has frequent middle ear infections, consider having him evaluated for allergies. *Food allergies* can cause swelling and inflammation of the eustachian tube and the nose and lead to fluid buildup in the middle ear. Omitting allergy-inducing foods can significantly reduce the frequency of ear infections. In one study, children who had persistent middle ear fluid and documented food allergies went on a diet by eliminating those foods for 16 weeks; middle ear fluid resolved in nearly 90 percent of them. When they were re-challenged with the problem food, 94 percent had recurrent middle ear fluid.[11] Please consult with a pediatric dietitian before starting any kind of restrictive diet to make sure your child continues to get optimal amounts of all essential nutrients.

Many parents reduce their children's intake of *dairy products* when they have a cold or ear infection, because they believe that milk increases mucus production. Also, some parents restrict *sugar, honey,* and sources of concentrated *fruit sugar* when their child has an infection, based on the theory that sugar inhibits the immune system. More research is needed to determine how effective these practices are before I routinely recommend them.

Do make sure that your child gets plenty of fluids; your child will feel even worse if she gets dehydrated.

Avoid vitamin D deficiency. If your child has frequent colds and ear infections, ask your health professional to check her vitamin D level and supplement if needed to get her level above 30 ng/mL.

Exercise

Because one of the underlying causes of ear infections is blockage of the eustachian tube, exercises that help open the eustachian tube may be helpful. How do you exercise the eustachian tube? Think of the ways you pop your ears after a sudden change in altitude: yawning, bearing down, blowing your nose with the nostrils pinched shut, and chewing gum. Probably the easiest of these for children is chewing gum or blowing up balloons. Obviously it's difficult, if not impossible, to have an infant chew gum, but most children over the age of about two-and-a-half are delighted to try. Sugarless gum is best. Beware: Too much sugarless gum can cause diarrhea. Don't let your child go to bed with gum, or he could awaken in the morning with a gooey mess in his hair. Is there any scientific evidence as to whether chewing gum is effective? Not yet, but it would sure be fun to find out!

Is *air travel* safe for children with ear infections? Yes. In a study of 14 children with middle ear fluid who flew in commercial, pressurized airplanes, none developed complications. Children who are on the verge of getting an ear infection or who have colds or allergies that intermittently block the eustachian tube may experience some pain, especially during landings. If your child has a cold and is prone to ear infections, let him suck on something or chew gum during takeoffs and landings to help keep the eustachian tube open.

It's also helpful to prop the child's head up, especially when he goes to sleep. Propping the head up also helps the eustachian tube drain the middle ear.

Environmental Therapy

For many years, people have used *heat packs and ice packs* to treat the pain of ear infections. Ice packs take the heat out of inflamed tissue; heat packs increase circulation and theoretically remove toxins more quickly. I have patients who use each. Some parents apply an ice pack or a bag of frozen peas or corn to a child's ear when the child complains of ear pain. Other parents put a hot water bottle, heating pad, or warm washcloth on the painful ear while waiting for the immune system to complete the healing process. If your child doesn't respond to warm compresses, you can try ice packs and vice versa. I don't know which of these is more effective (there are good reasons to believe in both) or if the main benefit is simply doing something for and spending time with the child. Please don't use heat packs or cold packs for more than 20 minutes at a time. Common sense precautions can prevent burns and frostbite.

Another time-honored technique is instilling *warm mineral oil* or *olive oil* in your child's ear. Heat the oil as you would for a baby's bottle and make sure it is not too hot before dropping in your child's ear. The heat speeds circulation and comfort, but no scientific studies have evaluated its effectiveness. Do not put any home remedies in your child's ear if there is pus draining out of the ear.

Do *not* try to remove earwax or anything else from your child's ear. Earwax does not cause middle ear infections. If you see pus or any kind of drainage from your child's ear, the eardrum may have ruptured. If there is drainage from the ear, do not put ANYTHING in the ear until you have consulted a health care professional.

A therapy that was popular a few years back was *ear candles*. These were special hollow candles used only to "help draw toxins out" of the ear. The bottom was placed in the ear canal (violating the old rule of never putting anything smaller than your elbow in your ear!), and the top was lighted. The theory was that the heat would melt earwax and allow it and other bad things to drain out. The problem was that they didn't work: Candle wax dripped into the ear instead, some kids got burned and had other injuries, and overall they were just much more trouble than they were worth. I do *not* recommend them.

Make sure your child's environment is free from tobacco smoke!

Mind-Body

Whenever a child is in pain, it is helpful to keep him calm. The best approach to keeping your child calm is to stay calm yourself. Visualize yourself as competent and able to care for and help your child; visualize your child as healthy and able to heal whatever is ailing him. Once you are in a calm frame of mind, you can distract your child from the pain by singing favorite songs, telling stories, playing quiet games, and reminding him of how much you love him.

Nathan was so distracted by the idea of having birds in his ear, he was eager for the previously dreaded ear exam.

BIOCHEMICAL THERAPIES: MEDICATIONS, DIETARY SUPPLEMENTS, HERBAL REMEDIES

Medications

A number of medications can either help the child feel more comfortable while he heals or help hasten the healing process. They are also used to reduce the risk of serious complications such as meningitis and mastoiditis. Even without antibiotics, serious complications are rare these days. This means that for the most part, treatments are aimed at keeping the child comfortable while his immune system fights the infection.

MEDICATIONS FOR EAR INFECTIONS

- Analgesics or pain medicines
- Antibiotics
- Steroids
- Antihistamines and decongestants

Pain is one of the worst symptoms for most children suffering from ear infections. Even the most powerful antibiotics take 24 hours or so to improve symptoms; in the meantime, you'll want to help your child manage his pain. The most commonly used pain relievers are *acetaminophen* (the active ingredient in Tylenol™ and other non-aspirin pain relievers) and *ibuprofen* (the active ingredient in Motrin™ and Advil™). Both are effective pain relievers and fever fighters; and both are available without a prescription. Acetaminophen takes about 45 minutes to start working; each dose lasts about four hours. Ibuprofen starts working a few minutes sooner and lasts six to eight hours. Ibuprofen can also decrease the inflammation in the ear that's causing the pain, not just cover it up. This may help decrease the amount of fluid in the middle ear and allow healing and normal hearing to occur faster. Do not give your child aspirin. Aspirin suppresses the immune system, and it can cause a sometimes fatal illness called Reye's syndrome.

Anesthetic eardrops can also help reduce pain in the ear. Commonly prescribed anesthetic eardrops include topical lidocaine, benzocaine, and a combination product, *Auralgan*™. These medications do not help the ear infection go away any faster, but they can help your child feel more comfortable while he fights off the infection. Auralgan is significantly more effective than olive oil drops, relieving pain within 30 minutes. Herbal eardrops can also help (see Herbal Remedies section). Do not put drops of any kind *in* your child's ear if there is anything draining *out* of your child's ear; if you see drainage, see your doctor. Allergies to topical anesthetics are possible.

Since the dawn of the *antibiotic* era, the frequency of complications due to ear infections has fallen dramatically. In 1938 (before antibiotics), 20 percent of cases of severe ear infections resulted in mastoidectomy (removal of the mastoid bone behind the ear). After the introduction of antibiotics, the rate of mastoidectomy has dropped to less than 1 percent. Is this because antibiotics work so well or have bacteria become more tame? No one knows for sure. The standard of care in the United States is to treat ear infections with antibiotics, particularly for children less than 6 to 12 months old and those with severe symptoms. On the other hand, doctors in the Netherlands, the UK, Sweden, and Iceland mostly focus on helping manage the child's pain, while they wait for the body's own immune system to take care of the problem.[12] Less than 2 percent of children with ear infections in these countries get prescriptions for antibiotics. Countries that use fewer antibiotics also have much lower rates of smart (drug resistant) bacteria; in fact, they have about 90 percent less drug resistant bacteria than countries that use antibiotics more freely.

Studies of the effectiveness of antibiotics have been conflicting. Some suggest that antibiotics are more helpful for younger children, who are more likely to have severe infections and less able to clear infections on their own. In the 1950s, studies in Europe and America showed that antibiotics were significantly more effective than placebo in decreasing drainage from the ear, resulting in fewer complications such as meningitis and mastoiditis. By the late 1960s, 90 to 95 percent of children treated with antibiotics improved in one to two days, compared with 75 percent of those who re-

ceived placebo medication. By 1997, one study showed that antibiotic treatment had no significant impact on pain at 24 hours and no impact on the number of subsequent bouts of infection or hearing loss a month later; the antibiotic group tended to have less pain at 48 hours, but in order to prevent one child from having pain, 17 children needed antibiotic treatment. This means that 16 children got antibiotics unnecessarily for every one child who benefited from them. Subsequently, a 2013 review of all published research concluded that about 20 children had to be treated to help one child have less pain at two to seven days after starting treatment (and no way to know which was going to be that one child), weighed against the risk of adverse events from antibiotics such as vomiting, diarrhea, and rash; antibiotics were most useful in children under two years of age, with infections in both ears or with a ruptured eardrum and pus draining out.

These findings led to a recommendation (for children over six months with mild to moderate symptoms, single ear infections, and no immune disorders) of delayed antibiotic treatment (waiting for 48 hours to see if the infection would start to clear on its own) for ear infections. In Switzerland, the Netherlands, and Britain, antibiotics are reserved for children who are not better in two to three days. In these countries, 92 to 96 percent of children suffering from ear infections reportedly recover within three days without antibiotic therapy. A review of 10 studies comparing different treatment strategies found no differences in complications; a benefit of somewhat earlier pain relief with antibiotics; but a huge reduction in the use of antibiotics in the delayed or no treatment group[13] American parents were slightly less satisfied with delayed versus immediate antibiotics (87 percent versus 92 percent satisfied). On the other hand, antibiotics wipe out a lot of beneficial gut bacteria, can cause upset stomach and yeast infections, and can lead to bacteria that learn to outsmart antibiotics (bacterial resistance).

I favor the European strategy of (a) managing pain and (b) empowering parents to make the choice about antibiotics, giving them an antibiotic prescription as a backup, but asking them *not* to fill it unless the mild to moderately ill child does not improve within 48 to 72 hours or if the child seems to be getting worse. Parents also need someone to call for help and reassurance if the child isn't improving. For children with severe symptoms that have lasted for more than 48 hours by the time I see the child, or children with fevers over 102°F (39°C), or with infections in both ears, or children less than six months old, I recommend that parents fill the prescription (typically for amoxicillin if the child isn't allergic to it) and start antibiotics that day; I also offer to see the child again if she's not improving within 48 to 72 hours of starting antibiotic therapy.[14] This approach protects the sickest children and empowers parents with the tools they need to help their child while providing support and backup.

If You Decide to Go for Antibiotics, How Do You Know Which One to Choose?

Overall, the commonly used antibiotics have similar effectiveness. That is, there is no major advantage of more expensive antibiotics over less expensive, generic antibiotics. Please don't ask your pediatrician for an expensive antibiotic just because it's new. Most pediatricians start with amoxicillin unless (a) the child has a penicillin allergy; or (b) the child has had amoxicillin for another infection within the previous 30 days.

Antibiotics can cause side effects in 20 to 30 percent of children: upset stomach, diarrhea, allergic reactions, and diaper rashes. The old standard was to give antibiotics for 10 to 14 days. Giving shorter courses (e.g., five days) appears to works just as well with fewer side effects.[15] If you're going to use antibiotics, use them for the shortest period of time required to fix the problem. A major problem with the common use of antibiotics is the development of bacteria that are resistant to them. The same bacteria that cause ear infections also cause sinus infections, bronchitis, and pneumonia. As antibiotics are used more widely, the bacteria start to figure out how to work around them and eventually become totally resistant to the medication.

For children who are vomiting or cannot take medicine by mouth for some other reason, a single shot of *ceftriaxone* is as effective as several days of antibiotics, though it costs more. This is an approach we often use in emergency settings if the child is quite ill, has an underlying immune problem, or it's clearly going to be a struggle to get the child to take oral antibiotics.

Please don't ask your doctor for a prescription for antibiotics to *prevent* ear infections. Antibiotics do *not* keep colds from turning into ear infections, and ongoing use of antibiotics promotes the growth of bacteria that are able to resist antibiotic therapy when it is needed.

I often recommend that parents give their child supplemental *probiotics or yogurt* while they are taking antibiotics for ear infections. The antibiotics kill many of the body's normal bacteria, and can lead to yeast infections in the mouth (called *thrush*), the diaper area, and the vagina (*Candidiasis*). Probiotics and yogurt help replace the body's normal bacteria and prevent yeast infections.

What Is Otitis Media with Effusion (OME)?

When fluid stays in the middle ear for six weeks or more following an acute ear infection, it is called a *chronic* ear infection or chronic otitis media with effusion (OME). Chronic ear infections impair hearing and predispose to repeated infections. In this situation, treatment with a *combination of an antibiotic* and *prednisone* (a steroid which decreases inflammation), followed, if necessary, by a second course of antibiotics may help clear up the middle ear.[16] While the antibiotic kills the bacteria, the steroid helps decrease eustachian tube swelling, thereby promoting middle ear drainage. Although steroids can have harmful effects when given over long periods of time, they have very few side effects when they are taken for less than a week.

What's the Safest Way to Use Steroids for Middle Ear Disease?

Steroid nasal sprays are more effective than placebo sprays in clearing middle ear fluid, and they avoid the potential body-wide side effects of taking steroids by mouth because very little of the medicine is absorbed into the bloodstream. However, compared with placebo, steroid nasal sprays do not appear to be a cost-effective approach to treating middle ear fluid.[17]

What about Eardrops to Treat Middle Ear Infections?

Antibiotic eardrops are *not* effective against middle ear infections (if the eardrum hasn't ruptured) because they do not cross into the middle ear space (the eardrum is in their way). Eardrops containing neosporin and polymixin may actually damage the inner ear if they are given to a child with a ruptured eardrum. I do not recommend antibiotic eardrops for children with ear infections unless (a) the infection is in the ear canal (external otitis or swimmer's ear); (b) the child has an ear tube that is draining pus; or (c) the eardrum has ruptured, and pus is draining out. When pus is draining out, this is called *suppurative otitis media* or *otor-rhea*. Please see a health professional to distinguish between swimmer's ear and suppurative otitis media.

What Is Swimmer's Ear?

Swimmer's ear is a painful infection of the ear canal. The middle ear is not infected, and this condition, though it is painful, does not cause long-term hearing loss or risk serious complications like meningitis. It is caused by different bacteria than those that cause middle ear infections. The bacteria that cause swimmer's ear love the moisture left in the ear after swimming and hate acid. This has led to the common practice of putting drops in the ear after swimming that contain alcohol (to speed evaporation) and vinegar or lemon juice (to make the ear canal slightly acidic). If prevention doesn't work, see your pediatrician for prescription eardrops.

Do Antihistamines and Decongestants Help?

Antihistamines and *decongestants* (common cold medicines) do not help drain middle ear fluid.[18] What's worse, they can cause side effects.[19] Antihistamines usually make children sleepy and dry their secretions. Paradoxically, some children react to antihistamines by becoming agitated. Decongestants can raise blood pressure and make children feel wound up, tense, and less hungry. I do *not* recommend them.

Dietary Supplements

Some practitioners recommend *dietary supplements* or foods containing high levels of *vitamins A, B-complex, C,* and *E, zinc,* and *evening primrose oil* to boost the immune system during an acute infection. Correcting deficiencies of vitamin A and zinc in poorly nourished children can help protect against the development of ear infections,[20] but most children in developed countries do not need them. On the other hand, many children have suboptimal *vitamin D* levels,[21] and optimizing vitamin D can help reduce the risk of ear infections. Breast-fed babies should receive vitamin D supplements—at least 400 IU daily for the first year of life. It is easy to test vitamin D levels with a blood test; if you're curious, ask your health professional to check.

I know it's unusual for a doctor to recommend sugar, but the sugar from birch trees, *xylitol,* has impressive effects in preventing ear infections. In fact, xylitol is so good at blocking bacteria that it is used in some chewing gums to help prevent cavities. In the 1990s two studies involving nearly 1,200 Finnish preschoolers showed that those who received chewing gum (kids old enough to chew gum) or syrup (younger kids) sweetened with xylitol (eight to nine grams daily) had 30 to 40 percent fewer ear infections than the children who received regular gum or syrup.[22,23] It doesn't work so well if you wait until the child is sick before offering xylitol gum.[24] For children who have frequent ear infections, I recommend preventing future infections by chewing at least three to five sticks a day of chewing gum sweetened with xylitol in children old enough to chew gum. Sorry, xylitol syrup doesn't work as well,[25] probably because it doesn't stay in the back of the throat and it skips the chewing action that can help pop open and strengthen the muscles around the eustachian tube. Read the labels; few sugarless gums are sweetened with xylitol, but you can find them if you look.

Everyone knows about using antibiotics to treat bacterial infections, but what about using *probiotics* to help the good bacteria replace the troublesome kind? A review of four studies concluded that using *Lactobacillus* GG (Culturelle™) compared to placebo reduced the risk of ear infections by 25 percent and reduced the need for antibiotic treatment by 20 percent without any serious side effects.[26] French researchers using a different mix of probiotic supplements failed to find a similar benefit.[27] So the benefits of probiotics depend on the type and dose of the strains used. While additional research is under way, I recommend *Lactobacillus* GG supplements to help reduce the risk of recurrent ear infections.

Herbal Remedies

Herbal eardrops can help soothe ear pain, but don't put them in an ear that has fluid draining out of it.

In 2001, a study of over 100 children 6 to 18 years old with ear infections found that herbal eardrops (*Otikon*™, Israel) containing garlic, calendula, mullein, and St. John's wort were as effective as anesthetic eardrops containing ametocaine and phenazone in relieving ear pain.[28] In a large follow-up study including children as young as five years old, the herbal eardrops were at least as helpful, if not more helpful, than conventional medications in improving ear pain.[29] Both studies were done in Israel, and it's not clear that American herbal products are as effective. Herbal products are regulated more like food than like medications in the United States, and there is a high degree of variability in herbal products.

Parents from many different parts of the world swear by *garlic oil ear drops* to treat their child's pain. To make garlic oil, soak four to five crushed cloves of garlic in one-quarter cup of olive oil overnight. *Strain out the garlic bits*. Please do *not* put chunks of garlic in the ear. Garlic pieces can get stuck in the ear canal. If your child's ear hurts when you wiggle the outer ear or there is any discharge from the ear, DO NOT place garlic oil or anything else in the ear canal without checking with a health care professional. Other herbal eardrops include *mullein oil* (mullein flowers, covered with olive oil overnight, then strained), *St John's wort oil*

(St John's wort flowers, mixed with olive oil and strained) or *tincture of plantain* or *witch hazel*. None of these individual products has undergone rigorous randomized controlled trials, so I cannot recommend specific formulations or doses, but I support families who want to use safe, culturally meaningful approaches to help soothe their children's pain.

Other home eardrop remedies I learned about from older pediatricians include supersaturated solutions of *sugar* or *Epsom salts*. The theory behind this practice is that the supersaturated solution will draw free water across the eardrum from the bulging side to the ear canal, decreasing pressure in the middle ear; that sounds good, but I don't know of anyone who has actually studied it. To make a supersaturated solution, heat a cup of water and stir in sugar (or Epsom salts) a teaspoon at a time until no more will dissolve. Gradually cool the solution until it is comfortable before putting it in the ear. If your child suffers *any* discomfort with any of these drops, flush the ear with warm water or hydrogen peroxide and discontinue their use. Again, there are no rigorous, randomized trials evaluating the safety and effectiveness of these eardrops, so I cannot recommend them, but I tolerate their use.

Several herbal *teas* have been used for children with ear infections, but none has proved effective and safe in preventing or treating ear infections so I can tell you what they are, but I'm not ready to recommend them until additional research establishes their role. These include *echinacea, goldenseal,* and *licorice root.* Other herbal teas or tinctures for children with ear infections include *calendula, chamomile, elderflowers, goldenrod, ground ivy, hops, hyssop, lobelia, peppermint, passionflower, red clover, skullcap, St. John's wort,* and *wintergreen.*

Herbs come in a variety of forms (dried, alcohol extracts, water extracts), and there is not much standardization in terms of dosage and frequency. This means that the product you buy may be a complete rip-off, containing little or none of the active herb. It might also be contaminated with pesticides, herbicides, metals, or the wrong herb or even medications! Although many of these preparations have been used for many years as part of traditional herbal healing, there is no scientific evidence evaluating their effectiveness in children with ear infections, and I can't recommend them as specific treatments to help cure ear infections. On the other hand, a nice cup of chamomile tea can be a gentle sedative for an upset child (or stressed out parent!). Do *not* give your child ephedra (ma huang) as a decongestant; it can cause heart problems, high blood pressure, and other serious side effects.

An unusual remedy recently reported from Italy is the combination of bee *propolis* and *zinc*. Propolis is a glue-like resin that bees collect from trees and sap, and use to seal small cracks in their hives. A combination propolis-zinc product given by mouth to children one to five years old who had suffered from recurrent ear infections showed a significant decrease in the risk of subsequent ear infections compared to a group of children who received a placebo.[30] Additional research is needed before I'd routinely recommend this preparation.

BIOMECHANICAL THERAPIES: MASSAGE, SPINAL MANIPULATION, SURGERY

Massage

Many parents naturally massage the area affected by illness or injury. Gentle *massage* of the upper back, neck, and around the ears and back of the head with *cocoa butter, camphorated oil, Tiger Balm,* or *Vick's VapoRub®* is soothing and relaxing. Some parents also gently massage the enlarged lymph glands in the neck. Massaging downward toward the chest may help drain the lymph glands that are swollen with white blood cells fighting the infection. Foot massage may also be relaxing and less uncomfortable if the child is very sensitive around the ears and neck. There are no studies evaluating the effectiveness of massage in treating children with ear infections, but I can't think of contraindications and would love to see a study evaluating it.

Spinal Manipulation

Osteopaths claim that their treatment, particularly cranial manipulation, is effective in both preventing and treating ear infections. Osteopathic manipulation was one of the most frequent recommendations made by the American psychic, Edgar Cayce, for children suffering from recurrent ear infections. The idea of treating a fundamentally anatomic problem (eustachian tube drainage) with an anatomic therapy (spinal or cranial manipulation) is appealing, but not well researched.

One of the most common reasons that families take their children to *chiropractors* is to help prevent recurrent ear infections. Chiropractors have reported numerous anecdotes about recurrent infections that finally stopped after a course of three to five sessions of chiropractic adjustments. A review of dozens of studies of spinal manipulation therapy (SMT) concluded that there was "limited evidence" that SMT was helpful; on the other hand, it is quite safe.[31] Until more data are in, I can't recommend chiropractic as a way to prevent recurrent ear infections, but I don't object to families trying it.

SURGERY: MYRINGOTOMY, TYMPANOCENTESIS, TYMPANOSTOMY TUBES, ADENOIDECTOMY

My grandfather practiced pediatrics in the days before antibiotics. He specialized in performing *myringotomies* on children with severe ear infections. A myringotomy involves taking a tiny scalpel and cutting a hole in the eardrum so that the pus from the middle ear can drain out. Myringotomies help relieve the pain-causing pressure, but they can cause permanent holes and scars in the eardrum. Studies performed since antibiotics became available show that antibiotics are a better choice, and now myringotomies are rarely done—gone the way of the horse and buggy.

Another one of the oldest treatments for ear infections is *tympanocentesis*, which is puncturing a hole in the eardrum with a needle and removing the pus. Unlike myringotomies, removing pus with a tympanocentesis needle rarely leaves a permanent hole in the eardrum or leads to scarring. This procedure is mostly done nowadays by ear-nose-throat (ENT) doctors and is rarely necessary except for infants less than a month old who develop ear infections (because they often have unusual bacteria that need different antibiotics) or for older children whose infections are not improving with the usual antibiotics.

Placement of *tympanostomy tubes* (ear tubes) is the most common operation performed on children. This procedure involves placing a tiny plastic tube (called a "grommet" by the British) through the eardrum. The tube allows drainage from the middle ear to the ear canal, functioning like a backup eustachian tube. Recurrent and persistent ear infections are frustrating, and placement of ear tubes offers hope for quick relief.

Ear tubes can provide some short-term benefits in terms of decreased fluid and improved hearing, but have no long-term benefit in terms of speech, language, or behavior. Ear tubes only work until the tube becomes plugged up or comes out (usually within 6 to 12 months). Ear tubes can scar the eardrum and cause other complications. Children who are almost certain to have frequent and severe ear infections, such as children with cleft palate, are definite candidates for ear tubes because they are clearly helpful for these children in maintaining hearing and developing speech. If a child does not have a facial abnormality like a cleft palate, I do not recommend ear tubes unless the child has already had a thorough evaluation and treatment for allergies, correction of vitamin D levels, pneumococcal and flu vaccine, and is still having three ear infections in six months, four ear infections in 12 months, or persistent ear fluid that is interfering with hearing and speech. If every pediatrician followed this protocol, many fewer children would face surgery.

Because swollen adenoids can block the eustachian tubes, *adenoidectomy* (removal of the adenoid lymph tissue) with or without *tonsillectomy* is another common surgical therapy for children with recurrent ear infections.[32] These additional surgeries done at the same time that ear tubes are placed are more effective than ear tubes alone in preventing recurrent infections and may reduce the risk of requiring subsequent surgery, but they also increase the risk of bleeding and other complications. Again, unless a child has facial abnormalities like a cleft palate or has had very frequent recurrent ear infections or persistent ear fluid despite optimal attention to allergies, immunizations, vitamin D levels, a trial of xylitol, and so on, I do not recommend this surgical approach.

BIOENERGETIC THERAPIES: ACUPUNCTURE, THERAPEUTIC TOUCH, PRAYER, HOMEOPATHY

Acupuncture

Acupuncture has been recommended for treating ear infections, but there are no studies evaluating the effectiveness of acupuncture treatment compared with antibiotics or other therapies for children with ear infections.

Therapeutic Touch and Prayer

I use Therapeutic Touch to help me diagnose ear infections before I look in the ear, and I have found it highly accurate in predicting what I find when I look with an otoscope. I also use Therapeutic Touch to soothe children with ear infections, but have not found it to be an instant cure or any better than the child's own immune system. There are no studies systematically evaluating the effectiveness of these "energy healing" techniques, but they are very safe and often valued by families, so I continue to use them until scientific studies have had conclusive results.

Homeopathy

Despite its years of historical use, the data from randomized controlled trials about homeopathy's effectiveness has only emerged in the last 25 years. In an early German study, children who went to homeopathic practitioners improved more quickly, had fewer recurrences, and used fewer antibiotics than children who saw ENT specialists.[33] However, this was a self-selected group of children whose average age was five years old, when the immune system is stronger than in infancy and most ear infections go away on their own without any treatment. It is possible (in fact, it seems pretty likely) that those who went to the specialists were sicker and less likely to do as well as those who went to homeopaths. An intriguing result, but a poor study design left more questions than answers. A British study reported that children who had fluid in their ears who were treated with homeopathic remedies did a little (but not significantly) better than those who didn't receive homeopathy.[34] Skepticism about homeopathy persisted.

More recent studies have been more rigorous and offered stronger evidence for the benefits of homeopathic treatments. The 2001 issue of the *Pediatric Infectious Disease Journal* rocked the world of conventional doctors by describing a randomized, double-blind trial comparing homeopathy to placebo in 75 children diagnosed with ear infections. Despite the expectation that homeopathy would not offer any benefits beyond sugar pills, the results indicated that homeopathy was better; there were fewer symptoms and fewer treatment failures with homeopathy than with placebo.[35] Ten years later, a Seattle study randomized 119 children to standard therapy alone or standard therapy plus homeopathic ear drops; again, the children who received homeopathy did better.[36] A randomized trial in India found that homeopathic treatment was faster than conventional pain medicines in reducing ear pain; homeopathic treatment was also associated with a marked reduction in the need for antibiotics.[37] A 2014 French study found that families who sought care from homeopathic practitioners ended up using far fewer antibiotics and analgesics than those who sought conventional care, though treatment outcomes were similar.[38] No large studies have shown homeopathy to be inferior to antibiotics for treating ear infections. Overall, the evidence for homeopathy supports the idea that using it for ear infections results in speedy symptom relief, less need for antibiotics, and fewer side effects than strategies that start with antibiotics.[39]

Homeopathic practitioners use a variety of remedies for children with ear infections depending on their exact symptoms. Most professional homeopaths rely on a single remedy, carefully selected for the patient's individual temperament and symptoms. The most frequently used remedies are *Pulsatilla, Chamomilla, Sulfur,* and *Calcarea carbonica.* Most parents, on the other hand, simply choose a remedy off the store shelf; many of these products contain combinations of the most commonly used individual remedies. For example, *ABC* is a homeopathic mixture containing minute amounts of *Aconite, Belladonna,* and *Chamomilla*; it covers all the major types of acute ear infections in children. Homeopathic remedies are typically given one to three times daily for less than five days.

Homeopathy looks like a promising approach for parents who are willing to hold off on antibiotics but who want to try *something* while their child is ill. Seeing a homeopathic practitioner takes longer than seeing a regular physician, and the costs may not be covered by your insurance unless your homeopathic practitioner is an MD or chiropractor. If you decide to use homeopathy and *if your child is not better after 48 hours of homeopathic treatment, please have him examined by another health care professional.* Also, if your child's earache is accompanied by a high fever (over 102), headache, or stiff neck, or the child is less than six months old, do not try homeopathic remedies or any other home treatment. Go directly to a health care facility to make sure your child does not have another serious infection that requires antibiotics.

WHAT I RECOMMEND FOR EAR INFECTIONS

PREVENTING EAR INFECTIONS

1. *Lifestyle—nutrition.* Breast-feed your infant for at least six to twelve months; do not feed your infant while he is lying on his back; do not put a child to bed with a bottle of juice or milk. Avoid giving cow's milk prior to four months of age. Make sure his vitamin D level is optimal. Minimize pacifier use; encourage sugarless lozenges or gum containing xylitol in children old enough to chew gum.

2. *Lifestyle—environment.* Do not smoke, and do not allow others to smoke around your child; avoid exposure to wood smoke, other irritants, and allergens. Avoid putting your child in day care until he's at least 12 months old; if your child is in day care, avoid crowded settings (more than six children per room), especially during the fall and winter when cold viruses abound. Advocate for clean air, and minimizing air pollutants such as nitrogen dioxide.

3. *Biochemical—medications.* Get your child immunized against pneumococcus and influenza. If you suspect allergies as a contributing factor, ask for a referral to a pediatric allergist for evaluation and treatment.

4. *Biochemical—supplements.* Consider probiotic supplements containing *Lactobacillus* GG to reduce recurrent infections.

5. *Lifestyle—exercise.* Have your child exercise his eustachian tube by blowing up balloons or chewing gum.

Take your child for a professional evaluation if he

- Is less than twelve months old

- Has a temperature higher than 102°F

- Refuses to drink liquids

- Has severe ear pain for more than one day

- Seems sicker than usual, or you are concerned about his appearance or behavior

- Has ear pain that has not improved within one or two days with home remedies

---

TREATING EAR INFECTIONS

1. *Biochemical.* Consider pain medication such as acetaminophen or ibuprofen; consider antibiotics for an acute infection, especially if your child is less than two years old. Consider standardized herbal eardrops (containing calendula, garlic, mullein, and St. John's wort) or anesthetic eardrops. If your child has recurrent ear infections, consider giving him probiotic supplements daily (*Lactobacillus* GG). Do not rely on antihistamine or decongestant medications to prevent or treat ear infections.

2. *Lifestyle—environment.* If your child has frequent colds and ear infections, have him evaluated and treated for allergies. If he has ear pain, try putting a warm pack, hot water bottle, or an ice pack over his ear for ten minutes to reduce ear pain. Do not allow anyone to smoke around your child! Do *not* use ear candles.

3. *Lifestyle—mind-body.* Comfort or distract your child (rocking, holding, telling stories, etc.).

4. *Lifestyle—exercise.* Have your child exercise his eustachian tube by blowing up balloons, chewing gum, or sucking on something. Prop his head up when he goes to sleep to minimize the pressure buildup in his ear.

5. *Biomechanical—surgery.* If other preventative and treatment strategies haven't worked within six months, it's time to consider surgery for ear tubes and a look at adenoid removal.

6. *Bioenergetic—homeopathy.* Consider a trial of a homeopathic remedy in consultation with a trained homeopathic physician.

17
ECZEMA

Wendy Fowler brought in her ten-month-old, James, because of his skin rash. He had been seen in the clinic four months ago by another doctor who had diagnosed eczema and recommended hydrocortisone cream. After the rash improved, they stopped the cream. Now the rash was back; James' itching was driving Wendy crazy, and she was concerned that the rash on his chest was becoming infected. James' father suffered from asthma and allergies, and James had had two ear infections already. He was otherwise healthy. Wendy wondered:

- What caused eczema?
- What else could she do to rid James of this rash?

Whether it's called eczema or *atopic dermatitis,* this rash is itchy, and it keeps coming back. Eczema is "the itch that rashes." Anything that provokes itching and scratching (such as an insect bite) can trigger an eczema flare-up. It starts out red, dry, and itchy. As it gets worse, it contains little blisters that break down into scabs, provoking more itching. Over time, the skin may become thickened and rough like tree bark, and the skin color may get darker or lighter than in unaffected areas. When the rash is well managed, the skin color and texture improve over several months. When it gets out of control, it easily becomes infected, becoming painful as well as itchy.

Eczema affects different areas of the skin as children grow and develop. In young infants, it usually appears on the cheeks. Drooling often irritates the skin around the mouth. During toddler years, the rash tends to occur on the backs of the arms and legs. In older children and adults, the worst symptoms are on the inside of the elbows and the backs of the knees. High-top sneakers and other shoes that make the feet sweat can aggravate eczema on the feet. The backs of the hands can be real trouble spots for dishwashers, cooks, and others who frequently have their hands in water.

Eczema is very common. By seven years old, about one in five children has been diagnosed as having eczema; 90 percent of cases appear by five years of age. Eczema most commonly starts between three and six months old, as many babies start consuming solid foods. As with James, it is more common in firstborn children, those who were especially large as newborns, and those who have allergies or asthma in the family. People who have eczema are prone to symptoms for their whole lives, but they often improve during adolescence and early adulthood.

As if itchy, dry red skin wasn't enough, children with eczema frequently suffer from other problems. *Asthma* and *allergies* are much more common in eczema sufferers. Infants with eczema are also prone to getting *cradle cap*. Children with eczema often put so much energy into healing their skin that there is not enough left over to grow well; thus, they may be *thin and short*. The *lymph nodes* in the neck, the armpits, and the groin can become quite *swollen* in children suffering from eczema. The neck lymph nodes swell when eczema affects the face; the armpit nodes swell when eczema affects the arms; and the groin nodes swell when eczema affects the legs. Miserable and itchy, children with eczema also frequently suffer from *sleep problems, behavior problems,* and *difficult temperaments.* Children with eczema are *not* at increased risk of developing skin cancer. They are prone to developing *skin infections* and *warts*. Scratched and irritated skin is easily infected with viruses (such as *Herpes*) and bacteria (such as *Staphylococcus*). If your

child's eczema suddenly takes a turn for the worse, especially if it becomes *painful,* take him to a health care professional to be evaluated for an infection.

It looked as if James had a typical case of hereditary eczema. He had scratched it vigorously, and a few spots were starting to look infected. The itch and pain were interfering with his sleep, and he was miserable.

OTHER ILLNESSES THAT CAN BE MISTAKEN FOR ECZEMA

Many things besides eczema cause childhood rashes. For example, newborn *seborrhea* (cradle cap) is frequently mistaken for eczema. Seborrhea is less itchy, and it usually appears before the infant is two months old, whereas eczema is very itchy and usually starts after three months of age. Like eczema, seborrhea tends to occur on the face and scalp, but unlike eczema, it often affects the diaper area as well. *Scabies* are tiny mites that cause a very itchy rash, which is quite contagious; its treatment is very different from the treatment for eczema. *Skin allergies* and *contact dermatitis* also look like eczema and are treated similarly. Children can have red, itchy skin from allergies to soap, detergent, fabric softeners, dryer sheets, and nickel jewelry. *Fungal infections* such as ringworm also look like eczema. *Rare immunological diseases* and serious nutritional deficiencies also cause symptoms that look like eczema, but they have other, more serious symptoms as well. If you aren't sure what's causing your child's rash, consult your health care professional.

HOW IS ECZEMA DIAGNOSED?

Eczema is diagnosed by a careful examination of the skin and listening to patients tell about their symptoms. There is no single blood test, X-ray, or other test to diagnose eczema. Measuring IgE and other biomarkers is not helpful.[1] Your health professional will look for itching and a rash in a typical pattern for the child's age. If there's any doubt about whether it might be another condition such as a fungal infection or scabies, a gentle skin scraping can be done to look under the microscope to rule them out. If your child is enrolled in a clinical trial, the health professional will likely use a formal scoring system such as Scoring Atopic Dermatitis (SCORAD) or Patient-Oriented Eczema Measure (POEM) to monitor progress; but otherwise, monitoring is usually done by taking photos at regular intervals to assess response to therapies.

Several factors increase the risk of developing eczema. As with most conditions, it's a combination of heredity, environment, and behaviors.

RISK FACTORS FOR ECZEMA

- Heredity: sensitive immune system
- Environment: over-bathing, drying soaps, allergies, irritants
- Diet: early feeding of solids, food allergies, obesity
- Inadequate levels of vitamin D
- Early exposure to antibiotics
- Mind-body: stress

Like asthma and hay fever, eczema tends to *run in families*. If one twin has eczema, chances are that the other twin has it, too. A particular problem on chromosome 11 is characteristic of the sensitive immune system found in eczema, asthma, and allergies. Many eczema sufferers appear to have a problem transforming the essential dietary fatty acids, such as linoleic and linolenic acid, to prostaglandin E_1. Prostaglandins affect the immune system and many other bodily functions. Abnormal immune function may account for the oversensitivity of children who suffer from eczema, asthma, and allergies.

Paradoxically, one of the most common triggers of eczema is overzealous *bathing* in hot water, if it's not followed immediately by applying emollient creams and ointments. Just as hot, soapy water helps remove grease from the dishes, it also removes the natural oils that protect the skin. They must be replaced immediately to avoid drying and cracking. Soap is generally unnecessary for babies, because babies don't make the smelly sweat that older children and adults do. Soap is only necessary for removing greasy dirt or sticky pulp. *Soaps* and *detergents* trigger eczema in many children. In my experience, Tide™ detergent and Ivory™ soap are two of the most common culprits. On the other hand, insufficient bathing is not the answer, and dirty skin can make the child prone to painful and aggravating infections.

I advised Wendy to switch to Dove™, Camay™, or another moisturizing soap. I also encouraged her to consider a milder laundry detergent without fragrances or softeners.

Other environmental triggers include *skin infections* (such as impetigo), *sweating, stress, irritants* (such as tobacco smoke and wool sweaters), and *allergies*. Allergies can be either from something that touches the child's skin or from something inhaled or eaten. Dust mites, cockroaches, house dust, molds, and grasses can all trigger eczema as well as sneezing, watery eyes, and coughing. Children who are allergic to animal fur, dust, and dust mites have fewer eczema symptoms when their environment is cleaned up.

Early introduction of a variety of *solid foods* before four months of age triples infants' risk of developing severe eczema.[2] A healthy baby does not need any food other than mother's milk in the first four to six months of life. Children can learn to like a variety of foods later in childhood; please do *not* start your baby on solids before he is four months old.

FOODS ASSOCIATED WITH ECZEMA

- Acidic foods, such as oranges, tomatoes, pineapple
- Proteins, such as eggs, nuts (especially peanuts), milk, soy, shellfish
- Wheat and corn
- Sweets, chocolate, soft drinks preserved with sulfur dioxide
- Artificial flavors[3]

Acidic foods such as citrus and tomatoes commonly cause a rash around the mouth in sensitive children.

Food allergies worsen symptoms in about 10 to 20 percent of children with eczema. The foods that trigger eczema (eggs, nuts, milk, soy, shellfish, wheat, and corn) are the same as those that trigger other allergic symptoms such as wheezing, runny nose, sneezing, and an upset stomach. For mothers unable to breast-feed their babies who have a high risk of eczema, offering a hydrolyzed formula for the first four months of life is a cost-effective approach to reducing the risk of developing eczema.[4]

How Are Food Allergies Diagnosed?

Unfortunately, skin tests and blood tests are not very helpful in diagnosing food allergies. Keep a diary to record your child's symptoms and any foods that seem to make them worse. If the child's previous reactions have been mild, parents can withhold a suspected food for several days, then give it and watch for 48 hours for symptoms. The problem with this approach is that the parents and child know that the child is getting the suspected trigger,

and negative expectations are powerful triggers of adverse effects.

It is difficult for families to test and treat their child for food allergies without professional assistance. A strong placebo effect comes into play whenever a challenging, costly treatment is recommended; this certainly applies to elimination diets. It is imperative that such an effort be undertaken with a professional experienced in dietary challenges in children such as a pediatric dietitian working with an allergist.

The best test is a double-blind challenge. In a double-blind challenge, the child avoids eating any of the suspected food for a week before the test. The health professional gives the child one of two different capsules—one contains the suspected allergen and the other contains an inert placebo. Neither the professional, the parents, nor the child know the contents of the capsule. Parents and professional watch the child for several hours (up to two days) to see if symptoms develop. If a child has had a previous severe reaction (including wheezing, shortness of breath, shock, or low blood pressure) to a particular food, the double-blind food challenge should only be done in a facility capable of managing pediatric emergencies.

Do Children with Food Allergies Have to Avoid Those Foods Forever?

Not necessarily. Even children who have proven severe food allergies often outgrow them within a year or two. No one knows exactly why, but you don't have to worry that a current sensitivity (or even a severe allergy) means a lifetime of restrictive diets.

What Other Factors Increase the Risk of Eczema?

Being *overweight* or obese also increases the risk of eczema.[5] Those chubby cheeks may be adorable, but they aren't necessarily protective against developing the itchy rash of eczema.

Low levels of *vitamin D* are also linked to an increased risk of eczema.[6] Maintaining healthy vitamin D levels is essential for promoting healthy immune function. It is possible that one of the reasons that eczema is more common in African American than Caucasian children and among urban compared with rural children is because of differences in vitamin D levels.

Early exposure to antibiotics increases the risk of developing eczema. It has been hard to determine whether the problem lies with the antibiotics or with the infection that led to the use of antibiotics. However, a German group of researchers studied 370 children over time; children who developed colds who were not treated with antibiotics had a 30 percent lower risk of developing eczema than children who didn't have colds, but those who received broad spectrum antibiotics for their colds had twice the risk of developing eczema as their healthy counterparts.[7] If your child requires antibiotics to treat a life-threatening condition, do not hesitate to use them. But please don't ask for antibiotics to treat minor illnesses; while killing bad bacteria, these antibiotics also wipe out many of the helpful bacteria living in our intestines that help regulate the immune system.

Stress triggers eczema symptoms in many children. On the other hand, having a chronic uncomfortable condition can be stressful. Breaking this cycle can help eczema sufferers, which is why we'll cover mind-body stress management skills later.

How to Treat Eczema

Let's tour the Therapeutic Mountain to find out what works. If you want to skip to my bottom-line recommendations on eczema, flip to the end of the chapter.

Lifestyle Therapies: Nutrition, Exercise, Environment, Mind-Body

Nutrition

Breast-feeding protects babies from developing eczema, and the protection lasts decades.[8] Another reason to nurse your baby! Nursing moms who eat more *fruits and vegetables* rich in vitamin C produce milk with higher vitamin C content and a 70 percent reduced risk of eczema developing in their babies.[9] These benefits are found only for children of women who got their nutrients through food, not through supplements.[10] One supplement that pregnant and nursing women can take that seems to protect babies against eczema is *probiotics*. Several studies have shown that moms who supplemented with *Bifidobacterium* or *Lactobacillus* had babies with up to 75 percent lower risk of developing eczema.[11,12] Remember that many of these healthy bacteria are found in fermented foods like yogurt, kefir, homemade sauerkraut, and kimchi. Finally, mothers who consume extra omega-3 fatty acids (like those found in salmon, sardines, and other fish) during pregnancy and while nursing have babies with a lower risk of developing eczema.[13]

Food allergies trigger symptoms in up to 35 percent of children who have eczema.[14] Allergies to eggs and cow milk protein are among the most common; other common triggers include peanuts, tree nuts, wheat, corn, soy, fish, shellfish, and acidic foods such as tomatoes and strawberries. Many food-sensitive children notice big improvements if they totally eliminate food triggers. Complete elimination of all possible food allergies (as in hospital programs relying on artificial food substitutes) can result in marked improvements in skin symptoms.

However, eliminating dietary staples runs the risk of nutritional deficiencies if done without careful planning or for excessive periods of time.[15] They are challenging to institute and maintain. The emotional and financial costs to the family of a child following a strict elimination diet are high. Even motivated families of children who have well-defined food allergies frequently find it impossible to maintain elimination diets. Consider a food elimination diet only if your child has severe, extensive eczema that is not controlled by other regimens *and* you seek the help of a nutritionist or dietitian.

Many children outgrow their sensitivity to foods. Even if your child's eczema is clearly worse with a certain food, it doesn't mean she can never have it again. You may carefully reintroduce it in 6 to 12 months. If there is still a reaction, wait another few months and try again.

We decided not to pursue possible food allergies at our initial visit and to wait to see how well James responded to other therapies. We held an elimination diet open as a possibility later if the initial therapies didn't do the trick.

Exercise and Sleep

Eczema not only interrupts *sleep* due to itching and scratching, but it is also worsened by insufficient sleep. Keep the bedrooms cool to minimize itching. Don't overdress the child. Make sure he has plenty of opportunities to sleep at night and even nap during the day. Sufficient sleep helps keep the immune system on an even keel.

Environment

In general, heat makes itching worse. Sweating washes the protective, lubricating natural oils off the skin. *Cold* relieves itch. I advise many of my patients with a small itchy area to rub an ice cube on it rather than scratching. Avoid overdressing your child, and avoid giving him baths in very hot water.

On the other hand, *sunlight* may improve eczema symptoms.[16] Ultraviolet light, UV-A and UV-B, are established treatments for psoriasis and eczema, but their long-term risks (excessive wrinkling, skin damage, and skin cancer) must be weighed against short-term benefits.

Prudent sun exposure (not enough to cause a sunburn) helps increase vitamin D production. On the other hand, sweating can aggravate eczema, and excessive sunlight increases the risk of developing skin cancer later on. Definitely avoid the most intense rays between 10 am and 2 pm.

Make sure your child's fingernails are cut short and kept clean to reduce the risk of infection from constant scratching. You may want to put a pair of gloves or socks over your child's hands at night to keep him from scratching himself in his sleep.

Too much bathing without sufficient lubrication afterward dries the skin and makes eczema worse. Although harsh soaps can remove protective oils from the skin, antibacterial soaps such as Hibiclens™ can rid the skin of bacteria that aggravate eczema.

BATH-TIME HINTS FOR CHILDREN WITH ECZEMA

- Give your child a bath one to three times weekly rather than daily.
- Use tepid (not hot) water.
- Avoid soaps or use only moisturizing types.
- Soap only the armpits, groin, feet, diaper area, and greasy, dirty areas.
- Put a tablespoon of vegetable oil in the bathwater to soothe the skin; *or* put two cups of oatmeal in an old stocking or pillowcase, tie the end in a knot, and throw it in the bathwater for a soothing soak; *or* add a cupful of baking soda to the bathwater.

- After the bath, pat, rather than rub, the skin dry.
- Apply an emollient immediately after bathing.

Avoid bubble baths because they can dry the skin and may cause an allergic reaction.

Many children with eczema have worse symptoms in the winter, when indoor air is dry. You might keep a *humidifier* going in the child's bedroom to add moisture to the air.

Studies show that some children with eczema had markedly improved symptoms when careful attention was paid to cleaning their environment.[17,18] The culprit here isn't dirt itself, but a microscopic critter called the *dust mite*. This bug also aggravates asthma and allergies in sensitive children. Ridding the house of dust mites by thorough dusting and using special hypoallergenic bed covers and high filtration rate vacuums can significantly improve eczema symptoms. Dust and dirt themselves do *not* cause eczema. In fact, children who grow up in farming families and who are exposed to lots of dirt and animals have about half the risk of developing eczema as their urban counterparts. Obviously, for some children, animal dander is associated with severe allergic symptoms; farm life is not for everyone, nor should itchy children take to sleeping with the family furry pet in hopes of driving the demon of itch away.

Avoid using detergents, soaps, fabric softeners, and dryer sheets that contain *perfume*. Nearly 4 percent of eczema patients are allergic to perfumes. Interestingly, those who react to the perfumes in dishwashing detergent seem to be most sensitive when the dishwater is very hot; washing in more tepid water causes less of an allergic reaction.

Wendy decided to decrease James' bathing from daily to twice a week, stop using soap, and use a milder detergent. She decided to add a cup of oatmeal to James' bathwater and to apply Eucerin and steroid cream immediately after the bath. She also went on a cleanup campaign to destroy dust mite strongholds in James' room.

Another environmental technique for treating eczema involves wet wrap therapy. You can do this with or without other treatments, such as medicated creams and with or without herbal remedies. Start with just one part of the body—say an arm or a leg. If you want to do it along with a cream, apply the cream before starting the wrap. Then soak some cotton flannel or old cotton diapers in water (or herbal tea such as chamomile or licorice) and wring it out so it's still damp, but not dripping. Wrap the moistened dressing around the limb. Then add one more layer of *dry* cotton. Leave this on for 20 minutes to two hours. Then move on to another affected area. Do *not* wrap the whole body at once, or your child may become overheated and uncomfortable; you may do both arms at once or both legs at once. Do *not* leave the wraps on overnight. Do *not* wrap too tightly—avoid cutting off the circulation! Do *not* use this technique if the eczema is complicated by a herpes infection or impetigo. Experiment to find out what combination works best for you and your child.

Mind-Body

Eczema can both contribute to stress and be worsened by stress. *Group educational visits* about eczema can help decrease stress while building understanding and social support, easing eczema symptoms.[19] *Hypnosis* can effectively alter skin sensations such as itch and pain. Hypnosis, autogenic training, and behavioral therapies that focus on relaxation and stress management may help children learn to manage the stressors that make their symptoms worse *and* help them cope with the stress of having an uncomfortable chronic condition. They can lead to significant improvements in itching and in the severity of the rash itself.

Biofeedback aimed at relaxation can also help people with severe eczema.[20] Other mind-body techniques such as visualization, progressive relaxation, meditation, or deep breathing that help children relax and manage stress may also be helpful. Such techniques affect the oversensitive immune system itself—the source of eczema symptoms.

I strongly recommend sound education, developing an eczema action plan, and mind-body approaches to help manage stress for patients with chronic conditions such as eczema.

BIOCHEMICAL THERAPIES: MEDICATIONS, DIETARY SUPPLEMENTS, AND HERBAL REMEDIES

Medications

Multiple medications are useful for eczema. Here are a few:

MEDICATIONS FOR ECZEMA

- Emollients or moisturizers
- Urea and coal tar
- Immunomodulators: steroids, tacrolimus, and pimecrolimus
- Antihistamines to relieve the itch
- Antibiotics to fight infections complicating the rash

The mainstay of medical treatment is to apply *emollients* or *moisturizers* to the skin several times daily. Avoid products that contain alcohol because alcohol can dry the skin and lead to more itching. Effective, inexpensive emollients include Eucerin™ ointment, Vaseline™ petroleum jelly, and even plain olive oil from your kitchen cabinet. Ointments and oils generally work better than creams or lotions because they don't wash off as easily. Using emollients daily decreases the risk of flares in eczema symptoms.

Urea and *coal tar* are old-fashioned skin remedies. Studies from the 1970s suggest that creams containing urea help soften and soothe dry skin in adults with eczema.[21] As awful as they sound, *coal tar* creams (Clinitar™) are also effective in treating chronic eczema when the skin has become thick and rough like tree bark;[22] these medications have fallen out of favor in treating eczema, as more powerful immunomodulator medications have been developed and marketed.

Immunomodulators are medications that affect immune system function. The most well known are corticosteroids (or steroids for short). A new class of medications, topical calcineurin inhibitors (TCI), include *tacrolimus* (Protopic™) and pimecrolimus (Elidel™); the newer medications can reduce reliance on topical steroids and even appear to be safe to use for infants.[23] They are generally the second choice after topical steroids. Newer medications are expensive and may cause a brief period of stinging, burning, and itching, and increase sun sensitivity; don't use them before sunbathing. Other potent immunosuppressive medications include *cyclosporine, methotrexate,* and *azathioprine* (Imuran™).

Steroid creams or ointments are the most widely used eczema medications. They come in many varieties and strengths. The mildest, 0.5 percent and 1.0 percent hydrocortisone cream, are available without a prescription. Most physicians recommend applying steroid ointments after a bath when the skin is warm and the pores are open to absorb the medicine. This application needs to occur within three minutes of stepping out of the tub in order to get through the top layers of the skin before they dry out. If you want to skip the stopwatch, just remember that you need to get the remedy on while the child's fingertips still look like wrinkled raisins. After applying the steroid, cover it with an emollient (such as petrolatum or olive oil) to seal it in. Stronger prescription steroids can cause side effects such as weakening and thinning of the skin. They should be used only under professional supervision and only as long as necessary to treat severe symptoms. When symptoms are under control, return to milder preparations or use several times weekly rather than daily. Your health care professional will help you determine the best regimen for your child.

Steroids often improve symptoms dramatically, but they do not "cure" eczema. When treatment stops, symptoms recur. On the other hand, there is *no* evidence that treating eczema with steroid creams (or any other treatment) "drives" the problem deeper into the body or that eczema treatment "causes" asthma or other problems.

Antihistamines, such as Benadryl™ (a non-prescription brand of diphenhydramine), reduce itching and scratching. Your doctor can prescribe stronger antihistamines to help your child sleep when the itching is particularly fierce. Daytime antihistamine use is for those children who can't stop itching and who are willing to be a little sleepy. *Note*: A few children have just the opposite reaction to antihistamines; they get hyper. Kids who react one way to one of these medicines won't necessarily react the same way to all of them. If you don't like the side effects with one antihistamine, bear in mind that there are several other options. Although antihistamines don't cure the underlying cause of eczema, they can help interrupt the itch-scratch-itch-scratch cycle and help improve symptoms.

Prescription *antibiotics* are useful for treating skin infections that frequently accompany eczema. Usually a five- to ten-day course of an oral antibiotic that kills *Staph* bacteria is sufficient to get the infection under control. Antibiotic ointments are useful if the infected area is small; they cause fewer side effects than antibiotics that are taken by mouth. Occasionally, eczema is complicated by yeast or a fungal infection. In this situation, the combination of a steroid and an anti-yeast, antifungal medication may be helpful. Remember that children with eczema are sensitive; some quickly become allergic to antibiotic creams common in first aid kits.

When they can't find anything else to put on eczema, some parents have resorted to *Calamine™* lotion. Though it may provide temporary relief, Calamine's drying effect makes it a poor choice for long-term use.

Dietary Supplements

HELPFUL SUPPLEMENTS FOR ECZEMA

- Essential fatty acids: gamma linolenic acid and omega-3 fatty acids
- Vitamin D
- Probiotics (yogurt, kefir, and supplements) and prebiotics

Overall, a review of 21 studies concluded that certain dietary supplements could help prevent or manage eczema.[24] The best evidence is for probiotics and essential fatty acids (gamma-linolenic acid and omega-3 fatty acids).

Supplemental *essential fatty acids* may be helpful. Evening primrose oil (EPO, Efamol™), borage oil, and black currant oil contain high amounts of one of these essential fatty acids, *gamma linolenic acid* (GLA). Back in the 1980s a randomized, controlled crossover trial in Italian children showed a significant improvement in eczema symptoms when the children received three grams daily of EPO compared with children who received placebo olive oil.[25] Since then, more than twenty controlled trials have shown that EPO improves eczema symptoms, but it takes at least four to eight weeks to see a benefit.[26] Although it is safe, EPO supplements are not as effective as prescription steroid preparations, and few dermatologists recommend them. You might want to try EPO supplements as a complementary therapy. If you don't see any improvement in eight weeks, stop the supplements and try another approach.

Other sources of *essential fatty acids* can be found in both the plant and animal kingdoms. Flax seed, herring, mackerel, and salmon contain omega-3 fatty acids, which can help decrease inflammation after four to eight weeks of daily use. I recommend that (unless they're allergic to fish) women who are pregnant or nursing and children old enough to eat solids eat these kinds of fish at least twice a week as part of a healthy diet;[27] if you hate fish, you can consider taking 1,000–2,000 milligrams daily of an omega-3 fatty acid supplement rich in EPA and DHA.

Ensure adequate vitamin D levels; low levels are associated with an increased risk of eczema. In a study of children with winter-related eczema, supplementation with vitamin D for one month produced a significant improvement in skin symptoms.[28] Other studies are starting to come in supporting the benefits of optimizing vitamin D levels for children with eczema.[29,30] There is preliminary evidence from one study in adults that adding vitamin E supplements to vitamin D supplements may be more helpful than vitamin D alone in improving eczema symptoms.[31] It is hard to tell what your child's vitamin D level is by just looking (although I've never found a deficiency in a child with tan lines in August), so if you want to know whether to supplement, ask your health professional to check a blood level.

Yogurt is good for the gut. Eating yogurt (and the healthy bacteria used to make it) also appears to be good for the skin. Our intestines are lined with millions of immune cells, ready to fight infectious diseases that might invade through the intestinal walls. It turns out that by keeping the intestines and their immunologic guards healthy with "good bacteria" such as *Lactobacillus GG*, *L. casei*, and *Bifidobacterium lactitis* (the same kinds of bacteria used to make yogurt, kefir, and other fermented dairy products), we can reduce the risk of food allergies, eczema, arthritis, and many other conditions associated with an imbalanced immune system. Mothers who consume these healthy bacteria for several weeks before delivery and then for the subsequent six months have half the risk of having a baby develop eczema as mothers who don't consume a diet rich in probiotics. This is because babies pick up the bacteria living on their mom's skin, and if the mom has good bacteria, chances are the baby will, too. If your child has been weaned or is drinking formula instead of breast milk, consider supplementing the formula with *Lactobacillus GG*, *L. casei*, and/or *Bifidobacteria* twice daily. Even if your baby already has eczema, these supplements may help soothe his skin and prevent future flare-ups.

Prebiotics may also be helpful for children with eczema. Prebiotics are compounds that nourish probiotic bacteria and yeast. Examples include inulin, fructo-oligosaccharides (FOS), galacto-oligosaccharides (GOS), and mannan oligosaccharides (MOS). Since these compounds make our microbiome happy, which makes our intestinal immune system happy, there is good reason to think they can help the entire body's immune response. A meta-analysis of four studies found a significant reduction in eczema among babies given prebiotic supplements in their formula.[32] Although prebiotics have exotic-sounding names, they are found in foods such as chicory root, Jerusalem artichoke, dandelion greens, garlic, onion, leeks, asparagus, wheat bran, and bananas.

I suggested that Wendy start giving James evening primrose oil and that she start feeding him yogurt every day and giving him Lactoba-

cillus GG, L. casei, or Bifidobacterium supplements twice daily. He was already eating soft solid foods, so I encouraged her to increase his servings of greens, garlic, onions, and asparagus. We checked his vitamin D level, which was low, and started him on supplements to boost his level above 30 ng/mL. To get his symptoms back under control, we also resumed his steroid cream and gave him a short course of antibiotics to treat his infection.

Despite initially hopeful case reports, subsequent studies have shown that supplementation with individual vitamins and minerals such as vitamins A and E, zinc, and selenium are not helpful for children with eczema. I do not recommend vitamin A, vitamin E, or zinc supplements for children with eczema unless they have a deficiency.

Herbal Remedies for Eczema

Randomized, controlled trials of *Chinese herbal tea* showed a significant improvement in eczema symptoms for both children and adults. In test tube studies, these herbs affect a variety of chemicals involved in immunologic reactions. Traditional Chinese medical herbal teas contain ten or more ingredients, and the potency of different ingredients can vary as much as fivefold between different batches. There are several reports of patients who developed severe liver toxicity while taking Chinese herbal remedies for eczema; the liver toxicity markedly improved once patients stopped taking the tea, but it returned when the tea was resumed. There is also a worrisome report of an adult who developed severe heart problems after taking Chinese herbs for eczema for two weeks. Many of these teas taste bad, and children cannot easily be persuaded to drink them. There is also a high rate of contamination and adulteration in Chinese herbal products. Despite the initially promising research,

I do not recommend them until quality control improves.

Licorice root's active ingredient, *glycyrrhetinic acid,* has anti-inflammatory properties similar to those of steroids. Several case reports in the 1950s British dermatology literature claimed benefits from ointments containing glycyrrhetinic acid. Unfortunately, other studies failed to reproduce these promising results, possibly because of the natural variation in the potency of licorice extracts. Consequently, licorice root never caught on as an eczema remedy. Still, some herbalists recommend compresses containing licorice root tea as a soothing home remedy for mild eczema. This is a safe home remedy, and many of my patients use licorice tea compresses instead of water-based wet wraps for children with severe eczema.

I'm much more comfortable recommending that patients drink familiar, recognizable teas rather than exotic supplements. That's why I was pleased to read a recent report in the *Archives of Dermatology* noting that 63 percent of adults with eczema who drank three cups daily of *oolong tea* reported moderate or marked improvements in their skin symptoms; the benefits were first noticeable within a week after starting to drink the tea regularly, and over 50 percent of persons continued to note benefits for up to six months.[33] Tea is a good source of antioxidants and is safe to drink if common sense precautions (avoid scalding anyone with excessively hot tea) are used.

Other herbs used to treat eczema include aloe vera, calendula, chamomile, lemon balm, coleus root (recommended for asthma and allergies as well), gotu kola, and St. John's wort oil (used for skin wounds and infections), yellow dock root, Oregon grape, and echinacea. Aloe vera helps heal wounds and kill bacteria that cause skin infections. Sage, nettle, and burdock tea added to the bath are drying teas said to be helpful for weepy, blistery eczema; additional research is needed to determine whether this means they are helpful for children with eczema. Poultices made from plantain, strawberry leaves, goldenseal, and violet are traditional cures for eczema. Whatever the cause of the itch, oatmeal compresses or oatmeal baths are soothing. Despite years of anecdotal experience, all of these herbs remain *untested* scientifically as treatments for childhood eczema; and I cannot recommend them except for trials of topical use until more research is done.

Don't assume that something is safe just because it's natural; St. John's wort can cause serious sun sensitivity. Beware of tea tree oil, *Melaleuca alternifolia*. Although it is a proven antiseptic, it is very irritating and can actually *cause* allergic, eczema-like rashes.

BIOMECHANICAL THERAPIES: MASSAGE

Massage is a natural way to apply emollients and other skin treatments for eczema. Vegetable oil, evening primrose oil, jojoba oil, and even plain old Vaseline™ and Eucerin™ have all been recommended as excellent emollients. Some herbalists recommend adding a drop or two of chamomile or yarrow oil to one teaspoon of vegetable oil to enhance its healing effects. There is one study suggesting that massage itself actually heals eczema.[34] We know

that massage benefits the immune system and helps children relax. Sounds to me like a win-win situation when it comes to treating eczema.

BIOENERGETIC THERAPIES: HOMEOPATHY

The most commonly recommended homeopathic remedies for eczema are: *Calcium carbonicum, Carcinosin, Housedust, Medorrhinum, Oleander, Pulsatilla, Sulphur,* and *Tub Bov.* In studies following children with eczema who were treated homeopathically over time, good results have been reported;[35,36] however, without a comparison group, it's hard to interpret these results. In a study comparing eczema severity among children whose families sought homeopathic care versus those whose families sought conventional care, there were similar rates of improvement for the two groups, though the homeopathically treated group incurred higher costs.[37] It is hard to know from this study whether homeopathy is equivalent to conventional care since the families in each group had their own reasons for choosing the treatments they did. Although homeopathic remedies may be safe, they are still unproved for eczema. I happily tolerate their use but until more research establishes effectiveness, I do not recommend specific products or doses.

WHAT I RECOMMEND FOR ECZEMA

PREVENTING ECZEMA

1. *Lifestyle—nutrition.* While you are pregnant and nursing, consume a diet rich in fruits and vegetables, fermented foods, and omega-3 fatty acid-rich fish. Do *not* introduce any solids in the first four to six months of life; after four months, introduce no more than one new food per week. Once your baby is eating solids regularly, make yogurt a daily part of the diet and gradually add fatty fish once a week. Avoid vitamin D deficiency. Supplement breast-fed babies with 400 IU daily of vitamin D.

2. *Lifestyle—environment.* Do not smoke, and do not allow others to smoke around your child. Minimize your child's exposure to allergens such as dust, dust mites, furry pets, and pollens. Do not use drying soaps on your baby or bathe her too frequently. Use tepid water in her baths.

> *Take Your Child to a Health Care Professional If*
>
> - You are concerned that the rash could be caused by something else.
>
> - The eczema takes a turn for the worse, becomes painful, or looks infected.
>
> - There is no improvement after four weeks of home therapy.

TREATING ECZEMA

1. *Lifestyle.* Enroll in an eczema education group with your health professional. If your child has moderate to severe symptoms, make sure you have an eczema action plan.

2. *Lifestyle—environment.* Use a moisturizer or emollient daily after baths, before bed, and before heading outdoors to help preserve skin moisture and protect the skin. Avoid drying or perfumed soaps. If you do use soaps, try a moisturizing type. Avoid strong, scented detergents. Make sure that all clothing has been rinsed well to remove soap residues before drying. Try putting two cups of dry oatmeal in an old stocking and drop it in your child's bathwater. Alternatively, put two tablespoons of vegetable oil in your child's bathwater. Let your youngster "scratch" with an ice cube. Use a high-filtration vacuum, install hypoallergenic mattress covers, and dust thoroughly to remove dust mites and other airborne eczema triggers. See a pediatric allergist and have your child evaluated for treatable allergies. Consider wet wraps

using water or tepid licorice tea. Talk with your pediatrician about prudent sun exposure.

3. *Lifestyle—diet.* Offer your child yogurt, kefir, or other fermented foods daily and eat fatty fish or flax seed at least once a week as well as vegetables rich in prebiotics. Drink oolong tea. If your child has severe, extensive eczema, talk with a pediatric nutritionist about trying an elimination diet.

4. *Biochemical—medications.* Talk with your health care professional about treatment with an immunomodulating medication such as a topical corticosteroid or steroid-sparing medication. Follow recommendations exactly. Consider using an antihistamine at night to reduce nighttime scratching. If the rash looks infected, try an antibiotic ointment or see your doctor. Do not use antibiotics to prevent skin infections; reserve them for active infections. Do not put Calamine™ lotion or other drying antihistamines on your child's eczema.

5. *Biochemical—nutritional supplements.* Consider supplementation evening primrose oil containing GLA. Consider giving your child *Lactobacillus* GG or *L. casei* or *Bifidobacterium* supplements daily. Provide vitamin D supplements if his levels are low.

6. *Lifestyle—mind-body.* Consider taking your child for hypnotherapy or biofeedback to help him learn to desensitize his skin and to learn to deal with stressful triggers. Consider using guided imagery or meditation techniques to learn to manage the stress of having a chronic, uncomfortable condition.

7. *Biomechanical—massage.* Give your child a massage at least three times a week using olive oil to help soothe the skin and the soul.

8. *Biochemical—herbs.* Do *not* use Chinese patent medicines because they have a high rate of contamination and adulteration. Consider using an herbal compress containing licorice, chamomile, calendula, or slippery elm bark tea.

18
FATIGUE

Alice Grey was a conscientious 15-year-old who was getting As during her freshman year of high school, ran on the track team, and babysat for her little brother, Matt. However, in the fall of her sophomore year, she developed mono, followed by a flu-like illness (she'd had her flu shot), and then couldn't quite get her energy back. She'd missed over 20 days of school by the time she came to see me in April, and she'd had to quit track and babysitting. She just couldn't concentrate, and she was tired all the time but didn't sleep well and didn't find that sleep was refreshing.

Back in the 1950s and 1960s, children and teenagers tended to complain about fatigue only after vigorous exercise or a sleepover, during a bout with infectious mononucleosis, or when battling a serious illness. That has changed. The last thirty years have seen the emergence in the late 1980s of chronic fatigue syndrome (CFS), also called myalgic encephalomyelitis (ME) and *systemic exertion intolerance disease* (SEID). Chronic fatigue may be accompanied by aching in the muscles and connective tissues (fibromyalgia, FM), depression, or other symptoms, but its main symptom is excessive, disabling fatigue after physical or mental exertion.

CFS usually includes more than just fatigue and achy muscles. It can affect just about every system in the body. In the brain, there can be problems with memory, concentration, fuzzy thinking, mood, and unrefreshing sleep as well as headaches. In the throat, there can be sore throats and tender lymph nodes. There is often pain in the muscles and joints, including the jaw joint and pelvis. Other common co-occurring symptoms include problems with standing up quickly (dizziness); rapid heart rate; allergies;

sensitivity to sounds, odors, and chemicals; and abdominal pain, bloating, nausea, constipation, and diarrhea.

CFS affects fewer than 3 percent of teens. It tends to occur more often in girls than in boys, and in my experience, tends to affect conscientious, high achievers like Alice more often than laid-back C students. It occurs in teens of all races and incomes. It can run in families, but it's not contagious. Chronic fatigue affects over one million Americans, and it often starts in adolescence.

Fatigue has a seriously negative impact on health and happiness. Kids who were at the top of their game academically, socially, and athletically often find themselves sidelined, unable to attend school, compete, practice, or socialize. This can lead to depression, which in turn is debilitating. It's a vicious cycle.

A person is only labeled as having CFS or one of these similar terms if the fatigue is not caused by another medical condition. What medical conditions cause fatigue? Just about all of them.

ILLNESSES THAT CAN CAUSE FATIGUE

Many kinds of illness—infections, cancer, organ failure, arthritis, and mental health challenges—can cause fatigue. Perhaps the most well-known cause of fatigue in teens is infectious mononucleosis (mono) caused by Epstein-Barr Virus (EBV). Just as adults battling cancer are often fatigued, so are pediatric oncology patients. Chronic pain is exhausting. Children with end-stage kidney failure, arthritis, and inflammatory bowel disease also often have severe fatigue. Iron deficiency anemia and B12 deficiency are classic causes of fatigue. Hypo- and hyperthyroidism can also cause fatigue. Mental health challenges such as depression and anxiety can appear as fatigue or inability to get out of bed and face the world.

WHAT CAUSES CHRONIC FATIGUE?

The truth is that scientists are still trying to determine what combination of genetics, infectious agents, lifestyle, toxic exposures, and other factors causes chronic fatigue. As in Alice's case, it's often triggered by a viral illness like mono or the flu. For other patients, the trigger may be persistent or recurrent viral infections.[1] Researchers have focused on understanding the relationships between oxidative stress, immunologic issues, endocrine problems, and fatigue.

DIAGNOSING FATIGUE

Blood tests can be used to rule out common causes of fatigue such as low iron, B12, or vitamin D levels and to test for thyroid problems. Your doctor may also order a white blood cell count or tests of inflammation to rule out hidden infections, malignancies, or inflammation contributing to fatigue; a blood sugar test to make sure it's not diabetes; and liver and kidney tests to make sure those organs are working normally. You may ask for tests to rule out Lyme disease and mono. Your doctor will likely ask about use of substances like alcohol, marijuana, heroin, and cocaine. Your doctor may also ask you to complete screening tests for depression and anxiety. But for the most part, the diagnosis of CFS is based on finding a classic constellation of symptoms rather than a particular finding in a blood or urine test.

CRITERIA FOR
CHRONIC FATIGUE SYNDROME

- New onset of severe fatigue lasting at least six straight months or longer, *not* related to exercise, *not* relieved by rest, and *not* due to other known medical conditions

- Fatigue severe enough to interfere with usual activities
- Four or more additional symptoms such as
  - Problems with memory or concentration
  - Sleep problems or unrefreshing sleep
  - Muscle or joint pain (myalgias or arthralgias)
  - Headaches
  - Sore throat
  - Sore lymph nodes in the throat or under the arms
  - Prolonged exhaustion and sense of sickness after physical or mental exercise
- *Optional* additional symptoms: brain fog; dizziness, fainting, or balance problems when upright; allergies; hypersensitivity to light, sounds, odors, or chemicals; bloating, abdominal pain, nausea, constipation, diarrhea; mood problems like depression or anxiety

Like fever, fatigue is a symptom, not necessarily a diagnosis.

Alice had classic symptoms for CFS. Her lab tests showed both low iron and low vitamin D levels, so we started replacements to get her back in the optimal range. She had normal B12 levels and no signs or tests positive for thyroid disease, cancer, diabetes, or serious arthritis. Her kidneys and liver were working well, and she had no signs of hidden infections. She was not pregnant. What else could she do?

How to Prevent and Treat Fatigue

Let's tour the Therapeutic Mountain to find out what works. If you want to skip to my bottom-line recommendations on fatigue, flip to the end of the chapter.

Lifestyle Therapies: Nutrition, Exercise, Environment, Mind-Body

The two therapies with the most evidence of effectiveness are both lifestyle therapies: mind-body (cognitive behavioral therapy [CBT]) and exercise (graded exercise therapy [GET]).[2] It's important for those with fatigue to set explicit goals for activities and then pace themselves to achieve them, balancing goal-oriented activity with rest. It's also important to create a supportive environment and eat nourishing food.

Nutrition

Good nutrition is the foundation for managing all kinds of chronic conditions, including chronic fatigue. Avoid eating junk foods and focus on foods with high nutritional content such as fruits, vegetables, seeds, nuts, legumes, whole grains, and fish. Try to use organic products when possible to minimize exposure to toxic chemicals, pesticides, and herbicides.

Exercise and Sleep

Graded exercise therapy is a form of physical therapy. Graded exercise means starting slowly (like five minutes of exercise daily) and gradually increasing the amount of exercise over time.[3] It doesn't work if the exercise is stopped, so regular exercise, even short amounts of stretching, needs to become part of a healthy lifestyle. Pace yourself. Start with slow stretching for each joint and stop before becoming completely fatigued, gradually increasing the duration of exercise (by five minutes per week, for example) as tolerated. Pay attention to your symptoms and stop when you're tired. You can look for a gentle yoga, QiGong, or tai chi class or YouTube video, or your insurance will likely cover a referral to a physical therapist.

QiGong is a Chinese mind-body practice that is closely related to tai chi. A 2013 analysis of seven trials involving nearly 400 patients concluded that QiGong was safe and possibly useful for patients with fibromyalgia.[4] A later study of 137 patients with chronic fatigue syndrome showed that 10 training sessions of QiGong over 5 weeks followed by 12 weeks of home practice improved both fatigue and mood.[5]

While it may be difficult to find QiGong classes in many American cities, you can easily find *tai chi* and *yoga* classes. As with QiGong, regularly practicing yoga, even for as little as 20 minutes daily, can help improve mood, sleep, pain, and fatigue.[6,7] Easy tai chi has proven helpful in improving fatigue in patients with cancer and multiple sclerosis.[8,9] Unless your child has a strong preference for another type of exercise, I recommend a mindful movement activity like yoga, tai chi, or QiGong.

Alice decided to enroll in a gentle restorative yoga class at her local YMCA. For the first two weeks, she could only participate for 10 minutes at each class, but gradually she felt her endurance improve and by the end of the eight-week session, she could participate for 30 minutes of activity. By the end of the next eight-week class, she was able to particulate for a full hour. She loved how the classes helped her feel calm, strong, flexible, and confident.

Restorative sleep is essential to recovery. Please encourage good sleep habits such as the following:

- Use the bed only for sleep and reading a book, not watching TV, using a computer, or other electronic equipment.

- Have a regular bedtime with a preparatory routine (bathing, tooth brushing, music, pajamas).
- Avoid naps after 3 pm.
- Avoid exercise within two hours before planned bedtime.
- Avoid caffeine after noon.
- Ensure the bedroom is dark, quiet, and cool.
- Avoid electronic screen time within an hour before bed.
- Spend some time in the morning in sunlight or exposed to full spectrum light.

See Chapter 24's section on dietary supplements (melatonin, valerian, herbal tea, theanine, 5-HTP, and magnesium).

Environment

In Europe and Japan, spa treatments like soaking in hot mineral water and using saunas are common approaches to many chronic health problems, including chronic fatigue.[10,11] An Italian group of researchers summarized more than half a dozen studies of spa therapy for adults with fibromyalgia and concluded that spa therapy improved pain, depression, and related symptoms.[12] Few studies have evaluated their effectiveness in children, and insurance is unlikely to pay for them. Nevertheless, if hot mineral baths and saunas are part of your culture and beliefs, I support using them with common sense precautions to avoid overheating.

Alice's grandmother was a big believer in hot baths with two to three cups of Epsom salts per bathtub. She called them Epsom soaks. She encouraged Alice to take Epsom soaks three nights a week before bed to help her relax. Alice said she slept better after these soaks and felt relaxed and stronger the next morning. I encouraged her to continue.

Mind-Body

Fatigue can contribute to stress and be worsened by stress. The most effective, proven therapy for chronic fatigue is a mind-body therapy: *cognitive behavioral therapy* (CBT).[13–15] CBT helps people understand the relationship between physical sensations, thoughts, beliefs, and emotions. CBT is generally safe and most insurance policies cover this well-accepted therapy. I recommend CBT as a first-line therapy for youth with chronic fatigue.

Mindfulness-based stress reduction meditation and mindfulness-based cognitive therapy are also emerging as popular mind-body strategies for adults with chronic fatigue. Preliminary data are promising,[16] and I support the use of safe mind-body strategies such as mindfulness for youth with chronic conditions such as fatigue. Insurance coverage varies.

Biofeedback is another mind-body approach to chronic conditions. In a case study, biofeedback was helpful for an adult with chronic fatigue syndrome.[17] Given the autonomic abnormalities present in chronic fatigue syndrome, I would like to see additional research evaluating the benefits of heart rate variability biofeedback. Insurance coverage for biofeedback varies. Biofeedback is very safe.

Alice started CBT a week or so after the yoga classes. She liked the therapist, who understood her well and didn't judge her at all. The therapist didn't think Alice was crazy, and helped her learn to use some new skills to explore the relationship between her physical sensations, negative thoughts (I'll never get better), and fears (I'll be lonely and dependent), and saw how dwelling on the negative thoughts and feelings sapped her energy and made it harder for her to recover. She learned to question her own thoughts and consciously choose to practice gratitude, which helped her feel more cheerful and optimistic.

BIOCHEMICAL THERAPIES: MEDICATIONS, DIETARY SUPPLEMENTS, AND HERBAL REMEDIES

Medications

The most commonly used medications to treat chronic fatigue are *antidepressant* medications. The evidence on their effectiveness in treating fatigue is mixed, but they can be helpful for those whose primary symptoms are fibromyalgia, depression, and sleep disturbances.[18]

Based on findings of lower cortisol levels in adults and children with chronic fatigue and the fact that many patients with Addison's disease (adrenal insufficiency) report nearly identical symptoms to chronic fatigue syndrome, some physicians have treated fatigued patients with low doses of hydrocortisone, prednisolone, fludrocortisone, and other *steroid* medications.[19–22] In fact, a growing number of specialists who treat children with chronic fatigue and pronounced autonomic problems (dizziness and fast heartbeat when standing up, nausea, and headaches), use fludrocortisone as an important part of therapy. However, not all studies have reported beneficial effects of steroid medications for patients with chronic fatigue.[23] Steroid medications can have serious side effects when used over a long period of time (thinning bones, increases in blood sugar, decreases in immune function, and more), and they should only be used when the patient is carefully and regularly monitored by a pediatric specialist.

Based on the observation that many children with chronic fatigue have increased autonomic nervous system activity in the sympathetic (fight or flight) branch, some researchers have given medications that are usually used to lower blood pressure (*clonidine*) in an attempt to "put the brakes" on sympathetic overactivity. While this approach may be understandable, in a large study evaluating the use of clonidine, it was not found to be helpful in improving chronic fatigue symptoms.[24] I do not recommend clonidine as a treatment for pediatric fatigue.

Alice did not have any symptoms that warranted medications other than her iron and vitamin D replacement. After our first visit, she read on the Internet that carnitine and coenzyme Q10 could help boost her energy. Could they? What other dietary supplements might be helpful?

Dietary Supplements

Because so many different nutrients are involved in energy production, a variety of supplements containing these nutrients have been tried to improve vitality for those suffering from fatigue. But not everything that is sold on the Internet has solid scientific evidence to support spending the purchase price.

DIETARY SUPPLEMENTS USED FOR FATIGUE

- B vitamins
- Vitamin D
- Carnitine (Acetyl-L-carnitine, ALC)
- Coenzyme Q10 (CoQ10, ubiquinone)
- Iron
- Magnesium
- Melatonin
- NADH
- Omega-3 fatty acids
- Probiotics

Among the most commonly used stress or energy supplements are those that contain *B vitamins*. B12 deficiency is uncommon in youth, but a well-known contributor to fatigue in the elderly. Because B12 deficiencies are generally due to inadequate intake or poor absorption, replacement is usually done with injections. Weekly B12 injections and *folate* supplements taken by mouth can improve symptoms in adults with chronic fatigue.[25] These supplements are safe, but please ask your health professional to check your child's B12 level before committing her to a regimen of B12 injections.

Low *vitamin D* levels are linked to muscle aches and depression, raising the question of whether supplementation can improve symptoms that are often part of the overall chronic fatigue picture. In a Japanese study, giving 100,000 units of vitamin D every two months for six months did not improve symptoms in adults with chronic fatigue syndrome.[26] In my clinical practice, about 65 percent of teens like Alice have suboptimal vitamin D levels, and gradual replacement (with 1,000–5,000 units daily of vitamin D supplements) has been associated with improvements in fatigue and muscle aches. This is consistent with findings from an American study of over 170 adults with fatigue (77 percent of whom had low vitamin D levels), whose fatigue significantly improved over five weeks of supplementation.[27] Ask your health professional to check your child's vitamin D level and supplement as needed to ensure optimal levels.

Carnitine is an amino acid found in protein-rich foods; it plays an important role in energy production and metabolizing fat. It is mostly used by older patients with mild memory problems and nerve pain related to diabetes or HIV infection. A randomized controlled trial in elderly patients with fatigue suggested that acetyl-L-carnitine (ALC) supplements helped reduce both physical and mental fatigue.[28] There are not enough studies showing that it helps improve fatigue in teens or young adults for me to recommend it. ALC is generally safe, but it can create a fishy odor to the breath, urine, and sweat.

Coenzyme Q10 (CoQ10, ubiquinone) is an antioxidant closely related chemically to vitamin K. It is essential for the cell's energy factories (mitochondria). It is often used to treat mitochondrial disorders, heart failure, high blood pressure, migraine headaches, Parkinson's disease, and muscle weakness. Because it's involved in 95 percent of the body's energy production, it has also been used as a remedy for chronic fatigue. A few preliminary studies support this rationale,[29,30] but definitive large randomized controlled trials have not established CoQ10's benefits in treating children with chronic fatigue. Patience is required. It takes about six months of supplementation for levels to build up enough to notice a benefit. CoQ10 supplements are generally safe, but product quality varies; check with an independent testing agency like ConsumerLab to ensure you are getting your money's worth if you try CoQ10 supplements.

Iron deficiency anemia is a well-known cause of fatigue. Please ask your doctor to check your child's ferritin level (an early marker of iron deficiency) and discuss supplements to optimize iron levels. Be careful and please do not start iron supplements on your own without a test. Excessive iron can be toxic.

Magnesium plays a key role in over 300 metabolic reactions in the body, including reactions involved in energy production. About half of teenagers do not consume the recommended dietary intake of magnesium (about 400 milligrams daily). Preliminary research suggests that restoring magnesium levels can help reduce fatigue.[31] Consider having your child's magnesium levels checked and provide food and supplements to correct any deficiencies.

Melatonin is a popular, effective, and safe sleep aid. In adults with Parkinson's disease–related fatigue, melatonin supplements have helped reduce fatigue as well as improving sleep and overall quality of life.[32] It has also proved helpful in other studies of adults with chronic fatigue and sleep problems.[33] Not all studies have had positive results. On the other hand, melatonin is low cost and low risk. If your child has significant problems falling asleep at night, consider a trial of supplementary melatonin (up to five milligrams nightly) for a week to determine how helpful it is for her.

NADH stands for nicotinamide adenine dinucleotide hydrate. NADH is an essential part of metabolism in every cell in the body. Small studies in adults with chronic fatigue have suggested that supplementation with 10 to 20 milligrams of NADH daily can help improve a sense of vitality and decrease fatigue, at least in the short term.[34,35] NADH is sometimes combined with CoQ10 and other dietary supplements. NADH has a low rate of side effects, but longer-term studies in children are needed before I routinely recommend it.

Omega-3 fatty acids can help reduce inflammation. A series of British patients treated with omega-3 fatty acid supplements rich in eicosapentaenoic acid (EPA) reported significant improvements in their energy levels;[36] however, this study did not include a comparison group, and it is hard to tell exactly how much of the improvement was due to EPA. Omega-3 fatty acids are essential nutrients that are generally safe and widely consumed. I have no objection to trying them, but until additional research is done in children with chronic fatigue, I cannot recommend specific doses for this condition.

Probiotics for chronic fatigue? Emerging research shows three abnormalities in the intestines of patients with chronic fatigue that suggest probiotics may be helpful: (1) inflammation, (2) changes in peristalsis (the waves of contractions that move food through the gut), and (3) differences in the types of bacteria that live in the gut (microbiome). Probiotic supplements could theoretically normalize gut bacterial colonies, decrease inflammation, and normalize peristalsis. Although probiotic supplements are generally safe, they have not undergone sufficient study as therapies for chronic fatigue for me to routinely recommend them for this condition. Stay tuned for emerging research.

Alice decided to stick with the vitamin D and iron supplements for a while to see how she felt.

She did buy some melatonin to use the nights she didn't have the hot Epsom salt baths to help her get restful, restorative sleep. And she decided to eat more yogurt (probiotics) and cold-water fish (omega-3 fatty acids) recommended by the Environmental Working Group.

Herbal Remedies for Fatigue

Coffee, tea, and *other caffeine-containing herbs* are widely used natural stimulants. Although they are generally safe, their benefits are short acting, and they may also have side effects such as racing heart, abnormal heart rhythms, and tremors.

Ginseng is a widely used stimulant herb with antioxidant effects. In a randomized controlled trial involving 90 Korean adults with chronic fatigue, *Panax ginseng* extracts helped reduce fatigue.[37] In another study, *Siberian ginseng* extracts were helpful for adults with less severe, longer lasting fatigue, but were not better than placebo for those with more severe or acute symptoms.[38] There are several species of ginseng that are used medicinally, and because they are so costly, there is great variability (including adulteration and dilution) in ginseng products. Until there are more studies in children and better quality control of available ginseng products, I do not recommend them as a treatment for fatigue in children.

Quercus robur is the Latin name of English oak or royal oak; it actually grows throughout most of Europe. Extracts of oak bark contain high levels of tannins, traditionally used to treat diarrhea. In a Scottish study, a proprietary oak bark extract (Robuvit™) significantly improved fatigue more than a placebo in otherwise healthy adults.[39] Because this is not a traditional remedy for fatigue and there is just one adult study, I'm not regularly recommending this herbal product as a remedy for fatigue, but I will pay attention to emerging research.

Alice was not a coffee or cola drinker and decided not to rely on caffeine to boost her energy since her heart rate was already a little on the fast side. She decided to stick with the yoga, CBT, vitamin D, iron, and hot soaks.

BIOMECHANICAL THERAPIES: MASSAGE

You might think that a relaxing therapy like massage would not be helpful for fatigue. However, at least one study has suggested that regular massage can help improve depression, anxiety, and pain in adults with chronic fatigue.[40] Consider offering your child a regular massage at home since professional therapy can be costly and is not usually covered by insurance companies.

BIOENERGETIC THERAPIES: ACUPUNCTURE, HOMEOPATHY

Acupuncture

Several studies have evaluated the benefits of acupuncture for treating chronic fatigue in adults, and most have shown significant benefits for both physical and mental fatigue.[41–43] Although acupuncture is seldom kids' first choice of treatments and few studies have specifically evaluated its effectiveness in young patients with chronic fatigue, acupuncture is quite safe, and I recommend it for teens who are not responding to graded exercise therapy and cognitive behavioral therapy.

Homeopathy

Homeopathic remedies are quite safe, but preliminary studies have reported mixed effects on fatigue symptoms in adults.[44,45] I happily tolerate the use of safe homeopathic products, but until more research establishes effectiveness, I do not recommend specific products or doses as treatments for fatigue.

Alice decided to keep acupuncture in reserve to see how well she did with the other therapies. After 12 weeks of yoga, eight weeks of CBT, the hot Epsom soaks, melatonin, and normalizing her vitamin D and iron levels, her energy levels were 90 percent back to normal. I encouraged her to continue her home yoga practice even after school started in the fall; eat a nourishing diet; and use good sleep habits. She sailed through her junior and senior years.

WHAT I RECOMMEND FOR FATIGUE

PREVENTING FATIGUE

1. *Lifestyle—nutrition.* Avoid deficiencies of essential vitamins and minerals like vitamin D, vitamin B12, folate, and iron. When in doubt, ask your physician to check levels.

2. *Lifestyle—exercise.* Learn to pace yourself: moderation in all things, including exercise.

3. Prevent infections that trigger CFS by washing hands frequently during cold and flu season and getting your flu vaccine.

Take Your Child to a Health Care Professional If You Are Concerned That

- The fatigue is lasting longer than the typical viral illness.

- Your child also has other symptoms such as fever, pain, loss of appetite, weight loss, rash, or seizures.

TREATING FATIGUE

1. *Determine the underlying cause.* Talk with your physician about testing for iron deficiency (ferritin), vitamin D deficiency, thyroid function (T4 and TSH), inflammation (CBC, sedimentation rate, CRP tests), vitamin B12 levels, a urine test to make sure it's not a bladder or kidney infection, and for teenage girls, a pregnancy test.

2. *Keep a symptom diary* to identify triggers and energizers. If your child has a rapid heartbeat, palpitations, faintness, or dizziness, ask for a referral to a cardiologist or neurologist.

3. *Lifestyle—mind-body.* Ask your pediatrician for a referral for cognitive-behavioral therapy (CBT). Consider a mindfulness-based mind-body program. Use safe, effective strategies to manage stress.

4. *Lifestyle—exercise.* Gradually increase the amount you exercise by five minutes daily per week to reach a goal of 150 minutes weekly. Consider gentle yoga (you can do it in a chair or on the floor), QiGong, or tai chi.

5. *Biochemical—nutritional supplements.* Provide B12, folate, magnesium, vitamin D, and/or iron supplements if her levels are low. Consider melatonin to help with sleep.

6. *Lifestyle—sleep.* Keep your bedroom cool and dark and your bed comfy. Go to bed earlier than usual. Develop a bedtime routine. Remove TV from bedroom. Keep a diary of things for which you are thankful. Possible dietary supplements to improve sleep (discuss with your physician): melatonin, 0.5 to 5 milligrams; valerian, 200–400 milligrams; herbal tea (chamomile, lemon balm, passionflower, hops); theanine, 25–50 milligrams (found in green tea); 5-HTP, 25–100 milligrams (an amino acid); or magnesium, 250–500 milligrams.

7. *Lifestyle—diet.* Avoid foods containing artificial flavors, colors, sweeteners, and MSG. Drink pure water, not soda. Avoid fried and processed foods. Avoid animal fats. Eat more fruits, vegetables, whole grains, seeds, beans, and nuts; minimize processed foods. Consider making a homemade beef broth to supply minerals and amino acids. Yogurt is a good source of probiotics. Fish contains omega-3 fatty acids that can help reduce inflammation (wild salmon, sardines, herring, and mackerel).

8. *Biochemical—medications.* Talk with your health professional about medications to treat related symptoms such as depression and autonomic imbalances. Do not ask for a clonidine prescription.

9. *Biofield—acupuncture.* Ask your health professional for a referral to an acupuncturist, and ask your insurance company to provide coverage for acupuncture if they don't already do so.

10. *Biomechanical—massage.* Consider offering your child a regular massage.

19
FEVER

Greg Edgecomb called me one Sunday evening because his three-week-old daughter, Nicole, had developed a fever of 102°F (39°C) and was not interested in breast-feeding. She was also sleepier than usual. I asked Greg to bring Nicole to the emergency room, so we could make sure she did not have a serious infection and not become dehydrated.

Sarah Klein called before work to ask if she should bring in Jacob, her two-year-old son, for an evaluation of his fever of 103°F (39.4°C). Jacob had had a cold for several days but had been drinking, playing, and sleeping normally until this morning when he awoke with a fever and rapid breathing. Sarah had given Tylenol and sponged Jacob, but his temperature was still 101°F (38.4°C). Sarah was concerned that Jacob might have pneumonia, and I agreed to see him as the first patient that day. On the way into the clinic, Jacob had a seizure. Sarah wanted to know if a seizure could be caused by a fever alone or if it meant that Jacob had meningitis.

Frank Little Bear brought in his eighteen-month-old, Tara, for a checkup and immunizations. The nurse noticed that Tara felt warm and checked her temperature: 101.5°F (38.7°C). Frank said Tara had a cold but had been behaving completely normally. He wondered if Tara could still get her scheduled immunizations or if they would have to make another visit.

Although they are frightening for many parents, fevers are a natural defense against infections. Even cold-blooded animals such as fish and reptiles seek warmer surroundings to boost their temperatures when they are fighting infections. Those that are allowed to raise their temperatures have higher survival rates than those that are kept cool. A fever can be a good thing!

Fevers are a leading cause of children's doctor visits. When winter's blizzards blow cold outside, fever reaches its peak season. Fever is not a disease itself, but it is a symptom of many common childhood illnesses. Most children who have fevers can be safely managed at home. Knowing a few basic facts about fevers may save you a trip to the doctor's office.

What Is a Fever?

Temperatures normally vary up to one full degree over the course of a day, with lower readings in the morning and higher ones in the afternoon. Children tend to run slightly higher temperatures than adults. Normal temperatures range from 95.7°F (35.4°C) to 99.5°F (37.5°C) when measured in the mouth or ear. Temperatures vary depending on when and where they are measured. Rectal measurements are generally higher than oral, underarm, or ear measurements. Temperatures are generally about 0.5° to 1° higher in the afternoon than in the morning.

TEMPERATURES CONSIDERED FEVERS

- *In the mouth or the ear*: above 99.5°F (37.5°C)
- *Under the arm (axillary)*: above 98.6°F (37°C)
- *Rectally*: above 100.4°F (38°C)
- *Low-grade fevers*: less than 102°F (38.9°C), measured rectally
- *Moderate-grade fevers*: between 102°F (38.9°C) and 104°F (40°C), rectally

- *High fevers*: greater than 104°F (40°C) measured rectally

Even among children with temperatures higher than 105°F (40.5°C), only about one-third have a serious bacterial illness. It is not until temperatures exceed 106°F (41°C) that the majority of children have a serious illness such as pneumonia. Infections rarely cause a fever of 106°F and *never* cause a fever over 107°F. Only heatstrokes or rare medical conditions cause temperatures higher than 107°F.

What Causes Fever?

Infections such as colds, flu, ear infections, and sore throats are the most common causes of fever. Hidden illnesses like bladder and bone infections can also cause fevers. Rheumatoid arthritis and cancer are uncommon childhood illnesses, but they can cause fevers. Teething does not raise the temperature above 101°F (38.4°C), though it can cause irritability, drooling, runny nose, and loss of appetite. Over-bundling can cause fevers in babies. The temperature in a closed car in summer sunshine can easily exceed 110°F (43.2°C) and cause heat illness, stroke, or even death in a child left inside while the parent makes a "quick stop" at the store. Do not leave your child unattended in a car.

Our temperature is normally controlled by a "thermostat" in the hypothalamus, a small area in the middle of the brain. When a child has an infection, the white blood cells fighting the infection send chemical messengers to the hypothalamus. The thermostat, normally set at 98.6°F (37°C), is raised to help the white blood cells fight the infection. The hypothalamus sends orders to the rest of the body to raise the temperature. The body responds by seeking a warm room, putting on more clothes, and shivering. When the set point returns to normal, new orders go out; the body reduces the temperature by sweating, flushing, throwing off covers, and drinking cool water.

This explains the "chills" when a fever builds and the sweating when a fever "breaks." When our body temperature is lower than the brain's thermostat, we feel chilled and try to warm up. Shivering is one of the most effective ways to warm up quickly. This is what happens when someone has the "shaking chills." When the set point returns to normal and our temperature remains high, we feel hot and start to sweat. Sweating is a sign that the set point has been lowered and the fever has "broken."

When a child is over-bundled, the temperature may rise even though the thermostat remains set at 98.6°F (37°C). Babies can't easily remove their own clothes, move to a cooler place, or sweat enough to cool off, so they can get overheated and have a higher temperature just from being overdressed or in a hot environment.

Fever raises the heart rate and breathing rate and increases the loss of fluids. Blood levels of zinc and iron fall, while copper and cortisol levels rise.

Do Fevers Cause Brain Damage?

There is *no* risk of brain damage unless the temperature rises above 107°F (41.7°C) and stays there. Fevers even from serious illnesses such as pneumonia and kidney infections do not cause temperatures high enough to cause brain damage. The main thing that causes that kind of brain-damaging temperature is *heatstroke*. Heatstroke occurs when someone is overexercised or overdressed in a hot environment and cannot cool off. During heatstroke, the internal thermostat remains set at 98.6°F (37°C), but the body can't dissipate the heat from exercise and the environment. Eventually it gives up.

I saw heatstroke bring down a North Carolina teenager who was playing basketball indoors in August and who didn't stop to rest or get a drink of water until he passed out with a temperature of 107.6°F (42°C). He was fine once we got some fluids into him and cooled him off.

Other than heatstroke, causes of very high temperatures (above 106.7°F) include rubeola, roseola, epsis, Kawasaki syndrome, and side effects of a few medications. Seek immediate medical care for any temperature over 106°F.

Many parents worry that fevers less than 102°F can cause brain damage. This fear has created an epidemic of overtreatment even when the child has a low-grade fever, which may be helping him fight an infection. Many physicians and pharmacists have perpetuated fever phobia by fixating on the child's temperature when inquiring about symptoms. Fever phobia contributes to a lot of unnecessary medication for children with low-grade fevers. Remember, care for the CHILD, not the temperature!

Do Fevers Cause Seizures?

Sometimes. Between the ages of six months and five years, about two to five percent of children have seizures when they have a fever; of those who have one fever-triggered seizure, most never have another, but about one-third have another seizure within two years. This is *not* the same as epilepsy. Seizures triggered by fevers last less than 15 minutes and typically involve jerking movements of the whole body, not just one side or one part.

Sarah's son, Jacob, fit this picture.

Seizures triggered by fevers run in families, are more common in children who have epilepsy, and are more common among children who attend day care. Why are they more common in day care attendees? Probably because children in day care get more infections and more fevers. In children who are otherwise healthy and developing normally, a typical fever-triggered seizure does *not* damage IQ or cause later school problems or epilepsy.

Seizures look very scary. Most parents who see their child have a seizure are afraid the child is going to die, even though the typical seizure stops within a few minutes and does not cause any long-term harm. If your child has a seizure with a fever, remain calm, check your watch to see how long the seizure actually lasts (even though it seems like hours, it probably will be less than five minutes), and make sure he is lying on his side on the floor or bed in a safe place away from sharp objects. Do not try to put anything in his mouth. Old-fashioned beliefs that seizures make people swallow their tongues are just plain wrong. If the seizure lasts longer than four minutes, call 911 or take him immediately to the nearest emergency room. If it lasts less than four minutes, take him to your regular health care professional or the emergency room.

When I examined Jacob fifteen minutes after his seizure, he was sleepy and had pneumonia. Sarah was scared by the seizure, which had lasted about two-and-a-half minutes. She said she finally understood what her mother had gone through when Sarah had fever-triggered seizures as a child. We treated Jacob with an injection of antibiotics for his pneumonia. He was better by the next day. As with most children who suffer a fever-triggered seizure, Jacob has never had another one. Note: Giving your child Tylenol at the first sign of illness will not necessarily prevent seizures.

Fevers are uncommon in babies under six months old. It's hard to tell just by looking at them whether or not young babies are ill; special tests may be needed, depending on your child's other symptoms. If your child is under three months old and gets any fever, have her evaluated by a health care professional. Do *not* try to treat a fever in an infant yourself without having her professionally evaluated. Fever-

ish infants under six weeks of age may need to be hospitalized while the cause of the fever is being sorted out.

Several tests used to be routine for children having their first seizure. Children up to a year of age will probably need to have a spinal tap if they have a fever-triggered seizure, just to make sure that meningitis hasn't caused the fever and the seizure. An EEG test and brain scans are *not* necessary for most children with their first uncomplicated fever-triggered seizures. Nor do most children need blood tests to evaluate electrolytes, blood sugar, or levels of different minerals or white blood cells unless there are specific findings on their physical examination that lead the doctor to suspect a problem in one of these areas.

Nicole's tests showed that she had a kidney infection. She was hospitalized and received antibiotics by vein. Later tests showed that the infection had been caught in time and there was no permanent kidney damage. Her parents were relieved that their quick response to Nicole's fever spared her problems she might have developed if they had delayed care.

Fevers can be signs of serious illness.

SEEK PROFESSIONAL HELP FOR FEVER IF THE CHILD

- Is less than three months old
- Is lethargic, refuses to drink fluids, is not interested in play
- Has a seizure, stiff neck, limb or abdominal pain
- Has trouble breathing or swallowing her own saliva
- Has pain when she urinates, is vomiting or not keeping down liquids
- Looks sicker than you'd expect from a viral illness

Do Fevers Do Any Good?

Yes. Fevers kill many germs. *Streptococcus pneumonia*, the bacteria responsible for many ear infections and pneumonia, is killed at temperatures over 104°F. In the pre-antibiotic era, practitioners often *caused* fevers in patients (e.g., with steam baths) to cure disease. I do *not* recommend that you give your child a fever on purpose, but you don't need to worry about the fever either. Instead, try to figure out why your child's body may be creating one. Treat the child, *not the thermometer*.

Can Children with Fevers Receive Their Regular Immunizations?

Yes. Children with fevers and minor illnesses such as colds and ear infections can receive immunizations safely.

Frank's daughter, Tara, could have her scheduled immunizations even though she had a fever.

What's the Best Way to Take a Child's Temperature?

Digital thermometers work quickly (in less than thirty seconds) and are accurate and easy to read. Many doctors' offices use electronic thermometers that measure the temperature in the ear. Even if your child has an ear infection, the reading in that ear will be less than half a degree higher than the child's real (core) temperature. Earwax doesn't affect the accuracy of temperatures measured by ear thermometers. *Glass thermometers* can be difficult to read and take several minutes to reach core temperature. They also contain mercury, which is poisonous for people and the environment (when the glass breaks and the mercury runs out). I do not recommend them.

Temperature *strips* and *temperature-sensitive pacifiers* are easy to use, but they tend to underestimate the true temperature. Newer models may be more accurate than the older ones available in the early 1990s, but they still tend to read about one-half a degree lower than rectal thermometers, and it may take several minutes to get a reading.

Our hands are not very accurate in judging fevers. Most parents and health care professionals tend to overestimate a child's temperature when feeling the forehead; on the other hand, they seldom miss a real fever. However, I still feel foreheads on my patients because I often get a sense of the child by touching him, and most children feel soothed when someone who cares about them strokes their forehead. And if I don't detect an increased temperature with my hand, a real fever is extremely unlikely.

Measuring the temperatures *under the armpit* is the safest and easiest for most parents. The temperature under the armpit usually is a little lower (about one degree lower) than the temperature inside the body. Do *not* leave your child alone with a thermometer under the arm. It can break (if it's an old glass one) or slide out, resulting in an inaccurately low reading.

Temperatures measured *rectally* are the most accurate, especially in children under five years old, but taking a rectal temperature can be tricky. Please ask your health care professional to show you how. Be sure to clean the thermometer carefully with rubbing alcohol and wash it in cool water before and after measuring a rectal temperature. Use a lubricant such as petroleum jelly on the thermometer before trying to insert it. The tip of the thermometer should be about one inch inside the rectum for the most accurate temperature; don't force it in. You will need to hold your child still with his "cheeks" (buttocks) together to hold the thermometer in place. Do *not* leave your child alone, even for a minute, with a rectal thermometer in place. The thermometer can slide in too deep or slide out completely. Glass thermometers can break if your child moves suddenly.

Temperatures can be measured in the

mouth (*orally*) with the thermometer under the tongue in older children. It is difficult to obtain an accurate reading in children whose noses are so stuffy that they have to breathe through their mouths. In these cases, an armpit (axillary) temperature may be better. For the most accurate reading with an oral thermometer, don't start to measure until several minutes after your child has eaten or drunk anything cold or hot. Be sure to clean the thermometer with rubbing alcohol and rinse it in cool water before and after using it. Let it dry thoroughly before putting it in the mouth. Place the tip of the thermometer under the tongue and have the child close her mouth, holding the thermometer in place with her hand, *not* her teeth. Glass thermometers need to stay in place for at least three minutes to be accurate. You can read an electronic thermometer as soon as it beeps (usually in thirty seconds or less).

Do *not* leave a child alone with an oral thermometer in place. Sick children may be distracted easily and forget to keep the thermometer under their tongue, resulting in an inaccurately low reading. Older children may have learned that they can create a "fever" (and thereby avoid an unwanted activity such as school) by holding a thermometer near a lightbulb, rubbing it between their hands, or running it under hot water. Don't tempt your child to use one of these common tricks by leaving her alone with the thermometer.

WHAT'S THE BEST WAY TO TREAT A FEVER?

Most parents can tell when their child is sick because the child doesn't eat, sleep, or play as she normally does. The child may whine, cling, and regress to more baby-like behavior or become quiet and withdrawn. If your child isn't acting sick, there's no reason to take or treat her temperature. The only reason to treat the child's fever is to help her feel more comfortable. Let's tour the Therapeutic Mountain to learn what works safely; if you want to skip to my bottom-line recommendations, flip to the end of the chapter.

LIFESTYLE THERAPIES: EXERCISE, ENVIRONMENT, MIND-BODY

Nutrition

Feed a cold and starve a fever? Most children with fevers are not hungry. Neither force-feed nor withhold food from a feverish child. Let appetite be the guide. Offer her foods that are easy to digest (such as rice, crackers, yogurt, and broth) several times a day, but let her appetite guide what and when she eats.

Do give your child *extra fluids* while she has a fever. Higher body heat causes extra sweating, and this fluid needs to be replaced. Fresh fruit juices are excellent. Carrot and beet juices are good choices as vegetable juices for feverish children. Some children like to suck on ice cubes or fruit juice popsicles when they have a fever. Others like to sip simple chicken broth.

Exercise and Sleep

Having a fever takes a lot of energy. Let your feverish child *rest*. Athletes also should take rest breaks and drink plenty of fluids to avoid becoming overheated on hot days. Gentle exercises, such as swimming, yoga, and tai chi, may be better exercise alternatives in August, and even these may be too strenuous for a child fighting a fever.

Environment

When a child has a fever, the body's thermostat is set higher temporarily, and the body will try to maintain the new higher temperature. Children maintain a higher temperature by shivering, moving to a warm room, or asking for extra blankets or to be held close. Removing all the child's clothes or giving sponge baths will make her uncomfortable and more likely to shiver because it does *not* reset the internal thermostat. Sponging does *not* affect the underlying cause of the fever. By artificially lowering body temperature without resetting the internal thermostat, sponging may make things worse. If your child is already sweating because her fever set point has been reduced (the fever has "broken"), sponging away the sweat with tepid water may help her feel more comfortable.

Do *not* sponge your child with rubbing alcohol. Rubbing alcohol can be absorbed through the skin, causing low blood sugar, seizures, and even coma.

Mind-Body

Remain calm yourself. Keep your child company. Read her stories while she is awake. Cuddle her and let her know she is loved. Remind her that she will be feeling better soon and that while she is sick it is okay for her to rest. The behavior of many children regresses when they are sick. Your child may want to cuddle an old favorite teddy bear or prefer stories that were outgrown months previously; she may start talking in baby talk again. This is normal; your child will be back to her usual self soon. You may want to play soothing music or recordings of children's books. You can play videos that you have approved as appropriate for your child. Do not give in to the temptation to let the TV be your sick child's babysitter.

BIOCHEMICAL THERAPIES: MEDICATIONS, DIETARY SUPPLEMENTS, AND HERBAL REMEDIES

Medications

Several medications can safely lower fevers, but all carry a small risk of side effects, so please do not exceed manufacturers' suggested dosages without specific advice from a health professional.

GENERALLY SAFE AND EFFECTIVE:

- Acetaminophen (Tylenol™, non-aspirin pain reliever)
- Ibuprofen (Pediaprofen™, Children's Advil™, Children's Motrin™)

How does *acetaminophen* (the active ingredient in Tylenol™ and other non-aspirin fever medications) work? It resets the brain's temperature thermostat. Acetaminophen also helps the child feel more comfortable with her other symptoms. It does *not* make the illness go away any faster. You can't tell how sick your child is (that is, whether you are dealing with a cold or pneumonia) by how much the temperature goes down when taking Tylenol™. A child with a high fever is unlikely to achieve a completely normal temperature from Tylenol™, but she may feel more comfortable. *Don't watch the temperature; watch the child.*

If you decide to use acetaminophen, read the package instructions carefully for the correct dosage for your child. Do not give it more often than every four hours. Overdoses can cause liver damage. There is no benefit to dividing doses every two hours. Keep medicine out of your child's reach (in a childproof or locked cabinet) between doses. It takes about 45 minutes to an hour for acetaminophen to start working. The effects wear off in four to six hours, so if your child is uncomfortable again in that time you can repeat the dose. Please do *not* wake up your child to give her acetaminophen. If she's sleeping, the medicine is much less important than the rest she needs to fight the infection.

Acetaminophen is very safe when taken in recommended doses. I've never seen a child experience side effects from it when it was taken as recommended. Acetaminophen does not prolong most illnesses or other symptoms or suppress the immune system. Generic or store brands are just as effective and safe as name brand acetaminophen.

Do *not* use *aspirin* without talking to your child's doctor. Giving aspirin when a child has influenza or chicken pox is associated with a sometimes-fatal condition called Reye's syndrome. Other side effects of aspirin include upset stomach, prolonged bleeding, and allergic reactions, especially in asthmatic children. It is not worth the risks when good alternatives like acetaminophen and ibuprofen are available.

Ibuprofen (Children's Motrin™ or Advil™) works a bit faster and a bit longer (six to eight hours instead of four hours for acetaminophen) and may reduce fever a bit more than acetaminophen (by about one degree Fahrenheit). However, ibuprofen is more likely than acetaminophen to cause an upset stomach. Because of its rapid onset and prolonged effects, ibuprofen has become my #1 choice when children need medication to treat their discomfort due to fever.

What about alternating acetaminophen and ibuprofen? You can give both at the same time or alternate them every few hours. If your child's discomfort is not better with acetaminophen and ibuprofen, it's time to call the doctor and get some help figuring out what's making the child so uncomfortable. Do not awaken a feverish child to give him acetaminophen or ibuprofen. Sleep is more valuable than a normal temperature. However, if your child is difficult to rouse or comatose, have him evaluated by a professional.

Dietary Supplements

No vitamins, minerals, or other dietary supplements are as effective as acetaminophen or ibuprofen in lowering a fever.

Chronic *vitamin A* overdoses can actually *cause* fevers. So can overdoses of a number of minerals.

Blood levels of *vitamin C* decline during fever. High doses (one gram or more in adult volunteers) of vitamin C temporarily *raise* the temperature by almost one degree. In animals, giving vitamin C supplements along with acetaminophen (Tylenol) boosts Tylenol's anti-fever effects and decreases its potentially toxic effects on the liver. It has not yet been studied as a fever fighter in children.

Children with Down syndrome have weak immune systems and tend to get a lot of fevers. Among children with Down syndrome, daily *zinc* supplements of 25 to 50 milligrams significantly reduced episodes of cough and fever compared with similar children given placebos. Zinc supplements have not been studied as preventive therapy in other children.

There are no studies evaluating ginger's effectiveness in treating fevers in children, but it is a helpful anti-nausea remedy. *Bee products* such as propolis have also gained popularity recently for treating a variety of illnesses, but studies have *not* demonstrated any anti-fever effects of propolis and there are *no* studies evaluating its effectiveness in feverish children. *Garlic* has been used to treat infectious illnesses since ancient times by Egyptians, Greeks, and Romans and is known as Russian penicillin. *Beet juice*, sometimes combined with carrot juice, is a British folk remedy for fever. There are *no* studies on the effectiveness of garlic or beets for reducing fever in children.

Herbal Remedies

A number of herbs used in other parts of the world have proven effective as fever fighters in animal studies, but none has proven as effective as acetaminophen or ibuprofen for children's fevers. A German combination containing ash tree, poplar, and goldenrod, known as STW 1, was shown in *animals* to be as effective as aspirin in reducing fever. I do *not* recommend these herbal remedies for treating feverish children until additional studies have evaluated their safety and effectiveness in humans.

White willow bark is the original source of common aspirin. It has been used since ancient times around the world to treat fever and the aches and pains of a variety of illnesses. Willow bark tea does not taste very good. Because the active ingredient in willow bark is metabolized in the body to salicylic acid (aspirin), it may cause the same problems as aspirin—upset stomach, suppressed immune system, possible allergies, and possible Reye's syndrome.

Other traditional remedies have not been evaluated in comparison studies, although they have been used for centuries. *Bitters,* such as Angostura bitters or gentian, have long been recommended to reduce fever, especially in illnesses characterized by upset stomach and diarrhea. *Boneset tea* was used by Native Americans to break the bone-breaking fevers and chills of influenza. It is bitter and may cause nausea and vomiting. *Catnip tea* is used in treating low-grade fevers to help a child relax and induce sweating. *Echinacea* helps bolster the immune system and is used to treat the common cold. *Elder flower tea* is sometimes used in combination with catnip or *peppermint* and *yarrow* to treat fevers. *Licorice root* has some antibacterial and anti-inflammatory properties and is used to treat fevers accompanying inflammation. Bitter-tasting yarrow flowers were said to have been used by the Greek hero Achilles to staunch the flow of blood from his battle wounds. Modern herbalists recommend a tea from its flowers to reduce fever and as a general tonic. There are *no* scientific studies evaluating the effectiveness of any of these herbal remedies in treating fever in children.

Herbal remedies are not necessarily safe just because they are natural. A traditional Asian remedy for fever, star anise (*Illicium verum*), proved an effective fever fighter in rodents but also caused seizures. *Feverfew* is used primarily as a painkiller for headaches and arthritis, but it can cause allergic reactions. Beware of possible contamination in herbal products. Most have been used for many years and generally are considered safe if given in small doses over a few days. If a child does not respond within a day or two to home remedies, her temperature exceeds 103.9°F, or she is lethargic, having trouble breathing, or acts very ill, take her to be evaluated by a health care professional.

Biomechanical Therapies

No biomechanical therapies have proven effective as fever fighters. Surgery may be needed if the underlying problem is appendicitis or an abscess.

Bioenergetic Therapies: Acupuncture, Homeopathy

Acupuncture

Acupuncture treatment effectively reduces fever in arthritis-like disorders in animals. Until such studies have been performed on human children, I do *not* recommend acupuncture as a primary treatment for typical fevers.

Homeopathy

Several remedies are traditionally used by homeopathic practitioners in extremely dilute doses (more concentrated doses can be poisonous) for treating various kinds of fever: *Aconitum, Arsenicum, Belladonna, Bryonia, Ferrum phos, Gelsemium, Nux vomica, Pulsatilla,* and *Sulphur.* The primary Bach flower remedy for fever is Rescue Remedy. There are insufficient scientific studies evaluating the effectiveness of homeopathic remedies for me to recommend them for treating children with fever. However, they are safe, or I tolerate their use for families whose culture relies on homeopathy.

WHAT I RECOMMEND FOR FEVERISH CHILDREN

Seek Immediate Professional Help If Your Child

- Is under three months old and has any degree of fever

- Is any age and the temperature is over 104°F

- Looks very sick

- Is more difficult than usual to awaken

- Won't stop crying with your usual loving care

- Is confused or delirious

- Loses consciousness or has a seizure

- Has a stiff neck

- Has painful or very frequent urination

- Is unable to keep down clear fluids

- Has trouble breathing or can't swallow her own saliva

- Cries when you touch or move her

- Has a rash with deep red or purple spots that don't fade or blanch when you press on the skin

WHAT TO AVOID WHEN YOUR CHILD HAS A FEVER

1. *Lifestyle—environment. Don't* sponge your child with alcohol or ice water. Sponge only when the fever is breaking, use tepid (not ice) water, and sponge only as long as it helps her feel comfortable. Neither over-dress nor completely undress your child. Her body will find its own proper temperature if you dress her as usual and offer a blanket or two.

2. *Biochemical—medications. Avoid* aspirin. Do *not* awaken your child to give anti-fever medicines.

When to Call or Visit Your Health Professional

- If your child is less than two years old and the temperature is above 102°F.

- If your child is urinating less often than usual (fewer than four wet diapers a day); this may be a sign that your child is becoming dehydrated. Offer additional fluids.

- If the fever lasts more than a day or two, especially if it's not clearly associated with a simple cold or flu.

TREATING A FEVER AT HOME

Remember. Treat the child, not the thermometer reading.

1. *Lifestyle—nutrition.* Give your child extra fluids to drink. Juices and broth are excellent and easily digested.

2. *Lifestyle—environment.* You may give a tepid water sponge bath if your child is sweating (the fever has "broken").

3. *Lifestyle—exercise.* Encourage your child to rest.

4. *Lifestyle—mind-body.* Stay with your child, offering love, support, and encouragement. Remember that most fevers resolve in two to three days on their own.

5. *Biochemical—medications.* Try acetaminophen or ibuprofen to reduce your child's discomfort. Do not exceed the doses recommended on the package label. Consider alternating acetaminophen and aspirin every two to three hours.

20
HEADACHE

Marvin Fahnestock called me shortly after he and his wife moved to town. Their 12-year-old daughter, Kim, had started having headaches a few weeks before they moved. She was now having them about twice a month. She had told her father that she knew when she was getting a headache because she saw flashing lights. Shortly afterward, she complained of terrible pounding pain, usually on the left side of her head. She had tried aspirin and acetaminophen, but that didn't help very much. She felt better after a long nap. She was otherwise healthy, though Kim had colic as a baby. Marv initially attributed the headaches to the stress of the move, and then he thought Kim might be getting migraines like her mother and grandmother. He called me to see if there were any new treatments for migraine and to make sure Kim didn't have something more serious, like a brain tumor.

Headaches are common, painful, and costly. Each year, Americans spend over $1 billion dollars for medical care to treat headaches, plus the cost of time missed from work and school. Although adults in their 30s and 40s are the most likely persons to have headaches, kids can have them, too; in any given month over 40 percent of teenagers report having had a headache. Most adults with headaches started suffering with them as teenagers, so addressing the problem early may prevent a lifetime of suffering. An estimated 10 to 15 percent of teenagers have migraine headaches. Children with migraines miss more than twice as many school days each year as other kids. Mondays are the most common day for migraines in school-age kids, with 50 percent more migraines reported on Mondays than on Tuesdays and twice as many on Mondays as on Saturdays.

Headaches run in families. For example, tension headaches are three times more common among those who have a close relative with frequent tension headaches than for kids who don't have headaches in the family. Migraine headaches often run in families like Kim's, too.

No age group is completely immune to having headaches. Recurrent vomiting may be the only signal that toddlers and preschoolers are having migraine headaches. Even infants as young as 12 months old have been diagnosed as having migraines.

Like Marvin, many parents worry that headaches are a sign of a brain tumor, but tumors are incredibly rare. About half of all headaches in children are due to muscle tension. About 25 percent are due to migraine, and the rest are due to fever, head injuries, sinus infections, dental problems, jaw problems, uncorrected vision problems, and ear infections. Less common causes of headache include very high blood pressure, meningitis, lead poisoning, carbon monoxide poisoning, eating very cold foods, altitude sickness, and overdoses of vitamins or medications. Withdrawing from certain foods, drinks, and herbs can also cause headaches. Overuse of headache medications can actually lead to headaches, too.

TENSION HEADACHE TRIGGERS

- Emotional stress
- Physical stress, lack of sleep, fatigue, heat, noise
- Poor posture
- Pain, pinched nerves
- Eyestrain, poor lighting
- Other illnesses, colds, flu

Muscular *tension headaches* are typically caused by emotional stress, anger, or de-pression. They can also be caused by physical stress (e.g., fatigue, cold, heat, irregular meals, noisy environment), poor posture, pain, pinched nerves, grinding teeth, eyestrain, poor circulation, and illnesses elsewhere in the body. Whew! Given this long list, it's surprising that not *all* kids get tension headaches. Tension headaches are usually most painful in the forehead or the back of the head. They feel like a band tightening around the head or a sense of pressure all over. They seldom throb, and are not aggravated by motion, light, or sound.

Migraine headaches can provoke pain, clenched muscles, and thus tension headaches. Often the two go together. Although migraine headaches and tension headaches can occur simultaneously, migraines have their own painfully unique characteristics. People who have migraine headaches are called *migraneurs*.

Famous migraineurs include President Thomas Jefferson; the painters George Seurat, Vincent Van Gogh, and Claude Monet; authors Virginia Woolf, Lewis Carroll, and Cervantes; actresses Elizabeth Taylor, Whoopi Goldberg, and Lisa Kudrow; musicians Carly Simon and Loretta Lynn; basketball players Kareem Abdul-Jabbar and Scotty Pippen; and numerous other athletes, artists, and business leaders.

CHARACTERISTICS OF MIGRAINE HEADACHES

- Run in families
- Recurrent headaches; pain-free between episodes
- Symptoms preceding the headache: aura
- Typically last two hours to seventy-two hours
- Pounding or throbbing on one side of head; can hurt all over
- Often very sensitive to light or sound
- Often have upset stomach, nausea, or vomiting

- Prefer to remain lying down in quiet, dark room
- Improved with sleeping

Marv was really on the ball in putting together his daughter's symptoms with his wife's history of headaches. Migraine headaches are also more common in women, adding more weight to the possibility that Kim was having migraines.

Like Kim, some migraineurs experience symptoms such as flashing lights or other images *just before* the headache starts. These symptoms are known as the headache's *aura*. Less often, migraineurs have other symptoms such as temporary blindness, tingling, or severe weakness in the arms or legs on one side of the body. Usually these symptoms last for a few minutes and go away when the actual headache begins, which is usually within an hour. Some people are so troubled by the aura symptoms that they actually look forward to the headache itself. For example, one of my colleagues had aura symptoms that included temporary blindness; when the headache started, he could finally see again—the pain offered a relief from blindness.

A migraine headache is typically a moderate to severe throbbing or pounding headache on one side of the head. However, many children experience migraine headaches as a kind of vague pain all over the head. Babies who have colic, like Kim, are more likely to grow up to have migraines than babies without colic. Most migraines are accompanied by nausea or a stomachache, making the sufferers truly miserable. When struck with a headache, most prefer to lie in a very quiet, dark room and to lie very still; moving makes the pain worse. Migraines typically last several hours, but they can last up to three days. Even after the headache is over, many sufferers feel wiped out and achy for a day or so. It's best to treat migraines as soon as symptoms start rather than trying to tough them out.

Like tension headaches, migraines can be triggered by many different things.

MIGRAINE HEADACHE TRIGGERS

- Stress, tension, or anxiety; being bullied or treated unfairly
- Being overtired, late nights; insufficient exercise or leisure time
- Diet: chocolate, cheese, aspartame, nitrites, MSG, caffeine
- Hormones: menstrual period, birth control pills
- Bright light, flashing lights
- Chinook winds

Children who suffer from migraines do not necessarily experience any more *stress* than other children, but they seem to respond differently to stress, often worrying more about challenges and hurting themselves in sports and games than other children. They also appear to be more sensitive to pain and more likely to develop other painful and fatiguing conditions like irritable bowel syndrome, chronic fatigue, and fibromyalgia; some might refer to this as "central sensitivity syndrome" or CSS. Studies using magnetic resonance imaging (MRI) indicate that the brain's cortex tends to be especially active and excitable in migraineurs. Regardless of sensitivity, being *bullied* or *treated unfairly* are common migraine triggers in teens.

Many migraineurs get their migraines after the tension has eased (e.g., after the exam is over) rather than while they are worried about it; thus some folks are most prone to getting a headache on the weekends or as soon as they've unpacked the suitcases on vacation. Sensitivity to changes or stress provides the rationale for mind-body therapies such as self-hypnosis, biofeedback, and meditation.

About 25 percent of families identify *fatigue*

as a trigger for migraines. Sleep problems like snoring, restless leg syndrome, grinding teeth, and difficulty falling asleep are also linked to an increased risk of migraines. Insufficient exercise and inadequate "down time" are also risk factors for migraines. Grandmothers who urge us to eat right, exercise, and get enough sleep probably had a pretty good strategy for minimizing headaches and a host of other health problems.

Missing meals or *eating certain foods* (such as chocolate, cheese, aspartame, nitrites, MSG, caffeine, and red wine) have also been blamed for migraines. Headaches are just one of many problems associated with underage drinking. Being overweight or *obese* increases the odds of developing migraines. *Chocolate* may be a trigger for some migraine sufferers, but in a randomized, double-blind crossover trial only *two* out of twenty-five adults who initially attributed their symptoms to chocolate had consistently negative reactions to it.[1] These results were duplicated in a study of 63 women who had thought their headaches were triggered by chocolate; but when they were actually tested blindly (they didn't know whether or not they were getting chocolate), it turned out that chocolate was not really the culprit.[2] If your child gets headaches after eating chocolate, try going six weeks without any chocolate and keep track of whether or not she has fewer headaches. Children do not *need* chocolate as part of a healthy diet, but they may not need to live a life of total deprivation to avoid headaches either.

The amino acid, tyramine, found in *cheese*, especially aged cheeses, has also been blamed for migraines. However, most people can eat cheese safely. Several studies in children have shown that tyramine does *not* trigger migraines, and restricting cheese does not reduce the risk of migraines.[3,4] This is why it's so important to keep a symptom diary to determine your child's unique triggers.

The artificial sweeteners *aspartame* (Nutrasweet™) and *sucralose* trigger some migraines.[5,6] There is increasing medical concern about the adverse effects of artificial sweeteners, which can also alter the balance of bacteria living in our intestines.[7] Your child does not require anything other than water as a beverage. If your child has frequent migraines, try omitting aspartame for six weeks and keep track of her headaches. Careful monitoring may uncover other triggers as well.

Nitrites and *nitrates* are used to cure and preserve many meats such as hot dogs, bacon, salami, and dried meat sticks. Nitrites and nitrates trigger migraines in some people. My stepfather's migraines were inevitably triggered when he took nitroglycerin for his angina. Again, try going several weeks without preserved meat products, and monitor your child's headaches carefully.

MSG is the key ingredient held responsible for the "Chinese restaurant syndrome": headache, dizziness, flushing, sweating, and abdominal cramps. When you eat out, ask to have your child's food prepared without MSG; most restaurants accommodate this request without a fuss. Because many people steer clear of MSG, manufacturers have started calling it by other names: hydrolyzed vegetable protein (HVP), hydrolyzed plant protein (HPP), and natural flavor enhancers. Read labels carefully. Sensitive children who consume small amounts of MSG in ordinary foods every day may end up with chronic headaches.

Caffeine, found in coffee, tea, cola, and chocolate, is an additive in many over-the-counter headache medications because it can help in the short term, but chronic use leads to caffeine addiction. Quitting coffee cold turkey not only results in more yawns, but in more headaches, irritability, and depression; caffeine withdrawal actually changes the EEG.[8] Likewise, the most common cause of postoperative headaches in adults is having abstained from coffee (as well as breakfast) prior to the operation! Children who drink two to three cans a day of caffeine-containing sodas are just as susceptible to caffeine withdrawal headaches as adult coffee drinkers who try to quit suddenly, who oversleep on the weekend and miss their morning dose, or who undergo surgery. If your child is a regular cola drinker, gradually reduce her intake over two weeks to avoid withdrawal headaches.

People who start taking over-the-counter medications every day for headaches and other aches and pains may end up addicted to caffeine and suffer severe headaches when they try to stop using them. This is called *medication-overuse headache*, and it can lead to chronic daily headaches.[9]

Menstrual periods and birth control pills are common triggers for migraines in teenage girls and adult women.

Bright or *flashing lights* trigger symptoms for some sufferers. Although some factors seem to trigger migraine headaches directly, others seem to work in combination. A teenager who usually tolerates nitrates without any problem could develop a raging headache if she has a bacon cheeseburger the day after a sleepless slumber party, especially if it's the wrong time of the month.

Many people feel their headaches are related to various changes in the *weather*. A Canadian neurologist listened to his patients complain that the warm, westerly Chinook winds in the Alberta province triggered their migraines; looking back over the headache diaries kept by 75 patients, he found that patients were much more likely to develop a headache in the two days before or during high Chinook winds than at other times.[10] If your child complains that certain kinds of weather trigger her migraines, well, it may not be all in her head.

Kim didn't notice that her headaches were triggered by weather, but did notice that they often appeared with her menstrual period and when she was feeling tired and stressed. She didn't miss many meals or stay out late on weekends, and she avoided processed meats and artificially flavored and sweetened soda.

Migraine sufferers who keep careful track of their headaches using a *headache diary* often find that their headaches occur more often at certain times of the day. In a study of fifteen meticulous migraine sufferers, the headaches occurred most often between 6 a.m. and 10 a.m. and were less likely between 8 p.m. and 4 a.m.[11] This is the same timing as heart attacks. It is also the same rhythm of the rising and falling of several stress hormones, suggesting a complex interaction between hormones, blood vessels, inflammation, and pain.

What Causes a Migraine?

Sophisticated imaging and biochemical studies are getting us much closer to answering that question. So far it looks like many migraine triggers focus in on the brainstem, the part of the brain closest to the spinal cord. Chemical imbalances in certain centers there, triggered by factors in the environment, start a wave of chaotic neural discharges firing across the brain, flooding the brain with chemical messengers, leading to leaky blood vessels, pain, and swelling.

Have you ever had an *ice cream headache*? Many people have. The problem is not the ice cream per se but the temperature. The headache occurs when cold food hits the nerves on the roof of the mouth, triggering a pain sensation that feels as if it comes from the forehead or behind the eyes. This type of headache may have a strikingly sudden onset, but it usually resolves within a minute. Your child can still eat ice cream and popsicles, but she may have to go slowly, allowing the ice cream to melt on her tongue before letting it hit the roof of the mouth.

DIAGNOSIS

There are no blood tests or X-rays that differentiate between tension headaches and migraines. X-rays may help make the diagnosis of a sinus headache. Expensive tests such as MRIs, CT scans, and EEGs are rarely necessary or helpful. The most helpful information is a complete picture of the child's symptoms. If your child has frequent headaches, keep a diary or calendar noting the headaches, when they start, what seems to trigger them, how long they last, what seems to relieve them, and any other symptoms. This kind of record is more useful than a hundred CT scans or MRIs, and it doesn't cost you a penny. The best diagnostic test is keeping track of symptoms and triggers and getting a thorough physical examination by your pediatrician.

WHAT IS THE BEST WAY TO TREAT HEADACHES?

The two most important aspects of any headache treatment are:

- Recognize and address the underlying cause of the headache.
- Prevent it or treat it early rather than wait it out.

The treatment for a sinus headache involves treating the sinus infection, while the treatment for a stress headache involves preventing or managing the stress. A headache signifying meningitis requires different therapy than a headache from a head injury. No matter what causes the pain, it's better to treat it early rather than wait until the body starts screaming even more loudly with worse symptoms.

This chapter does *not* cover the specific treatments for all causes of headaches. It focuses on treatments for tension headaches and migraine headaches.

SEE YOUR HEALTH CARE PROFESSIONAL IF YOUR CHILD HAS A HEADACHE THAT

- Is very severe
- Is accompanied by fever or stiff neck (possible meningitis)
- Also has vomiting following a blow to the head (possible brain injury)
- Has seizures, dizziness, confusion, double vision, or trouble hearing
- Is worse with "bearing down," coughing, or sneezing
- Is worse in the morning and gets better during the day
- Lasts longer than two days or gets worse over time

Also seek professional care if you suspect that your child might have a sinus infection, a dental problem or jaw problem, lead poisoning, high blood pressure, or some other serious problem underlying her headaches.

Let's tour the Therapeutic Mountain to find out what works for tension and migraine headaches. If you want to skip to my bottom-line recommendations, flip to the end of the chapter. We'll start with Lifestyle Therapies.

LIFESTYLE THERAPIES: NUTRITION, EXERCISE, SLEEP, ENVIRONMENT, MIND-BODY

Nutrition

Food allergies and food sensitivities are commonly blamed for migraine headaches. A British study showed that 93 percent of children suffering from weekly migraines and other allergy symptoms (e.g., skin rashes, diarrhea, wheezing) markedly improved when they consumed a low-allergen diet.[12] Another study by the same group of researchers showed similar improvements among children who had both migraine headaches and seizure disorder when they tried a low-allergen diet.[13] A Turkish study confirmed the importance of eliminating allergenic foods (diagnosed by IgG blood tests) in reducing headache frequency and severity.[14] Other research has confirmed these findings in adult migraine sufferers.[15,16]

What is this powerful "oligoantigenic" diet? It allows only one type of meat, one fruit, one vegetable, one carbohydrate, water, and vitamins for three to four weeks, gradually adding one new food each week.[17] Any foods provoking headaches are withdrawn. The foods that provoke symptoms most often are cow's milk, egg, chocolate, orange, wheat, corn, cheese, tomato, cane sugar, benzoic acid, and the yellow dye, tartrazine.[18]

Not all children with frequent migraines that are thought to be triggered by foods actually have symptoms when given them under double-blind study conditions. Before you begin severely restricting your child's diet, consider working with a pediatric dietitian, neurologist, and allergist team.

Avoiding allergenic foods is rarely successful because it is very difficult to change eating habits. In a study of adults with chronic headache, 75 percent identified one or more foods as headache triggers. However, most of them did not act on this knowledge. They ate just like people who didn't suffer from chronic headaches, except they drank a little less red wine.[19] This may be one reason physicians rarely pursue food sensitivities in the treatment of headache; even when extensive tests are done, patients rarely follow through on the rigorous diets required.

There are no reliable tests that can tell you in advance which foods will cause problems for a particular child, so if you want to find out whether your child has a true sensitivity, you need to start with the bare bones and gradually add things. An alternative approach is to eliminate those foods most commonly associated with headaches in other children and then slowly add them back, one at a time. Some clinicians advocate for blood tests measuring IgG or IgE antibodies to commonly consumed foods, but others question the reliability of this approach. Many children who are sensitive to particular foods do eventually "outgrow" their sensitivity and tolerate these foods later without any symptoms. Besides being extremely difficult to institute, these diets run the risk of nutritional deficiencies unless supervised by a pediatric nutrition expert. *See your health care professional before starting any special diet.*

Exercise and Sleep

Being overly tired is a setup for developing a headache. Make sure your child gets *plenty of rest*, especially during periods of increased stress. Going to bed in a quiet, dark place is useful even after one feels early warning signals for a migraine starting. For many people, going to bed is the main treatment for migraines. Usually by the time the sufferer awakens, the headache is gone. Rest is also helpful for many of the illnesses (such as influenza, intestinal flu, strep throats, and sinus infections) that cause headaches.

Vigorous *exercise during the day* (when the child is not having a headache) is one of the best remedies for stress. Aerobic exercise also seems to be a natural antidepressant. Exercise programs can be very useful in preventing migraine headaches as well as relieving the stress that triggers tension headaches.[20,21]

Environment

Being in a *quiet, dark environment* is soothing for many headache sufferers. Often the stress that precipitates a headache is due to loud noises (e.g., construction noises right outside), bright lights, or overstimulation. Relaxing in bed in a quiet, dark room helps reduce the environmental stress level.

Pressure, heat, and *cold* have all been reported to ease headache pain. Some folks feel better if they stand under a hot shower or apply a heat pack to their neck and shoulders. Others prefer *ice packs* to the back of the neck or forehead. Nowadays you can purchase a gel-filled headband that can be tightened around the head with a Velcro strap; the headband can either be heated in the microwave or cooled in the freezer and then strapped around the head, applying pressure and the most favored temperature for the headache sufferer. This device was actually studied in 15 adults with severe headaches and found to be helpful and have fewer side effects than many headache medicines.[22] Other home remedies include a ten- to fifteen-minute *hot foot bath or soak.* The heat dilates the blood vessels of the feet and promotes general relaxation.

Transcranial magnetic stimulation (TMS) has become increasingly popular as a way to treat a variety of neurologic conditions in adults. Preliminary studies suggest it may be helpful for patients with intractable headaches that have not responded to other therapies.[23,24] However, there is insufficient research for me to recommend TMS for pediatric headache patients.

Exposure to *toxic chemicals, fumes,* and *carbon monoxide* is a hidden cause of many headaches. When I was studying hypnosis, we learned of a girl with severe recurrent headaches who had been evaluated for several serious illnesses. Her headaches responded briefly to hypnotherapy but then recurred. Finally her parents looked for an environmental cause. They found it in their own garage (which was next to the girl's bedroom); a faulty furnace was spewing carbon monoxide into the air and poisoning their daughter. When the furnace problem was corrected, their daughter's headaches disappeared. Carbon monoxide poison-

ing can occur wherever there is incomplete combustion such as from a leaky car exhaust pipe, smog, and fumes from charcoal being burned or a portable electric generator operating indoors during a power outage.

Mind-Body

Stress is a definite trigger for headaches. Children who report bullying or other distressing events at school are much more likely to suffer from headaches than other kids. If your child suffers from recurrent headaches, keep a *headache diary* or log of her symptoms. You can use a regular calendar, a smart phone app, or make your own. Write down the date and time of the headache: Was it a school day or a weekend? Was it the first thing in the morning or after school? Was it just before or during a menstrual period? Was it after she witnessed a fight at home or in the neighborhood or after a fight with a friend? Next, write the food or other factors that seemed to trigger it, what you did to treat it, how long it lasted, and if anything seemed to make it worse or better. By keeping careful track of your child's symptoms, you will be well on the way to determining their cause and the best ways to prevent and treat them.

Mind-body therapies such as *biofeedback, hypnosis, cognitive behavioral therapy,* and *progressive relaxation* effectively prevent headaches for many children and have fewer side effects than most medications.[25] Remember, no method is 100 percent effective for all children. However, mind-body therapies can help 50 to 75 percent of headache sufferers.[26] They have become mainstream therapies at the neurology and headache clinics of leading children's hospitals because they are safe and effective. The combination of *deep breathing, autogenic training, biofeedback,* and *progressive muscle relaxation* can even ease headaches after they've started as well as reduce the frequency of getting them in the first place.[27] These techniques require training and regular practice. Most teens require about two months of practice before they notice fewer symptoms. I usually refer patients with frequent disabling headaches to the pediatric psychologist in the headache clinic for eight sessions of mind-body skills training.

EFFECTIVE MIND-BODY THERAPIES FOR PREVENTING MIGRAINES

- Hypnosis
- Autogenic training
- Biofeedback: thermal and EMG
- Progressive muscle relaxation
- Mindfulness-based meditation
- Cognitive behavioral therapy (CBT)

Several studies have demonstrated that the most effective preventive therapy for migraines in children is *relaxation* whether through self-hypnosis, the relaxation response (meditation), autogenic training, biofeedback, or progressive muscle relaxation.[28–32] As with other therapies, one type may work better for your child than another. A review of many studies found that relaxation therapy alone improved symptoms for 38 percent of adults with migraine headaches and 45 percent of those with tension headaches. Unlike the benefits of medication, which tend to wear off soon after the medication is stopped, the benefits of relaxation training last for several *years* beyond the initial training period. Some of these therapies can be learned and practiced at home after a few training sessions with a professional, which leads to lower cost than continuous treatment with prescription medications.[33]

Self-hypnosis is even more effective than the most commonly prescribed medication, propanolol, in preventing migraines in children.[34] Children as young as six years old can learn hypnosis to help them prevent headaches. Hypnosis is simply being in a very relaxed state of mind in which the child's attention is focused on pleasant, healing images. Applications are available for smart phones and tablets, and CD recordings and MP3 downloads are available from Health Journeys (www.healthjourneys.com).

You can help your child relax through *deep breathing exercises* (in which the child watches her belly rise and fall with each breath), *progressive muscle relaxation,* or repeating a favorite phrase, affirmation, or prayer. Relaxing, pain-reducing images for your child might include the following:

1. Imagine that the pain is like a bunch of bubbles. Take a deep breath in. Each time you breathe out, blow the bubbles out of your head and watch them disappear as they float up into a soft, blue sky.
2. Imagine yourself in your favorite place. It may be home in bed with your parent reading you a story. It may be at the beach or in the woods or Disneyland. Imagine yourself there feeling very safe and happy.
3. Imagine the sun warming your hands and feet.
4. Imagine the pain as a point on a dial inside your head. Reach over and gradually turn the dial down until it is at zero.
5. Imagine the pain as a thermometer reading. Watch the thermometer go down and down as the headache disappears. Imagine putting an ice cube at the bottom of the thermometer so it goes down even faster.

I'm sure you and your child can imagine even more ways of blowing away or turning down the pain.

Autogenic training is a kind of self-hypnosis in which the person tells herself several relaxing statements over and over. Autogenic training is an effective treatment for chronic tension headaches. Check out the autogenic training recording at the Ohio State University Center for Integrative Health and Wellness website (http://go.osu.edu/guidedimagerypractices).

Biofeedback is particularly useful in preventing pediatric migraine headaches, and I almost always refer my migraine headache patients to a psychologist who can help train them in biofeedback techniques. The simplest example of *thermal* biofeedback is the old-fashioned mood ring. More sophisticated devices use thermistors to measure temperature and display the results in colors, lights, or sounds. Thermal (fingertip temperature) feedback has proved especially helpful in preventing migraines and drastically reducing the need for medication; for many children, biofeedback is even more effective than medication.[35] Benefits last at least six years in those who have participated in just eight to ten weeks of biofeedback training.[36] Fewer formal training sessions are necessary if children continue to practice relaxation techniques at home;[37] this makes biofeedback a more financially attractive treatment option. Children can learn thermal biofeedback very readily, and many science museums around the country now contain exhibits using thermal biofeedback devices to teach kids about biofeedback.

Electromyographic (*EMG*) *biofeedback* provides information about muscle tension to help one learn to relax specific muscles. By relaxing tense muscles, EMG biofeedback has been especially helpful for those with chronic tension headaches.[38] Controlled studies document the effectiveness of EMG biofeedback training in reducing the pain and frequency of muscle tension headaches in children as young as six years old.[39]

Progressive relaxation exercise combined with *diaphragmatic breathing* can help reduce the frequency and intensity of headaches for many children. As preventive therapy for headaches, they can be as effective as biofeedback training and less costly.[40] Relaxation exercises increase the child's sense of mastery, decrease her sense of distress, decrease her muscle tension, and distract her from pain. The benefits of progressive relaxation exercises, self-hypnosis, and autogenic training can persist for years in children with both tension headaches and migraine headaches.

Mindfulness-based meditation programs are gaining ground as ways to help cope with chronic recurrent pain, and in the process, reduce the pain itself. My colleagues at Wake Forest conducted a randomized controlled trial and showed that mindfulness was not only safe, it also helped reduce the severity of migraines and their impact on daily life.[41] Although more research is always helpful, mindfulness has made its way into many schools as a way to improve concentration, behavior, and educational outcomes; and I'm in favor of more mindfulness in health care settings, too. Check with your insurance to see if group training led by a psychologist certified in teaching mindfulness is covered.

Cognitive behavioral therapy, CBT, is a solution-focused concise counseling technique that is helpful for many kinds of pain. In a study comparing a standard medication plus education to medication plus CBT for teens with migraines, the CBT combination won hands down.[42] CBT is usually provided by a licensed psychologist and covered by most major insurers.

Mind-body techniques are most effective if practiced regularly. Children benefit from home practice and the encouragement of supportive parents. Most successful programs combine biofeedback training with imagery, breathing, or relaxation techniques. I recommend professional relaxation, hypnosis, and/or biofeedback training as first-line treatments for children with frequent or severe headaches. Even those children who end up requiring medications often benefit from these non-drug therapies.

I referred Kim to our clinic's psychologist, who quickly taught her three new skills: (1) deep slow breathing; (2) autogenic training; and (3) thermal biofeedback. Kim found the sessions easy and enjoyable. She started practicing 10 minutes daily at home.

BIOCHEMICAL THERAPIES: MEDICATIONS, DIETARY SUPPLEMENTS, HERBAL REMEDIES

Medications

For regular everyday *tension headaches,* all your child may need is a non-prescription pain reliever and a nap in a dark, quiet room.

NON-PRESCRIPTION PAIN RELIEVERS

- Aspirin
- Acetaminophen (Tylenol™ and other brands)
- Ibuprofen (Advil™, Motrin™, and others)
- Naproxen (Aleve™ and others)

Aspirin is inexpensive and widely available. It has even proven effective in treating some migraine headaches. Aspirin should *not* be taken when your child has influenza or chicken pox because of the risk of developing Reye's syndrome, a potentially deadly disorder of the liver and brain. Because of this risk, many physicians now steer clear of aspirin; I also recommend that you get the influenza and chickenpox vaccines. Aspirin can cause an upset stomach and bleeding of the stomach lining; frequent or high doses can cause ringing in the ears.

Acetaminophen (Tylenol™ and other brands) is the most widely used non-prescription pain reliever. Generic brands are just as effective as name brands. Acetamino-

phen is inexpensive and is very safe if taken in recommended doses. Acetaminophen starts to work in thirty to forty-five minutes; benefits last for about four hours. Excessive doses damage the liver. Do not exceed recommended doses.

Ibuprofen (Advil™, Motrin™, and other brands) is my personal favorite non-prescription, over-the-counter non-steroidal anti-inflammatory drug (NSAID) for headaches. It starts a bit sooner and lasts about two to four hours longer than acetaminophen or aspirin. Ibuprofen can cause upset stomach. With chronic use, particularly on an empty stomach, it can lead to bleeding of the lining of the GI tract. If used as recommended, it is generally safe.

Naproxen (Aleve™) is another NSAID like ibuprofen, but it lasts up to twelve hours. It can cause stomach upset and may only be tolerable if taken with food. It is effective for migraines but is probably overkill for simple tension headaches. I recommend that migraine sufferers keep some on hand to treat severe headaches.

Combining *caffeine* with any of these non-prescription pain relievers boosts their effectiveness. You can either buy a combination product (such as Extra Strength Excedrin™) or have your child take a pain reliever with a cup of coffee or tea.

Be aware that daily use of headache medications, even over-the-counter products, can lead to medication overuse headaches. If you find yourself reaching for headache medicines more than once a week, please see your health professional for a thorough evaluation and discussion of other treatment options.

For *migraine headaches,* additional medications have proved effective in prevention and treatment. If your child has migraines more than once a month, talk with your health care professional about preventive medications. There is a *high placebo response rate* to migraine medications. This means that many children will improve with a medication or supplement, regardless of which one is started. This is why most pediatricians recommend starting with the safest treatments before trying riskier options.

PREVENTIVE MEDICATIONS FOR MIGRAINES

- Beta-blocker medications: Propranolol (Inderal™), Metoprolol, Timolol, and others
- Blood pressure medications: Calcium channel blockers: (Flunarizine™, Nimodipine, and others); ACE inhibitor (Lisinopril™ and others); Clonidine (Catapres™, Kapvay™, and others)
- Antidepressant medications: Amitriptyline (Elavil™), Trazodone (Desyrel™, and others), pizotifen
- Antihistamines: Cyproheptadine (Periactin™), Diphenhydramine
- Seizure medications: Phenobarbital, Phenytoin (Dilantin™), Carbamazepine (Tegretol™), Topiramate (Topamax™), Divalproex, and others
- Other: Botulinum toxin (Botox™)

Beta-blocker medications such as propranolol (Inderal™) have been the medication of choice for preventing migraine headaches for many years. In adults, propranolol reduces migraine headaches an average of 43 percent—about the same reduction achieved with biofeedback training.[43] Propranolol also prevents pediatric migraines.[44] Propranolol should *not* be used in children suffering from asthma, heart failure, or depression, because it can worsen these disorders.

Calcium-channel blocking drugs were originally used to treat high blood pressure. Not all blood pressure medications prevent migraine headaches, and it may take several months before taking them starts to reduce headache frequency. They are not a perfect cure; even after six months, calcium channel blockers such as flunarizine only cut headache frequency in adults by about 50 percent. Angiotensin-converting enzyme (ACE) inhibitors, such as lisinopril, are also effective for some patients in cutting the frequency of migraine episodes.[45] Nimodipine proved successful in one randomized control trial in children in reducing the number of headache days. Additional research is needed to compare the effectiveness of calcium channel blockers to other treatments in preventing pediatric migraines.

Certain *antidepressant* drugs including tricyclic antidepressants (such as amitryptiline), selective serotonin reuptake inhibitors (such as fluoxetine), and others (including trazodone) have successfully reduced migraine frequency more than placebo even in children. They may be useful to try in children who have migraines and depression, but I do not recommend them as first-line migraine prevention for children who are not depressed. *Trazodone* has proved helpful in preventing more headaches than placebo medication in at least one pediatric study; additional research is needed to compare its safety and effectiveness to other medications.[46]

Periactin™, a widely used prescription antihistamine, also prevents migraines;[47] its side effects include drowsiness and increased appetite. Periactin is used both to prevent and to treat pediatric migraines.[48] Its effectiveness is similar to propranolol, amitryptiline, and flunarizine.[49]

In the past, the seizure medications *phenobarbital* and *phenytoin* were used to prevent migraines, but they have substantial side effects and are less commonly used nowadays. *Valproate* prevents migraines in older children and adults nearly as well as propranolol, but it is rarely used unless the migraineur also suffers from seizures because of the potential for serious side effects.[50] More recently, *topiramate* has become a mainstay of migraine management; it significantly reduces headache frequency and headache-related disability in children with frequent migraines.[51]

OnabotulinumtoxinA (*Botox™*) was approved by the U.S. FDA for use in preventing headaches in adults in 2010. Preliminary studies suggest it can be modestly useful in preventing pediatric headaches, but Botox™ injections have not become mainstream preventive treatments for children.

In general, most pediatricians start with propranolol as a preventive medication for pediatric migraines in children for whom it is not contraindicated; an increasing number of neurologists turn to topiramate and trazodone as second-line preventive therapies. Remember, there is a high placebo response rate, so start with the safest options. No medication offers 100 percent protection, and most can cause serious side effects. I recommend using them for patients with frequent migraines in conjunction with healthy lifestyle choices, massage, acupuncture, and mind-body therapies such as biofeedback.

What Prescription Medications Are Used to Treat Migraine Headaches?

The most common recommendation for treating migraines is to start with non-prescription ibuprofen. (Other non-steroidal anti-inflammatory medications include naproxen and aspirin, but these are less commonly recommended than ibuprofen). If your child has frequent (twice a week or more often), moderate, or severe headaches that do not respond to ibuprofen, your pediatrician may recommend a higher dose of ibuprofen or a prescription medication such as a triptan, a nausea reliever, or an ergotamine preparation.[52]

TREATMENT MEDICATIONS FOR MIGRAINE (PRESCRIPTION ONLY)

- Triptans: Sumatriptan (Imitrex™) and others
- Antihistamines such as Pizotifen
- Anti-nausea medications: Metoclopromide (Reglan™), prochlorperazine (Compazine™), and chlorpromazine (Thorazine™)
- Ergotamine and dihydroergotamine (DHE)
- Topical anesthetics: intranasal lidocaine
- Sedatives and narcotics (butalbital, Valium™, Demerol™, morphine)
- Steroids (dexamethasone and prednisone)
- Combinations: Midrin™ and Fiorinal™

Sumatriptan (Imitrex™) was the first triptan medication used to treat migraine headaches. It is available for use by mouth, by nose spray, and by injection. The triptan family includes almotriptan, eletriptan, naratriptan (Amerge™), rizatriptan (Maxalt™), and zolmitriptan (Zomig™). A sumatriptan/naproxen combo is also available. Triptans start to work in minutes, relieving the pain, the oversensitivity to lights and sounds, and the nausea and vomiting of migraine headaches. It works best when given at the beginning of headache symptoms and usually works with just one dose; it is *not* a preventive medication, which means you take it when you feel a headache coming on, not daily when there are no symptoms. Sumatriptan causes a brief rise in blood pressure, which is safe for most adolescents but may be hazardous in those with severe heart disease or those with high blood pressure. Other side effects include experiencing a bad taste and tingling in the mouth, flushing, tingling all over the body, warmth, and light-headedness. Triptans have proven useful in teenagers as well as adults, and have become one of the most commonly used migraine medications for children for whom ibuprofen alone is not enough.[53]

Antihistamines usually lead to drowsiness and help with nausea. *Pizotifen* is an antihistamine that is also chemically related to triptans; it is one of the secondary choices of therapies for children whose main headache symptoms include nausea. Chronic use leads to increased appetite with weight gain.

Nausea and vomiting are among the most troublesome migraine symptoms. *Anti-nausea medications* not only treat nausea, but by promoting normal intestinal movement, they help other pain medicines stay down and get absorbed. These medications are usually used in emergency room settings rather than at home. These medications may cause drowsiness, dizziness, and a drop in blood pressure, but including them in a combination treatment regimen in addition to a pain medicine can help reduce the time in the emergency room and the likelihood of requiring an admission to the hospital.[54]

Although it is one of the oldest medications used to treat migraines, *ergotamine,* which is derived from the ergot fungus that grows on wheat, is rarely used as a first-line medication since triptan medications became widely available. Ergotamine is better absorbed and more effective when *caffeine* is added. If it is taken as soon as the child feels the headache coming on, Cafergot™ (caffeine and ergotamine) is frequently successful in minimizing subsequent symptoms, and was the medicine I was first prescribed when I developed migraine headaches during my pediatric residency. (I haven't needed it in many years now.) Cafergot™ has significant side effects including cold hands and feet from blood vessel spasms; it can also cause nausea and a sense of anxiety. Because it causes blood vessel spasms, it should not be used during pregnancy. Synthetic *dihydroergotamine* (DHE) is available by prescription for injection (in the emergency room, for example) and by nasal spray (Migranal™) and suppositories. The injection and nasal spray cost about 5 to 10 times as much as ergotamine pills, but if nausea is a big problem with your child's migraines, he may not be able to take a pill in the midst of his headache, and it's good to have some other options. DHE is mostly used nowadays in emergency rooms when other treatments have failed to improve the migraine enough to send a child home.

Remember the shot of numbing medicine you got last time you went to the dentist? Well, a related numbing medicine or topical anesthetic, *lidocaine,* can help relieve headache pain for over half of adults suffering from migraine if a dose is squirted up the nose soon after the headache starts.[55] It has been used to treat cluster headaches in adults, but it is seldom used to treat pediatric migraines.

Many of the *sedative* medications used to treat migraine work by promoting sleep, which is the way people historically handled migraines. Sedatives such as butalbital and diazepam work solely by putting the patient to sleep; they do not address the underlying cause of the headache, and I do *not* recommend them. *Narcotic* pain relievers (such as morphine and codeine) are widely used in emergency rooms as migraine treatments, but they carry the risk of addiction. Emergency room personnel have become very wary of persons showing up in the ER complaining of headaches and demanding narcotic medications. None of these medications actually addresses the cause of headaches, but they have been used for many years when no other effective medication was available.

Midrin™ is the brand name of a combination medication: acetaminophen, isomethephene, and dichloralphenazone. It is rarely used in children. *Fiorinal*™ is the combination of aspirin, caffeine, and butalbital (a sedative). Both Midrin™ and Fiorinal™ can be addicting and are not any more effective than other headache medications. I do *not* recommend them.

Dietary Supplements

Have any dietary supplements become mainstream recommendations for preventing migraines? Yes, the supplements most commonly recommended by pediatric neurologists include riboflavin (vitamin B2), magnesium, and coenzyme Q10, and the herbs feverfew and butterbur.[56]

DIETARY SUPPLEMENTS FOR HEADACHES

Yes: Riboflavin (B2), Magnesium, coenzyme Q10

No: Vitamin A, zinc, niacin

Possible: Folate, vitamin B6

Although it is generally better for your child to get her essential nutrients from foods rather than from pills, sometimes pills are necessary to achieve the high levels necessary to achieve a specific health goal.

For example, *riboflavin* (vitamin B2) is typically required in levels of 0.3 to 1.6 milligrams daily, which is readily available in foods. However, much higher doses (200 milligrams twice daily or 400 milligrams daily) that help prevent migraine headaches can only be found in supplements.[57] Riboflavin is quickly eliminated through the kidneys, turning urine a bright yellow. Riboflavin is generally safe. It does not help relieve pain once a headache starts, but it can reduce the frequency of headaches when taken daily in these extraordinarily high doses. Lower doses have not proven helpful.[58]

Magnesium supplements have long been used as laxatives and to treat pre-eclampsia. More recent studies have reported benefits of magnesium for asthma and to prevent migraines. Although the recommended intake of magnesium is 100 to 400 milligrams daily, about 50 percent of teens fail to consume that much. Children who suffer from migraines have lower levels of magnesium than healthy children, with even lower magnesium levels during headaches. I recommend that your child eat plenty of magnesium-rich foods such as leafy green vegetables, legumes, nuts, seeds, and whole grains, but these foods are seldom favorites among children and teens. Theoretically, it is possible to consume enough magnesium to reduce the risk of migraines (400 to 600 milligrams daily) through food alone, but for most youth, supplements are an easier, more reliable strategy.[59] The main side effect of excessive magnesium intake is diarrhea.

Coenzyme Q10 (COQ10) is a vitamin-like nutrient that is present in high levels in the heart, kidneys, and pancreas. It is essential for energy production in human cells and also helps reduce the frequency of migraine headaches by about 30 percent. CoQ10 supplements are most useful for children who have low blood levels of CoQ10.[60] Typical doses are 100 to 300 milligrams daily. CoQ10 is generally safe.

Avoid excessive intake of Vitamin A, zinc, and niacin. Excessive *vitamin A* causes a buildup of pressure inside the brain that produces such severe symptoms that it has been mistaken for a brain tumor. Excessive *zinc* can cause headaches as well as stomachaches, nausea, vomiting, and decreased appetite. *Niacin* injections were used in the 1940s to treat headache pain, but later studies showed that they were no better than a placebo for curing headache pain. I do *not* recommend them.

The data on *folate* and *vitamin B6* are mixed. The higher the dietary intake of folate (found in leafy greens), the lower the risk of migraines;[61] individuals with genetic differences in folate metabolism are at even higher risk of migraines if they consume insufficient dietary folate, and they may benefit from folate supplements. There is not enough research yet for me to routinely recommend folate supplements as a proven preventive strategy for pediatric migraine, but I do think that eating leafy green veggies daily is a good idea. For many women, headaches are one of many troublesome symptoms of the premenstrual and early pregnancy periods and are sometimes a side effect of oral contraceptive pills, which can deplete vitamin B6 levels. *Vitamin B6* has been suggested as a safe supplement to prevent hormonal headaches, but I will wait until further research is published before I routinely recommend B6 supplements to prevent pediatric migraines.

Herbal Remedies

Although many herbal remedies have been used globally to prevent and treat headaches, the best modern research supports two: *butterbur* and *feverfew.*

Butterbur is the common name for *Petasites hybridus,* a European and North American shrub that grows in wet, marshy ground. Historically, its large leaves were used to wrap butter in warm weather. Used daily, it can help prevent migraine headaches.[62] Typical doses used in adults are 50 to 150 milligrams daily.[63–65] The raw plant contains potent liver toxins, pyrrolizidine alkaloids. If you decide to try butterbur as a preventive therapy for your child's migraines, be sure to look for a product that is pyrrolizidine-free.

Consuming as little as two to three *feverfew* leaves a day (or 25 milligrams twice daily of a freeze-dried preparation) can also dramatically decrease the frequency and severity of migraine headaches.[66,67] Feverfew is used preventively; you cannot wait until your child has a headache and then expect feverfew to relieve it. It must be taken daily to work. Suddenly stopping it or missing several doses can cause rebound headaches. About 10 percent of those who take feverfew develop sores in their mouths. The amount of active ingredient in different commercial products varies widely, and some contain no active ingredients at all. Herbal products are not well regulated in the United States. There is great variability in the strength, potency, and purity of commercially available feverfew preparations. So, even though it may be helpful for European adults in several studies, I do not recommend feverfew to children here in the United States until more research has established safety, reliability, and effectiveness of American feverfew products in preventing migraines.

Capsaicin is the spicy ingredient in hot peppers. It appears to decrease the nerve transmitters responsible for pain. Hot peppers have long been a folk remedy for painful conditions. Capsaicin has *not* been studied as a treatment of childhood headaches. It can certainly sting and cause severe irritation if it gets in the eyes. Do not spray pepper up your child's nose. If you want to try capsaicin, use it in food or use one of the commercial capsaicin products as a rub over painful areas of the scalp away from the eyes.

Ginger is a traditional remedy for headaches and upset stomachs. It is widely used in East Africa and in Ayurvedic medicine as a headache cure. Certain compounds in ginger appear to work the same way as aspirin does, blocking inflammation and pain. There are several case reports that taking raw ginger or ginger powder daily helped prevent migraines in adults.[68] An Iranian study suggested it could be as effective as a triptan medication in halting headaches.[69] There aren't enough studies for me to recommend ginger as a preventive treatment for pediatric migraine headaches. Nevertheless, ginger is safe, inexpensive, and widely available; it can also help with nausea, and I have no objection to a trial of home remedies of ginger.

Tiger Balm is a popular non-prescription salve used for all kinds of aches and pains, including headache pain. It contains a number of herbal extracts including menthol, cajuput, camphor, and clove oil. In a randomized, controlled crossover trial among 57 adults who came to clinic suffering from a severe tension headache, Tiger Balm was significantly more effective than placebo rubs in relieving headache pain; in fact, it worked faster and was as strong as acetaminophen (non-aspirin pain reliever) taken by mouth.[70] Although this salve may cause a little tingling and tickle the nose, Tiger Balm is very safe. I like to massage it into the temples, the neck, and the shoulders as soon as a headache starts.

Herbal remedies are sometimes *combined* with each other and with dietary supplements. For example, in a French study, the combination of feverfew and white willow (*Salix alba*) effectively reduced the frequency and intensity of migraine headaches.[71] Another project compared a moderate dose of riboflavin to a combination of riboflavin, magnesium, and feverfew, finding that there were similar significant improvements with both kinds of treatment;[72] unfortunately, this study lacked a true placebo, so it is hard to tell whether the response to both interventions was anything other than a placebo effect. Another study compared placebo treatment to a combination of under-the-tongue (sublingual) tablets of ginger and feverfew; in this study, the herbal combination resulted in significantly quicker and more reliable resolution of migraine symptoms than the placebo treatment.[73]

Herbal remedies can also be combined with music therapy, acupuncture, and massage to help ease headaches. For example, one study found that the combination of feverfew and acupuncture was more effective than either alone in reducing headache pain and the impact of headaches on the overall quality of life.[74]

Other herbal headache remedies include: angelica, balm mint, chamomile, cowslip flowers, ginseng, feverfew, hawthorn, lavender, peppermint, rosemary, skullcap, valerian, violet flowers, and white willow bark. *Willow bark* is the original source of aspirin (now made synthetically); the bark itself contains very small quantities of the painkiller. *Valerian* and *skullcap* are traditional sedative herbs, but there is little research on their benefits for preventing or treating headaches. Lavender, chamomile, and *mint* are used in many stress-related conditions because they are so calming. Unfortunately, there aren't enough studies evaluating herbs' effectiveness in treating adults or children suffering from headaches.

An herbal bath may be soothing. Herbs traditionally added to the bathwater include balm mint, chamomile, hops, St. John's wort, lavender, rosemary, peppermint, and rose. Put the dried herbs in an old pillowcase or a stocking to make cleanup easier. Alternatively, you can add a few drops of the essential oils of these herbs to the bathwater. There aren't enough scientific studies to recommend herbal baths as pediatric headache remedies, but you can use these time-tested techniques if they are consistent with your values and culture.

Headaches can be a *side effect* of herbs. *Ginkgo* causes headache and upset stomach even at typical doses. *Hops* (even in the innocuous sleep pillow) have also been accused of causing headaches and nausea. *Ephedra*, commonly used to treat congestion and colds, can cause high blood pressure, rapid heart rate, and headache.

BIOMECHANICAL THERAPIES: MASSAGE, SPINAL ADJUSTMENTS, SURGERY

Massage

Massage is one of the most widely used headache remedies. It is a perfectly natural response to headache pain, and it has proven effectiveness. In one study, adults with chronic tension-type headaches received either placebo ultrasound or massage focusing on trigger points twice weekly for six weeks; both groups had improvements in headache frequency, but those who received real massage had a significantly better improvement in pain severity, too.[75] A similar study compared placebo ultrasound to Thai massage for adult headache patients; the massage group had a significant increase in pain threshold compared with the placebo group.[76] In another study, adult headache sufferers who came to the doctor looking for stronger medication were offered either massage or acupuncture; both massage and acupuncture relieved pain for most patients, but massage appeared to have a slight edge.[77] In a German study of children with tension-type headaches, trigger point massage therapy led to significant reductions in pain frequency, severity, and duration.[78]

Massage is one of the most underutilized therapies in medicine. It can be expensive when provided by a professional if insurance doesn't cover the cost. Contact your insurance carrier and urge them to cover the cost of therapeutic massage for headaches! Also, it's an easy skill for parents to learn. Just about every child would benefit from having a parent provide massages on a regular basis, and this is particularly true for kids with recurrent headaches.

Recipe for a Heavenly Headache Massage

1. Start by massaging your child's forehead. Put your thumbs together at the center and top of her forehead and slowly stroke away from the center. Repeat this movement, moving gradually down toward the eyebrows with each repetition.
2. Next, massage her temples slowly using tiny circular motions.
3. Move to the back of the skull at the place where the skull meets the spine.

Rub from the center away toward the ears and make slow, gentle circles in all the tender spots.
4. Then massage the back of the neck, moving from the back of the skull downward toward the shoulders.
5. Before doing the shoulders, try a brisk scalp rub. Use your fingertips or knuckles to lightly "scrub" your child's scalp as if you were giving her a shampoo: first one side, then the other.
6. Gently knead the muscles of the shoulders and upper back. Experiment with what feels best for your child.
7. Finish the head and neck massage by massaging your child's face. In general, it feels best if you use your thumbs, moving from the center toward the sides of the face. Slide along the bones that form the rim of the eye socket. Rub from the bridge of the nose down to the tip. Be sure and massage the jaw from the chin up toward the ears. Gently tug and wiggle her earlobes.
8. If your child is too tender to be touched on the head, neck, or shoulders, try giving her a foot rub instead. It may help relax her, and in one study of adults, reflexology (specialized foot rubs) effectively reduced headaches for 81 percent of patients.[79]

Be very gentle. There are often sore spots where the muscles have been tightly tensed. It is better to massage these spots lightly, focusing on the less painful areas and coming back to the tight spots, gradually increasing pressure as your child tolerates it. If your child's head or neck is very tender, you may want to give her a foot massage first to help her relax. If your child's headaches are particularly severe or recurrent and you want professional assistance, see a massage therapist or physical therapist to learn how to help.

Spinal Adjustments

Cranial and spinal adjustments or manipulation are widely used to treat headaches and neck aches, and many chiropractors claim that this is the treatment of choice for headaches, but more well-controlled clinical trials are needed to verify exactly what type and frequency of treatments are helpful.[80] If you've found help with chiropractic or osteopathic treatments, there is no need to stop them because they are generally safe and covered by most insurance carriers.

Surgery

There is no need for surgery to treat most childhood headaches. However, surgery can be lifesaving for severe head trauma.

BIOENERGETIC THERAPIES: ACUPUNCTURE, THERAPEUTIC TOUCH, HOMEOPATHY

Acupuncture

Over the past twenty years, evidence has accumulated that supports the effectiveness, safety, and relative cost-benefit of acupuncture as a preventive treatment for recurrent tension-type and migraine-type headaches.[81,82] I refer more and more of my migraine patients to my acupuncture colleagues as do many pediatric pain treatment specialists. Most children are afraid of needles, so acupuncture is seldom the families' first choice of headache remedies in the United States. However, if your child's headaches have not responded to other remedies, acupuncture therapy is worth trying.

Therapeutic Touch

My introduction to Therapeutic Touch (TT) came when I was a pediatric resident and a colleague cured my migraine headache with a brief TT treatment. I didn't learn how to provide this helpful therapy myself for another five years. Research supports my personal experience. TT successfully treats headaches. In a comparison trial, TT effectively relieved tension headaches in 90 percent of recipients, and relief was sustained for over four hours.[83] A related therapy, Healing Touch, has similar benefits. Therapeutic Touch is safe, and I often use it in my practice for both patients and colleagues who have headaches.

Homeopathy

The studies on homeopathy for headaches have shown mixed, but mostly negative results. The most commonly recommended homeopathic headache remedies are: *Belladonna*, *Bryonia, Gelsemium,* and *Nux Vomica.* There aren't enough studies evaluating the effectiveness of homeopathic remedies in treating headaches in children. They are certainly safe, but I do not routinely recommend them.

COMBINATIONS

An increasing number of clinicians use what is called a multimodal approach to healing headaches naturally.[84,85] This means combining careful tracking, mind-body strategies to reduce stress, meditative exercises like yoga or tai chi, dietary supplements, and manual therapies. In my experience, most families of children with chronic conditions like headache have already tried one medication after another, and they're ready for a comprehensive approach to healthy living to improve headaches and life in general.

WHAT I RECOMMEND FOR HEADACHES

PREVENTING HEADACHES

If your child has frequent headaches, *keep a diary or log of the headaches, headache triggers, and remedies.*

1. *Lifestyle—exercise and sleep.* Make sure your child gets plenty of regular sleep at night and vigorous exercise during the day.

2. *Lifestyle—nutrition.* If your child suffers from other allergy symptoms (runny nose, diarrhea, skin rashes) as well as frequent headaches, see a pediatric nutritionist or allergist to discuss the possibility of a low-allergenic diet. Try avoiding nitrates, nitrites, aspartame, MSG, chocolate, and aged cheeses, and monitor your child's headaches. Avoid sudden changes in caffeine consumption. Give your child folate and magnesium-rich foods. Also consider including ginger and hot peppers as part of her diet.

3. *Lifestyle—environment.* Allow your aching child to rest in a quiet, dark room. Avoid known allergens and toxins such as carbon monoxide and lead. Try using a gel-pack filled headband that can be tightened around the head to apply pressure to the painful areas. These headbands can be cooled or heated, as preferred by the child. Or try a hot footbath.

4. *Lifestyle—mind-body.* Help your child practice relaxation exercises or meditation regularly to help minimize stress and build a sense of mastery. Consider professional mind-body therapies such as self-hypnosis, guided imagery, or biofeedback training.

5. *Biochemical—medications, herbs, and supplements.* See your physician about preventive prescription medications. For migraines, consider giving your child dietary supplements containing vitamin B2 (400 milligrams daily), magnesium (check with your doctor about the proper dose for your child), and/or coenzyme Q10 (100–300 milligrams daily) to prevent migraines. Consider a trial of the herb butterbur to prevent migraine headaches.

6. *Biomechanical—massage.* Give your child a massage several times weekly.

7. *Bioenergetic—acupuncture.* If it's acceptable to your child and a pediatric acupuncturist is available, consider acupuncture therapy.

8. Avoid excessive vitamin A, zinc, or niacin; avoid ginkgo, hops, and ephedra.

See Your Health Care Professional If Your Child Has a Headache and ...

- You are concerned about possible meningitis (fever, stiff neck, spotted rash), or your child is lethargic or irritable.

- Your child has recently suffered a head injury *and* vomits more than three times, vomits more than one hour after the injury, has difficulty hearing or seeing properly, has clear or bloody fluid coming out of the nose or ears, has a seizure, or becomes more clumsy, off-balance, or dizzy.

- The headache is severe and won't go away with simple home remedies.

- The headache is worse in the morning and gets better during the day.

- You are concerned about a possible sinus or tooth infection: facial pain, runny nose or thick mucus from the nose, fever, or pain in the jaw, teeth, ears, or eyes.

- The headache lasts longer than two days or gets worse over time.

- You are concerned that your child has an underlying serious illness: frequent recurring headaches despite home remedies; the child seems unwell, tired, has a poor appetite, is not growing well, has a change in personality, or becomes clumsier.

TREATING HEADACHES

1. *Biochemical—medications.* If your child's migraines are only moderately severe, try a non-prescription pain reliever such as acetaminophen (Tylenol), ibuprofen (Advil or Motrin), or naproxen (Aleve) in combination with caffeine. For children with more severe migraines, see your physician about a triptan medication to have on hand to treat the headache in its early stages. You are much better off starting with something strong than starting weak and working your way up.

2. *Biochemical—herbs.* Try rubbing a little Tiger Balm salve into the temples, the neck, and the upper shoulders at the first sign of a headache. Consider ginger to help manage nausea. Consider a combination of ginger and feverfew to abort a migraine.

3. *Lifestyle—environment.* Have your child rest quietly in a dark room; try an ice pack on the back of her neck. Alternatively, try a hot footbath or warm towels around her feet.

4. *Biomechanical—massage.* Try a head, neck, and shoulder massage. If she is too sore, try a foot massage instead. Consider adding oil of peppermint or eucalyptus to the massage oil. Consider seeking professional help from a licensed massage therapist or physical therapist.

5. If available, try Therapeutic or Healing Touch.

21
JAUNDICE

Steven Chen called me just as the office opened one morning about his newborn daughter, May. Steven's mother, Rose, was in town helping with the baby and had noticed May was becoming jaundiced. May was a little yellow in the face the previous evening, but now the jaundice was noticeable on her chest as well. Rose told Steven that when he was a baby, he had jaundice and underwent many blood tests. She had been told to stop nursing him, and he had to stay three extra days in the hospital under special lights. Rose had continued to be extra cautious with his health throughout his infancy. She wanted him to make sure that his bad luck wasn't starting in his daughter.

Steven's wife, Bonnie, was much less concerned. Her pregnancy and delivery had been completely normal. May had weighed 7 pounds, 6 ounces at birth, and had gone home 20 hours after her delivery. Bonnie's milk supply was starting to come in, and she was nursing six or seven times a day. A public health nurse was coming to visit them later that afternoon for a routine checkup following the early discharge, but Rose was urging Steven to take the baby to the hospital right away to get May started on therapy. He felt trapped between his mother's concern and his wife's confidence, and he wanted some professional advice about what to do.

WHAT IS JAUNDICE?

Jaundice is a yellow color of the skin and the whites of the eyes. It is caused by a buildup of the yellow pigment *bilirubin* in the blood. Bilirubin is a normal breakdown product of the hemoglobin in red blood cells. Every day about 1 percent of our red blood cells die and are replaced with new cells. Red blood cells contain iron and hemoglobin. Rather than waste the useful iron in the dead red blood cells, the body saves and recycles it. Hemoglobin, the iron-carrying molecule, is easily replaced, so while the iron is saved, the old hemoglobin is broken down and metabolized in the liver.

When hemoglobin breaks down, it becomes *biliverdin,* which is a green pigment. Biliverdin is transformed into *bilirubin,* which is yellow. (See Figure 1.) This is why bruises change color over several days from reddish purple (fresh blood) to green (biliverdin) to yellow (bilirubin). The liver processes bilirubin to prepare it for excretion in the stool. If the liver gets backlogged, some of the bilirubin spills over into the bloodstream, turning the skin and the whites of the eyes yellow. Once the liver has finished processing the bilirubin, it passes it on to the gallbladder (which functions as a kind of holding tank), which in turn passes it on to the intestines for excretion.

Before it passes out of the body, bilirubin faces another hurdle. An enzyme in the intestines can free the bilirubin to be reabsorbed into the blood. The longer it takes for the intestinal contents to be excreted, the more chance there is for bilirubin to be reabsorbed into the blood. It then has to return to the liver to be reprocessed and then on to the gallbladder and intestines.

Jaundice is not an illness; it is simply a sign

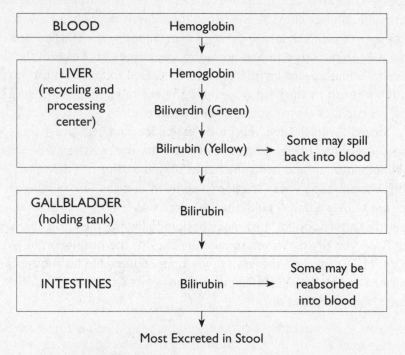

Figure 1: Hemoglobin's Jaundiced Journey

that bilirubin has built up in the blood. For reasons that no one quite understands, levels tend to be higher in the morning than in the evening—just the opposite of temperatures, which tend to be highest in the afternoon and evening and lowest in the morning. Babies tend to have higher bilirubin levels than adults.

WHY DO BABIES HAVE HIGHER BILIRUBIN LEVELS THAN ADULTS?

Now that you know where bilirubin comes from, you can easily see the different reasons it builds up in babies.

WHY BABIES HAVE HIGH BILIRUBIN LEVELS

- Higher turnover of red blood cells.
  - Fewer red blood cells needed after birth
  - Bruising during birth (forceps or vacuum-assisted deliveries)
  - Maternal-infant blood type incompatibility (Rh or ABO)
  - Genetically fragile blood cells like G6PD deficiency or hereditary spherocytosis
- Slower liver metabolism or excretion
  - Babies are slower than adults; genetic differences in metabolism
  - Diseases that slow down liver metabolism
  - Diseases that block secretion from the liver to the gallbladder
- Slower movement through the intestines, allowing time for reabsorption

First, babies *have more red blood cells* for their size than adults. They need more red blood cells to help carry oxygen before they are born. Once they are breathing on their own, babies do not need so many red blood cells. The extra blood is broken down and re- cycled to supply future needs. The extra work of breaking down blood cells creates a backlog for the liver, and some bilirubin tends to spill over into the blood.

To compensate for thinner air (less oxygen), babies born at high altitudes have higher hemoglobin levels than their sea level counterparts; babies born at high altitudes are also more likely to be jaundiced than babies born at sea level. Babies who get bruises on their head from pushing during delivery (or from the forceps or vacuum extractors used to assist delivery) are also more likely to become jaundiced because of the extra work for the liver in breaking down the blood in the bruises.

Certain *diseases* can also cause a higher turnover of red blood cells. In the bad old days, the most feared cause of jaundice was *Rh disease*. Rh disease occurs when the *mother* has Rh-*negative* blood and the *baby* has Rh-*positive* blood. Rh-negative mothers make antibodies to the Rh factor in their baby's blood cells, breaking them down rapidly, leading to anemia, very high bilirubin levels, and sometimes death or permanent disability, called *kernicterus*. Fortunately, scientists found a way to block this reaction: *Rhogam*. Rhogam derails the destructive reaction against babies' blood so Rh-negative mothers can have normal, healthy Rh-positive babies without breaking down their red blood cells.

Rh disease was terrible and frightening. High bilirubin levels meant the possibility of kernicterus, which was to be avoided at all costs. The memory of kernicterus has been linked to jaundice in the minds of older parents and physicians alike. The fear of kernicterus is why many people fear jaundice. The good news is that kernicterus has virtually disappeared in full-term babies with the advent of Rhogam, so younger parents and physicians are less likely to panic at the first tinges of yellow skin.

Other *blood group differences* (such as *A, B, O, AB blood types*) between mother and baby can also lead to breakdown of the baby's red blood cells. These differences are rarely as severe as Rh disease. Mothers with *type O blood* are the most likely to have an immune reaction to their babies' blood. Mothers who have O-positive blood make antibodies to blood types A, B, and AB. If the mother has blood type O and the baby has one of the other types, it is called an *ABO set-up* or *ABO incompatibility*. This is why physicians check mothers' blood types during pregnancy. If the mother is type O, her baby's blood type is checked soon after delivery. Careful monitoring and early treatment can prevent damage from ABO incompatibility.

Some *genetic diseases* result in fragile blood cells, which are easily destroyed. These diseases have long names such as *glucose-6-phosphate dehydrogenase* (G6PD) *deficiency* and *hereditary spherocytosis*. G6PD is common in certain racial groups such as Sephardic Jews, Greeks, and Nigerians, but it is rare in most other peoples. Hereditary spherocytosis is more common among people with northern European ancestry. Genetic diseases causing fragile blood cells can be diagnosed with simple blood tests.

Even under normal circumstances, the newborn's *liver* has a lot of work to do metabolizing broken-down blood cells. However, babies' livers are not as speedy or developed as adults'. They just aren't as adept at the hemoglobin recycling program. Bilirubin builds up as it waits for the liver to metabolize and excrete it. *Low–birth-weight* and *premature* babies are more likely to become jaundiced because their livers are even less fully developed and ready to metabolize bilirubin than full-term babies.

The rate at which the liver gets up to full speed varies slightly from baby to baby and in different ethnic groups. *Asian* babies have a higher rate of jaundice than other races. Hispanic babies are more prone to jaundice than Caucasians, and African-American babies are least likely to be jaundiced. These differences are not just because it's easier to see jaundice in certain racial groups; the bilirubin levels really differ in different races. The differences are probably due to genetic differences in liver metabolism in different races. For unknown reasons, boys are more likely to become jaundiced than are girls.

May Chen had no problems with her blood type, and she was not premature or low birth weight. Her Asian background did put her at slightly higher risk of jaundice, but not necessarily a worrisome predisposition to G6PD or spherocytosis.

If there are *liver diseases* such as hepatitis or other illness that slow down the liver, a backlog of bilirubin accumulates, spills into the blood, and causes jaundice. Any major illness, such as *sepsis* or *meningitis*, can also slow liver metabolism. *Hypothyroidism* slows metabolism throughout the body, including the liver. Blockages in the pathway from the liver to the intestines by way of the gallbladder can also lead to bilirubin buildup.

May's newborn screening test showed no signs of hypothyroidism or other genetic problems, and she had no symptoms of sepsis or meningitis.

Once bilirubin has made it to the *intestines*, you'd think the way would be clear to excretion in the stool, but there is another hurdle before bilirubin's final exit. The longer the bilirubin stays in the intestines, the more chance it will be reabsorbed into the blood. From there it has to go back to the liver to be re-metabolized and re-excreted. This process is known as the *entero*(intestinal)-*hepatic* (liver) *circulation*.

Anything that *slows down the intestines* can slow the excretion of bilirubin and increase its entero-hepatic circulation. Everyone who has changed a newborn baby's diaper knows that newborn stool is dark, thick, and sticky. This thick, sticky stool, known as *meconium,* builds up before the baby is born. It needs to be cleared out before normal stool can pass. Every time a baby nurses, the intestines contract, moving the meconium and stool along on their way out. The less often the intestines contract, the more slowly meconium is excreted. Babies who are constipated or who don't pass a stool within the first day of life are more likely to develop jaundice than those who pass a stool in the first 24 hours.

The most common cause of slowed excretion is not consuming calories frequently. Formula-fed babies get a full complement of calories at every meal from day one. Breast-fed babies have smaller meals until their mother's milk supply comes in. The less frequently breast-fed babies nurse (for example, if the baby is fed only every four hours or if water is substituted for a feeding), the less the intestines are stimulated and the more slowly the mother's milk supply comes in. Overall, breast-fed babies are about twice as likely to be jaundiced as formula-fed babies. In the first few days of life, they tend to take in fewer calories, have less stimulation to move their bowels, and have fewer bowel movements.

May was nursing well, and her mother's milk supply was slowly starting to increase.

For reasons no one quite understands, there is an enzyme in some women's breast milk that enhances bilirubin's reabsorption from the intestines. Families that have this particular enzyme tend to have babies who become mildly jaundiced after they're a week old and they tend to stay mildly jaundiced for about a month. But they don't seem to be at higher risk of developing kernicterus or any other problem—except for being slightly less photogenic for the first few weeks!

How Common Is Jaundice?

In light of all the factors that can cause increased bilirubin, it's not surprising that so many babies become jaundiced. Jaundice is very common even among perfectly normal healthy babies with perfectly normal deliveries. About 80 percent of babies have visible jaundice, but only about one to two per hundred have bilirubin levels above 15, and only about one to two per 1,000 babies reach bilirubin levels of 18–20 or more. With modern screening and early treatment, the frequency of bilirubin levels over 30 is much lower than that—about one in 10,000 babies, and the frequency of kernicterus is about one in 50,000 babies.[1]

Jaundice typically becomes noticeable around the third day of life as the red blood cells break down and the liver falls behind on metabolizing it. Bilirubin levels are highest on the fourth or fifth day of life. They gradually decline over the next several days as the liver gets into gear and the mother's milk supply comes in.

Because newborn jaundice doesn't show up right away, pediatricians recommend that all babies have a screening test in which a special, highly reliable device reads the baby's bilirubin level in the skin (transcutaneous bilirubin) before she goes home from the hospital.[2] If the level is high, she can stay and start treatment; if it's moderate, the level can be rechecked in a few hours. If the level is low, she can be rechecked in 24 to 48 hours to see how fast the bilirubin level is rising. This is the reason so many hospitals and health plans are moving toward having nurses visit new babies at home. Even if phototherapy is needed, it can usually be set up at home. This home-based approach to care is much more convenient while still making sure that babies get the therapy they need.

May had typical newborn jaundice. Because she was breast-feeding, Asian, and jaundice ran in her family, she was at higher risk than the average baby of being visibly jaundiced, but she was not necessarily in any danger. Her transcutaneous bilirubin level was fine before she left the hospital for home, and the visiting nurse would check her bilirubin that afternoon. She was behaving normally. She did not need to come to the emergency room.

These could be signs that your baby has a serious problem as a cause for the jaundice or that your child needs additional treatment.

May first became visibly jaundiced when she was three days old. She had no signs of serious illness; it was safe for her to be evaluated and treated at home.

TAKE YOUR CHILD TO A HEALTH CARE PROFESSIONAL FOR AN EVALUATION IF

- The mother is Rh-negative or blood type O.
- Abnormal or fragile blood types or severe jaundice run in your family.
- Jaundice is visible within the first 48 hours of life.
- The jaundice lasts longer than 10 days.
- The jaundice extends all the way down to the chest or the belly button.
- The baby is more than three weeks premature.
- The baby develops dark urine or pale stools.
- The baby acts sick, is very sleepy, or is not interested in nursing.
- The baby has trouble breathing or is breathing very fast.
- You are concerned about the baby in any way.

DOES JAUNDICE CAUSE BRAIN DAMAGE?

In the days when *Rh disease* caused devastating red blood cell destruction, anemia, and even death, autopsies showed bilirubin deposits (or staining) in the brain. This brain staining accompanied by either death or permanent disability was attributed to bilirubin. Many studies over the past 40 years have looked at whether bilirubin causes brain damage in babies who are *not* affected by Rh disease.

High bilirubin levels do affect *hearing*. However, these hearing deficits resolve as soon as the bilirubin levels drop to normal. Many babies also become *sleepier* when they are jaundiced, but when the jaundice resolves, their sleepiness disappears. Long-term follow-up studies of thousands of otherwise healthy, full-term babies with normal to moderately high bilirubin have not shown any long-term adverse effect of jaundice on IQ, development, or behavior. These studies do not necessarily apply to very premature babies or babies whose jaundice is due to other serious problems such as hypothyroidism. Sick babies tend to have persistent problems following high bilirubin levels, but among healthy babies, the long-term effects are not serious if the jaundice is monitored and treated. Some researchers have even suggested that bilirubin itself is not toxic to the brain; in fact, it may be a protective antioxidant that simply marks injured areas of the brain. For the vast majority of babies, jaundice is not dangerous, not brain damaging, and not life threatening. On the other hand, for a few, it can be a sign of a serious problem that requires prompt medical attention. So how do you know whether or not to be worried?

DIAGNOSIS

Talking to parents about family history of jaundice, liver disease, blood problems, and how the baby is currently doing (feeding, sleeping, peeing, and pooping) can uncover clues to the diagnosis. A physical examination can point the way to other conditions contributing to jaundice. The bilirubin level in the blood can be determined easily by a blood test. Along with a bilirubin level, your health care professional will check the baby's and mother's blood types and may test the baby's blood cells. Depending on the results of the initial discussions, examination, and blood tests, other laboratory studies may be helpful. For example, a different set of tests are needed if the jaundice lasts longer than two weeks or the baby had extensive bruising at birth or missed her vitamin K injection.

Because so many babies are out of the hospital in less than 48 hours, doctors recommend that bilirubin levels be checked before they go. It turns out that babies who have bilirubin levels less than six mg/dL have a very low risk (about 2 percent) of developing significant jaundice later on. Early screening is reassuring. For those babies who are discharged before three to five days of age, it's also a good idea to get rechecked as we had planned for May. The policy of pre-discharge screening with a recheck within a few days after discharge and prompt treatment if needed has led to a dramatic reduction in the prevalence of the kinds of high bilirubin levels linked to kernicterus. Screening is a good thing.

What Is the Best Way to Prevent and Treat Jaundice?

The best way to *prevent* jaundice is to feed your baby early and often—at least eight to ten times daily during the first week of life. Also make sure your baby is exposed to some sunlight every day. Let's consider all your options to find out what works best in treating jaundice; if you want to skip to my bottom-line recommendations, flip to the end of the chapter.

Lifestyle Therapies: Nutrition, Exercise, Environment, Mind-Body

The most effective and commonly used therapies to treat jaundice are lifestyle therapies—nutrition and environmental interventions.

Nutrition

Feeding frequency. As they say about Chicago voting, do it early and often. Frequent feeding, especially for breast-fed babies, results in lower bilirubin levels and less jaundice. Nursing mothers should breast-feed their newborn babies at least eight to ten times daily. Jaundiced babies may be a little sleepy, but wake them up to nurse every two to three hours. Frequent nursing in the first week of life helps mothers produce more milk more quickly, stimulates the babies' intestines to move things along, and helps babies eliminate bilirubin. Breast milk also helps protect babies against infectious diseases, allergic diseases, colds, and asthma. Keep on nursing!

Formula and water supplements. Supplementing with formula seriously sabotages the eventual success of breast-feeding. Because jaundice is generally so benign and breast-feeding is so beneficial, most pediatricians do not recommend formula or water supplements for jaundiced breast-fed babies. Rather than stopping or interrupting breast-feeding, I recommend that you nurse your baby *more* often, at least eight times daily. If your baby is very jaundiced and seems to be hungry even after she nurses, please discuss the situation with your health care professional or a lactation (breast-feeding) specialist.

If you are already feeding your baby formula, you may want to choose a special *casein-hydrolysate formula* rather than a standard cow milk formula. At least one study suggests that babies who drink these special formulas have a significantly lower risk of developing jaundice than those who drink regular formula.[3] If you receive food supplements from the Women, Infants, and Children (WIC) program, you may need a prescription from your pediatrician for this special formula.

Environment

Light therapy (*phototherapy*) is the main-stay of jaundice therapy. Phototherapy was discovered years ago by a nurse and physician who observed that infants who were taken out-side on bright, warm days or whose cribs were near the window were much less likely to be jaundiced than babies whose cribs were away from the window. Astute pediatricians since then have noted that jaundice occurs more commonly during the fall and winter months when babies are exposed to less sunlight.

In the presence of sunlight, our skin changes bilirubin to another form, which is much more easily excreted by the kidneys. This change al-lows bilirubin to bypass the liver and intestines to be excreted directly by the kidneys without the possibility for reabsorption.

Phototherapy is recommended at different bilirubin levels depending on the baby's age (Table 21.1). Remember, most jaundice is not harmful and most babies do not need multiple daily bilirubin tests or hospitalization.

May's bilirubin level was 15 at four days of age, and both she and her mother had type A (Rh-positive) blood. The nurse encouraged more frequent breast-feeding and two outdoor strolls per day to help bring down the bilirubin. However, it was overcast and rainy, and May never made it outside. When the nurse returned the following day, May's bilirubin had climbed to 18, so she was started on home phototherapy.

Phototherapy can be provided by stan-dard fluorescent lights, special blue-green spectrum fluorescent lights, halogen lights, light-emitting diodes (LED lights), or special blankets containing fiber-optic lights ("bili blankets"). Sandwiching the baby between lights is called double phototherapy, and it may be even more effective than single photo-therapy.[4] Single phototherapy, such as the bili blanket alone, is also effective, especially if the baby is close to the light source.[5] Newer prod-ucts containing LED lights may be even stron-ger than the old standard of blue-spectrum fluorescent lights.[6] With bili blankets, the baby can be snuggled by the parents rather than being confined to an incubator. For otherwise healthy full-term babies with responsible par-ents and moderate bilirubin levels (that don't require double phototherapy), there is no evidence that home phototherapy is less ef-fective than phototherapy delivered in a hos-pital setting.[7] Phototherapy is typically used until bilirubin levels fall to 13–14 in full-term babies.

Phototherapy has few side effects. Because of the light intensity, the baby's eyes need to be protected to prevent eye damage. The room temperature may need to be turned up if the baby is lying naked on top of a bili blanket. The baby may also need some extra milk to compensate for the increased evaporation of fluid from the skin during phototherapy. Long-term follow-up studies have not shown any increased risk of skin cancer or other skin dis-

Table 21.1: Bilirubin Levels Leading to Phototherapy at Different Ages	
Baby's Age	Bilirubin Level for Phototherapy
Less than 24 hours old	Any jaundice
25–48 hours old	12–15 or higher
49–72 hours old	15–18 or higher
73 or more hours old	17–20 or higher

eases in children who received phototherapy as infants.

By the time May was a week old, her bilirubin level was down to 13; she was nursing well and her mother's milk was in. She had a total of two skin tests, three blood tests, three nurse visits, a few calls to the physician (me), two days of home phototherapy, and had avoided extra trips to the office and an expensive hospital stay. Her grandmother, Rose, was very impressed at the changes in health care in the 23 years since Steven was born. Bonnie was confident that May was basically a healthy baby and was glad NOT to have had to make repeated trips to the clinic in the first week after her delivery. In the old days, the treatment for jaundice was so aggressive, it made moms worry that their babies were especially fragile and increased the number of doctor visits they made in the first few months to make sure everything was okay. With modern screening and treatments, babies have fewer visits and moms are more confident.[8]

BIOCHEMICAL THERAPIES: MEDICATIONS, HERBS, NUTRITIONAL SUPPLEMENTS

Medications

Although a number of medications have been tested over the years, none has caught on as the mainstay of treatment for newborn jaundice.

MEDICATIONS TO PREVENT OR TREAT JAUNDICE

- Activated Charcoal (no longer used)
- Agar (rarely used as an adjunctive treatment)
- Phenobarbital (effective, but has side effects; rarely used)
- D-penicillamine, clofibrate, gemfibrozol, metalloporphyrins (rare)

Activated charcoal binds bilirubin. It is not absorbed into the bloodstream, so it passes through the intestines, absorbing bilirubin and toxins on the way out. It prevents bilirubin's reabsorption. A Minneapolis study from the 1960s showed that feeding babies activated charcoal from the time they were four hours old decreased their bilirubin levels somewhat. However, the charcoal was not very helpful if it was started later when the babies were actually jaundiced. It is also messy. Activated charcoal never really caught on as a jaundice treatment, but it is used to treat some cases of poisoning.

Agar, an extract of seaweed, also absorbs bilirubin in the intestines and helps prevent its reabsorption. Different kinds of agar have varying abilities to absorb bilirubin. However, even the highest medical grade agar is not very potent. In a study my colleagues and I did at Yale in the 1980s, the most absorbent agar lowered bilirubin levels only a couple of points. This difference was statistically significant, but it probably had little real impact on the health of the babies over the long term. High-grade agar is used in some medical centers as an adjunctive treatment to light therapy, but in most medical centers, it's never used and is not even available.

Phenobarbital, the well-known seizure medication, lowers bilirubin levels by revving liver metabolism. Because of its sedative effects, it is rarely used nowadays to treat typical newborn jaundice. Several other medications and immune globulin infusions were used historically, but have not found much favor in the phototherapy era.

Several medications *increase* bilirubin levels. They should *not* be used in newborns because of the risk of shifting jaundice from a benign condition to a worrisome problem. For this reason, medications such as *sulfa* drugs and *ceftriaxone* are not used in babies less than two weeks old.

Dietary Supplements

Fearing that the baby might be dehydrated, many mothers give *water supplements*. Jaundice is not caused by dehydration. Water supplements do not improve jaundice.

No vitamins or minerals have been proven helpful in treating newborn jaundice. Megadoses of *niacin* can actually cause jaundice in older children. I do not recommend any dietary supplements as treatments for newborn jaundice.

Herbal Remedies

No herbs have proven safe and effective in treating newborn jaundice. Herbal remedies traditionally used for adults with liver disease or jaundice (such as *aloe vera gel, cascara bark, dandelion root, parsley,* and *milk thistle*) have not been evaluated in babies and may not be safe. Traditional herbal remedies used in other parts of the world have not been tested for safety and effectiveness in the United States, and I don't recommend them (examples include zizyphus jujuba; camel's thorn; flixweed; and Yinzhihuang or Yin Zhi Huang).

Several herbal remedies can actually cause liver damage and jaundice. For example, the Chinese herbs *Yin-Chen* and *Cheun-Lin*, which are traditionally used to treat jaundice, displace bilirubin from the blood to other tissues such as the brain! Other herbs, such as chaparral and *Jin Bu Haan* (*Lycopodium serratum*) actually cause liver toxicity and jaundice. Goldenseal, which is sometimes recommended as an herbal antibiotic, can bump bilirubin off of its carrier protein, making it easier for the bilirubin to pass into the brain; it should never be used in babies less than three months old. I do not recommend any herbal remedies to treat newborn jaundice.

BIOMECHANICAL THERAPIES: MASSAGE AND SURGERY

Massage

In many parts of the world, infant massage is a routine part of baby care. Although gentle belly massage may prompt more bowel movements,[9] there is limited evidence that it affects bilirubin levels in a clinically significant way. Nevertheless, when common sense is used, massage is safe, and if it's part of your culture and family tradition, I encourage you to provide this kind of comforting care for your baby.

Surgery (Transfusions)

If a baby's bilirubin level climbs into potentially dangerous levels despite changes in feeding and despite light therapy (phototherapy), *blood transfusions* can be helpful. This is called an *exchange transfusion* because some of the baby's blood (containing high bilirubin levels) is removed and exchanged for low-bilirubin blood. This procedure must be done in the hospital by physicians who are very experienced in the care of newborn infants. It rapidly reduces bilirubin levels, but the procedure itself causes significant side effects. Because of the risks of this potentially lifesaving procedure, most physicians prefer to use it as a last resort, relying on *frequent feeding* and *phototherapy* as the mainstays of treatment.

BIOENERGETIC THERAPIES: ACUPUNCTURE, THERAPEUTIC TOUCH, PRAYER, HOMEOPATHY

There are no studies establishing the effectiveness of acupuncture, prayer, Therapeutic Touch, or homeopathy in treating jaundiced newborns.

WHAT I RECOMMEND TO TREAT NEWBORN JAUNDICE

Take your baby to your health care professional if

- The mother is Rh negative or blood type O.

- Abnormal or fragile blood types or severe jaundice run in your family.

- The baby becomes excessively sleepy or unresponsive (sleeping for more than four or five hours in a row in the first week of life).

- The baby develops a fever or acts sick.

- Jaundice is visible within the first 48 hours of life.

- The jaundice extends all the way down your child's chest to the belly button.

- The jaundice lasts longer than 10 days.

- You are concerned about her in any way.

PREVENTING AND TREATING JAUNDICE

1. Get good *prenatal care* so you know what your blood type is and can take preventive measures early if you are at risk for having a jaundiced baby.

2. Get good *pediatric care.* Have your baby's bilirubin checked with a skin test (transcutaneous bilirubin) before you leave the hospital or birthing center and again within two to three days.

3. *Lifestyle—nutrition.* Feed your baby at least eight to ten times daily. If you need help, ask for a lactation consultation.

4. *Lifestyle—environment.* Take your baby outside to get some sunlight every day. Avoid the high-intensity hours between 10 a.m. and 2 p.m. If your baby develops higher bilirubin levels, use professional phototherapy.

5. *Biomechanical—surgery.* If your baby develops very high bilirubin levels despite other therapies, she may need an exchange transfusion, exchanging her high-bilirubin blood for lower bilirubin blood. Seek care from an experienced pediatrician.

6. *Avoid* stopping breast-feeding, using water supplements, or herbal remedies to treat jaundice.

22

PAIN

with Nisha Rao

Maggie Murphy brought her six-year-old son, Matthew, for evaluation of leg pain. For the past few weeks, he had been waking up in the middle of the night complaining of aches in his legs, and sometimes it was so severe that it made him cry. During the day, Matthew's behavior was completely normal. He was able to run and play with his friends like he always had. Maggie was concerned that Matthew's leg pain could be the beginning of something more serious like arthritis, which ran in their family.

Children of any age—even babies—can feel pain.[1] Pain can occur anywhere. The most common sites of pain in children are the head, the belly, the back, and the arms and legs. Children of any gender and age can experience pain, but girls and teens are more likely to report pain than boys and younger children.[2]

The most common *causes* for immediate, *short-term (acute) pain* are traumas like stubbing a toe or bruising a shin. But there are lots of other triggers, too. Intestinal gas and anxiety can trigger bellyaches, and stress can trigger headaches. Strong emotions can lead to chest pain that has nothing to do with having a heart attack.[3] Weather changes worsen pain for some children, especially those with joint pain.[4]

Many factors increase the risk of developing *chronic* or *recurrent pain*. Among these, the most common include stress and depression. Like many other conditions, some types of pain run in families. For example, some forms of back pain, arthritis, and migraine headaches are more likely in children whose parents have these conditions. Rapid growth spurts are often associated with leg pain. Certain kinds of sports can increase the risk of painful trauma, repetitive use injuries, bruises, strains, and sprains. On the other hand, being sedentary can lead to deconditioning, weakness, and general achiness. Poor posture while watching television or working at a screen for prolonged periods can also increase the risk of back and neck pain. Smoking is bad for general health and is another risk factor for developing chronic pain. Don't smoke and discourage your child from taking up this horrible habit.

SUMMARY OF PAIN TRIGGERS AND RISK FACTORS

- Family History of Pain
- Gender (girls)
- Age (older more than younger children)
- Stress and negative emotions
- Strenuous sports or being sedentary
- Weather changes
- Rapid growth
- Poor posture
- Smoking

Maggie had been trying to think about possible triggers for Matthew's pain. He liked to run around the backyard with his friends and jump on the trampoline, but he hadn't been involved in any strenuous activities lately. He hadn't had a fall or bruised his legs recently. The only risk factor she could think of was their family history of arthritis. Maggie's mother had a bad knee because of osteoarthritis. Maggie wondered if Matthew was developing something similar. Most of all, though, Maggie was worried that the pain was affecting Matthew in subtle ways. She began wondering whether pain poses problems in addition to discomfort.

Yes, pain is not only uncomfortable; it interferes with sleep, studying, school, socializing, and hobbies. Pain can make kids irritable, forgetful, and moody and interfere with their appetite.

Pain is a signal in the same way a red traffic light or a fire alarm are signals. We don't just unplug the light or ignore it. Whenever possible, we don't just treat or cover up the pain; we try to determine the cause and address that. This means your health professional will try to make a diagnosis.

DIAGNOSING THE CAUSE OF THE PAIN

Diagnosing the cause of pain depends on where the pain is located, what it is like, how long it has been there, the age of the child, and other factors.

For belly pain, blood tests might be used to check the function of the liver, pancreas, or kidneys, or to check for celiac disease. Urine tests are often ordered to determine whether the child has a urinary tract infection or diabetes causing belly pain. A lactose breath test is used to test for lactose intolerance causing belly pain. Stool tests are used to test for parasites or malabsorption. Imaging studies such as ultrasound, X-rays, CT scans, or MRI scans are used to test for appendicitis, twisted testicles or ovaries, and other problems that can cause severe pain.

Similarly, diagnosing the cause of joint pain and back pain may involve multiple tests and multiple steps. After identifying the location of your child's pain, your doctor may look for other physical signs like swelling, redness, warmth, and changes in the way your child is able to move the area. Depending on the findings, he or she may recommend certain blood tests such as rheumatoid factor, anti-ds-DNA antibody, anti-Smith antibody, erythrocyte sedimentation rate, complement reactive protein, or others. Your doctor may also recommend imaging in the form of X-ray, CT, or MRI. Sometimes, doctors may also recommend draining a swollen joint and evaluating the fluid for signs of infection or inflammation.

Matthew's pain was in his shins, not the knee or ankle joints. He had classic symptoms of grow-

ing pains and did not require any blood tests or imaging studies.

Back pain is common in adults, but less so in children. When children have back pain, we don't just think about bruises or muscle sprains and strains (though they are the most common types of pain). We also do diagnostic tests to make sure there isn't a tumor or a problem in the bones or spinal cord. Kidney problems can also be associated with back pain, so health professionals may also order a urine test for a child with back pain.

What Is the Best Way to Treat Pain?

The best way to treat pain is to:

- Recognize and address the underlying cause of the pain.
- Track symptoms, triggers, and relievers.
- Prevent it or treat it early rather than waiting it out.

The treatment for appendicitis may involve surgery, while the treatment for an anxiety-related tummy ache may involve learning and practicing mind-body skills. The treatment for gluten intolerance is a change in diet. No matter what causes the pain, it's better to treat it early rather than wait until the body starts screaming even more loudly with worse symptoms.

This chapter does *not* cover treatments for *all* causes of pain. It focuses on treatments for the most common kinds of abdominal pain and joint/muscle/bone pain. See Chapter 20 for headache pain.

Also seek professional care if you suspect that your child might have some other serious problem underlying his abdominal, muscle, bone, or joint pains. Let's tour the Therapeutic Mountain to find out what works for common

---

> ### SEE YOUR HEALTH CARE PROFESSIONAL IF YOUR CHILD HAS PAIN THAT
>
> - Is very severe
> - Is accompanied by a fever or pain with urination or defecation
> - Is located in the right lower portion of the belly
> - Lasts longer than two days or gets worse over time
> - Is accompanied by swelling, stiffness, or weight loss
> - Interferes with daily activities like eating, sleep, or attending school or sports

kinds of abdominal, muscle, bone (growing pains, overuse pains), and joint pains (arthritis). If you want to skip to my bottom-line recommendations, flip to the end of the chapter. We'll start with Lifestyle Therapies.

Lifestyle Therapies: Nutrition, Exercise, Sleep, Environment, Mind-Body

Nutrition

When a child has a tummy ache, the first suspect is something the child ate. For example, lactose (milk sugar) intolerance is a common cause of belly pain, gas, and bloating in older children and teens. Children with lactose deficiency can experience a remarkable improvement in their tummy aches when they stop consuming milk, cheese, and ice cream or start using lactase enzyme supplements or lactose-free dairy products. A pediatric gastroenterologist can do a formal test to determine whether your child is lactose intolerant, or you can find out for yourself informally through trial and error.

A diet deficient in vegetables, fruits, and other high-fiber foods may lead to painful constipation. See Chapter 12 for more details about including high-fiber foods and fluid to combat this common problem.

For children with irritable bowel syndrome (IBS), a special diet called the FODMAP diet helps relieve symptoms, including belly pain.[5] FODMAP stands for **f**ermentable **o**ligosaccharides, **d**isaccharides, **m**onosaccharides, and **p**olyols. What a mouthful! FODMAPS are carbohydrates that are easy for bacteria to digest or ferment but harder for humans to break down and absorb. As the bacteria break them down, these carbohydrates produce gas and other by-products, leading to abdominal pain and diarrhea. Improvement can start as quickly as 48 hours after avoiding FODMAP foods.[6] Table 22.1 shows FODMAP foods, and you can find more detail on many Internet sites. However, given the number of foods high in FODMAPS and the extent to which you are likely to change your family's diet, if you decide to try it, please seek the help of a registered dietitian.

Can Nutrition Affect Inflammation and Pain?

Yes. A diet rich in vegetables, fruits, olive oil, and fish (sometimes called the Mediterranean diet) tends to decrease inflammation throughout the body. Because of this anti-inflammatory tendency, the Mediterranean diet can help control symptoms due to inflammation like asthmatic wheezing, allergies, joint pain, and back pain.[7] A Mediterranean diet is generally healthy for adults, too. It's a better bet than the standard American fare of processed, fried, and sugary foods.

Although everyone can benefit from a healthy diet filled with fruits and veggies, please talk with your health care professional about any radical change in your child's diet.

Sleep and Exercise

Pain makes it hard to *sleep*. Poor sleep makes us more sensitive to pain, and children who experience poor sleep have a harder time performing their usual activities because of their pain.[8] It's a vicious cycle.

Vigorous *exercise during the day* (when the child is not having pain) can help release the body's own painkillers (endorphins). Exercise is one of the best remedies for stress and pain. Aim for at least one hour of moderate to vigorous exercise daily, but be careful not to overdo it. Avoid painful overuse injuries.

Tai Chi can help control pain in adults with arthritis, and may also be helpful for kids with pain, inflammation, and sleep problems. For example, in a study of older women with osteoarthritis, regular Tai Chi practice improved joint pain, balance, and general activity level over 12 weeks.[9] A program that taught Tai Chi to middle school children for five weeks improved the children's well-being, calmness, relaxation, and sleep.[10] By decreasing stress and improving overall well-being and sleep, Tai Chi could decrease your child's pain.

Yoga exercises can also decrease the intensity and frequency of abdominal pain in school-aged children and teens with functional abdominal pain.[11] For example, Iyengar yoga helped a 14-year-old girl who had persistent, disabling abdominal pain resume her daily activities.[12]

Environment

Toxic elements in the environment can make pain worse. For example, carbon monoxide from a poorly vented furnace, heater, or engine exhaust can cause many symptoms including headache, dizziness, abdominal pain, diarrhea, vomiting, and flushing.[13] Carbon monoxide poisoning can kill. If you suspect carbon

Table 22.1: FODMAP Diet for Irritable Bowel Syndrome—Dos and Don'ts		
Food Category	OK to EAT (DO eat)	AVOID (Don't eat)
Fruit	Banana, blueberries, citrus fruits, cranberries, cantaloupe, grapes, melons, kiwi, papaya, pineapple, rhubarb, strawberries	Apples, apricots, blackberries, cherries, coconut, nectarine, mango, pear, peaches, plums, prunes, watermelon, dried fruits; fruit juices
Vegetables	Avocado, alfalfa sprouts, arugula, bean sprouts, bell peppers, carrots, corn (small amounts), celery (small amounts), cucumber, eggplant, endive, ginger, green beans, kale, parsnip, potato, radish, spinach, squash, chard, sweet potato, tomato, turnips, zucchini	Artichoke, asparagus, brussels sprouts, broccoli, beets, cabbage, cauliflower, garlic, leeks, lettuce, okra, onions, shallots, snow peas
Beans/Legumes	Tofu	Chickpeas, lentils, kidney beans, soybeans, edamame
Grains	Oats, polenta, quinoa, rice, gluten-free pasta, gluten-free bread, gluten-free baking mix	Wheat, rye, triticale
Dairy	Coconut milk, rice milk; lactose-free milk; lactose-free yogurt; small amounts of hard cheese such as cheddar, Swiss, parmesan	Milk and milk products, yogurt, sour cream, ice cream, cottage cheese, ricotta
Nuts/Seeds	Almonds, chia, flax, macadamia, peanuts, pecans, pine nuts, pumpkin seeds, sesame seeds, sunflower seeds, walnuts	
Meat/Fish/Eggs	Beef, chicken, eggs, fish, pork, tofu	
Sweeteners (read labels!)		Agave nectar, honey, high fructose corn syrup, mannitol, sorbitol, xylitol, isomalt
Other		Coffee substitutes, energy bars, inulin

monoxide exposure, seek immediate medical care, and then make sure that the source of the carbon monoxide is identified and repaired.

If your child's pain seems to be worsened by *weather* changes, try to keep note of these patterns in a diary. This could help you to anticipate and respond to changing weather.

On the other hand, pain can be improved by environmental factors ranging from temperature and weather to music and spending more time in nature. Novel environmental therapies for pain include magnets (electro-magnetic therapy), low-level lasers, and ultrasound. Let's look at them one by one.

A simple approach to helping with pain is to control the microclimate. *Ice packs* can help ease the pain of a recent injury or a sudden toothache, while warm water bottles or *heating pads* can soothe many chronic aches and pains.

Music can help distract people from pain. One study found that children who underwent blood draws (venipuncture) in the presence of musicians trained to work in medical settings showed less distress before, during, and after the blood draw.[14] Similar benefits have been found using music therapy for other procedures.[15] Music therapy can also help hospitalized children and adolescents receiving palliative care near the end of life.[16]

Spending unstructured time outdoors in *nature* (playing in a park or recreational area, for example) has numerous health benefits—decreasing stress, instilling a sense of well-being, and improving attention.[17] This brings a sense of comfort and ease. In one study, hospitalized patients with a view of deciduous trees took fewer doses of pain medication than adults viewing a brown brick wall. In another study, adults undergoing flexible bronchoscopy were given a tape recorder with nature sounds to listen to before, during, and after their procedure, and a nature scene mural was placed at their bedside; those with views and sounds of nature reported less pain than those without views and sounds of nature.[18]

New and emerging environmental therapies for pain include the use of low-level lasers and electromagnetic therapy. Unlike hot lasers used for precision cutting, *low-level laser therapy* (LLLT, aka cold lasers, soft lasers, or laser acupuncture) is used in veterinary medicine and physical therapy to speed wound healing and ease joint pain. The use of LLLT for pain relief in adults with rheumatoid arthritis, neck and low back pain, and frozen shoulders makes me guess that it won't be long before enterprising therapists start using LLLT to help children with many kinds of pain.[19–22]

Similarly, static and pulsed *electromagnetic fields* are useful in some adults with joint and muscle-related pain,[23–25] and someone is bound to start trying them for children with arthritis. Until more research is done, it's premature for me to recommend low-level lasers or electromagnetic therapies for children with painful conditions.

Transcutaneous electrical nerve stimulation (TENS) is another therapy that has been used to control pain. TENS involves sending low-voltage electrical currents into the skin. There is growing evidence that TENS reduces pain in adults.[26] A small study showed that it has helped control the pain of dental procedures in children.[27] But TENS has not been studied enough as a therapy for abdominal and joint pain in children for me to routinely recommend it yet.

Therapeutic *ultrasound* is a physical therapy technique that involves sending high-frequency vibrations to certain parts of the body to help control pain and inflammation. Therapeutic ultrasound has successfully been used to help manage whiplash pain, osteoarthritis, and rheumatoid arthritis in adults.[28–30] Therapeutic ultrasound has a good safety record. More research is needed on therapeutic ultrasound for pediatric pain.

Mind-Body

It is stressful to have chronic, recurrent, uncontrolled pain. Stress also increases sensitivity to painful stimuli and promotes worrying thoughts about what the pain means. It's another vicious cycle. Stress management can help control pain, especially pain associated with irritable bowel syndrome (IBS) and juvenile idiopathic arthritis (JIA).[31–33] A comprehensive approach to stress management might include identifying and eliminating a child's stressors, using a pain diary, counseling with a mental health practitioner, and learning to practice effective mind-body strategies.

If your child suffers from recurrent pain of any kind, keep a *pain* (or comfort) *diary* or *log* of his symptoms, triggers, and relievers. You can use a regular calendar, a smart phone app, or make your own diary or log.

Other mind-body strategies such as *deep*

breathing, progressive relaxation, biofeedback, guided imagery, hypnosis, mindfulness-based meditation, and cognitive behavioral therapy can also help manage pain.

EFFECTIVE MIND-BODY THERAPIES FOR PAIN

- Tracking symptoms and stress
- Slow, deep breathing
- Progressive muscle relaxation
- Guided imagery and hypnotherapy
- Biofeedback
- Mindfulness-based meditation
- Cognitive behavioral therapy (CBT) and support groups

Slow breathing can be very effective in helping to manage pain because people often take shallow breaths or hold their breath when they are in pain. Breathing deeply can induce a relaxation response, and relaxation can reduce sensitivity to pain.

Progressive muscle relaxation can help children recognize the body tension that comes with pain and gradually learn to modulate that tension. This technique can also decrease anxiety and discomfort, especially when used with other techniques like deep breathing or guided imagery.[34]

Guided imagery with progressive muscle relaxation decreases the frequency of recurrent abdominal pain and helps children resume their normal activities.[35] During this technique, a child is guided to imagining a favorite place or activity. Focused imagination promotes relaxation and often relieves pain.

One study found that biofeedback, hypnosis, and relaxation were effective for controlling headaches, abdominal pain, and fibromyalgia in children.[36] Physiologically, hypnotherapy and guided imagery modify the brain's perception and response to pain. Clinically, hypnotherapy has proven helpful for children with

irritable bowel syndrome and functional abdominal pain.[37–39]

Mindfulness-based meditation is effective in reducing stress and pain.[40] Mindfulness means paying attention intentionally and nonjudgmentally to present moment experience. Several mind-body practices fall under the umbrella of mindfulness-based practices—the body scan, yoga, mindful breathing, mindful walking, and mindful eating. Mindfulness-based meditation practices can help manage your child's pain and the stress that feeds pain.[41]

Cognitive behavioral therapy (CBT) improves pain and activity in children.[42–44] It can be provided in individual sessions, groups, or even over the Internet. For example, a study of 11–18-year-olds showed that CBT sessions held over the Internet were effective in controlling pain in both the short-term and three months later. Cognitive behavioral therapy for the child's family also helps improve pain in children.[45]

Mind-body techniques are most effective if practiced regularly. Children benefit from home practice and the encouragement of supportive parents. I recommend stress management, deep breathing, progressive muscle relaxation, biofeedback, guided imagery, hypnotherapy, mindfulness-based meditation, and cognitive behavioral therapy for children with frequent or severe pain. Even children who use medications often benefit from these non-drug therapies.

Maggie and Matthew asked for a referral to our clinical psychologist to learn more about mind-body techniques and how to use them. They learned the technique of mindfulness, and Matthew began breathing slowly and deeply into his belly, exploring the sensations in his leg with friendly curiosity instead of fear. He learned that pain shifted, moved, and gradually subsided as he breathed into the area where the sensation felt strongest.

BIOCHEMICAL THERAPIES: MEDICATIONS, DIETARY SUPPLEMENTS, HERBAL REMEDIES

Medications

There is a *high placebo response rate* to pain medications. This means that many children will improve with a medication or supplement, regardless of which one is started. This is why most pediatricians recommend starting with the safest treatments before trying riskier options.

MEDICATIONS FOR ABDOMINAL PAIN

- *Acid neutralizers, blockers, or preventers*: Tums™, Pepcid™, Zantac™, Prilosec™, Prevacid,™ and others
- *Antidepressant medications*: Elavil™ and others
- *Anti-inflammatory medications*: Prednisone™, Remicade™, and others
- *Beta-blockers and triptans*: Inderal™, Imitrex™, and others
- *Narcotic drugs* (for very severe abdominal pain): Morphine and others

Most medications for belly pain (such as antacids) work by neutralizing or reducing stomach acid. *Antacids* include medications like Tums™. If excess stomach acid isn't causing the pain, they aren't much more helpful than powerful placebos. Antacids are widely accessible without a prescription. However, they are not always effective or safe. With prolonged use, some can cause aluminum toxicity, and others can cause milk-alkali syndrome. I do not recommend them for chronic treatment of abdominal pain unless you have an established diagnosis of excessive stomach acid and you've already tried safer lifestyle approaches.[46]

Other medications known as proton pump inhibitors (PPIs) block the production of stomach acid. These include cimetidine (Tagamet™), famotidine (Pepcid™), lansoprazole (Prevacid™), omeprazole (Prilosec™), and ranitidine (Zantac™). Again, if the problem isn't excessive acid, these medications are no more effective than placebo. They are generally well tolerated but can sometimes have side effects like headache, diarrhea, constipation, and nausea.

Another type of medicine pushes food faster through the gastrointestinal tract. One example is metoclopramide (Reglan™). This medication can have serious side effects, which led to an FDA-mandated black box warning. These drugs should not be used routinely and should be reserved for severe cases that have not improved with other treatments.

Prescription antidepressants such as amitriptyline (Elavil™) are sometimes used to treat pain. Studies on their effectiveness in managing abdominal pain in children have had conflicting results. One small study suggested that a low dose of tricyclic antidepressants may help control pediatric abdominal pain.[47] But a review analyzing two trials of antidepressant medications for IBS concluded that they were no better than placebo.[48] Since antidepressant medications can have serious side effects (dry mouth, blurry vision, constipation, difficulty urinating, memory problems, and heart arrhythmias), don't ask for a prescription for them unless you've already tried safer lifestyle strategies.

Immunosuppressive and immune modulating drugs like azathioprine, prednisone, cyclosporine, methotrexate, Remicade™, and others are used to treat Crohn's disease, ulcerative colitis, and other forms of inflammatory bowel disease and some types of severe arthritis.[49] These drugs effectively reduce inflammation. However, they have significant side effects. In suppressing the immune system, they can increase the risk of serious infections and cancer.[50]

Medications like *beta-blockers* (propranolol or Inderal™) and *triptans* (Imitrex™ and others) are usually used to prevent migraine headaches. (See Chapter 20.) These medications can also be used to prevent a specific type of abdominal pain called an abdominal migraine. Beta-blockers can cause a slow heart rate, lower blood pressure, and worsen asthma. Triptans can actually cause headaches, chest pain, weakness, dizziness, and drowsiness. Be sure you have an accurate diagnosis before trying any new medication.

Finally, *narcotics* like morphine are sometimes given to children with very severe abdominal pain, usually in an emergency department or hospital units while waiting for or recovering from surgery.

What Are the Most Commonly Used Medications to Treat Other Kinds of Pain?

For everyday muscle, bone, and joint *aches and pains*, the only medication your child may need is a non-prescription pain reliever.

NON-PRESCRIPTION PAIN RELIEVERS

- Aspirin
- Acetaminophen (Tylenol™ and other brands)
- Ibuprofen (Advil™, Motrin™, and others)
- Naproxen (Aleve™ and others)

Aspirin is inexpensive and widely available. Aspirin should not be taken when your child has influenza or chicken pox because of the risk of developing Reye's syndrome, a potentially deadly disorder of the liver and brain. Because of this risk, many physicians now steer clear of aspirin. (To help guard against influenza and chicken pox, I also recommend the influenza and chickenpox vaccines.) Aspirin can cause an upset stomach and bleeding of the stomach lining; frequent or high doses can cause ringing in the ears.

Acetaminophen (Tylenol™) is the most widely used non-prescription pain reliever for children. Generic brands are just as effective as name brands. Excessive doses damage the liver. Acetaminophen is slightly less powerful than ibuprofen as a pain reliever, but it is less likely to cause stomach irritation.[51] *Ibuprofen* (Advil™, Motrin™) is my personal favorite non-prescription, over-the-counter non-steroidal anti-inflammatory drug (NSAID) for muscle, bone, and joint pain. It starts a bit sooner and lasts about two to four hours longer than acetaminophen or aspirin. However, ibuprofen can cause stomach irritation and even bleeding. If used for short periods as recommended, it is generally safe.

Naproxen (Aleve™) is similar to ibuprofen, but it lasts up to twelve hours. Naproxen is used to treat juvenile (pediatric) arthritis and other forms of muscle, bone, and joint pain. However, in addition to sometimes causing stomach upset (it should be taken with food to minimize this side effect), naproxen has been associated with cardiovascular toxicity and kidney failure in children.[52]

If these over-the-counter medications aren't strong enough, higher doses of ibuprofen are available by prescription.

As with abdominal pain, pain in the joints and muscles is sometimes treated with immune-suppressing medications like prednisone.[53] For pediatric arthritis, steroids may be injected directly into the joint, sparing body-wide side effects like the increased risk of infections and cancer. These medications are powerful, and if your child receives a prescription for them, he should be closely monitored by a pediatric specialist.

Dietary Supplements

Dietary supplements commonly used for abdominal pain include magnesium, guar gum, and probiotics. Dietary supplements used for bone and joint pain include *vitamin D, omega-3 fatty acids,* and *S-adenosylmethionine* (SAM-E). Let's review them one at a time, starting with supplements for abdominal pain.

DIETARY SUPPLEMENTS FOR ABDOMINAL PAIN

- Magnesium for constipation
- Guar gum for irritable bowel syndrome
- Probiotics

Magnesium helps ease uncomfortable constipation. (See Chapter 12.) It is available over-the-counter in products such as magnesium hydroxide (Milk of Magnesia™) and magnesium citrate. Magnesium is generally considered safe when taken at or below recommended dosages. It can cause upset stomach, nausea, vomiting, and diarrhea as well as low blood pressure, sleepiness, and weakness when excessive doses are chronically consumed.

Guar gum is a fiber supplement that reduces the frequency of irritable bowel syndrome symptoms, including pain.

There are many species of the millions of healthy bacteria (*probiotics*) that live in our intestines. Probiotics containing *Lactobacillus rhamnosus* (LGG) alone and combinations of *Lactobacillus and Bifidobacteria* can improve abdominal pain associated with chronic constipation.[54,55] Combinations including *Lactobacillus, Bifidobacteria*, and *Streptococcus thermophilus* have also proven helpful in reducing the pain associated with functional abdominal pain, irritable bowel syndrome (IBS), and ulcerative colitis in children.[56–60] Research in this area is just beginning, and doses range from 100 million for the pain for newborn colic to over 350 billion to treat inflammatory bowel disease (IBD). Probiotics are generally safe, and are even used in premature babies to reduce the risk of serious infections in newborn intensive care units. Side effects can include

gas, bloating, and general abdominal discomfort, so start with a low dose and work your way up. Check with an independent testing group like ConsumerLab to find a product that meets your child's needs in terms of formulation, avoidance of dietary triggers like gluten and dairy, and cost.

DIETARY SUPPLEMENTS FOR BONE AND MUSCLE PAIN

- Vitamin D
- Fish oil and other sources of omega-3 fatty acids
- S-Adenosylmethionine (SAM-E)

Although it is generally better for your child to get his essential nutrients from foods rather than from pills, sometimes pills are needed to achieve the high levels required to achieve a specific health goal.

Vitamin D deficiency is often linked to achy muscles and bones. In fact, many patients I see with vague muscle and bone aches have vitamin D deficiencies that had gone undiagnosed.[61] I recommend regular safe (non-burning) exposure to sunlight for all children and regular intake of vitamin D supplements if necessary. These therapies can help prevent the vague achiness that sometimes results from a vitamin D deficiency.

Fish oil, which contains omega-3 fatty acids, has anti-inflammatory properties. Fish oil helps reduce symptoms caused by inflammation such as adult rheumatoid arthritis[62] and pediatric asthma.[63] Fish oil and other omega-3 fatty acid supplements are quite safe. Over 90 percent of the fish oil products tested by ConsumerLab are free of environmental toxins such as mercury and dioxins. Prescription omega-3 fatty acids are available, but they are expensive and may not be covered by your insurance.

S-adenosyl-methionine (SAM-E) supplements

are often used by adults to help manage the pain of arthritis in the hip and knee. A review of several studies suggests it can be both safe and effective in adults with osteoarthritis.[64] However, additional research is needed to determine whether this expensive supplement is helpful for the kinds of arthritis that afflict children.

Herbal Remedies

Herbal remedies for abdominal pain include chamomile, ginger, peppermint, and combinations such as Iberogast™. Herbal remedies for muscle, bone, and joint pain include turmeric, valerian, willow bark, and menthol or camphor rubs. Let's start with supplements for abdominal pain.

HERBAL REMEDIES FOR ABDOMINAL PAIN

- Chamomile
- Ginger (for nausea); also may help with muscle and joint pain
- Peppermint
- Iberogast™

Chamomile is often used in combination with other herbs such as peppermint, anise, and fennel to treat belly pain, gas, indigestion, and bloating. Chamomile reduces inflammation and prevents spasms in the gastrointestinal tract. Even though it is widely used to help control abdominal pain in children, there are no studies proving its benefit for this specific purpose.[65] On the other hand, chamomile is generally considered safe.

Ginger contains both anti-nausea and anti-inflammatory compounds. It has proven effectiveness as a nausea remedy. (See Chapter 26.) Evidence of its helpfulness in controlling muscle and joint pain is conflicting. Some researchers suggest that patients may have to take ginger for several weeks to months before it improves their arthritis.[66] Ginger is safe

and inexpensive. Until more studies are done to evaluate ginger's benefits in treating muscle and joint pain in children, I recommend it for nausea, but not necessarily for joint pain.

Peppermint is used to help reduce indigestion, gas, and gastrointestinal spasms. It is especially helpful in treating IBS.[67–71] Enteric-coated peppermint oil is generally well tolerated; potential side effects include abdominal pain, heartburn, perianal tingling, and allergies.[72] Peppermint oil should be avoided in children who have gallbladder or bile tract problems, gallstones, liver disease, and reflux because it can make these problems worse.

Herbal remedies are sometimes *combined* with each other and with dietary supplements. For example, the European product *Iberogast™* is a combination herbal remedy that improves general gastrointestinal uneasiness (functional dyspepsia) as well as abdominal pain associated with IBS and chronic abdominal pain in children.[73,74] Iberogast™ contains extracts of several herbs: candytuft, angelica root, milk thistle, greater celandine, licorice root, chamomile, lemon balm, caraway seeds, and peppermint leaves. These herbs are generally safe; potential side effects include abdominal cramps, diarrhea, nausea, and dizziness.

Next, let's review common herbal remedies for muscle, bone, and joint pain.

HERBAL REMEDIES FOR MUSCLE, BONE, AND JOINT PAIN

- Turmeric
- Valerian
- Willow bark
- Menthol or camphor balms

Turmeric is the yellow spice in curry powder. It has anti-inflammatory properties. Turmeric is used by adults as a treatment for arthritis in doses of 400–600 mg three times daily.[75] In fact, turmeric is probably as effective as ibuprofen in controlling the pain of osteoarthritis in adults.[76] Turmeric also benefits adults with abdominal pain due to inflammatory bowel disorder (IBD). Turmeric can prevent IBD relapse and improve patient symptoms when used with other anti-inflammatory medications.[77] Turmeric is generally safe. Additional research is needed on using turmeric or its active compound (curcumin) for medicinal purposes in children.

Valerian is used to ease insomnia. Valerian may help improve adults' reactions to stressful situations.[78] Easing stress and improving sleep may make it easier to cope with pain. Additional research is needed on valerian's benefits for children with painful conditions.

Willow bark has anti-inflammatory properties and was the original source of aspirin. Both contain acetyl-salicylic acid. Willow bark is considered to have fewer side effects than aspirin. It is generally a safe remedy for everyday aches and pains, but, like aspirin, willow bark should not be used when your child has influenza or chicken pox due to the risk of Reye's syndrome.

Menthol and camphor rubs (which are found in creams like Icy Hot™ and Tiger Balm) can be applied to the skin to soothe muscle soreness. They work by stimulating and then numbing nerve cells that detect pain and are often very effective pain relievers.[79] Some forms of Icy Hot also contain methyl salicylate; like acetyl-salicylic acid from willow bark, methyl salicylate is also useful in reducing musculoskeletal pain (especially acute pain).[80] Don't get them in your eyes; they sting!

Maggie decided to start treating Matthew's growing pains with Tiger Balm. When Matthew woke up in the middle of the night complaining of leg pain, she reminded Matthew to breathe into the pain and then she massaged Tiger Balm into his legs. Most of the time, he fell asleep within ten minutes.

BIOMECHANICAL THERAPIES: MASSAGE, SPINAL ADJUSTMENTS, SURGERY

Massage

Massage is highly effective for helping children in pain. One study showed that children with juvenile rheumatoid arthritis who were massaged by their parents for 15 minutes a day for 30 days had less stress and anxiety immediately after the massage and decreased pain after the 30-day period; their activity levels also improved.[81] Massage is safe. I recommend regular massage for children with chronic pain.

Spinal Adjustments and Joint Manipulation

Although case reports claim that chiropractic manipulation improves growing pains,[82] these benefits have not been reliably duplicated in randomized controlled trials. Clinical trials evaluating the use of chiropractic manipulation for abdominal pain in children have not yet been performed. Additional research would be helpful.

Surgery

There is no need for surgery to treat most childhood pain. However, surgery can be lifesaving for appendicitis or twisted testicles or ovaries. Surgery can also be an important therapy for fractures, bone tumors, and scoliosis. Optimal treatment of an infected joint or abscess generally includes surgical drainage, too.

BIOENERGETIC THERAPIES: ACUPUNCTURE, THERAPEUTIC TOUCH, PRAYER, HOMEOPATHY

Acupuncture

Over the past twenty years, increasing evidence has supported the effectiveness and safety of acupuncture as a pain treatment. For example, one study of 6- to 18-year-old children with chronic pain showed that six weekly sessions of acupuncture combined with hypnosis resulted in decreased pain and improvement in functioning.[83] Another study showed that most children and their parents who had completed acupuncture treatments felt that acupuncture had improved their symptoms and that acupuncture was a pleasant experience for them.[84]

Maggie decided that if Matthew's pain wasn't better in six weeks, they'd give acupuncture a try. Because he improved so much with his mindful breathing and the nighttime massage (and our old favorite for growing pains, tincture of time), he never needed the acupuncture.

Therapeutic Touch, Healing Touch, Reiki

Therapeutic Touch, Healing Touch, and Reiki are safe and widely used to restore harmony and a sense of balance, comfort, and relaxation.[85] These techniques decrease stress and anxiety and improve sleep. They have been used to relieve the pain of arthritis, burns, and surgeries.[86] They can be used along with other therapies like music therapy and guided imagery, and can help to reduce anxiety and pain. I nearly always offer Therapeutic Touch to children in pain.

Prayer

Prayer is widely practiced and very safe. Many different forms of prayer can be comforting. Additional research is needed on the benefits of prayer for children in pain, but if prayer is part of your family's values, beliefs, and culture, I encourage you to pray.

Homeopathy

Homeopathic treatment is certainly safe, but there are no large randomized trials showing it is more helpful than placebo in treating children's abdominal or joint pain.

COMBINATIONS

An increasing number of clinicians use what is called a multimodal approach to healing pain naturally. This means combining careful tracking; healthy lifestyle, including mind-body strategies to reduce stress; meditative exercises like yoga or Tai Chi; manual therapies; acupuncture; and medications.

WHAT I RECOMMEND FOR PAIN

FOR PAIN IN GENERAL

If your child has frequent abdominal pain, *keep a diary or log of the abdominal pain, abdominal pain triggers, and remedies that seem to help.*

1. *Lifestyle—exercise and sleep.* Make sure your child gets plenty of deep, restorative sleep at night and at least one hour of vigorous exercise during the day. Consider introducing your child to exercises like Tai Chi and yoga, which can help reduce stress and pain.

2. *Lifestyle—environment.* Avoid toxins like carbon monoxide. Help your child manage his or her pain with music. Surround your child with reminders of nature and incorporate plenty of outdoor activity into your child's days. Use ice packs or heating pads if they are helpful.

3. *Lifestyle—mind-body.* Help your child practice slow, deep breathing; meditation; and progressive muscle relaxation exercises regularly to help minimize stress and build a sense of mastery. Consider counseling to further help your child cope with the stress of having a painful condition and to learn cognitive behavioral therapy strategies. Visit your health care professional to learn more about mind-body therapies such as biofeedback, guided imagery, and hypnotherapy. Remember that these therapies are most effective when practiced regularly.

4. *Bioenergetic—acupuncture, Therapeutic Touch, prayer, homeopathy.* If it's acceptable to your child and if a pediatric acupuncturist is available, consider acupuncture therapy. Consider learning more about or asking a health professional to provide Therapeutic Touch. Incorporate prayer into your child's life if it is consistent with your values and beliefs.

FOR ABDOMINAL PAIN

1. *Lifestyle—nutrition.* Monitor food-related triggers (like lactose) for your child's abdominal pain. Encourage more fruits and vegetables as part of a healthy Mediterranean-style diet. Visit a dietitian if you are considering special diet plans, such as the FODMAPS diet for IBS.

2. *Biochemical—medications, herbs, and supplements.* Consider supplements with magnesium for constipation. Consider using guar gum, probiotics, and Iberogast™ for IBS. Consider a cup of chamomile or peppermint tea. Ginger can reduce nausea. Talk to your health care professional if your child has frequent abdominal pain for a diagnosis before starting antacids, proton pump inhibitors, or other medications. Prescription medications may be useful for more serious causes of abdominal pain.

3. *Biomechanical surgery.* If you suspect appendicitis or a problem with your child's ovaries or testicles, seek professional help immediately. Surgery can be lifesaving.

See Your Health Care Professional If Your Child Has Abdominal Pain That ...

- Is very severe
- Is accompanied by fever
- Is located in the right lower portion of the belly
- Lasts longer than two days or gets worse over time

FOR PAIN IN THE MUSCLES AND JOINTS, LIKE GROWING PAINS

1. *Lifestyle—nutrition.* A Mediterranean diet can help control inflammation. Consider visiting a registered dietitian if you are thinking of making significant changes to your child's diet.

2. *Lifestyle—environment.* Stay tuned to emerging research on using electromagnetic fields, ultrasound, low-level lasers, and infrared therapies for pediatric pain in the joints and muscles.

3. *Biochemical—medications, herbs, and supplements.* Make sure your child gets safe (non-burning) sun exposure every day to build up healthy levels of vitamin D. Consider a vitamin D supplement if necessary because deficiencies of vitamin D can lead to muscle and bone aches. Fish oil, willow bark, and turmeric are safe, and may help reduce joint pain. Massage menthol and/or camphor balms (like Icy Hot™ or Tiger Balm) onto painful muscles and joints. Non-prescription pain relievers like aspirin, ibuprofen, and acetaminophen can also offer short-term relief. See your health professional for stronger or longer-lasting medications.

4. *Biomechanical—massage.* Gently massage the area that your child finds painful. Consider seeking professional help from a licensed massage therapist or physical therapist to learn more about how to provide a massage safely and effectively. Advocate with your insurance carrier to cover this valuable professional service.

See Your Health Care Professional If Your Child Has Musculoskeletal Pain That…

- Is very severe

- Is accompanied by fever, weight loss, or rashes

- Lasts longer than two days or gets worse over time

- Is accompanied by swelling, stiffness, or difficulty moving the area normally

---

23
RINGWORM AND OTHER FUNGAL INFECTIONS

Marie Washington brought in her sons, Irving and Michael, to be treated for ringworm and athlete's foot, which they'd picked up at school. Eight-year-old Irving had classic ringworm—silver dollar–size scaly patches on his arms and legs. Michael, the sixteen-year-old star of his basketball team, was complaining of athlete's foot—itching and burning on his feet, especially between his toes. Marie wanted to know:

• Is ringworm really caused by worms?
• Can only athletes get athlete's foot?
• What is the best way to treat both boys?

Despite its name, ringworm is caused by a fungus, *Tinea corporis,* not worms. Ringworm is the most common fungal infection in children over a year old. Similar fungal infections also occur on the feet (athlete's foot, or *Tinea pedis*), in the groin (jock itch, or *Tinea cruris*), and in the scalp (*Tinea capitis*). *Tinea* simply means fungal infection. The second part of each medical name refers to the body part affected (*capitis* = head, *corporis* = body, *pedis* = feet). Several types of fungus cause infections.

They have long, poetic names such as *Trichophyton, Microsporum,* and *Epidermophyton.* Despite their beautiful names, these fungi cause a rash of rashes.

Tinea corporis is one of the most common skin conditions of childhood. It is passed from person to person. When this happens to wrestlers, it's called *trichophytosis gladiatorum.* Kids can also pick up these infections from the clothing or equipment of infected persons.

Tinea capitis (fungal infection of the scalp) affects about 5 percent of school-aged children under nine years old. It is usually caused by *Trichophyton tonsurans* in the United States, but it may be caused by other species in tropical countries. Although most fungal infections are passed from person to person, scalp infections can be caught from dogs or cats. These fungi can cause big pimple-like swellings, mushy areas, and crusting on the scalp, called a *kerion*. Kerions can look so serious they have fooled experienced doctors into thinking that something other than a simple fungal infection was responsible, leading to costly treatment errors and even hospitalization. The infection can also look like plain old ringworm or it might cause patchy bald areas on the head (alopecia) and swollen lymph nodes in the neck. When the fungi cause the scalp hairs to break off, the resulting bald patches with dots of broken hairs are called *black dot ringworm*. Swollen lymph nodes in the back of the neck and behind the ears are other common symptoms of fungal infections on the head. Tinea capitis is common in school-age kids, but becomes much less of a problem once kids enter puberty and their scalps become oilier and less hospitable to fungus.

Unlike tinea capitis, which fades as adolescence begins, *Tinea pedis* (athlete's foot) becomes more common as adolescent hormones (and high-top sneakers) kick in. In one study of marathon runners, about one in ten had symptoms, and twice that many carried the ringworm fungus. This finding points out the need for careful hygiene in the locker room. Even if your buddy doesn't have any symptoms, he could still be carrying the fungus that will make you miserable. Athlete's foot occurs even among those whose only exertion is Monday morning quarterbacking. Once the fungus has set in and the skin starts to scale, itch, and break down, the scene is set for bacterial in-fections to move in, creating an even bigger problem.

Fungal infections of the fingernails and toenails are known as *onychomycosis*. These are much less common in children than in older adults. They are difficult to treat. The nail may become thick and white or yellow, with ridges or curls. Treatment nearly always requires antifungal medications taken by mouth.

Those who have fungal infections of the feet or nails can spread the fungus to others. For example, a teenager with athlete's foot can spread the fungus to a sibling who ends up with a fungal infection elsewhere on the skin (ringworm).

It's possible Irving picked up his infection from Michael, who probably got it from a teammate who had no idea that he was spreading a fungal infection.

Other skin infections include *pityriasis* (*tinea*) *versicolor*, which is a yeast (*M furfur*) infection of the skin (the infected areas fail to tan in the summer, so the patches look pale), and *candidiasis*, which frequently causes diaper rashes, vaginitis, and oral thrush.

ARE YOU SURE IT'S A FUNGUS?

Fungal infections can mimic several other skin conditions such as eczema, impetigo, insect bites, lupus, psoriasis, pityriasis rosea, plain old dandruff, and allergic reactions. Special tests may be needed to make the right diagnosis. Some fungi glow under black lights; your doctor may use a special light (known as a Wood's lamp) to see if the infected skin glows in the dark. Alternatively, she may scrape a tiny bit of skin onto a slide and examine it under a microscope to be sure it's a fungus and not an imitator. Sometimes a culture is necessary to be certain. Culture results can take several weeks. Most folks are eager to take action much sooner, so you may start therapy even without these diagnostic tests.

WHAT'S THE BEST WAY TO TREAT FUNGAL INFECTIONS?

All therapies take several weeks to eradicate fungus; there are no quick cures, but there is relief! Let's tour the Therapeutic Mountain to find out what works best. If you want to skip to my bottom-line recommendations, flip to the end of the chapter. Let's start with healthy lifestyle approaches before diving into medications and other treatments.

LIFESTYLE THERAPIES: NUTRITION, EXERCISE, ENVIRONMENT, MIND-BODY

Nutrition

There are no studies suggesting that any particular diet prevents or eradicates fungal infections.

Exercise and Sleep

Exercise common sense. If your child is in gym class at school and takes showers in a common area, make sure he has shower sandals to keep his feet out of direct contact with the inevitable fungus on the floor. If your child has an infection, it's even more important to wear shower shoes to prevent the infection from spreading to others. Have him dry his feet well after a shower to make them less hospitable for fungi. After showering is the perfect time to apply antifungal medication. Even if your child doesn't have athlete's foot, he may want to dust his feet with antifungal powder several times a week in the locker room to prevent picking up a case.

Adequate sleep is always important, but there is no evidence that extra sleep speeds healing from fungal infections of the skin.

Environment

Fungi love to live in warm, damp environments. High-top sneakers are perfect breeding grounds for athlete's foot fungus. If your child is prone to athlete's foot, have him forget the high-tops for a while and go with more breathable shoes such as sandals or thongs when he's not playing basketball. It takes a day or so for regular shoes to dry out from normal sweat. Changing shoes and socks twice daily helps reduce moisture; try not to wear the same shoes two days in a row, and look for breathable socks. Also, put those sneakers in the washing machine with the white clothes, detergent, and bleach every week. You may find that as the shoes are cleaned and dried and the rash clears, there is a remarkable improvement in foot odor.

Children *can* catch fungal infections from pets and farm animals. If your pets or livestock have skin problems, have them checked by your veterinarian and treated promptly to reduce your child's chances of catching a fungal infection.

If your child has a fungal infection of the scalp, he can return to school after he's been taking medication (both by mouth and therapeutic shampoo) for at least a day. Please do not allow your child to share helmets or participate in sports with head to head contact (e.g., wrestling) for at least six to eight weeks after starting treatment to avoid spreading the infection.

BIOCHEMICAL THERAPIES: MEDICATIONS, DIETARY SUPPLEMENTS, AND HERBS

Medications

Medical treatments are similar for most fungal infections *except* for those affecting the scalp. Scalp infections require medication that reaches deep beneath the skin surface because they usually involve the deep hair shafts. Most other fungal infections on the skin respond to medicated creams, lotions, or sprays. Even for scalp infections that require oral medications, using a topical cream or shampoo can help cut the risk of spreading the infection to others.

NON-PRESCRIPTION MEDICATIONS FOR FUNGAL INFECTIONS

- Undecylenic acid: (Desenex™, Pedi-Dri™, Cruex™)
- Miconazole: (Micatin™, Monistat™); Clotrimazole: (Lotrimin™, Mycelex™)
- Tolnaftate: (Tinactin™, and others)
- Terbinafine (Lamisil™)
- Selenium sulfide shampoo: (Selsun™ and others)
- Zinc pyrithione, 1 percent for pityriasis versicolor or adjunct for tinea capitis

Undecylenic acid (Desenex™ and other brands) has been used to treat athlete's foot for over thirty years. The most commonly recommended antifungal agents in our clinic are undecylenic acid, clotrimazole, and miconazole because they have proven safety and effectiveness and are inexpensive. They are applied twice daily until after the rash has been gone for at least one week to ensure that all the fungi have been eliminated. I prefer creams and lotions to powders and sprays because they tend to stay on better, but you can experiment to find out what works best for your child.

Tolnaftate (Tinactin™) and *Terbinafine* (Lamisil™) are effective antifungal medications. They are traditionally used to treat athlete's foot and jock itch, and they are effective against ringworm, too. Improvement is usually apparent within ten days, although complete cure may take a month or more.

Selenium sulfide shampoo (Selsun™, Exsel™) is an adjunctive therapy to treat fungal infections in the scalp. It does not eradicate the infection, but it helps prevent it from spreading to others. Non-prescription strength selenium sulfide shampoo appears to be as effective as the more expensive prescription formulations. Selenium sulfide is also an effective treatment for pityriasis versicolor; the shampoo or lotion is applied for 10 minutes daily before being rinsed off.

PRESCRIPTION MEDICATIONS FOR FUNGAL INFECTIONS

- Griseofulvin (Fulvicin™, Grisactin™, Grifulvin V™)
- Imidazoles: Econazole (Spectazole™), Fluconazole (Diflucan™), Itraconazole (Sporanox™), Ketoconazole (Nizoral™), Oxiconazole (Oxistat™), Sulconazole 1% (Exelderm™), Butenafine 1% (Mentax™)
- Allylamines: Terbinafine (Lamisil™), Butenafine, Naftifine (Naftin™)
- Others: Halprogin (Halotex™), Ciclopirox (Loprox™)

A variety of prescription medications have been developed to fight fungal infections. Some are taken by mouth while others are applied directly to the rash. Mild infections typically respond to topical treatments, but deep-seated or widespread infections require a systemic approach that includes oral medication. Many of these medications have not been thoroughly tested or approved for use in children, but that hasn't stopped doctors from recommending them when a stubborn fungal infection fails to improve with older, approved drugs.

For fungal infections in the scalp (*tinea capitis*), the oral medicine *griseofulvin* is the treatment of choice because of its low toxicity and low cost; it has been around since 1958, so we know a lot about how safe it is. Like penicillin, griseofulvin is derived from our old friend, the *Penicillium* mold. It must be taken daily for at least six to eight weeks to eradicate fungal infections. Because fungi have gradually adapted to medications, the dose required to knock them out has gradually increased. Griseofulvin is one of the rare medicines that actually works better if it is taken with food, especially fatty foods. Try giving it to your child with a glass of whole milk. It can interfere with the effectiveness of birth control pills and it should not be used during pregnancy; children taking seizure medications may need to take higher than usual doses of griseofulvin to achieve the same effect as children who are not taking other medications. Griseofulvin can cause headaches, nausea, and rashes. Because fungi have developed resistance to griseofulvin, a growing number of other medications are used to treat fungal infections of the scalp.

Econazole is a prescription imidazole medication that works quickly; it is effective within two weeks against ringworm and within one month against athlete's foot. *Fluconazole* is used to treat children with systemic (not just on the skin) fungal infections, but it is sometimes used to treat tinea capitis, too, as an alternative to griseofulvin. It should not be used by children with liver or kidney disease. Because of its potential toxicity, liver and kidney functions are usually tested with simple blood tests before starting fluconazole. *Ketoconazole* and *itraconazole* can eradicate symptoms and even prevent outbreaks in kids who are frequently exposed to fungi, such as wrestlers; ketoconazole should not be used by mouth to treat tinea capitis because it can cause liver damage, but it can be used safely in shampoo and other hair products. *Itraconazole* has become a treatment of choice for serious lung infections with a nasty fungus called *Aspergillus* as well as fungal infections affecting the fingernails or toenails—sites that are notoriously

difficult to treat. It is not officially approved for use in young children. The prescription imidazole medications are effective, but they are expensive. They can also cause an upset stomach and markedly interfere other medications, including allergy medicines, seizure medicines, asthma medicines, heart medicines, and others.

The *allylamine medications, terbinafine* and *naftifine,* are among the most effective antifungal medications.[1] *Terbinafine* (Lamisil™) is very effective in treating even difficult cases of athlete's foot, toenail fungus, and tinea capitis.[2–4] Terbinafine creams are available over-the-counter without a prescription. However, taking terbinafine by mouth may cause side effects when given to patients who are already taking other medications such as cimetidine, rifampin, or phenobarbital. Taking terbinafine by mouth means it has to be metabolized by the liver, so patients with liver disease should avoid it. Terbinafine is approved for use by children four years old and older; treating tinea capitis takes about six weeks of daily medication. Terbinafine packets are generally sprinkled on non-acidic food like mashed potatoes or pudding. *Naftifine* is another very powerful antifungal medication for treating athlete's foot.[5] Like all fungal medicines, it can cause burning and stinging when applied to already irritated skin.

Cicloprix shampoo is as effective as selenium sulfide shampoo when combined with oral griseofulvin to treat fungal infections of the scalp. Do not rely on shampoos alone to treat fungal infections of the scalp; they are too deep-seated for shampoos to eradicate them.

The most cost-effective medical approach to the typical fungal infection of the skin is to start with an inexpensive but effective non-prescription medication (such as miconazole or terbinafine); if there is no improvement after a month of twice daily use, see your physician about a more expensive and powerful prescription medication.[6] Fungal infections of the scalp respond best to oral medications combined with antifungal shampoos.

Both Irving and Michael started treatment with terbinafine cream.

How Long Should You Treat a Fungal Infection?

A good rule of thumb is to treat for at least seven days after the rash has cleared. Usually the fungus remains on the skin even after the rash is gone, so additional treatment is needed to prevent a recurrence.

What about Treating an "Id" Reaction?

An *id reaction* is an immune response to a fungal infection. The immune system over-reacts and sets up an itchy rash on the face, hairline, trunk, arms, and legs, usually just after medication has been started to treat a fungal infection of the scalp. It is not a reaction to the antifungal medication, so you can continue the treatment; your health professional may recommend a mild steroid cream (like hydrocortisone) or an antihistamine (like diphenhydramine) to help with the itching while the immune system stabilizes.

How Are Medications Best Delivered?

Some medications (like griseofulvin) are usually taken by mouth, whereas others are applied in creams, lotions, shampoos, or sprays. For fungal infections of the skin (other than the scalp and nails), most clinicians advise direct application of the medication to the rash. A novel approach to delivering medication right to the problem (at least when the problem is athlete's foot) is through special medicated toe socks. In one study comparing control treatment to the medicated (with clotrimazole) toe socks (each toe is enclosed separately), the medicated socks group had a significantly higher cure rate than the control group that wore similar socks without the medication.[7]

Nutritional Supplements

Vitamin C in powder or crystal form has been recommended as topical treatment for fungal infections of the skin, but there are no studies evaluating its use.

Herbal Remedies and Essential Oils

Plants have evolved many tricks over the ages to ward off fungal infections. Several strong-smelling undiluted essential oils are known to kill fungi and bacteria in test tubes and petri dishes.[8–12] The most commonly used herbs to treat yeast and fungal infections include angelica, basil, bergamot, cinnamon, eucalyptus, garlic, green tea, lemon, lemongrass, mint, tea tree oil, rosemary, thyme oil, and bee propolis. However, some essential oils reduce the effectiveness of standard antifungal medications.[13] Essential oils can irritate skin when applied straight (undiluted). These oils should not be used as the sole treatment for serious or widespread fungal infections or those affecting the scalp or nails.

Garlic extracts have proven antifungal effects in test-tube studies and have been tried in adults with fungal skin infections with some success.[14,15] Garlic can synergistically improve the effects of antifungal medications.[16] Garlic extracts have not yet been tested in children in well-designed randomized controlled trials to treat fungal skin infections. Garlic poultices can cause burns and severe skin irritation if left on too long. I do not recommend garlic applications for children when safe, effective topical medications are readily available to treat fungal skin infections.

Green tea extracts kill Trichophyton in test tubes. In a Japanese study, daily footbaths with green tea extracts for 12 weeks helped decrease the size of adults' fungal skin infections better than footbaths with water.[17] However, green tea has not been compared to standard non-prescription antifungal medications and has not been tested in children. It is generally safe as long as common sense precautions are used (do not immerse your child's skin in very hot tea!), but I do not recommend green tea as a replacement for antifungal medications.

Tea tree oil has antibacterial and antifungal properties in test tubes and horses.[18–20] In an Australian study it was compared to tolnaftate to treat athlete's foot. Although patients reported improved symptoms with both preparations, tea tree oil failed to kill the fungi.[21] Tea tree oil is safe when applied to non-irritated skin, but it should not be ingested because it can be toxic. As with any treatment, if the skin becomes more irritated, stop using it.

Ringworm of the scalp is very rare on the Indian subcontinent, where many people's hair dressings contain vegetable oil. A test tube study showed that the most popular Indian hair oils (mustard oil, coconut oil, and Amla oil) had antifungal properties.[22]

These preliminary results are intriguing. Vegetable oils are safe and inexpensive, but do *not* forsake medications that have proven effectiveness for treating fungal infections that have already taken hold in the scalp.

Some herbalists recommend teas to help the immune system fight fungal infections. Teas reputed to have this effect include burdock, cleavers, dandelion root, echinacea, nettles, peppermint, and red clover. Teas used as washes for fungal infections are agrimony, burdock, and marigold. There are no studies evaluating the effectiveness of any of these teas or tinctures as treatments for fungal infections. I do *not* recommend them as primary treatments.

Lemon juice and cider vinegar have both been recommended to make the skin more acid (lower pH) and less hospitable for fungi. There are no scientific studies documenting their effectiveness in treating children with fungal infections, but they are inexpensive and safe when kept out of sensitive tissues like eyes.

BIOMECHANICAL THERAPIES: MASSAGE, SPINAL MANIPULATION, SURGERY

None of these therapies has scientifically proven benefit in treating children with ringworm or other fungal infections. I do *not* recommend them for childhood fungal infections.

BIOENERGETIC THERAPIES: ACUPUNCTURE, THERAPEUTIC TOUCH, PRAYER, HOMEOPATHY

None of these therapies has scientifically proven benefit in treating children with ringworm or other fungal infections. Until more research confirms their safety and effectiveness, I tolerate them, but I don't recommend them as treatments for fungal infections.

WHAT I RECOMMEND FOR FUNGAL INFECTIONS

PREVENTING FUNGAL INFECTIONS

1. *Lifestyle—environment.* Use good hygiene. Prevent athlete's foot by using shower shoes, sandals, or flip-flops in gym showers and locker rooms. Avoid sharing helmets. Keep the feet clean and dry. Avoid high-top sneakers. Wear breathable shoes. Change socks and shoes frequently. Wash sneakers in the washing machine and dry them thoroughly every week. If your pets or livestock develop skin problems, have them evaluated and treated promptly to prevent possible spread to your children.

See Your Health Care Professional If

- You suspect a fungal infection of the scalp. This requires prescription medication. It may also benefit from regular shampooing with antifungal shampoo.

- The rash is not improving after ten days of therapy or is getting worse with home treatment.

- The rash is spreading.

- The rash involves the toenails or fingernails.

TREATING FUNGAL INFECTIONS

1. *Biochemical—medications.* Treat ringworm, jock itch, and athlete's foot twice daily with non-prescription antifungal medications containing clotrimazole (Lotrimin™, Mycelex™), miconazole (Monistat™), terbinafine (Lamisil™), tolnaftate (Tinactin™), or undecylenic acid (Desenex™, Cruex™, Caldesene™) for at least four weeks. Continue treatment for at least one week after all symptoms have resolved to make sure the fungi have been eradicated.

2. *Biochemical—medications.* If the rash is not gone within a month with non-prescription treatments, see your health care professional to evaluate the need for prescription medication.

3. *Biochemical—medications.* For fungal infections of the scalp, use a prescription medicine by mouth plus an antifungal shampoo daily.

24
SLEEP PROBLEMS

These are some of the challenges with regard to children and sleep:

- Getting your infant to sleep through the night
- Getting your toddler to bed
- Nightmares and night terrors
- Insomnia

"Our baby is four months old and still waking up to nurse at 2 a.m. and 5 a.m. When will he outgrow the need for middle-of-the-night feeds so we can get some sleep? Will giving him rice cereal at bedtime help him sleep better?"

"Our three-year-old is driving us nuts. Every night it's the same thing. We put her to bed, and then she wants a glass of water; next it's a story; by that time she has to go to the bathroom again. We get her in bed and five minutes later she's up again; if we don't read her another story, she starts crying. Bedtime started out as a 15-minute process, and now it takes an hour. How can we get back to a reasonable routine?"

"Our four-year-old has been waking up a few hours after we put him to bed, screaming like someone is killing him. We rush into the room to find him sitting up in bed, staring blankly. He doesn't seem to know we're there and won't talk. In the morning, he doesn't even seem to remember it. This has happened several times. Should we take him to a psychiatrist?"

"Our teenager has 7:30 a.m. classes every day, but he has a hard time getting up, especially on Mondays. He comes home from school and sleeps on Friday afternoons, and then stays up playing computer games online with friends until 2 or 3 in the morning, then sleeps late Saturday and Sunday. This seems like a terrible habit. What can I do to make sure he gets enough sleep?"

One of the most common and frustrating problems parents face is getting their children to sleep through the night. Babies learn sleep and wakefulness cycles just as they learn hunger cycles, bladder control, language, and walking—over time. Children's sleeping patterns change throughout infancy and early childhood, from initially waking every few hours to nurse, to sleeping longer stretches, to intermittently waking again in later infancy, to full night's sleep plus naps at toddler age, to school-age regular nighttime sleep, to adolescents who may require more sleep as they go through growth spurts (Table 24.1). Annual surveys conducted over the last twenty years have shown that America's youth are sleeping much less than they did before the advent of computers, cell phones, and other electronic devices.[1] Altogether we have an enormous national sleep deficit.

It takes the average two-month-old over 25 minutes to fall asleep, while the average nine-month-old needs just over 15 minutes to drift off. Most infants awaken every few hours to nurse until they are three to four months old, and some continue to do so for the first year and a half. By six months of age, over 80 percent of infants sleep at least five hours in a row. Believe it or not, in pediatric circles five to six straight hours of sleep is considered "sleeping through the night."

Table 24.1: Typical Sleep Patterns

Age	Total Sleep	Naps?
Newborn	16–17 hours	2–3
6–12 months	14–15 hours	2
Toddlers	12 hours	1
School-age	10 hours	none
Older school-age	8 hours	none
Teenagers	8–10 hours	none or weekend

Almost all children give up naps by first grade; some outgrow naps by the time they are three years old. School-age children typically need fewer than 10 hours of sleep, but should get at least 8 hours a night. Adolescents often sleep more during periods of intense growth and maturation. During illnesses and recovery from injury, children sleep more. Each child is different. But many children would benefit from more sleep than they're getting.

The two main phases of sleep are rapid eye movement (REM) and non-REM sleep. Non-REM sleep is quiet sleep during which the child moves from wakefulness into deeper sleep with slower brain waves. During deepest sleep the brain waves are very slow, and the child is very still. This is when growth hormone secretion is at its maximum. REM sleep is fairly light and physically active sleep. Following REM sleep the child may briefly waken before returning to quiet sleep and repeating the cycle. Like cats, infants can enter REM sleep almost immediately. Older children and adults sleep quietly (non-REM) for about 70 to 90 minutes before REM sleep starts. A teenager who sleeps eight hours goes through about four or five repetitions of this cycle every night.

If your child has sleep difficulties, you are not alone. Neither is your child. Almost half of American children are reported to have some kind of sleep difficulty. The most common problem is getting toddlers to go to bed. Ten to 15 percent of kids have a hard time falling asleep or awake in the middle of the night at least once a week. Fifteen to 20 percent have troubles waking up in the morning or feel fatigued during the day from inadequate sleep. Over one-third of children experience nightmares or night terrors by age two; over 10 percent walk in their sleep.[2] Other sleep-related problems include bedwetting and teeth grinding. Children's sleep problems frequently disrupt parents' sleep, resulting in parental fatigue, frustration, anger, and guilt. Adults with sleep problems have a greater risk of anxiety, depression, low work productivity, high blood pressure, diabetes, pain, and unintentional injuries.

Parents aren't the only ones who are stressed, exhausted, and depressed when their children have trouble sleeping. The kids suffer, too. Children and adolescents who don't get enough sleep are less able to pay attention in school, more irritable, less happy, and less likely to comply with parental or teacher requests. Many kids who are inattentive or impulsive are actually sleepy; they tend to fall asleep more during the day and to fall asleep more quickly at night than their peers. Anytime a child is being evaluated for attention deficit hyperactivity disorder (ADHD), depression, anxiety, or oppositional behavior, the clinician and parent should pay close attention to the child's sleep. Poor sleep may be a precursor to other behavioral problems.

FACTORS THAT INTERFERE WITH NORMAL SLEEP

- Illness, pain, allergies
- Anxiety, depression, tension, or stress
- Change in routine, travel
- Being overtired
- Too much TV or exposure to handheld electronic devices
- Caffeine-containing beverages or foods
- Stimulant medications
- Difficult temperament
- Bedtime habits—falling asleep in parents' arms
- Nightmares

Ear infections are perhaps the best example of a painful illness that awakens children in the middle of the night. *Teething* and *itchy rashes* can also keep a child from a sound sleep as can breathing problems from congestion or asthma. *Diabetes* and *bladder infections* increase urination and thereby increase nighttime awakening. Strangely enough, cow's milk *allergies* often interfere with sleeping as well.[3,4] There is at least one study showing that sleep dramatically improves when allergic children stop drinking cow's milk.[5] (See Chapter 5.) If your child has other allergic symptoms, consider a trial off of cow's milk to see if sleep improves.

Like adults, children can suffer from *stress, anxiety, depression,* or *tension* that keeps them awake. Preschool-age children may be concerned about parental disputes or disruptions in normal routines. School-age children and adolescents may worry about upcoming tests, peer pressure, dates, or scary movies. Major catastrophes and natural disasters often trigger nightmares. Following the 1989 San Francisco earthquake, nightmares were eight times more frequent in Bay area students than among students living in Tucson, Arizona. Try to keep evening time quiet and relaxed. Discuss gruesome news (if necessary) and family disagreements during the day when coping and communication are at their best rather than before bed when they tend to be strained.

Paradoxically, being *overtired* is a common cause of sleep problems. When a child becomes overtired, he compensates for his fatigue by increasing levels of stress hormones (such as epinephrine) that keep him awake. It takes a certain calm focus to allow the events of the day to slide away and let sleep set in.

School-age children who watch more television, especially those who watch violent shows, are much more likely to have sleep problems than children who watch less.[6] Having televisions in their rooms makes it far too easy for kids to overindulge in TV, decreasing healthy physical activity and delaying bedtime. If your child has trouble sleeping, remove the TV from his room and limit him to less than 90 minutes daily of television, preferably nonviolent shows. When he watches, stay with him; you might be surprised how much violence there is in kids' shows, and how much it helps your child to talk with you about what he's viewing.

Adults are not the only ones who are kept awake by *caffeine*. Its stimulant effects are found in tea, cola, and chocolate as well as coffee. I saw a teenager recently who complained of being unable to fall asleep before 11:30 p.m. and then having trouble getting up for his 6 a.m. alarm. It turned out he was drinking massive quantities of cola after supper while doing his homework. His ability to fall asleep dramatically improved when he decided not to drink any cola after 4 p.m.

Common cold medications and herbal remedies also contain potent *stimulants*: ephedrine, pseudoephedrine, phenylpropanolamine, ephedra, and ginseng. If you are nursing your baby, check with your health care professional before taking non-prescription cold medications because some of them end up in breast milk and may keep your baby awake.

Infants who have trouble falling asleep or who waken frequently in the middle of the night are more likely to be viewed as having *difficult temperaments* than children who sleep soundly. Children with persistent sleep problems are also more likely to have temper tantrums and other behavior problems. It is difficult to sort out whether less adaptable children are more likely to have sleep problems or whether children who awaken their parents in the middle of the night are labeled as troublesome.

Poor bedtime sleep habits can also lead to difficulties in learning to fall asleep without parents' help. Infants who are *put into bed after they have fallen asleep* or *who fall asleep in the presence of a parent* are more likely to awaken in the middle of the night (and wake their parents) than infants who fall asleep by themselves in their crib. This is a strong argu-

ment for putting your child down to sleep while he is still awake. If your child is already used to falling asleep while you are holding him or rocking him, try making the change first at naptime. When he learns to fall asleep on his own at naptime, try putting him down at night while he is still awake. By learning to fall asleep on his own, he can later put himself to sleep when he awakens in the middle of the night without waking you.

Sleep patterns can be substantially improved with a few straightforward measures. In a study of breast-fed infants and their parents, one group was taught how to train their infant to sleep through the night; the other group was put on a waiting list and not given any special instructions on infant sleep until the baby was eight weeks old. The techniques included: (1) late night feeding between 10 p.m. and midnight; (2) gradually lengthen time between late night feeding and next feeding; (3) alternative responses to crying including diapering, swaddling, walking; and (4) maximizing differences between daytime and nighttime environments. Those who were taught the techniques were able to train their infants to sleep better through the night (100 percent were sleeping at least five consecutive hours by the time they were eight weeks old compared to only 23 percent of control group infants). The trained group also rated their babies' temperaments as significantly easier than the untrained group whose babies were still waking frequently.[7] Parents who are taught to train their infants to sleep through the night also report an increased sense of competence as parents.[8] The advantages of early sleep training may extend beyond peaceful parental slumber to benefit infant temperament and parental confidence.

Nightmares are most common between the ages of three and six years old. Children who awaken with nightmares are generally fully awake, can be comforted by parents, and can tell their parents details about the bad dream.

Children who suffer from nightmares are not maladjusted, more anxious, or under more stress than children who don't have nightmares as frequently. Nightmares are not due to behavior problems, seizures, or brain wave abnormalities. They are simply bad dreams. Adults who suffer from nightmares tend to be more sensitive and creative than those who aren't beset by troubling images at night.[9] Children, of course, have more vivid imaginations than adults, and have active fantasies about imaginary friends, ghosts, monsters, and all kinds of creatures. It's no surprise that many of them turn up in dreams. Nightmares can be inspired by anything from television violence to shadows cast by a nightlight to actual scary experiences.

FREQUENT NIGHTMARE TOPICS

- Scary animals
- Monsters
- Strangers attacking or chasing them
- Being abandoned or helpless in a strange place
- Being shot or attacked

Night terrors are different from nightmares and occur most commonly in children between the ages of one and four years old. Unlike dreams or nightmares, they do not occur during REM sleep. Rather, they occur in the deepest non-REM stages of sleep. Night terrors tend to occur within the first two hours of sleep whereas nightmares occur during dream sleep in the later part of the night (between 3 a.m. and 6 a.m.). During a night terror episode, the child utters terrifying screams, sits bolt upright, and stares ahead with a glassy gaze. His eyes are open and he appears to be awake, but he does not notice or respond to his parents. This is very unnerving for the parent who is trying to comfort an obviously distraught child who is so soundly asleep that he really doesn't hear anything. Often, the child doesn't even recall the incident the following morning. The episodes typically last fewer than 15 minutes. Children with sleep terrors are more likely than others to become sleepwalkers later in childhood. Although they are known to run in families, the precise cause of night terrors is unknown, but children usually outgrow them by school age. They are *not* signs of a psychiatric problem. However, if your child has frequent night terrors (more than twice a week), see your health care professional for advice. Some prescription medications are helpful for this condition.

WHAT IS THE BEST WAY TO TREAT SLEEP PROBLEMS?

The best treatments for all children's sleep problems are to:

- Prevent them in the first place.
- Treat the underlying problem if there is one.
- Reassure yourself and your child that good sleep is attainable.
- Use proper sleep habits, such as bedtime routines.

Let's tour the Therapeutic Mountain to make sure we cover all the bases. We'll start with lifestyle strategies, which are the most important factors to consider in improving sleep. If you want to skip to my bottom-line recommendations, flip to the end of the chapter.

LIFESTYLE THERAPIES: NUTRITION, EXERCISE, ENVIRONMENT, MIND-BODY

The primary treatment for childhood sleep problems is creating a healthy lifestyle. Paying attention to your child's diet, exercise patterns, and sleep environment is key to improving sleeping patterns.

Nutrition

Anyone up for a glass of warm milk before bed? Some foods historically have helped with sleep. Which ones does science support?

FOODS TO HELP WITH SLEEP

- High-protein foods: yes
- Rice cereal: no
- Sweets: always brush teeth afterward

Formula-fed babies generally fall asleep faster and sleep longer between feedings than breast-fed babies in the first six months of life; this means that during the first six months, breast-fed babies tend to wake more often in the night to nurse, though they have lower risks of developing wheezing, coughing, snoring, or other breathing problems.[10] Between ages six and twelve months, formula-fed babies do not waken in the middle of the night any less than breast-fed babies.[11] Breast-feeding for the first year is better for your baby's health, though it may mean more night-awakenings during the first six months of life.

Protein-rich foods like milk contain a rich mixture of amino acids, including tryptophan. Tryptophan is an important precursor to hormones that promote healthy sleep and mood. Dieters who cut back on protein may find themselves having trouble sleeping at night and focusing during the day. Both research and my grandmother suggest that eating a protein-rich meal late in the day helps you sleep better at night. I often recommend a late night snack of milk, peanut butter, or a turkey sandwich for kids who have trouble falling asleep. Be sure to brush teeth before bed.

Hot beverages are traditionally used before bedtime to induce a state of relaxed drowsiness in older children and adults. You might try a cup of warm (non-caffeinated) herbal tea or a glass of *warm milk*. There are many other com-pounds in milk that may contribute to sleepiness in addition to the tryptophan. I've been unable to find a study comparing the impact of hot to cold beverages on sleep. On the other hand, unless your child has problems with bed-wetting, it's worth a try.

Despite common myths to the contrary, adding high-carbohydrate foods like *rice cereal* to your baby's evening bottle does *not* help him sleep better through the night. Randomized, controlled trials clearly demonstrated that adding rice cereal to the evening feeding did not improve infants' sleep.[12] It turns out that many parents become frustrated with night awakenings and add cereal just about the time the baby is learning to sleep through the night on his own anyway. The cereal erroneously receives the credit for the baby's accomplishment in learning to fall asleep and stay asleep.

Sweets are often called "comfort foods." *Honey* is a traditional ingredient in home sleep remedies. To avoid the risk of botulism, do *not* give honey to infants under a year of age. If you give your toddler or older child something sweet to help him sleep, do so *before* he brushes his teeth!

Do NOT put your child to bed with a bottle in his mouth. Falling asleep with a bottle of milk or juice will rot your baby's teeth. Decay can start as soon as the teeth begin to emerge (usually between five and seven months of age). If your child is already in the habit of going to bed with a bottle and can't seem to fall asleep without one, here's a trick that will eliminate that habit within a week. The first night, dilute the milk or juice (whatever the child has been taking to bed) with one-eighth water. The next night, go to one-quarter water; the third night, half; the fourth night, three-quarters; and the fifth night, nothing but water in the bottle. Most children don't like the taste of water as much as whatever they were drinking before. Some may protest at the watering down, but many will not even be aware of the gradual change. By the time you've reached pure water, most children willingly release the bottle altogether. If not, at least the water will not harm your child's teeth.

Exercise

Many physicians (and grandmothers) recommend that vigorous exercise be avoided in the hour just before bedtime. Multiple studies in adults show that exercises such as yoga and tai chi can help decrease inflammation and pain and improve sleep.[13–17] Just about any regular daytime exercise can certainly contribute to sound sleep,[18] but avoid running, dancing, spinning, and other frenetic exercises in the hour before bed. Better bets are reading stories, listening to quiet music, and taking a hot bath.

Environment

Bed and bedtime should be as pleasant as possible. Do not send your child to bed during the day as a punishment. Associating bedtime with punishment will only make him want to avoid it.

SLEEPING ENVIRONMENT

- Light (dark), sound (quiet), temperature (cool)
- With parents vs. alone
- Hot bath before bed
- Music: Bach vs. Barney
- Favorite stuffed animal, blanket, pacifier
- Sleep position (face up vs. face down)

Emphasize the day-night differences in the environment. Pay attention to light, sounds, aromas, and temperature. Most children sleep better in a *cool* environment. *Quiet and dark* are conducive to sleep, but are not essential. Many children feel more comfortable with a

nightlight or having their bedroom door open so they feel safe.

What Is SIDS?

Sudden Infant Death Syndrome (SIDS) is the sudden, unexpected death during sleep of a baby, usually between one and five months old. It is the second leading cause of death among infants. The number of SIDS deaths could be cut in half if all babies were put to sleep on their backs (supine) rather than on their tummies (prone). *Lay your baby down to sleep on his back, NOT face down.* SIDS deaths have declined dramatically since widespread publicity has changed sleep habits from tummy sleeping to sleeping on the back. The American Academy of Pediatrics, the National Institute of Child Health and Human Development, the Indian Health Service, the Consumer Product Safety Commission, and the Bureau of Maternal and Child Health all recommend that you put your baby to sleep on his back or side to reduce the risk of SIDS. This is one time you should pay attention to medical authorities; rarely are we in such strong agreement.

TO REDUCE THE RISK OF SUDDEN DEATH DURING SLEEP, DO NOT

- Put your infant to sleep on his stomach.
- Smoke cigarettes or use heroin, crack, or cocaine.
- Overheat or over-bundle your baby.
- Put your infant to sleep on a sheepskin or waterbed.

Parental *smoking* and *drug use* markedly increase the risk of SIDS. Several studies have shown an increased risk of SIDs among babies who are *over-bundled, have their heads covered,* or who sleep in *overly warm* rooms.[19] Although sheepskin and lambskin are soft and cuddly, they may increase your child's risk of suffocation. Waterbeds may also be problematic. Unlike firmer foam bedding, natural fibers and waterbeds tend to conform to the baby, making it difficult for the baby to turn his head away from possible suffocation if he lies on his tummy with his nose facing into the mattress.

Should a Baby Sleep in the Same Bed with His Parents?

This question must be answered by parents according to their own values and beliefs. The American Academy of Pediatrics opposes co-sleeping on safety grounds because there is an increased risk of Sudden Infant Death Syndrome (SIDS) in babies who co-sleep with their parents, particularly if parents smoke, drink alcohol, or use drugs, or if the baby was born prematurely.[20]

CO-SLEEPING: BENEFITS

- Cultural norm
- Easier breast-feeding

In many cultures it is unthinkable for a baby *not* to sleep with his parents. Also, most nursing mothers find it convenient to have the baby in bed with them so when the child awakens to feed there is no need to get up. On the other hand, there are some drawbacks.

CO-SLEEPING: DRAWBACKS

- Increased risk of Sudden Infant Death Syndrome (SIDS)
- Baby learns to depend on parental presence to fall asleep
- Parents awaken every time baby awakens
- More sleep problems reported by parents
- Challenges to parental intimacy

The major drawback of co-sleeping is the increased risk of SIDS. Another less serious, but more common consequence is that the baby learns to fall asleep only with parents present, and doesn't learn to fall asleep by himself; it is harder to get the baby to fall asleep by himself in his own bed later on when parents want the privacy of their own bed back. Some parents are light sleepers who awaken every time the baby moves or awakens. For these parents, it may be preferable to have the baby in a crib nearby for the first few months when the baby requires a nighttime feed, and then move the baby's crib to another room when a night feeding is no longer needed. *Parents must be attuned to and honor their own needs as well as the baby's needs.* Parents who are exhausted and irritable from insufficient sleep are prone to making poorer parenting decisions than parents who are well rested.

Flexibility is important when the toddler or infant is acutely ill. Children with chronic conditions such as asthma don't necessarily have more sleep problems than other children. However, children with acute illnesses such as colds, sore throats, and ear infections often feel more comfortable with a parent nearby. Many parents choose to allow the child to sleep with them for a few nights. Children often want to continue this pattern after the illness has resolved. In these cases, the parents will need to help the child reestablish a bedtime routine for falling asleep in their own beds and staying there. Some parents bring a mattress into the child's room when the child is ill so the parents can sleep on the floor near the child without having the child come into the parents' room.

Mind-Body

The most important thing you can do to help your child learn to fall asleep at night is to establish a bedtime routine. The routine should suit your needs as a parent and be consistent with your family beliefs.

COMMON ELEMENTS OF A BEDTIME ROUTINE:

- Same bedtime every night
- Going through the same routine of bathing, brushing teeth, toileting, hot bath, cool room; putting on pajamas, stories, songs, lights out, shades drawn every night
- Wearing special clothes (pajamas) to bed; bedclothes should not be the same as day clothes
- Special bedtime stories or consistent prayer
- Special bedtime lullabies or other music
- Bedtime massage—back rubs, stroking the forehead, foot rubs, etc.
- Favorite blanket, stuffed animal, pacifier, or pillow in bed (transitional object)
- Same words for goodnight
- A hug and "I love you" from parents

By establishing the *same bedtime* and *same routine* every day, you help your child establish a regular sleep-wake cycle. Our innate circadian rhythms are actually slightly longer than twenty-four hours, so it is natural for all of us to stay up a bit later each night and wake a bit later each morning. If you allow your child or teen to stay up later on weekends and sleep in the next morning, you may have a real struggle on Mondays when you try to revert to the old schedule and your child's body wants to continue the later pattern. The same thing can happen during vacations.

Children who go to sleep early tend to awaken early, and those who stay up later tend to sleep later. You may need to adjust your child's bedtime to reflect your family's needs.

If you are a night owl and like to sleep late in the morning, you may want to establish a late bedtime for your child. On the other hand, if you have to be up early for work, you may want your child asleep earlier. Children who have long afternoon naps are less sleepy for early bedtimes than those whose afternoon naps are shorter. Adjust bedtimes and nap times to find the schedule that works best for you. Try to follow the same pattern at least three days in a row. If you switch sleep patterns every day, your child will just be confused and irritable.

It is very important to *put the child down to sleep while he or she is still awake*. Studies using all nighttime lapse photography show that children as young as three weeks old who are put to bed while they are still awake can learn to put themselves to sleep. In fact, infants who learn to put themselves to sleep in their crib more easily learn to fall asleep on their own when they waken in the middle of the night.

If you allow the child to fall asleep in your arms, he will learn that he can *only* fall asleep in your arms. He will then fuss and cry until you come pick him up and he will not fall asleep until you are holding him. Babies are often drowsiest right after a meal. Try putting your infant down as soon as you've finished the last feeding of the evening. Falling asleep is a skill and a habit. Teach your child early on that he can fall asleep on his own, and you will save yourself many later struggles.

Please put your child to sleep in the *same place* every night. If you are out visiting friends or family for the evening when your child's bedtime comes, try to put your child to sleep using as many of the same routines as possible—pajamas, story, song, favorite blanket or stuffed animal, and so on.

A *hot bath* before bed is a time-honored way of relaxing for sleep. Again, make sure the bath time is relatively quiet and calm. Too much excitement can rev a child up just when you want him drowsy.

Music is a key component of many families' bedtime routine. Some families start with soothing classical music or lullabies when the child is still an infant. Keeping to a routine of the same music every night helps the child use it as a stimulus to sleep. One of the advantages of using the same music is that it can be brought along in the car or on trips so that at least that aspect of the bedtime routine can be maintained no matter where the family is. *What music works best?* Common sense says that if the music is calming for parents, it will contribute to a calm, relaxing atmosphere for the child. One of my physician colleagues swears that Gregorian chants have miraculous sleep-inducing effects on her infant daughter. Others prefer Bach and some like Barney. You'll have to experiment to find out what works best for your family.

Many children grow attached to *certain objects* such as a favorite blanket, stuffed animal, pacifier, or pillow. In a survey of toddlers' parents in a Cleveland area clinic, over half reported that their child needed a transitional object (blanket, pacifier, toy, thumb-sucking) to help them fall asleep.[21] Having this object along can help ease the way to sleep when the family is in the car, at grandparents', or otherwise outside of the usual sleep environment.

WHAT TO DO WHEN YOUR BABY WAKES UP IN THE MIDDLE OF THE NIGHT

- Check the baby: Does he need to be fed? Does the diaper need changing? Is he in pain (e.g., an open diaper pin, a hair wrapped around a finger or toe, ill)?
- Attend to his needs for nourishment, hygiene, or pain relief.
- Reassure him that you love him and are nearby with a few brief words, a pat on the tummy, stroking his head, or patting his hand.

- When you finish, leave the room.
- If he keeps crying, recheck him in five minutes; repeat the above, then leave the room.
- If the baby keeps crying, recheck him in 10 minutes; repeat the above, then leave the room.
- If the baby keeps crying, recheck him in 15 minutes; repeat the above, then leave the room; and so on.
- Reward yourself for hanging in there and teaching your child a new skill!

Make your middle-of-the-night visits as *brief* and *boring* as possible so your baby isn't interested or stimulated by them. Make sure his basic needs are met. Reassure him by speaking a few loving words and patting him on his tummy or stroking his head. If you have a music tape that helps him fall asleep, turn it on. Make sure he has his favorite blanket or toy nearby where he can see and touch it. Do not get your baby out of bed except to change diapers, swaddle, or feed him. If you rock, cuddle, and walk your baby back to sleep, he will grow dependent on your behaviors. If you teach him to need this kind of attention in the middle of the night, he will continue to demand it long after you want to go to sleep. If this is okay with you, offer him all the cuddling and entertainment you want.

If your baby is used to you getting up and playing with him or holding him and rocking him back to sleep and you stop doing those things, *he will probably cry* so you will do them again. This is the most difficult part for parents, and the part requiring the most self-control. If you want him to fall asleep on his own, leave the room even though he cries for you to return. It seems very cruel to leave the baby alone and crying, but remind yourself of your long-term goal. If he's still crying after five minutes (he probably will be), go back and recheck him. It may help for you to set a small timer for yourself so you know when five minutes are up. Go through the same routine of checking and brief reassurance. Then leave. Next time wait 10 minutes. Repeat this procedure, waiting an extra five minutes each time before going back to recheck, so that the intervals between rechecking lengthen to 15 minutes, then 20 minutes, then 25 minutes, and so on. You don't need to recheck him if he's not crying. This technique is called "Ferberizing" in honor of Dr. Richard Ferber, the Boston pediatrician who developed it. Ferberizing has been successfully used by thousands of American parents.

The first few nights this process may take longer than it would have taken simply to go in, pick him up, and rock him or cuddle him until he falls back asleep. That's okay. Within a week, there will be much less protest and fewer night awakenings because your child will have learned to put himself back to sleep. Even after your child has learned this skill, things may briefly fall apart the next time he becomes ill. Children who are ill often have legitimate needs for their parents in the middle of the night, so feel free to go comfort your child as needed during illnesses. After your child has recovered, he may want to continue the habit of extra attention. Then you may be called upon to repeat this routine until he has relearned to put himself back to sleep.

Acknowledge yourself for having the discipline to teach your child to put himself to sleep. This is one of the most difficult parenting tasks; many parents give in night after night rather than listen to their child cry. Give yourself some small reward for hanging in there for the first five minutes, then ten minutes, and so forth. Go ahead and brag to your friends, family, and colleagues. You've earned it! And you can be proud that your child has learned a new skill.

Many parents just can't stand to hear their baby cry, even for five or ten minutes at a stretch. There is an alternative to help your infant learn to sleep through the night called *scheduled waking*.

SCHEDULED WAKING

- Keep a diary of your child's night awakenings.
- Note what time your child usually wakes up.
- Set the timer for 10 minutes before your child usually wakes.
- Awaken your child *before* he wakes himself.
- Feed or soothe him.
- Repeat for each of the usual night awakening times.
- Every few days, reduce the number of times you waken your child.

Scheduled awakenings may seem bizarre, but they work. It takes a bit longer than Ferberizing, but it generally works within six weeks. It may be worth trying for parents who don't want to ignore their child's cries, but who do want to eliminate the nighttime awakenings. This strategy has also proven successful in reducing *night terrors* in children who tend to have them at the same time each night.

Truthfully, I did not Ferberize my child nor do scheduled awakenings. When he awoke and wanted me, I was there. I cuddled him; I rocked him; I sang to him. I got less sleep, but I didn't feel guilty, either. It wasn't a problem for either of us. He's gone off to sleepovers at friends' houses and away to camp for weeks at a time without any sleep problems. Managing your child's sleep is an individual choice every family must make (and live with).

WHAT TO DO IF YOUR TODDLER KEEPS GETTING OUT OF BED

- Follow bedtime ritual. (Be firm about good-night.)

- Explain consequences of getting out of bed.
- Follow through with consequences.

For parents of many toddlers, the most difficult part of getting them to bed is the seemingly endless requests for more water, another story, another back rub, and so on. What starts out as a 15-minute routine turns into a nightly hour-and-a-half battle. The parents wearily give in to each request, not wanting to be cruel and hoping that this is the last thing the child wants. Parents may warn the child, "This is the last . . . ," but give in to one more thing to avoid a tantrum or crying. This is a tough situation. Fortunately, it responds well to tough love. Tough love in this situation means setting limits, establishing clear consequences, and then sticking to them. It sounds easy, but it is very challenging. If it were easy, you wouldn't be facing the problem now. Be firm, be consistent, and be willing to put up with (and ignore) some serious crying for the first few days.

First, establish a bedtime routine that all adults in the house can live with. Make sure all adults know what's going on and agree with the plan so the child can't play one against another. Decide in advance how long the routine will take. Plan ahead. If you want the real bedtime to be 9 p.m. and the routine involves a bath, brushing teeth, a story, and a prayer, you'll need to start the routine by 8:15. Warn the child at 8 p.m. that it's time to start wrapping up the evening's activities and getting ready for bed. Go through the bedtime routine you have agreed on in advance. Praise your child for cooperating with each phase of the routine (bath, tooth-brushing, picking the bedtime story, lying quietly while the story is read, prayer, and so on). Positive reinforcement for following each part of the bedtime routine is very effective in getting children to cooperate with bedtime, reducing tantrums, and staying in bed once lights are out.

Decide in advance what the *consequences* of getting out of bed will be. For most parents, the consequence is simply leading the child back to bed, having them get into bed, and repeating that it is now bedtime and the child is expected to remain in bed. The consequence must be immediate (not after five more minutes of TV or cuddling), loving, and firm. You may decide that the consequence for getting out of bed a second time is that the child's bedroom door is closed or the music is turned off. You can anticipate that the child will get out of bed to test these limits, and most likely he will cry. This is normal. Your child is not being bad; he is simply testing the rule to see if it's a real rule or not. It's up to you to make it clear that it is a real rule, and you will enforce it by an immediate return to bed. The consequence must be immediate to be effective. Telling a three-year-old that the consequence is no TV the following day is nearly meaningless; many adults can't remember a consequence for that long!

This tough love, consistent consequences technique is effective, even for children who are learning disabled, autistic, or have genetic developmental differences.[22–24] Most children get the idea within a week. Be prepared to be firm for five to seven days in a row and be prepared for a temporary escalation in whining, complaining, and crying. Don't try to start this rule/consequences routine when you are dealing with another major stressor or disruption in routines such as a new baby in the house, or you will doom yourself to frustration and failure. By and large, the only times this technique does not work is when parents are unable to follow through with enforcing the limits they wanted to set.

There is an alternative for getting your toddler to bed. This is a sneakier way of introducing limits. Instead of forcing your child to go to bed at your preset time immediately, let your child stay up as late as he wants. Go through your bedtime routine in a calm, relaxed fashion. The next night, start the routine 15 minutes earlier. Be sure to give your child 15 to 20 minutes of warning so that he can wrap up his activities and be ready for bedtime routines. Every few days, push the routine back 15 minutes earlier so that your child's biological clock is gradually reset to your desired bedtime. This technique takes a bit longer, but it may save you a few tantrums, and it works.[25-26]

For older children and teens with insomnia, a professionally provided mind-body therapy called *cognitive behavioral therapy* (CBT) can be helpful. CBT is particularly useful for those whose insomnia is related to anxiety, depression, or pain. As few as six to ten CBT sessions can have better and longer lasting effects than prescription sleep medicines.[27-28] CBT is generally provided by a licensed psychologist or social worker and is generally covered by insurance.

Nightmares and Night Terrors

Do not talk about your child's fears or anxieties right before bed. Talking about the monsters in the closet or anxieties about an upcoming test may temporarily comfort your child, but it may also keep the anxieties spinning in his mind as he tries to fall asleep. Instead, reassure your child that everything is all right and that you can talk about such issues in the morning. If you postpone talking about your child's concerns, make sure that you really *do* set aside time the following morning to discuss them. Alternatively, you can talk about them early in the evening, well before bedtime so that fears are resolved before the child gets in bed. Skillfully switch the conversation to focus on things for which you are grateful or things that amused you during the day.

Being wakened by a terrified, screaming child is a frightening event for most parents. First reassure yourself that the house is not on fire, and there is no immediate emergency. *Re-main calm* as you enter your child's room. Ask to hear about the scary dream. Listen and be sympathetic, letting your child know you hear him and understand his fears. While you listen, calmly stroke his back, forehead, or hand; then reassure him that you are there and he is loved. Many children are comforted by hearing that parents, too, have frightening dreams sometimes, but that the dreams aren't real and that everything is really okay. Reassure the child that you can talk more about the dream in the morning if he would like to discuss it further.

Hypnosis and *guided imagery* can help reduce the frequency of insomnia, nightmares, and night terrors. The noted child hypnotherapist, Dr. G. Gail Gardner, describes a simple technique that parents can use to help children reduce the frequency of nightmares and improve self-esteem. During the daytime when your child feels relaxed and comfortable, begin to discuss the nightmare. Have the child tell you the story of the bad dream from beginning to end. When he has finished, ask him to pretend that the nightmare is a scary story that you read together and that he can make up a new, happy ending. Have him start at the beginning of the dream again, and this time, when the scary part comes (e.g., big animals chase him), instead of feeling frightened and running, he can be brave (e.g., turn around and ask the animals why they chase him) or ask for help from a superhero or fairy godmother or see the situation transformed (the monsters turn into butterflies) and see how things turn out differently (e.g., the monster becomes his friend, he is a hero, or the butterflies fly away). When he has completed the story to his satisfaction, tell him that he can do the same thing with his dreams at night *while he is dreaming them*. It is true. Becoming involved with dreams and changing them at the time one is dreaming is called *lucid dreaming*. Children who learn to become involved in their dreams in this way develop a powerful sense of self-esteem. This technique of rewriting the nightmare is also known as *rehearsal,* and it has proven effective in adults and teens with severe nightmares.

Professional *hypnotherapy* has also been used to help children with *night terrors.* After getting the night terrors under control with medication, school-age children trained in self-hypnosis gradually learn to control their night wakening with hypnosis alone, eliminating the need for medication.[29]

Another therapy, *recording* developed at the University of New Mexico, has proven very effective in adults suffering from severe, recurrent nightmares. In recording therapy, the child recalls the nightmare while he is relaxed and comfortable in a safe place. He then writes down the nightmare or draws it in great detail. Although this technique is very simple, it has proven very helpful. Getting something out on paper is empowering.

Several of these techniques can be combined for children who suffer from severe nightmares. One ten-year-old boy who suffered from recurrent nightmares after a car ran into his house benefited from such a combined program: progressive relaxation, reminding himself that "it is only a dream" and reconstructing positive endings for his nightmares. Within weeks he was able to sleep through the night without waking his parents.[30]

Insomnia

For *insomnia* in older children and teenagers, the problem is likely to be *anxiety* about upcoming events or *overstimulation* from activities or caffeine. Make sure the child uses his bed only for sleeping. Have him listen to music, talk on the phone, do his homework, and do other activities somewhere else. Have him engage in relaxing activities such as reading or listening to quiet music for the hour before he goes to bed.

Meditation, autogenic training, and *self-hypnosis* have proven helpful for adults suffering from insomnia.[31–33] Autogenic training is a form of self-hypnosis in which a series of six phrases are repeated over and over: (1) My feet and legs are heavy and warm; (2) My hands and arms are heavy and warm; (3) My heartbeat is calm and regular; (4) My breathing is easy and free; (5) My belly is soft and relaxed; (6) My forehead is cool. Repeating these phrases (even silently to oneself) results in relaxation. An easy technique is to use a recording that includes calming sounds and music as well as a monotone voice and a repetitive, relaxing message that emphasizes the child's ability to fall asleep and stay asleep. This kind of recording can be used either for problems falling asleep or for problems of waking up in the middle of the night and not being able to fall asleep again. At The Ohio State University's Center for Integrative Health and Wellness, we recommended these techniques so often that we put recordings on our web page for free, easy access for our patients.

Progressive relaxation is an easy technique for older children and teenagers to learn, and it is remarkably effective in allowing a child to relax so he can fall asleep. With this technique, the child focuses on one part of the body at a time, starting with the feet. As he focuses on that part of the body, he intentionally tightens the muscles there as tight as he can, holds it for a few seconds, and then releases. Attention moves from the feet to the lower legs, upper legs, hips, belly, chest, lower back, upper back, hands, arms, neck, and head. By the time the child tightens and relaxes the facial muscles, the whole body is generally quite relaxed and ready to fall asleep.

Combining progressive relaxation and pleasant imagery has also proven useful in children who are afraid of the dark or who are reluctant to go to bed. Confronting oneself with potentially scary thoughts (such as fear of the dark, monsters, or tests) while one is very relaxed gradually drains the fear from the scary thought. This technique is called *desensitization* when used by professional therapists. It is very effective in reducing fears of the dark and nightmares.

Biofeedback training can help insomniacs learn to affect their brain wave activity. By giving patients information about their brain waves and muscle tension, biofeedback devices help patients progress from normal waking rhythms (beta activity), through light relaxation (alpha rhythm), and into the deep theta and delta states normally achieved only with deep sleep or meditation. Even young children can learn to use biofeedback to improve their sleep.[34] Biofeedback is usually taught initially in an office with help from a psychologist or licensed professional counselor or social worker, but devices and smart phone applications (apps) are available for home use so the child can practice at home on his own.

BIOCHEMICAL THERAPIES: MEDICATIONS, DIETARY SUPPLEMENTS, AND HERBAL REMEDIES

I do not recommend biochemical therapies for children who simply don't want to go to bed when their parents want them to sleep. Nor are they appropriate therapy for infants who still waken in the middle of the night to nurse or feed. The only time I recommend such therapies is for children who have undergone a significant life change or disruption in their sleep-wake cycle, such as those who have been in a natural disaster, witnessed a death, traveled across three or more time zones, have a chronic health condition that affects sleep, or are hospitalized. Even under these circumstances, biochemical remedies should supplement, not replace, parental reassurance and lifestyle strategies. Do not use any medication, herbal remedy, or dietary supplement daily for more than five days without consulting your health care professional. Tolerance to sleeping medications begins to develop within a week, and higher doses may be needed to achieve the same effect.

Medications

The medication most commonly recommended for children suffering from temporary sleep disruption is diphenhydramine (Benadryl™). Benadryl is an antihistamine whose main side effect is drowsiness. Benadryl is very safe, inexpensive, and may help your child sleep after a long flight across several time zones. Benadryl is very safe when taken as recommended. However, some who take Benadryl become more aroused and excitable. It's impossible to predict in advance if your child will react this way. Benadryl given 30 minutes before bedtime to children with troubled sleep significantly reduces the time it takes to fall asleep and decreases episodes of awakening in the middle of the night, but some children who take it feel drowsy and "out of it" the next day. These hangover effects typically disappear within a few days of regular use—not a habit I'd encourage.

MEDICATIONS USED FOR CHILDREN'S SLEEP

Antihistamines: Diphenhydramine (Benadryl™), Hydroxyzine (Vistaril™), Trimeprazine (Temaril™)

Benzodiazepines: Diazepam (Valium™), Lorazepam (Ativan™), Alprazolam (Xanax™), Temazepam (Restoril™)

Sedatives/Barbiturates: Phenobarbital, Secobarbital, Chloral Hydrate, Zolpidem (Ambien™), Zaleplon (Sonata™), and others

Vistaril™ (hydroxyzine) is a prescription antihistamine similar to Benadryl. It is frequently used to help hospitalized patients fall asleep. Another prescription antihistamine, *trimeprazine* (Temaril™), was studied in three-year-old children with sleep problems. The usual doses were not very helpful, but higher doses (three times the normal antihistamine dose) knocked the toddlers out and kept them out throughout the night. The medication is not a long-term cure, but it has few side effects. Most physicians prefer behavioral (lifestyle) changes rather than relying on antihistamines for typical toddler sleep problems.

Other sleeping medications are available with a prescription from your health care professional. Most are not approved for use in children, so pediatric prescribing is considered "off label." I do *not* recommend sleeping pills for children unless you have tried safe home remedies without success and you have your child thoroughly evaluated by a professional. Many sleeping pills make children feel groggy or hung-over the next day. Some can lead to addiction. A few can be fatal if an overdose is taken.

The most commonly prescribed sleep medications for adults are benzodiazepines and sedatives. The benzodiazepines help reset the biological clock for those suffering from jet lag. Benzodiazepines also suppress deep sleep, the time when sleep terrors are most likely to occur. Other prescription medications have been useful in treating children suffering from sleep terrors. I do not routinely recommend them because of the risk of serious side effects, because safe alternatives (such as hypnosis) are available, and because most children outgrow night terrors without any treatment.

Dietary Supplements

Our own bodies make internal sleep promoters using amino acids, vitamins, and minerals. A shortage of any of the raw ingredients can lead to sleep problems. This has led to a large industry selling dietary supplements to ensure an adequate supply for successful sleep.

SLEEP-PROMOTING SUPPLEMENTS

- Amino Acids (tryptophan; 5-HTP)
- Minerals (calcium, magnesium)
- Vitamins (B vitamins)
- Hormones (melatonin)

The amino acid, *L-tryptophan,* is a building block for the nerve transmitters serotonin and melatonin, which play important roles in regulating sleep and moods. It is an effective sleep promoter in adults, infants, and children.[35–38] It was one of the most popular supplements sold until 1989 when it was banned due to its association with the deadly disease *eosinophilia myalgia syndrome* (EMS). Before it was banned, L-tryptophan was linked to over 5,000 cases of EMS, including 27 deaths. Because all of the EMS cases could be traced to the L-tryptophan from a single manufacturer, a search was made for possible contaminants. It turns out that the problem was probably *not* due to L-tryptophan, but to a contaminant introduced in the manufacturing process. Since the problem was identified and corrected, L-tryptophan is back on the market, and typical adult doses are 400 to 1,000 milligrams an hour before bed. Breast milk contains substantially more L-tryptophan than formula. Rather than taking supplements, I recommend that you breast-feed your baby to promote healthy sleep.

Another amino acid closely related to tryptophan is *5-hydroxytryptophan* (5-HTP). Because it was never linked to EMS cases, many adults switched from tryptophan to 5-HTP supplements. With the help of an enzyme and vitamin B6, 5-HTP is broken down in the body to serotonin. Beware of combining tryptophan or 5-HTP with antidepressant medications such as selective serotonin reuptake inhibitors (SSRIs) and MAO-inhibitors.

Calcium and *magnesium* supplements are frequently recommended to adult sufferers of insomnia, sometimes in combination with melatonin supplements. However, there are no scientific studies evaluating the safety and effectiveness of such supplements as sleep promoters in children. Children generally get all the calcium they need from milk, but about one-half of American youth consume less than the recommended amounts of magnesium. If your child does not consume foods rich in calcium and magnesium, talk with your health professional about using a supplement, at least until the diet is optimized.

B vitamins are widely consumed to help manage stress. Methylated *vitamin B12* supplements have been reported to be useful in a few adults and teenagers suffering from sleep problems,[39–40] but B vitamin supplements have not been compared with other therapies as sleep aids for children. They are generally safe.

Melatonin is a hormone produced by the pineal gland, deep in the brain. Melatonin supplements are a popular over-the-counter sleep remedy, especially for adults who have to work the night shift and airline personnel who have to combat jet lag.[41] A review of several studies concluded that melatonin can improve the time it takes to fall asleep and is generally safe in doses of 0.5 to 7.5 milligrams an hour before desired bedtime.[42]

Herbal Remedies

Traditional folk medicine has long relied on herbs as sleep aids. Most, such as *chamomile,* are given as tea; some, such as *valerian,* are taken as tablets or capsules. Still others, such as *lavender,* may be included in bathwater, massage oil, or as a few drops of essential oil on a favorite pillow or blanket. An herbal "sleep pillow" filled with a combination of dried sedative herbs can be tucked under a sleepless child's regular pillow.

HERBS TRADITIONALLY USED FOR INSOMNIA

- Chamomile
- Balm mint (also known as lemon balm) and catnip
- Hops
- Kava kava
- Lavender
- Passionflower, skullcap, and valerian
- St. John's wort

Chamomile flowers are among the safest and most widely used herbal remedies for children. Chamomile is calming and was given to Peter Rabbit following a harrowing escapade in Mr. MacGregor's garden. Chamomile is widely available in commercial herbal teas in your local grocery store. It has a pleasant aroma and taste. It is easily grown and often can be found growing wild along roadsides.

Balm, also known as *lemon balm* because of its refreshing fragrance, is included in herbal tea blends used to treat insomnia, anxiety, and "nerves." It is also commonly used to calm an upset stomach. Lemon balm is often cultivated in household gardens because it attracts bees and is a tasty addition to summertime teas and salads. *Catnip* is from the same family of plants. Not only is it attractive to cats, it is also said to be soothing and sedating, allaying anxiety and preventing nightmares. There are no modern studies comparing the effectiveness of traditional herbs like chamomile, balm, or catnip to dietary supplements or medications in children with sleep problems, but the herbs have a long historical track record of safety.

Hops have been traditional herbal sedatives as least as far back as Roman times. They are genetically related to the *Cannabis* (marijuana) family of plants. Hops are traditional ingredients in European "sleep pillows," having been used by King George III to cure his insomnia. Tea made from hops can also be added to bathwater for a relaxing soak. Hops are best used soon after they are picked. Within a year after drying, they lose 90 percent of their effectiveness.

Kava kava is a traditional beverage in the Pacific Islands, where it has been used historically on social occasions to induce a carefree state of conviviality and lack of anxiety. It has become more popular in the United States and Europe as an anti-anxiety herb and to help adults with a lot on their minds to relax and fall asleep.[43,44] There are no studies evaluating it in children, and there is some potential for misuse by teens who may make the mistake of driving while under the influence of kava (yes, it can impair driving performance). Modern preparations of kava have also been linked to liver problems. I don't recommend kava for children.

Lavender flowers are often included in bedtime herbal tea blends. A few drops of essential oil of lavender can also be added to the child's evening bath or to a massage oil to help him relax and sleep.[45] Lavender has a very pleasant, soothing scent. As more hospitals adopt an integrative approach to care, lavender aromatherapy, which has been studied in several controlled trials,[46] is even making its way into inpatient and intensive care units to help patients sleep.

Passionflower is a common herbal sleep remedy that is included in several herbal tea blends. Its name is not derived from its use as an aphrodisiac, but from the flower's imagined resemblance to the crucifixion scene (the passion of Christ). There are no studies evaluating its safety or effectiveness in children.

Despite its creepy name, *skullcap* is a common ingredient in herbal bedtime teas. It is a traditional sedative that was historically used to treat rabies. A potential problem with skullcap is liver toxicity. I do not recommend it, although you often find it in commercially available teas marketed to improve sleep.

The name *valerian* comes from the Latin word *valere*, meaning to be strong or well. Valerian root was used as a sedative in ancient Greece, Rome, China, and Persia and was also used by Native Americans. Valerian decreases the length of time it takes to fall asleep and improves the quality of sleep without leaving that drowsy morning-after feeling that is common with many sleep medications.[47] In adults suffering from insomnia, it improves the quality of sleep and reduces nighttime awakening as effectively as several prescription sleep medicines. Because it improves nighttime sleep, it may reduce daytime drowsiness.[48] Valerian's major drawback is its unpleasant scent, which can be described as musky, stale, sweaty gym socks. Phew! A more serious drawback is valerian's toxic effect on the liver. There is substantial variation in the amount of the sedative compounds in different varieties and preparations of valerian. However, I have used valerian myself when I've had to deal with jet lag; and I've recommended it to adolescents having trouble falling asleep in the hospital. The combinations of valerian and lemon balm and valerian with hops are widely used in Europe, where herbal products are more commonly used and better regulated.[49]

The leaves and flowers of *St. John's wort* have been used since ancient Greece, and they remain one of the most commonly recommended herbal remedies for a variety of sleep disturbances, including nightmares. St. John's wort is widely used to treat depression and is a reasonable choice for children whose sleep is disturbed due to a recent loss, such as the death of a grandparent or pet. One potential side effect of St. John's wort is photosensitivity; that is, when a child takes St. John's wort, his skin may become more easily sunburned. Another problem is that St. John's wort may cause the body to metabolize other drugs more quickly so they are less likely to benefit the child. If your child takes any other medication, be sure to talk with your doctor before starting St. John's wort.

Combinations of the above herbs are widely available in commercial teas such as Nighty Night™, SleepyTime™, Surrender to Sleep™, Tension Tamer™, Calming Tea™, and others. This is certainly a convenient way to go because the herbs have been selected, combined, and are available in easy-to-use teabags. No matter what blend you choose, avoid herbal teas before bed if your child is prone to bed-wetting. Also *avoid stimulating herbs*: coffee, tea (with caffeine), cocoa, chocolate, cola, ephedra (Ma huang), guarana, ginseng, and mate. They can all keep your child awake long after he should be finished counting sheep.

Caution: Herbal products are not regulated by the federal government with the same degree of safety considerations as drugs are. Herbal products may not contain any of the active ingredients we associate with the herb on the label; or herbal products may be contaminated with heavy metals (such as lead, cadmium, and mercury), pesticides, herbicides, etc.; or they may *also* be "spiked" with drugs. So, herbs may cause some toxicity themselves, and commercially marketed products may present additional hazards. The most commonly reported problems involve liver damage; if your child develops diarrhea, dark urine, or pale stools while taking an herbal remedy, stop the herbs and see your doctor immediately.

BIOMECHANICAL THERAPIES: MASSAGE, SURGERY

Massage

Many parents and children enjoy a relaxing *massage* before falling asleep. In a study of adolescent psychiatric patients, a nightly back rub was more effective than watching relaxing videotapes in helping the children fall asleep. The positive effects of massage carried over into the day, helping the children manage their anxiety and depression and decreasing their levels of stress hormones.[50] Massage is safe, even for premature infants.[51] It can help them gain weight and leave the hospital faster.[52] Massage can be as simple as a back rub, stroking a child's forehead, or gently rubbing his hands or feet. Scented oils containing essential oils like lavender may also help the child relax and feel drowsy. If you feel uncertain about the best way to massage your child, check out one of the many books on massage from your local library, or schedule a visit with a local massage therapist who treats children and ask her to show you effective strategies.

Surgery

If your child suffers from *severe snoring*, consider taking him to an ear/nose/throat (ENT) doctor for an evaluation for *sleep apnea*. Sleep apnea or severe breathing obstruction is a chronic condition that affects about 10 percent of children with severe snoring. In this condition the child's adenoids and tongue block airflow during sleep, resulting in loud snoring; long, frequent pauses in breathing; and increased work to breathe that results in poor growth, daytime sleepiness, and hyperactivity. It can also lead to high blood pressure. After removing the obstruction, the child's sleep returns to normal, resulting in improved wakefulness during the day and an improved appetite, growth, and general disposition.

BIOENERGETIC THERAPIES: ACUPUNCTURE, THERAPEUTIC TOUCH, PRAYER, HOMEOPATHY

Acupuncture

While most kids (and some adults) are afraid of needles, acupuncture is actually helpful in improving sleep. An analysis of 33 studies involving over 2,000 participants showed that compared with no treatment or placebo needles, real acupuncture significantly improved sleep.[53] Be sure to ask if your acupuncturist has experience treating children. Acupuncture is most likely to be accepted by children whose sleep problems are related to chronic pain. Insurance may cover acupuncture as a treatment for pain and secondary insomnia; check with your carrier and advocate for acupuncture.

Therapeutic Touch and Reiki

Therapeutic touch has proven benefits in helping patients relax and in decreasing anxiety.[54,55] A study in nursing home patients showed that Therapeutic Touch significantly improved sleep in elderly residents.[56] Reiki can also help children relax and feel less stressed. I regularly use Therapeutic Touch and Reiki to treat hospitalized children.[57] Many fall asleep during the treatment. They become very relaxed, and several older children have been able to stop using sleeping pills.

Prayer

This chapter would not be complete without including my grandmother's favorite advice about falling asleep: *counting your blessings*. I know that the standard advice is to count sheep, and that counting sheep has worked for thousands of people. It is very boring and therefore very effective. However, counting your blessings, enumerating all the things for which you are grateful, has the added benefit of generating feelings of gratitude. This is one of the most potent anti-anxiety and antidepressant techniques I know.[58] Since we know that anxiety is a major contributor to sleep problems, my grandmother again proved she was ahead of her time.

Homeopathy

There are no scientific studies evaluating the effectiveness of homeopathic remedies as sleep aids in children, though preliminary research suggests some benefits for total sleep time, non-REM sleep, and number of nighttime awakenings.[59] Traditional homeopathic remedies for sleep problems include: *Arsenicum* (nightmares or other symptoms worse after midnight), *Belladonna* (for sudden frights), *Calcarea carbonica* and *phosphorica*, *Chamomile* (for teething or irritable infants), *Coffea cruda* (treating sleeplessness with extremely dilute concentrations of coffee, another perfect example of the homeopathic principle of like cures like), *Ignatia* (for insomnia due to grief), *Kali phosphoricum* (for night terrors), *Lycopodium*, *Natrum muriaticum* (nightmares), *Nux vomica* (for insomnia due to overindulgence in caffeine or overstimulation), *Passiflora* (for insomnia due to an overactive mind in older children and teenagers), *Pulsatilla*, *Rhus toxicum* (for restless sleep), *Silica*, and *Sulphur*. Homeopathic remedies are certainly safe and most are inexpensive, but their effectiveness for treating pediatric sleep disorders has not been established in large randomized controlled trials in children. If they've worked for you, keep using them.

WHAT I RECOMMEND FOR SLEEP PROBLEMS

Take Your Child to a Health Care Professional If

- Your child's sleep does not improve with lifestyle changes or home remedies.

- Your child's nightmares interfere with daytime activities or his ability to sleep by himself.

- Your child has other symptoms, and you are concerned that something more serious is going on (such as an ear infection, allergies, diabetes, or a bladder infection).

- You think your child may need medication (long travel, severe stress).

- You think your child may need surgery (loud snoring, or sleep apnea).

- You are becoming increasingly frustrated by your child's wakefulness.

1. To help an infant learn to go to sleep:
- Feed thirty minutes before bed; do not put infant to bed while nursing.
- Put him to sleep in the same place every night while he's still awake.
- Create a consistent bedtime routine. Put him to sleep on his back to minimize the risk of SIDS.
- Stop smoking.
- Do not rely on rice cereal to help your baby sleep through the night.
- Do not put the baby to bed with a bottle.

2. To help decrease awakenings in the middle of the night:
- Check on your child to make sure he's not hungry or in pain. Attend to his needs. Reassure him.
- Leave the room. If he still cries, recheck him at longer and longer intervals.
- Or use the scheduled waking technique.

3. For children who won't go to bed on time or those who keep getting out of bed:
- Establish a brief, consistent bedtime routine.
- Set firm limits, state consequences, and stick to them *or* let the child stay up as late as he wants; gradually move bedtime earlier by 15 minutes every few nights.
- Do not let him watch more than 90 minutes of television daily. Bedtime should not be delayed to watch "special" programs.

4. For children with nightmares:
- Prevent nightmares by minimizing your child's exposure to scary or violent movies, television, and news before bedtime.
- Reassure your child at the time he awakens.
- Briefly listen to him if he wants to describe the nightmare; discuss the nightmare with him the next day; have him record the details of the nightmare by keeping a dream journal or drawing a picture of the scary dream; give him the opportunity to rewrite the nightmare with a different ending.

5. For a child who has night terrors:
- Reassure yourself that your child is normal; most children outgrow them.
- Do not try to waken your child or to soothe him unless he wakens; do stay in the room with him in case he wakes up and wants you.

6. For the school-age children or teenagers with insomnia:
- Encourage vigorous exercise during the day, not before bed; consider yoga or tai chi.
- Remove the television from the bedroom.
- Keep the room dark and cool; try a warm bath before bed; encourage relaxing music or activity before bed; use the bed only for sleep (not for homework, TV, telephone).
- Consider mind-body strategies like CBT, self-hypnosis, autogenic training, guided imagery, and biofeedback; counting blessings; practicing meditation; progressive muscle relaxation, or other relaxation techniques.
- Avoid caffeinated beverages and food. Avoid stimulating cold medicine. Consider a glass of warm milk, peanut butter, or turkey for bedtime snacks.
- Discuss stresses during the day when coping skills are highest rather than the evening when coping skills ebb.
- Ensure adequate intake of essential nutrients including amino acids, vitamins, and minerals. Consider supplements with melatonin. Consider aromatherapy with lavender.
- Consider a back rub before bedtime to help him relax. If your child snores loudly or stops breathing in his sleep, have him evaluated for obstructive sleep apnea.
- Consider a Therapeutic Touch or Reiki treatment to help him relax. If insomnia is related to pain, consider acupuncture.

25
SORE
THROATS

Peter Phillips brought in his six-year-old daughter, Angela, first thing one cold winter morning. Angela had been ill for a day with a very sore throat, swollen, tender glands in her neck, and a fever. Her throat was so sore she couldn't even drink orange juice. Peter had been suffering from laryngitis himself; his voice was raspy and faint. He had heard there was a flesh-eating strain of strep that could kill you, so he wanted Angela treated as rapidly as possible. He also didn't want to miss a day of work, so he wanted to take her directly from the office to school.

WHAT CAUSES SORE THROATS?

Sore throats are the third most common cause for doctor visits in the United States. There are different kinds of sore throats and different causes for each kind (Table 25.1). Viruses cause most of the problems, but overuse, irritants, bacterial infections, and a variety of other things can all make for a painful throat.

It sounded as though Peter had laryngitis, and Angela had tonsillitis or pharyngitis.

Pharyngitis can be caused by viruses, bacteria, allergies, irritants, overuse, or emotions.

Only 15 to 20 percent of the time does the bacteria group A beta-hemolytic *streptococcus* cause the pharyngitis; another 45 percent of cases are caused by viruses, and the rest are a mixed bag.

Tonsillitis is usually caused by a viral or bacterial infection. It can be caused by strep.

Viruses, irritants, and overuse can also cause *laryngitis*.

CAUSES OF SORE THROATS

- Viruses: Epstein-Barr virus (infectious mononucleosis), adenovirus, cocksackievirus (hand-foot-mouth disease), herpes virus (cold sores), canker sores, others

- . Bacteria: Various kinds of *streptococcus, staphylococcus, mycoplasma, diphtheria* (rare in vaccinated people), and other bacteria
- Allergies: Dust, pollen, animals, or pollutants
- Irritants: Tobacco smoke, air pollution
- Miscellaneous: Overuse; grief, anger, and other strong emotions

The most famous sore throat *virus* is the *Epstein-Barr virus* (EBV), the cause of infectious mononucleosis, or mono. Teenagers with mono typically suffer from sore throats, swollen lymph glands, a swollen spleen (a giant way station for lymph cells in the upper left side of the belly), and fatigue. Mono usually lasts for two weeks. Unfortunately, the fatigue can last weeks or even months, causing kids to miss major amounts of school. The sore throat can be so severe and the tonsils so swollen that it is hard to swallow anything. This makes it easy for mono sufferers to become dehydrated. De-

hydration makes them feel worse and even less like drinking, a downward spiral of symptoms that is best avoided by drinking plenty of fluids throughout the illness. Mono is uncommon before school age. Despite its reputation, the "kissing disease" is rarely caught from kissing.

It didn't sound as if anyone in the Phillips family was suffering from mono.

Preschoolers' and summertime sore throats are usually caused by *adenovirus*. Adenovirus sore throats are often very painful and may be accompanied by conjunctivitis or pinkeye.

Adenovirus was not a likely cause for either of the Phillips' symptoms.

Enteroviruses cause about 10 percent of sore throats in children. *Coxsackievirus, Type I Herpes virus* (cold sores), and *canker sores* can all cause red spots or blisters in the mouth and throat. Coxsackievirus also causes tiny blisters or red spots on the palms of the hands and the soles of the feet, which is why it's sometimes called hand-foot-and-mouth disease. Children typically feel ill and complain of pain a day or two before the blisters appear and for several days afterward. *Herpes virus II, chlamydia,* and *gonorrhea* can cause sore throats in teenagers experimenting with oral sex. And *AIDS* can also cause a sore throat and other flu-like symptoms soon after infection occurs.

Group A beta-hemolytic streptococcal bacteria (strep) are responsible for only 5 percent of pharyngitis cases in adults, but up to one-third of sore throats in school-age children, especially in the winter and early spring. True strep throat is very rare in children under three years old outside of crowded day care settings. Infections are spread from one person to another by mucus. Symptoms appear within a week after exposure. About 2 to 5 percent of children carry strep bacteria without having any symptoms. Strep carriers are *not* at risk of developing the most feared consequences of

Table 25.1: Types of Sore Throats

Type of Sore Throat	Body Part Affected	Symptoms
Pharyngitis	Pharynx, back of throat	Pain, fever, redness, sometimes ulcers or red spots on palate or back of throat
Tonsillitis	Tonsils	Pain, fever, red, swollen tonsils, sometimes with pus or white spots
Laryngitis	Larynx, voice box, lower throat and upper windpipe	Hoarseness, whisper-voice, cough, sometimes pain; usually sound worse than they feel

strep infection—rheumatic fever and kidney disease—but they can infect others. Strep's symptoms typically last about three days, but can go on for a week. Strep infections are easily treated with antibiotics. Because antibiotics prevent strep's serious consequences, *strep infections are important to diagnose and treat with antibiotics*. If someone ever develops rheumatic fever, it is strongly recommended that antibiotics be taken regularly to prevent recurrences.

Other kinds of strep (*group C* and *group G beta-hemolytic streptococcus*) cause sore throats that look just like Group A strep. Infections caused by these kinds of strep also yield to antibiotics. Group C and group G strep do not lead to rheumatic fever or kidney disease, so antibiotics are not strictly necessary.

Mycoplasma pneumonia bacteria cause sore throat and "walking pneumonia" in school-age children. The symptoms caused by mycoplasma look just like those caused by strep except that mycoplasma also causes coughing, while strep rarely does. A few other bacteria can also cause sore throat symptoms; these included diphtheria, plague, syphilis, tularemia, and other nasty bugs.

Prior to the start of widespread immunization, *diphtheria* caused serious, sometimes fatal, cases of sore throat. Diphtheria is the "D" in DPT and TDaP vaccines. Thank goodness for vaccines making diphtheria a thing of the past. Prevent diphtheria from returning; make sure your child is immunized.

Allergies cause sore throats as well as itchy eyes and watery noses. The most common and easily avoided throat irritant is tobacco smoke. Ironically, in the 1950s menthol cigarettes were advertised as medically approved sore throat remedies. Now we know that nothing could be further from the truth. The menthol may be soothing, but the tobacco smoke makes things much worse. Children whose mothers smoke have more frequent and more severe sore throats.

Throat irritants include air pollution, fumes, smoke from wood stoves, dry air, mouth-breathing, and the nasal drainage that drips down the back of the throat from colds or allergies.

A sore throat secondary to *overuse* is a common affliction of cheerleaders, singers, and sports enthusiasts. Although adults may recognize that a "lump in the throat" is due to *grief, anxiety, anger, or other strong emotions*, children may not be able to interpret this feeling except to say they have a sore throat.

Peter admitted that he smoked about a pack a day. His wife had quit smoking about a month earlier and was trying to get him to quit, but he said he just didn't have time, and there was no lung cancer in his family anyway. Peter also reported that since the furnace had been on this winter, the air in the house was very dry. They had a humidifier, but it was up in the attic, gathering dust. I advised him to stop smoking now and to clean the humidifier and get it up and running. He agreed, but really wanted antibiotics for Angela, assuming she had a strep throat.

How Can You Tell a Strep Throat from Other Types of Sore Throat?

Even experienced clinicians are accurate only about 50 percent of the time when they guess whether or not strep is the culprit behind a sore throat based on symptoms and physical exam alone (Table 25.2).

Diagnosis

Most children who have other symptoms such as a runny nose or a cough do *not* have strep or mono and do *not* need any diagnostic tests. Although many doctors have tried to develop scoring tests to predict which patients need antibiotics and which ones don't, none of these clinical scoring systems is as accurate as some simple, quick lab tests.

Table 25.2: Symptoms and Types of Sore Throats

Symptoms	Strep	Viruses	Irritants	Allergies
Rapid onset of symptoms	✓			
Fever	✓	✓		
Red throat	✓	✓	✓	✓
Pus or white patches on tonsils	✓	✓		
Swollen glands in neck	✓	✓		
Trouble swallowing	✓	✓	✓	✓
Headache	✓			✓
Nausea, vomiting	✓			
Bad breath	✓			
Red, sandpapery skin rash	✓			
Hoarse voice		✓	✓	✓
Cough		✓	✓	✓
Watery or red eyes		✓		✓
Runny nose		✓		✓
Fatigue	✓	✓		✓
Under three years old		✓	✓	✓

LABORATORY TESTS FOR SORE THROATS

- Infectious Mononucleosis—blood tests
- Strep throat—throat swab for rapid test and/or culture

Blood tests are useful for diagnosing *infectious mononucleosis*. The test doesn't become positive until the disease has been present at least four or five days. A negative test in the first day or two of illness could be wrong. Children who have mono can also have a strep infection.

The only way to be absolutely sure an infection is caused by strep is with a laboratory test. This means your child will need a swab of the back of the throat. The swab is tested either with a rapid test that checks for signs of strep (like a tracker following an animal's footprints) or a throat culture that checks for the presence of the bacteria themselves (like seeing the animal directly).

A rapid strep test can be ready within minutes. If the test is positive (meaning strep is present), you can start antibiotics the same day. The bad thing about rapid tests is that a negative result doesn't necessarily mean that your child doesn't have strep. If a tracker doesn't see footprints, it doesn't mean there is no animal—just that the signs weren't visible. Rapid tests miss up to 10 to 30 percent of cases. If the rapid test is negative, many physicians perform a throat culture to make sure an infection isn't missed.

On the other hand, some doctors just get a throat culture and don't bother with rapid tests at all. Throat cultures are the gold standard for diagnosing strep throats. They are very reliable. But it takes at least 24 to 48 hours for the results to come back.

If you want to minimize your child's chances of being overtreated with antibiotics, insist that a throat culture be performed *before* filling a prescription for antibiotics. Do *not* treat your child with leftover antibiotics before getting a throat culture. You can treat your child's discomfort with analgesics or other home remedies. It takes a while for rheumatic fever or kidney disease to follow a strep throat, so antibiotics are not an emergency.

I prefer to do a rapid test, and process the test while the patient is still in the office. If the test is positive for strep, we start antibiotics that day. If the test is negative, we do a throat culture and wait for the results before starting antibiotics. With this approach, most patients who need antibiotics get them right away. Those who don't need them avoid unnecessary antibiotics (allergic reactions, upset stomachs, medication costs, and the killing of helpful bacteria in the gut—the microbiome). My only exceptions to this approach are children who have:

- Symptoms of scarlet fever as well as sore throat
- The classic symptoms of a strep infection during a strep outbreak
- A history of rheumatic fever

These children all get a throat culture to confirm the diagnosis *and* antibiotics even before the culture results are back because of their high risk of serious disease. Early treatment (within the first two days of symptoms) helps children feel better faster than delaying treatment for a day or two while waiting for the cultures to come back.

The rapid strep test for Angela was positive, so no culture was needed. Peter's rapid test and his throat culture were negative for strep.

WHY IS A STREP THROAT WORSE THAN ANY OTHER KIND OF SORE THROAT?

The two reasons strep throat is more serious than other kinds of sore throat are *rheumatic fever* and *glomerulonephritis* (kidney disease). Rheumatic fever follows strep infections about two to three weeks after the sore throat is gone. Just after World War II, before antibiotics were widely available, strep throat was followed by rheumatic fever in 2 to 3 percent of soldiers. Treatment with ten days of antibiotics reduces this risk tenfold. Treatment can be delayed as long as nine days after the first sore throat symptoms appear and still effectively prevent rheumatic fever.

Many people think rheumatic fever is a disease of the past, but outbreaks still occur. In the 1980s, outbreaks were reported in Ohio, Pennsylvania, Salt Lake City, San Diego, New York, and Nashville. Most of these outbreaks were in suburban or rural middle-class neighborhoods, and the initial illnesses were so mild that many people had not even visited their doctors. Strep comes in slightly different varieties—some of which cause very nasty infections and others of which are more benign—and these varieties shift and change in subtle patterns year to year, leading to differences in disease severity over time.

One to three weeks after some strains of strep infection, the immune reaction between the strep bacteria and the child's antibodies (immune molecules) can settle in the kidneys, resulting in a serious kidney disease known as *post-streptococcal glomerulonephritis*.

If not treated, the bacteria from a strep throat sometimes spread, leading to ear infections, lymph node infections, and abscesses in the tonsils and throat. Strep bacteria can also cause serious, life-threatening infections such as shock, septicemia, and infections of the skin, connective tissues, and muscles. These complications, though well publicized, are extremely rare. Some strains of strep are more aggressive than others. Some are also easier to spread than others. It is impossible to sort out all of these strains of strep without special laboratory tests.

WHAT'S THE BEST WAY TO TREAT A SORE THROAT?

The best treatment depends on what's causing the sore throat. Let's tour the Therapeutic Mountain to find out what works. If you want to skip to my bottom-line recommendations, flip to the end of the chapter.

LIFESTYLE THERAPIES: NUTRITION, EXERCISE, ENVIRONMENT, MIND-BODY

Nutrition

As with other infectious diseases, drinking *plenty of fluids* is the order of the day. Your child's fluid intake is sufficient if it necessitates diaper changes or visits to the bathroom every two to three hours.

One of the most time-honored sore throat remedies is the saltwater gargle: one-half teaspoon of salt (or one-quarter teaspoon of salt plus one-quarter teaspoon baking soda) in 8 to 16 ounces of warm water.

Other home remedies are:

- Hot tea with honey
- Honey mixed with roasted onion
- Honey and lemon juice (or cider vinegar) in six to eight ounces of hot water
- Apple juice and clove tea (spiced apple cider)

There are no scientific studies evaluating the effectiveness of any of these home remedies (and none are cures for strep throat or mono), but I encourage families to use home remedies that are safe, don't cost much, and seem to make everyone feel better. Remember chicken soup, too! It can help your child stay hydrated, and many children find warm broths soothing as well as nourishing.

Exercise and Rest

Rest is indicated for every infectious disease. Rest allows the body to concentrate its energy on healing rather than exercise. Teenagers suffering from mono may be fatigued for months after the acute infection has resolved. They should *not* participate in contact sports such as football or soccer for at least a month to protect their spleens. There's no point in forcing a fatigued teenager out onto the football field or gymnastics floor anyway. Continue mild, regular exercise such as yoga, tai chi, or walking to keep the muscles from becoming weak and wasted. This is not the time for wind sprints.

Rest and patience are the mainstay of therapy for children whose throats are sore from too much talking, shouting, or singing.

Peter realized that Angela needed her rest. He decided that he would take the day off to stay home with her, rent comedy movies, drink tea, and take naps.

Environment

For sore throats due to allergens and irritants, the best treatment is environmental: Avoid the irritant. Do *not* smoke and do not allow others to smoke around your child. Avoid other sources of smoke, exhaust, and chemical fumes.

Humidity is the main treatment for laryngitis. Humidity also reduces throat irritation caused by dry winter air.

I suggested Peter get the humidifier out of the attic, clean it out, and get it going.

Cold foods and drinks are also soothing to sore throats. Even kids who won't eat their normal food when they are sick can often be coaxed to try ice cream, sorbet, popsicles, or even plain ice cubes.

Mind-Body

There are no studies evaluating the effectiveness of *mind-body* techniques such as hypnosis, biofeedback, or meditation in treating children with sore throats. On the other hand, they may help those whose sore throats are due to an emotional "lump in the throat."

BIOCHEMICAL THERAPIES: MEDICATIONS, DIETARY SUPPLEMENTS, HERBAL REMEDIES

Medications

Some medications just help you live with the uncomfortable symptoms while the body does the healing work. Others attack the root problem (if the problem is a bacterial infection).

MEDICATIONS USED TO TREAT SORE THROAT SYMPTOMS

- Analgesics: acetaminophen, ibuprofen (non-prescription)
- Throat lozenges (non-prescription), antacid gargles (non-prescription)
- Antibiotics (prescription only): penicillin, erythromycin, others
- Steroids

Analgesics, for example, soothe sore throat pain, but they don't wipe out viruses or bacteria. The most widely used analgesics are *acetaminophen* (Tylenol™ and other aspirin-free pain relievers) and *ibuprofen* (Advil™ or Motrin™). *Aspirin* is not recommended because of its association with the sometimes fatal Reye's syndrome. Acetaminophen and ibuprofen are both effective pain remedies. Ibuprofen works a bit sooner and lasts a bit longer than acetaminophen and may be slightly more effective against pain. That's what I take, and that's what I give my son when he has pain.

Similarly, some *throat lozenges* containing *benzocaine* (such as Cepastat™) help numb sore throat pain, but they don't do anything to treat an underlying infection. Some children have had serious side effects from benzocaine. I do not recommend lozenges for children under four years old because of the risk that they will choke on them. Chloraseptic™ throat spray significantly reduces sore throat symptoms compared with a placebo. I use throat sprays when I have a severe sore throat, but since my son hates the taste and temporary stinging, I haven't been able to convince him that they're worth it.

Antacids (such as Maalox™) can coat and soothe not only the stomach lining but also the throat. For children with frequent mouth and throat irritation from cancer chemotherapy, canker sores, or fever blisters, I often recommend Magic Mouthwash—a combination of Tylenol™, Maalox™, and Benadryl™. This is a soothing combination of safe ingredients that can be repeated several times daily. Ask your physician to help you calculate the right dose for your child.

What Other Medications Help Ease the Pain of a Sore Throat?

Chlorhexidine is a non-prescription ingredient in several throat sprays, mouthwashes, and lozenges that can help ease the pain of a sore throat while also serving as a mild disinfectant.[1] *Benzydamine* (Difflam™) is another non-prescription pain reliever that can be gargled or sprayed on the throat.[2] Some preparations combine chlorhexidine and benzydamine; allergies and side effects are rare with short-term use of these products. Prescription *lidocaine* lozenges also reduce pain temporarily while the body heals itself; allergies are possible.[3]

What Antibiotics Are Best for Strep Throats?

Antibiotic treatment reduces strep throat symptoms, consequences, and the chance that the sufferer will pass the illness on to someone else. Antibiotics are *not* useful for treating sore throats caused by viruses, allergies, irritants, overuse, or emotions.

ANTIBIOTICS TO TREAT STREP THROAT (ALL REQUIRE A PRESCRIPTION)

- Penicillin; amoxicillin (Amoxil™); amoxicillin-clavulanic acid (Augmentin™)
- Erythromycin, azithromycin (Zithromax™), clarithromycin (Biaxin™)
- Combinations: Erythromycin-sulfisoxazole (Pediazole™)
- Cephalosporins: cefadroxil (Duricef™), cefaclor (Ceclor™), cefuroxime (Ceftin™), cephalexin (Keflex™), cefpodoxime (Vantin™), and others
- Clindamycin (Cleocin™)

Our good old friend, *penicillin* (actually derived from the *Penicillium* mold—does that make it a natural remedy?) remains the official treatment of choice for children with true strep throats. Penicillin decreases the length of symptoms by about a day. More importantly, it dramatically reduces the subsequent risk of developing rheumatic fever. Penicillin can be given as an injection, pills, or syrups. If taken by mouth, it must be given at least twice daily for at least ten days to eradicate strep bacteria. Some parents quit giving medication by the sixth day because the child is better and the need for medication seems less pressing, but penicillin is less effective in preventing rheumatic fever when taken less than ten full days. It also requires refrigeration, which is a drawback for families who are about to go on camping trips or doing extensive travel. For maximal absorption and effectiveness, penicillin should *not* be given within an hour before or two hours after mealtime—pretty inconvenient for today's busy families. I give parents the choice of whether they want their child to receive an injection or oral medication. As much as most people hate needles, I'm surprised how often families choose to "take the shot and get it over with."

Amoxicillin is a derivative of penicillin that is also inexpensive and effective. It tastes better than penicillin, and it can be given as little as once a day and still work.[4,5] Like penicillin, it requires refrigeration. If amoxicillin is given to patients with mono, they break out in a rash. This is another one of the reasons physicians don't just give antibiotics to everyone who has a sore throat. Sadly, after nearly 50 years, some strains of strep have developed resistance to penicillin and amoxicillin, forcing physicians to rely on secondary choices of antibiotics to reduce the risk of rheumatic fever.[6]

The combination of amoxicillin and clavulanic acid (Augmentin™) may have less resistance than amoxicillin alone, but it has a higher rate of side effects (upset stomach, diaper rash, thrush, etc.) and is more likely to wipe out healthy bacteria in the intestines.

For children who are allergic to penicillin, the antibiotic of choice has historically been *erythromycin*. However, more and more strains of strep have also become resistant to erythromycin.[7] It has to be given three or four times daily, and it often causes an upset stomach. About a third of patients stop taking it because of side effects. Antibiotic relatives of erythromycin, *azithromycin* (Zithromax™) and *clarithromycin* (Biaxin™), are more effective treatments for strep throat. Azithromycin is even effective with only five days of treatment. Both of these antibiotics cost more than penicillin or erythromycin—you pay more for convenience.

The *combination antibiotics* and the *cephalosporins* (such as Ceclor™, Duricef™, and Vantin™) are also more effective and expensive than penicillin, amoxicillin, or erythromycin. Some are more convenient because they only require dosing once or twice a day and they do not require refrigeration. Cefpodixime (Vantin™) taken twice daily for five days or cefuroxime (Ceftin™) taken twice daily for four days is as effective as penicillin taken three times daily for ten days. Most combination medications or cephalosporins cost three to four times as much as penicillin or erythromycin. They can also cause upset stomach and diarrhea and wipe out the healthy bowel bacteria, leading to yeast infections.

Clindamycin is a powerful antibiotic that effectively eradicates strep. It is my choice for *Strep* throats that are resistant to penicillin and cephalosporins because it also does an excellent job of clearing *Strep* from carriers who keep unconsciously reinfecting family members. Unfortunately, it also kills many of the healthy bowel bacteria, leading to diarrhea. If clindamycin is recommended for your child, be sure to stock up on extra yogurt and supplemental probiotics (such as *Bifidobacteria* and *Lactobacillus*) to replace the healthy bacteria in the intestines that have been wiped out by antibiotics.

Trimethoprim-sulfamethoxazole (Bactrim™ or Septra™), *tetracycline,* and *sulfa* drugs are *not* effective in eradicating strep from the throat. They are *not* recommended for strep throat, though they can be very helpful in other infectious illnesses.

It doesn't hurt to wait a day or two to start antibiotics, but typically the sooner treatment starts, the sooner the child will feel better. Delaying treatment for 24 to 48 hours may allow the body to build up its own antibodies (immune defenses against this and future attacks) and prevent subsequent infections. One study found that children who were treated immediately had an eight times higher risk of developing another strep infection within the next four months than children whose treatment was delayed for two days.[8] Other studies have found *no* difference in recurrence risks between those treated immediately and those whose treatment is delayed by 48 hours.[9] I do *not* withhold antibiotic therapy from a child who is sick with a laboratory confirmed strep throat. I also don't rush to treat a child who has a sore throat before appropriate tests have established a diagnosis.

Pretty much whenever I recommend antibiotics, I also recommend that the child get extra helpings of yogurt or kefir containing active cultures; also, there are a number of products containing the healthy bacteria (probiotics: *lactobacillus, bifidobacterium,* and *saccharomyces*), which normally live in the intestines. Providing probiotics helps reduce the diarrhea that often accompanies administration of potent antibiotics.[10] They are very safe, and they're natural.

After weighing all the options and considering Angela's lack of allergies and recent local patterns of antibiotic resistance, we settled on using penicillin to treat her strep. Peter also decided to stop at the grocery store on the way home to stock up on some organic yogurt to help replace the healthy bacteria killed by the antibiotic and reduce her risk of developing antibiotic-associated diarrhea.

When Can Your Child Return to School or Day Care?

Children are no longer contagious to others when they have completed 24 hours of antibiotic therapy. Even children who have received an injection should wait 24 hours before returning to school or day care to make sure the medicine has had a chance to begin eradicating the strep bacteria and reducing the risk of spreading the infection.

Do Antibiotics Always Work against Strep?

No. No antibiotic is 100 percent effective.

REASONS FOR ANTIBIOTIC FAILURE

- Failure to take medication
- Reinfection from other family member or peers
- Not a bacterial infection
- Resistant bacteria, requiring different antibiotic
- Weak immune system

The main reason antibiotics fail is that people fail to take them often enough and long enough. Even if the child takes all of her doses, if someone else in the family is infected and not treated, that person can re-infect the child. Infection can go back and forth between family members like a Ping–Pong ball. For repeated, recalcitrant infections, it's worth testing and treating every infected person in the household. Children can also get re-infected by close peers in school and day care. In some cases, a second, more powerful antibiotic is needed. Antibiotics are *not* effective against viruses; if your child's sore throat is not due to bacteria, antibiotics won't help. Some children's throats are just more susceptible to the strep bacteria. Antibiotics can't do the job alone; they require a healthy immune system to wipe out bacteria and keep them out. If antibiotics haven't worked for your child, see your health care professional to discuss alternative antibiotics or other measures to support the immune system.

Lest you think that modern medicine can only treat bacterial sore throats, here's some reassuring news. The antiviral medication *acyclovir* is very effective in reducing children's mouth and throat pain caused by cold sores if it is started within three days of the start of symptoms (which are caused by Herpes vi-

ruses).[11] If your child suffers from severe cold sores that also cause sore throat, don't just wait around hoping it will get better.

Is There a Role for Steroids in Healing Sore Throats?

Yes. Steroids can help ease recovery after tonsillectomy when the throat is very sore and there's a high rate of nausea.[12] Steroid medications are also sometimes used alone or in combination with acyclovir to treat severe sore throats caused due to infectious mononucleosis.[13,14] German pediatricians sometimes recommend steroids as part of the overall treatment for severe sore throats, but other European pediatricians do not.[15,16] Please don't ask your pediatrician for steroids to treat run of the mill sore throats.

Dietary Supplements

Vitamin C lozenges may help relieve sore throat pain, but there are no studies comparing their effectiveness to other throat lozenges or hard candy.

In an analysis of adult studies evaluating the impact of *zinc lozenges* on cold symptoms, zinc helped reduce scratchy throat by 33 percent and sore throat by 18 percent, but had no effect on fever or headache.[17] Lozenges are generally safe for children four years and older because they are less likely to choke on them than younger children. Many children, however, complain about the taste of zinc lozenges or report nausea or upset stomach with zinc supplements.[18] Until more research is done on dietary supplements as a treatment for sore throat in children, I do not recommend any specific vitamin or mineral lozenge over any other kind of lozenge. If your child uses lozenges for symptomatic relief, be sure to brush teeth to reduce the risk of cavities.

Herbal Remedies

No herbal remedy replaces antibiotics for treating strep throat. However, herbal remedies have a long history of soothing irritated, sore throats. See the box that lists some of the most commonly used herbal treatments that can help your child deal with the discomfort:

HERBAL REMEDIES TO SOOTHE SORE THROATS

- Demulcents (throat soothers): *aloe vera, horehound, licorice, mullein, slippery elm bark*
- Immune boosters: *echinacea*
- Inflammation easing: *chamomile, licorice*
- Bacteria fighters: *garlic, onions, goldenseal*
- Astringent: *gum myrrh, mullein*
- Miscellaneous: *honey, sage, thyme, oil of eucalyptus, mustard, essential oils of chamomile, lavender, and lemon*

Aloe vera is used to heal a variety of minor skin irritations, burns, canker sores, and ulcers in the mouth and throat.[19,20] There are few studies published in English evaluating its effectiveness in treating sore throats in adults or children, but pure aloe juice is generally safe, particularly if used as a mouthwash (swished and spit).

Horehound has historically been used as a demulcent and expectorant (loosens mucus) ingredient in herbal cough drops. Look for horehound as an ingredient in Ricola Natural Herb Cough Drops™.

Licorice root contains compounds (such as carbenoxolone and glycyrrhetinic acid) that ease inflammation. Licorice tea may be especially useful in treating sore throats due to cold sores or canker sores.[21] Be careful not to overdo it with licorice. Taking too much for too long can cause problems with the balance of blood salts, blood pressure, and heart function. But it is safe for short-term use and is fairly sweet, so kids do not usually object to the taste. When I have a sore throat, I enjoy a cup of tea combining licorice root and slippery elm bark.

Chamomile helps ease inflammation, reducing pain and stress. Drinking or gargling chamomile tea may help ease a painful sore throat. It has very few side effects; it's readily available, inexpensive, and very few people are allergic to it.

Mullein leaves and flowers are boiled into tea as a demulcent and astringent used for coughs, colds, and sore throats.

Slippery elm bark is another of my favorite remedies because it is so soothing to mucous membranes. It is the main ingredient in Throat Coat™ tea. Slippery elm bark lozenges (Fisherman's Friend™) are also available as sore throat soothers.

Comfrey tea is a demulcent (throat soother), but contains dangerous pyrollizidine alkaloids that can cause serious liver problems; I do not recommend comfrey tea to treat sore throats in children.

Red clover blossoms and leaves are another traditional sore throat remedy, but they can cause bleeding problems, so I do not recommend red clover for symptomatic relief of sore throats.

Echinacea. A Nebraska farmer stole the idea for this remedy from local Native Americans who had long used it as an anti-infective agent. Echinacea is widely used in Germany to treat the common cold and sore throats. In a randomized controlled trial conducted in teens and adults in Switzerland, an herbal throat spray containing echinacea and sage was as effective as the combination of chlorhexidine and lidocaine (both numbing medications).[22] Taking echinacea daily for two months may reduce the risk of catching a cold or sore throat.[23–25] Echinacea is very safe, but most herbalists recommend that it be used on a short-term basis (less than six to eight weeks at a time) only. Echinacea does not kill strep or prevent rheumatic fever.

Pelargonium sidoides is a member of the geranium family that is used to treat the common cold and sore throats. In a controlled trial conducted in the Ukraine, a standardized pelargonium extract was just as safe, but more effective than placebo for children with non-strep sore throats.[26] Pelargonium is sold in the United States as the brand names Umcka™ or Umckaloabo™.

Raw garlic and *onions* have antibacterial properties and are common remedies for respiratory infections in Europe, though they have never been specifically tested as treatments for sore throats. You might try giving your child raw garlic blended with mashed potatoes or in a blender with vegetable juices. Some of my European colleagues swear by this remedy. Having tried it once, I cannot in good conscience recommend that you try to force garlic or onion juice down the throat of any child who is already unhappy. It tastes terrible.

Goldenseal root contains the active ingredient berberine. Berberine kills many bacteria (including strep) in test tubes studies. Goldenseal also stimulates the immune system. There are no studies evaluating its effectiveness in treating children with strep throat, and it is going extinct from overharvesting; I do not recommend it.

Gum myrrh is the gum resin of trees indigenous to East Africa and Arabia. It has been used since ancient times as a perfume and incense. It is also an ingredient in mouthwashes and gargles for its fragrance and its mildly astringent properties.

Other traditional herbal remedies are compresses of *sage* and *thyme,* flannel mufflers of *Eucalyptus oil,* and *mustard* poultices. Some families add *chamomile, lavender,* and *thyme* to steam baths to soothe the mucous membranes of the nose and throat. Others massage the child's neck and throat with massage oil and added essential oils of *chamomile, lemon,* and *thyme*. I encourage parents to follow their families' safe healing traditions

In a German study involving over 200 adults with sore throats, a throat spray containing *sage* was significantly better than placebo in reducing sore throat pain.[27] Combination sprays containing sage, aloe, and other herbs are available from American suppliers. I have not used them yet myself, but after reviewing these studies, I may try one the next time I have a sore throat.

Though not technically an herb, *honey* is traditionally added to tea or lemonade to help soothe sore throats. Honey can help heal wounds, including speeding recovery and easing pain from tonsillectomy.[28,29] To avoid the risk of botulism, do not give honey to infants under one year of age. I use and recommend honey for older children to help soothe coughs and sore throats.

Peter decided he would get some slippery elm bark lozenges for Angela and make hot licorice and slippery elm bark tea with lots of honey for both of them. He wanted to talk to his wife about aloe vera and sage before purchasing them.

BIOMECHANICAL THERAPIES: MASSAGE, SPINAL MANIPULATION, SURGERY

Massage and Spinal Manipulation

There are no studies evaluating the effectiveness of either massage or spinal manipulation in treating children with sore throats. Massage can be very soothing and if done by parents, it is also inexpensive.

Surgery

At one time, *tonsillectomy* was the most common operation performed on children. Tonsillectomy is less common now. However, there still are times when tonsillectomy is beneficial. Carefully performed studies in Pittsburgh have shown that some children with recurrent infections do benefit from surgery. Those who had *seven or more* culture-proven infections with strep in one year and underwent a tonsillectomy had only half the number of infections in the next year compared with similar children who did not have surgery. The benefits only lasted a year or two. Three years later, the number of infections was identical in both groups.[30] A subsequent analysis of several studies comparing surgical and nonsurgical treatments for children with frequent strep throats (seven or more documented strep infections in the prior year; five or more in each of the past two years; or three or more in each of the past three years) also reported a modest reduction in the number of strep throats in the year following surgery among the children with the most infections prior to surgery, but the benefits waned over time as many children got better at resisting infections without any specific treatment.[31] Tonsillectomy is generally a safe procedure for children between the ages of two and twelve years, but there is a risk of bleeding and pain. In Europe, a partial tonsillectomy (called a tonsillotomy) has become a popular procedure because it saves some of the lymph tissue and reduces postoperative pain, the risk of postoperative bleeding, and the number of days of hospitalization, and has the same benefits on recurrent infection as full tonsillectomy.

This was Angela's first strep throat, so there was no need to discuss surgery.

BIO-ENERGETIC THERAPIES: ACUPUNCTURE, THERAPEUTIC TOUCH, PRAYER, HOMEOPATHY

Acupuncture

There is an acupuncture point in the muscle between the base of the thumb and first finger (known as LI4) that helps relieve pain in the head, mouth and neck. You can find it on yourself by massaging the area deeply. It is the point where deep massage feels a bit uncomfortable or odd. In a randomized trial of 60 adults with newly diagnosed sore throats, those treated with acupuncture had an immediate significant drop in their pain levels.[32] You can try massaging this point on both of your child's hands for several minutes as an experiment to see if it helps decrease his pain. But remember that acupuncture is not a cure for strep and will not prevent rheumatic fever.

Therapeutic Touch and Prayer

There are no scientific studies evaluating the effectiveness of Therapeutic Touch or prayer in treating children with sore throats. Because both are so safe, they are certainly worth trying if they are consistent with your beliefs.

Homeopathy

In a randomized, controlled trial, a complex homeopathic remedy safely helped improve pain and redness in children 6 to 12 years old with viral sore throats.[33] In another study, homeopathic treatment was no less effective than conventional care for children with colds, coughs, and sore throats.[34] Homeopathic remedies should not be used to treat strep throats. Homeopaths caution that effective remedies often make the patient feel worse before they feel better. Homeopaths recommend that if you are going to use a homeopathic remedy, avoid giving your child cough drops or other remedies that contain eucalyptus, menthol, camphor, or other strongly aromatic substances.

The most commonly recommended homeopathic remedies for sore throats are: *Aconitum, Belladonna, Bryonia alba, Hepar sulph, Ignatia, Lachesis, Lycopodium, Mercurius vivus, Nux vomica, Phytolacca, Pulsatilla, Silica,* and *Sulphur.*

WHAT I RECOMMEND FOR SORE THROATS

PREVENTING SORE THROATS

1. *Lifestyle—environment.* Minimize throat irritants. Do not smoke and do not allow others to smoke around your child. Avoid allergens. Minimize your child's exposure to other children with sore throats. Wash your hands and encourage your child to wash his hands to prevent spreading viruses and bacteria that cause sore throats.

2. *Lifestyle—exercise.* Avoid overusing the throat.

3. *Biochemical—medications.* Have your child immunized against diphtheria.

4. *Biochemical—herbal remedies.* Consider using echinacea daily for two months during cold season to prevent colds and sore throats.

See Your Health Care Professional If You Suspect Strep or If Your Child

- Is not better within forty-eight hours of starting treatment

- Has trouble swallowing her own saliva or starts drooling

- Develops signs of dehydration (such as fewer than four wet diapers within twenty-four hours or no tears when crying)

- Has severe pain that interferes with sleeping or drinking

- Has trouble breathing

TREATING SORE THROATS

1. *Biochemical—medications.* For sore throat pain, try analgesic medications, anesthetic throat sprays, or (for children over four years old) lozenges. If your child has a strep infection, I recommend a full course of antibiotics as prescribed by your health professional. Steroid medications can be useful in recovery from tonsillectomies and severe cases of infectious mononucleosis.

2. *Lifestyle—nutrition.* Encourage your child to drink plenty of fluids. Saltwater gargles may be helpful for children who are old enough to gargle. Yogurt and kefir may help replace healthy bacteria wiped out by antibiotics.

3. *Biochemical—herbs.* Try sipping hot tea with honey (slippery elm bark, licorice, or sage tea). *No honey* for infants less than one year of age. Consider an herbal throat spray or gargle containing aloe, licorice, and/or sage. Consider aloe vera for sore throats due to canker sores.

4. *Lifestyle—environment.* Offer cold non-carbonated drinks, popsicles, sorbet, ice cream, ice cubes, or frozen yogurt. Humidifiers may help those with laryngitis.

5. *Lifestyle—exercise.* Rest.

6. *Biomechanical—surgery.* If your child has had many recurrent strep throats, having her tonsils out may be helpful.

7. *Lifestyle—mind-body.* Relaxation and stress management strategies may help those whose painful "lump in the throat" is due in part to stress or strong emotions.

8. *Biofield—acupuncture.* Acupuncture can help ease sore throat pain, but is not a cure for strep throat.

9. *Biofield—homeopathy.* Consider homeopathic remedies to relieve sore throat pain, but not as the only treatment for strep throats.

26
VOMITING AND NAUSEA

Jack Burnside called me about his 18-month-old daughter, Patrice, who had started vomiting the day before. She had thrown up about three times, most recently in the car on the way to the grocery to get some home remedies his mother had given him when he was as a child to settle his stomach—cola and chicken soup. Jack had stopped giving her milk, and started giving her clear liquids such as flat soda. His wife was concerned that their four-month-old baby, Anna, was also spitting up after nursing. He had several questions:

- What caused the girls' vomiting?
- Should he bring in Patrice, Anna, or both for evaluations and treatment?
- What were the safest and most effective remedies for vomiting?

WHAT CAUSES VOMITING?

Vomiting is a distinctly uncomfortable experience, but it can also help eliminate toxins from the body before they do serious damage. Vomiting is generally preceded by nausea, the feeling that one is about to vomit. Children who are nauseated generally avoid food, although most will still sip liquids if they are offered. Children vomit more easily than adults.

Gastroenteritis (also known as the intestinal flu) is the most common cause of nausea, vomiting, diarrhea, and abdominal pain in children. Technically, this illness is not the "flu" because it is not caused by the influenza virus, but this doesn't stop health professionals and parents alike from calling it the "flu." Gastroenteritis is normally caused by a viral infection and is over within 48 hours.

Patrice's symptoms sounded as if they were due to plain old gastroenteritis. We agreed that Jack would try some simple home remedies and would bring her in the next day if she was still vomiting or developed any signs of dehydration.

In addition to gastroenteritis, which is usually caused by viruses, there are several other common causes of nausea and vomiting.

COMMON CAUSES OF VOMITING

- Gastroenteritis (intestinal flu)
- Spitting up (regurgitation or gastroesophageal reflux, GER)
- Food poisoning
- Swallowed phlegm dripping into the throat from the nose or sinuses
- Overeating, food intolerance, or food allergies
- Side effect of severe coughing
- Side effect of other treatments (chemotherapy, other medications, herbs, vitamins, surgery)
- Motion sickness
- Pregnancy
- Symptom of other illnesses (e.g., appendicitis, strep throat, bladder infection, migraine headache, hepatitis, and others)

All babies *spit up* milk from time to time. In a Chicago area study of nearly 1,000 babies, reflux was reported in 50 percent of infants less than three months old, and climbed to 67 percent (two-thirds) at four months. Most babies had outgrown the problem by the time they were seven to eight months old. The milk that is spit up can come from the mouth, the back of the mouth, the throat, the esophagus, or the stomach. If it comes from the esophagus or stomach, it is called *gastroesophageal reflux.* Regurgitation is common in babies who are fed too much too quickly. Sometimes the baby swallows what she regurgitates, and sometimes she spits it out.

Usually it looks like the baby is spitting up a lot, but appearances can be deceiving. If you are concerned about how much milk is being spit up, try this simple test. Take a measuring tablespoon of milk and spill it on the floor, an old rag, or a towel. You'll probably be as surprised as I was the first time I tried this experiment at how much milk it looks like even when you know perfectly well that it's only one tablespoon.

Gastroesophageal reflux can cause heartburn pain, though it most often poses more of a laundry problem than a medical problem. Severe, persistent reflux can cause inadequate nutrition and poor growth if the baby is not keeping down enough milk to grow. Severe reflux can cause irritability, poor growth, and breathing problems such as coughing or turning blue. If your baby is spitting up so much that she's not growing or she appears to be in pain, irritable, or having breathing problems, take her for a professional evaluation.

Anna had been drinking avidly. She was gaining weight normally and had no other symptoms. She was so eager to eat that she seldom paused to burp. She was so tired by her vigorous feeding that she often fell asleep immediately after eating; when she was laid down, she spit up undigested milk. When Jack and his wife actually measured the amount she was spitting up, it was only about one-and-a-half teaspoons. Anna didn't need any medical therapy, just a bit more burping.

If severe vomiting occurs in the first few days of life, it may be due to *intestinal obstruction,* which requires immediate professional evaluation and treatment. If the baby or child develops a yellow tinge to her eyes or skin along with vomiting, take her for an evaluation for liver problems like *hepatitis*.

Vomiting can be a useful reaction for *eliminating tainted food* (food poisoning) before it is absorbed into the system. If your child is vomiting because of food poisoning, it is better not to suppress the symptoms, but to allow Nature to rid the body of the poisons.

Children also vomit if they *swallow a lot of phlegm*. Children suffering from colds, ear infections, bronchitis, and particularly strep throat and sinus infections are prone to vomiting because the excessive phlegm irritates the stomach. The best way to reduce this kind of vomiting is to treat the underlying condition and to remove extra phlegm with frequent nose blowing, a bulb syringe, or nasal aspirator.

Overeating frequently causes indigestion, heartburn, and nausea, but it rarely causes vomiting in adults. Children seem to be more sensitive than adults to this manifestation of overindulgence. A trip to the ballpark or circus, complete with hot dogs, candy, soda pop, and so on, sometimes ends unpleasantly in a rushed trip to the bathroom. *Certain foods*, such as green apples, have a well-deserved reputation for causing nausea and stomachaches.

Food allergies, such as *cow milk* and *egg allergy* can also appear as vomiting. However, vomiting is seldom the only sign of a food intolerance. There may be gas, rashes, swelling, headaches, or other symptoms, too.

Severe coughing can trigger vomiting. Vomiting triggers a reflex that actually relieves the coughing spasm. In the old days, ipecac (the medicine used to induce vomiting in case of accidental poisoning) was used to treat children with asthmatic coughs. Nowadays effective asthma medicines don't necessarily cause vomiting.

Speaking of ipecac, many common *medications, herbs, vitamins,* and even the anesthetics used during *surgery* induce nausea and vomiting. *Aspirin* and *ibuprofen* commonly cause a queasy stomach. The antibiotic *erythryromycin* and its relatives, *clarithromycin* and *azithromycin,* frequently cause nausea and vomiting, although smaller doses are sometimes used to help speed stomach emptying. *Chemotherapy* for cancer is notorious for causing severe nausea and vomiting. Several *Chinese herbs*, notably *chuanwu* and *caowu* (the roots of Aconitum, used as anti-inflammatory therapy) and *bajiaolian* (used to treat snake bite, tumors, and swollen lymph glands) have caused serious nausea and vomiting. *African herbal remedies* have also been reported to have serious side effects, including vomiting and dehydration. Herbs high in *tannins* (such as uva ursi and even black tea) can also cause nausea. *Vitamin A* overdoses can cause vomiting, headaches, and other serious neurologic problems. *Zinc* can also cause nausea, and overdoses can lead to vomiting.

Anyone who has had major *surgery* can tell you that one of the worst parts of an operation is feeling sick to your stomach right after you wake up from the anesthesia. Depending on the type of operation, anywhere from 30 to 80 percent of children experience nausea and vomiting within 24 hours of surgery that requires anesthesia.

Children are also more susceptible than adults to *motion sickness*. Motion sickness can be brought on by the motion of a car, boat, airplane, or amusement park ride. Exhaust fumes and overexcitement make motion sickness worse.

Other conditions also include vomiting and nausea as key symptoms. Morning sickness is a classic early sign of *pregnancy*—an all too common condition in American teenagers. About 50 percent of women have nausea and vomiting in early pregnancy, and an additional 25 percent have nausea alone. During pregnancy, nausea can be triggered by specific odors, fumes, and foods; the risk can be reduced by an adequate intake of B vitamins.[1]

If *constipation* continues long enough, the child may become nauseated, and may even vomit. Children suffering from *migraines* often experience nausea and vomiting as cardinal symptoms of their "sick headaches." Abdominal pain, nausea, and vomiting are also symptoms of severe *lead poisoning*.

Other serious causes of vomiting include appendicitis, bladder or kidney infections, and meningitis. Rare genetic diseases such as hereditary fructose intolerance or severe vitamin B12 deficiency can also cause vomiting. Vomiting is also one of the signs of liver disease. Children who have suffered a *head injury* often vomit once or twice within the first hour following the injury.

TAKE YOUR CHILD TO BE EVALUATED IF SHE:

- Spits up more than a tablespoon of milk with each feeding
- Is losing weight or not gaining weight as well as predicted
- Is less than ten weeks old and has very forceful (projectile) vomiting
- Is vomiting even small amounts (less than a tablespoon) of water
- Has violent retching
- Vomits up bile (greenish fluid) or blood or coffee-ground looking material
- Has a temperature over 100.5°F if she is less than two months old
- Has a temperature over 103.9°F at any age
- Has severe abdominal pain, especially the lower right side (could be appendicitis)
- Vomits more than three times or more than one hour following a head injury
- Vomits continuously for more than 24 hours
- Seems uninterested in drinking or appears to be dehydrated
- Develops a yellow tinge to her eyes or skin
- Has symptoms or appearance that cause you concern

DEHYDRATION

Dehydration can occur within a day, especially if your child has diarrhea as well as vomiting. If your child has any of these signs, she needs to be evaluated by a health care professional to determine the cause of the vomiting and the potential need for fluids by vein (intravenous).

SIGNS OF DEHYDRATION

If your child:

- Voids (pees/urinates) less than four times per day or less than half of what is normal for her
- Doesn't have tears when she cries
- Loses weight
- Has sunken eyes or a sunken fontanel (soft spot on baby's head)
- Has dry lips and tongue or stringy saliva
- Has hands and feet that are much cooler than her arms and legs

WHAT CAN BE DONE TO PREVENT VOMITING?

Prevent *food poisoning* by keep hot foods hot and cold foods cold. Cook all meat, poultry, and eggs thoroughly to kill bacteria that cause vomiting and diarrhea. Wash all poultry products well before cooking. Wash the cutting board, knives and other utensils, and your hands after preparing poultry. Do not leave foods prepared with mayonnaise (such as chicken or tuna salad) out of the refrigerator for more than an hour before eating them. Do not let your child eat cookie dough or cake batter made with raw eggs that have not been cooked.

Avoid *motion sickness* by having the child sit in the front seat, keeping a window open, and encouraging her to look out at the horizon rather than reading.

Avoid spreading infectious *gastroenteritis* by frequently washing your hands and your child's hands. Treat underlying illnesses promptly. Keep your child's nose cleaned out when she has a cold or sinus infection to prevent her from swallowing lots of phlegm.

Prevent spitting up by slowing feedings, frequent burping, thickening feedings with rice cereal, or keeping your child upright for 20 to 30 minutes after feeding.

I suggested that Jack and his wife try slowing down Anna's feedings, having her burp at least twice each feeding and once afterward. I also suggested that they keep her upright for half an hour after she fed to give the milk a chance to pass on through the stomach, making it less likely that it would come back up when she lay down. Anna continued to grow well and stopped spitting up when she was about six months old.

WHAT IS THE BEST WAY TO TREAT VOMITING?

The best way to treat vomiting depends on what is causing it. The treatments for motion sickness are not necessarily the same as for food poisoning or vomiting after surgery. Let's tour the Therapeutic Mountain to find out what works for different kinds of vomiting. If you want to skip to my bottom-line recommendations, flip to the end of the chapter.

LIFESTYLE THERAPIES: NUTRITION, EXERCISE, ENVIRONMENT, MIND-BODY

Nutrition

Giving *small amounts* of *clear fluids frequently* is the *mainstay* of treatment for vomiting because it's important to prevent dehydration. "Small amounts" means one to two tablespoons at a time. Frequently means every five to fifteen minutes or so. To give your child's stomach a chance to settle, wait for 15 to 30 minutes after a vomiting spell before trying again. *Avoid large volumes* at one time. Often a child who vomits after four ounces of soda guzzled quickly can easily tolerate a tablespoon repeated every five minutes over an hour, which is equivalent to the same four ounces.

To replace fluid losses, especially if your child has diarrhea as well as vomiting, use a commercially prepared rehydration fluid (such as Pedialyte™) or make your own.

HOMEMADE REHYDRATION SOLUTION

One quart of clean water
One-half teaspoon salt, or one-quarter
  teaspoons salt + one-quarter teaspoon
  salt substitute (potassium)
Eight teaspoons of sugar or one cup of
  infant rice cereal

Use measuring spoons to measure precisely. Mix thoroughly. If you prefer to avoid straight sugar, you can substitute rice cereal. The sugar or carbohydrate helps the body absorb the water and salt. Don't omit it.

A traditional Japanese home remedy for upset stomach is *miso soup.* Some children love it, but others really don't care for the flavor. You can find miso soup in most health food stores. This is one of my favorite home remedies for an upset stomach. A more American version is chicken bouillon or chicken broth. Both contain salt.

If your baby is *breast-feeding*, continue to do so, even if the baby spits up every time she nurses. If she becomes uninterested in breast-feeding or has fewer wet diapers than usual, she needs to be evaluated professionally for dehydration and possible additional therapy.

Avoid solid food until the vomiting has stopped for at least two to four hours. Most children won't be hungry for solids until their stomachs have settled a bit anyway. Start cautiously with small amounts. Resume with bland foods such as rice, bananas, dry toast, dry cereal, teething biscuits, or crackers. Start with small amounts given frequently rather than one large meal. For at least a day or two, avoid greasy foods such as meats, fried eggs, or anything else that is fried. If your child is prone to motion sickness, give her a few crackers or some toast before leaving on a trip. An empty stomach can actually make her nausea worse than having a little something on board.

Avoid very cold foods and beverages as these can be hard on the stomach.

Avoid very sweet beverages. Lower sugar solutions (6 percent glucose) are better absorbed and cause less nausea than higher sugar beverages (12 percent glucose). If your child craves fruit juices, consider diluting them with water and offering them along with a non-fried salty snack. Or alternate salty beverages like miso or chicken broth with diluted fruit juice.

Tannins are chemical compounds found in many foods, including tea, wine, sorghum, barley, cocoa, and some berries. Tannins give food and beverage a dry, astringent quality. Avoid foods and beverages rich in tannins when your child is nauseated. Consuming tannin-rich foods like tea on an empty stomach causes nausea in some people.

Avoid giving *coffee* to a nauseated child or teenager. Though it can promote downward movement through the colon and help correct constipation, it can also worsen reflux and heartburn.[2]

Thickening feedings may help minimize spitting up in babies with mild to moderate reflux.[3–5] You can thicken feedings by adding one to two tablespoons of infant rice cereal to each ounce of formula or purchase pre-thickened formula such as Enfamil AR™. Commercial thickeners may contain carob or locust bean gum, tapioca, and other compounds.[6] Thickened feedings are most effective in combination with keeping the baby upright for 20 to 30 minutes following a feeding.

If your child has chronic problems with vomiting or serious spitting up, consider having her evaluated for a possible allergy to cow's milk. In one study of children who had been referred to gastroenterology specialists for reflux, 41 percent turned out to be allergic to cow's milk, which is the main ingredient in most formulas.[7–9] Some children will improve markedly when they are switched to soy, and some will require a more elemental formula. See your doctor if you think this is the problem.

Exercise

Although exercise helps build strong muscles, lungs, and heart, it has some side effects on the intestines. Most long distance runners are familiar with runner's diarrhea. Vomiting is also common after long runs or intense bicycle rides, possibly because the blood that normally supplies the stomach and intestines is diverted away to supply the muscles, interfering with normal gut function. Exercise also sometimes induces gastroesophageal reflux, resulting in heartburn and nausea.[10] Your child is not likely to suffer these symptoms unless she is exercising very vigorously as part of a track or cross-country team. If your child does have heartburn with exercise, consider cutting back on the intensity of her workout. Light exercise actually encourages the stomach to empty properly. Adequate training minimizes diarrhea, cramping, and nausea associated with intense exercise.

As with most illnesses, rest is beneficial. If your child is tired, encourage her to rest. Try lying down with her to read a story or listen to quiet, soothing music.

Environment

Avoid stale, stuffy air. Do not let your child become chilled, but do keep fresh, cool air flowing around your child's face.

Keep an empty receptacle (a pot, bowl, or basin) near your child's bed or wherever she is resting so that if the urge strikes, she does not need to run all the way to the bathroom and worry about losing it on the way.

Aromatherapy is the use of aromatized essential oils to achieve health goals. An increasing number of hospitals use aromatherapy to help patients manage symptoms like nausea and insomnia. In a review of nine controlled trials, aromatherapy using isopropyl alcohol was more effective than saltwater placebo in reducing nauea.[11] In another study, a combination of peppermint, ginger, lavender, and spearmint was more effective than placebo in reducing postoperative nausea.[12] Similarly, combining a mind-body approach (controlled, slow, deep breathing) with peppermint aromatherapy offered significant benefits in reducing postoperative nausea and vomiting.[13] On the other hand, one study found that peppermint aromatherapy was much less effective than medication and no more effective than placebo

in reducing postoperative nausea in women;[14] similarly, another study found no benefits of ginger aromatherapy for nausea secondary to chemotherapy.[15] Aromatherapy is certainly safe (if allergens are avoided), and I readily tolerate its use, but I cannot recommend a specific type or dose of aromatherapy to treat nausea in children until additional research clarifies its benefits.

Some children are soothed by a cool, damp cloth placed on the forehead and over the eyes or the back of the neck.

Gently clean your child's face after she vomits.

If she is old enough, have her swish some water in her mouth and spit it out afterward to clean the taste of vomit out of her mouth.

Mind-Body

Hypnosis can be helpful in several situations: reducing intractable vomiting; for children struggling with the severe nausea and vomiting due to cancer chemotherapy; for cutting back on the gagging children sometimes have when trying to swallow pills; and for eliminating habitual vomiting.[16–20] *Desensitization therapy, relaxation therapy,* and *biofeedback* have also proven useful in treating air sickness in pilots and nausea in patients undergoing chemotherapy.[21,22] If your child has recurrent vomiting from chemotherapy or motion sickness, it is probably worthwhile to consult a hypnotherapist. If your child has a simple case of food poisoning or gastroenteritis, the vomiting will probably be over by the time you get an appointment to see a psychologist for hypnotherapy or biofeedback.

For most people, there is something vaguely shameful about vomiting. It is important to reassure your child that you love her and will stay with her, that it's okay to vomit if she needs to, and that vomiting will help clear the toxins from her system.

BIOCHEMICAL THERAPIES: MEDICATIONS, DIETARY SUPPLEMENTS, AND HERBAL REMEDIES

Medications

Because vomiting is often the body's way of eliminating toxins from the system, and because young children are more prone to suffering side effects, medications to suppress nausea and vomiting are *not* recommended for children under two years old. If your young child is undergoing chemotherapy or requires some other therapy that causes nausea, ask your health professional about safe treatments to minimize symptoms.

NON-PRESCRIPTION ANTI-NAUSEA MEDICATIONS

- Meclizine (Antivert™; Bonine™)
- Diphenhydramine (Benadryl™)
- Dimenhydrinate (Dramamine™)
- Cyclizine (Marezine™)
- Doxylamine (Unisom Sleep tablets™)
- Emetrol™; Coca-Cola™ syrup
- Pepto Bismol™

Antivert™ (Meclizine) is commonly used to prevent and treat *motion sickness*. It starts working within one hour and lasts for 12 to 24 hours. It is not recommended for children younger than 12 years old. It can cause drowsiness, restlessness, blurred vision, and a drop in blood pressure. It should not be taken if the teenager is planning to drive or operate dangerous equipment. Overdoses can cause seizures. Although it sounds scary, meclizine is actually safe enough to be used during pregnancy to treat morning sickness.

Benadryl™ (diphenhydramine) is one of the most commonly used medications for motion sickness, nausea, and vomiting. It is often used in combination with more powerful drugs to prevent nausea due to chemotherapy. Benadryl is fairly safe even for young children. It causes drowsiness, increased thirst, and dry mouth in most children, and irritability and excitement in a few.

Dramamine™ (dimenhydrinate) is the classic medication used to prevent motion sickness. It is chemically very similar to Benadryl and has similar side effects: drowsiness or irritability, confusion, and dry mouth. It is not approved for use in children less than two years old, but it is the number one choice for treating motion sickness in children between 2 and 12 years old.

Cyclizine (Marezine™) is approved for children of all ages. It starts working within 30 to 60 minutes. However, after the death of an Alaskan teenager was attributed to cyclizine abuse, enthusiasm for this drug has waned.

Doxylamine is an effective nausea remedy; its main side effect is sleepiness, and it is usually marketed as a non-prescription sleep aid.

Emetrol™, *Nausetrol*™, and *Naus-A-Way*™ are all carbohydrate (sugar) solutions with phosphoric acid. Although they are marketed as anti-nausea medications, they are actually pretty similar to *Coca-Cola*™ *syrup*. There is no scientific evidence that any of these medications are any more effective than flat soft drinks in reducing nausea and vomiting. None have as much scientific evidence as ginger (see the section on Herbal Remedies). In addition to non-prescription nausea remedies, a number of prescription medications have proven effective for severe nausea associated with chemotherapy, surgery, or other conditions.

PRESCRIPTION MEDICATIONS FOR VOMITING

- Prochlorperazine (Compazine™), Promethazine (Phenergan™)
- Hydroxyzine (Atarax™, Vistaril™)
- Metoclopramide (Reglan™)

- Trimethobenzamide (Tigan™)
- Ondansetron (Zofran™)
- Scopolamine (Transderm Scop™)
- Others

Compazine™ and *Phenergan*™ are chemically related treatments for vomiting due to other medications and surgery. Compazine is a second line treatment for pregnancy-related nausea (following vitamin B6 and doxylamine) and one of the first-line treatments for other kinds of nausea because it is safe and available as a tablet that can dissolve between the cheek and teeth. These medications can be very sedating; Compazine is less sedating than promethazine. Phenergan™ helps prevent motion sickness. Neither is typically used for children less than two years old.

Hydroxyzine is an antihistamine like diphenhydramine. It is a sedative that also helps control nausea and vomiting as well as hives and other allergic reactions. Tigan™ is commonly used in hospitalized adults and is approved for use in children as young as two years old. Vistaril™ is often given in combination with other anti-nausea medications. Hydroxzyine should not be used during pregnancy. Side effects include sleepiness, dizziness, blurred vision, dry mouth, confusion, and tremor. Don't allow your teenager to use hydroxyzine before driving.

Reglan™ is less commonly used to treat gastroesophageal reflux and, in combination with other medications, to prevent chemotherapy-induced vomiting. It should *not* be used in babies who simply spit up a bit after feeding if they are growing well and do not have other problems. Reglan is a third-line treatment for pregnancy-related nausea. It has become less popular since a growing number of side effects have been reported. There is a black box warning about the potential for a side effect called "tardive dyskinesia"—involuntary movements of the face, jaw, and tongue, which may not resolve even after treatment is stopped.

Ondansetron (Zofran™) is a powerful medicine used to help prevent nausea and vomiting in patients undergoing chemotherapy.

Transdermal scopolamine is chemically related to poisonous belladonna plant; however, it is a useful treatment for motion sickness in teenagers and adults.[23] Scoplamine is available as tablets, capsules, oral solutions, and an adhesive patch to prevent motion sickness. Because younger children are more susceptible to its side effects (dry mouth, drowsiness, blurred vision), it is not recommended for children under 12 years old.

Other medications that treat serious nausea and vomiting in hospitalized patients are steroids (methylprednisolone or Medrol™), sedatives, and droperidol. Droperidol has fallen out of favor since it can cause cardiac arrhythmias. Olanzapine is a medication primarily used to treat schizophrenia, but has been tested as an anti-nausea remedy for children undergoing chemotherapy.[24] These medicines are sometimes combined with others to enhance their anti-emetic (anti-vomiting) effect. These combinations have the risk of severe sedation and other side effects. They should only be used under the supervision of a physician with extensive experience in treating children.

Medications are not indicated for the vast majority of children suffering from gastroenteritis, spitting up, or food poisoning. Neither Patrice nor Anna needed medication for their symptoms.

Dietary Supplements

Vitamin B6 (pyridoxine) has proven effective in both preventing and managing the nausea and vomiting of pregnancy, and it's the usual first-line recommendation because it is so safe and effective. The dose is 10 to 25 milligrams every eight hours until symptoms are controlled. It is also combined with the medication doxylamine (Unisom™ or Diclectin™).

Pyridoxine has also been used to treat the nausea caused by radiation therapy.[25] It is not helpful in treating gastroenteritis-associated nausea. Because pyridoxine is generally safe, you can try giving your child 10 milligrams of vitamin B6 an hour before traveling to help minimize motion sickness.

Probiotics can also help reduce the risk of nausea associated with stress and with antibiotics.[26] For example, for children treated with antibiotics to improve gastritis due to *Helicobacter pylori* infection, probiotics help reduce nausea and vomiting, improve medication adherence, and improve eradication of the harmful bacteria.[27–29]

Herbal Remedies

Ginger is my favorite home remedy to prevent and treat nausea. I recommend it regularly to friends who are pregnant and to patients who require chemotherapy or those recovering from surgery. I keep some candied ginger stocked in my office and at home as part of my first aid kit. I take it with me on airplanes, too. Research has shown it to be helpful in treating the morning sickness of pregnancy, seasickness, and nausea caused by chemotherapy.[30,31] In a study of postoperative nausea, ginger was as effective as prescription Reglan™.[32] Ginger is very safe. The total dose of powdered ginger root is typically divided into four doses over the course of a day (for an adult, a one gram daily total). The tea can be sipped throughout the day. I usually combine ginger and peppermint tea, sometimes adding a little chamomile as well to the recipe.

GINGER TEA

One quart of water
Three to five slices (about two inches chopped) of ginger root

Simmer together for 10 to 20 minutes. Tea can be poured over peppermint leaves or chamomile flowers and allowed to steep for another 10 minutes. Sweeten as desired and sip as needed.

You can also try ginger soda, but check to make sure the ginger ale you buy contains real ginger and not just ginger flavoring. You can also add freshly grated ginger (one-eighth teaspoon at a time) to your child's juice, applesauce, or hot cereal. Some teenagers prefer to chew on candied ginger or candies with ginger flavoring. Younger children sometimes complain that ginger is too spicy, so I usually don't recommend it for kids less than six years old.

Other traditional tummy-settling teas contain *chamomile* and *peppermint*. You can find this combination under several brand names (such as Celestial Seasonings' Grandma's Tummy Mint™ tea). Sometimes *catnip* and other members of the *mint* family are added or substituted for peppermint. Common additions include *anise, dill, fennel, lemon balm, lime flowers,* and *meadowsweet*. Some herbalists recommend bitter herbs (such as *gentian*) to treat stomach upsets and stimulate appetites. These herbs don't taste very good, and your child may resist them. The combination of *chamomile, vervain, licorice,* and *balm mint* has proven useful in treating infants suffering from colic, a common newborn condition often thought to be due in part to an upset stomach.[33]

This is the combination that Jack chose to try: ginger, mint, chamomile, and lemon balm for Patrice. He also decided to change from cola to real ginger ale. We agreed that he would alternate ginger ale with chicken broth so that Patrice would get the balance of salt (from chicken broth) and sugar (from ginger ale) that her body needed.

Castor oil packs. For mild gastroenteritis, you may want to try an old folk remedy which was frequently recommended by Edgar Cayce: soak a piece of flannel cloth in *castor* oil and lay the cloth over your child's abdomen. Cover with a towel, then with a hot water bottle or heating pad set to low, and let the child rest for 30 to 60 minutes before gently rinsing off with a baking soda and water solution. There are no scientific studies evaluating this remedy for nausea, but it is safe and inexpensive. Do not drink the castor oil—it is a potent laxative.

Other common herbal remedies that have less scientific support in treating gastroenteritis include basil, clove tea, cinnamon tea, goldenseal (*Hydrastis canadensis*), and barberry (*Berberis vulgaris*). Because goldenseal is going extinct from overharvesting, herbalists seldom recommend it anymore. The Japanese herbal remedy *rikkunshito* has been tested in preliminary studies as a remedy for regurgitation and reflux in babies, but additional research is needed before it becomes a standard remedy in the United States.[34]

BIOMECHANICAL THERAPIES: MASSAGE, SURGERY

Massage

Massaging the belly could trigger more nausea in sensitive children, and I do not recommend it as a treatment for routine gastroenteritis, regurgitation, or reflux.[35] One study tested the effects of massage and healing touch in comparison with therapeutic presence alone or standard care for Minnesota cancer patients; in this study, both treatments were helpful for pain, fatigue, and mood, but not for nausea.[36] However, in another study, massage therapy significantly reduced nausea and vomiting in children undergoing chemotherapy.[37] If your child is undergoing chemotherapy, ask if her nurse can provide massage or if there is a massage therapist on staff who can help address your child's symptoms.

Surgery

Surgery is unnecessary to treat the flu, but it can be lifesaving if the vomiting is due to appendicitis. If your child has severe abdominal pain, especially if it localizes on the right, lower side where the appendix is located, please seek medical care immediately. Surgeons are probably the most skilled physicians around when it comes to figuring out belly pain and vomiting.

BIOENERGETIC THERAPIES: ACUPUNCTURE, THERAPEUTIC TOUCH/REIKI, HOMEOPATHY

Acupuncture

There are numerous studies evaluating the effectiveness of *acupuncture* and *acupressure* in treating nausea and vomiting.[38] The point most often used is Pericardium 6 (P6), located about one inch above the wrist crease, between the two tendons leading to the palm. This is about where your watch clasp falls on your wrist.

In over 40 comparison studies, acupuncture therapy given before surgery significantly reduced postoperative nausea and vomiting in adults as effectively as anti-emetic medications.[39] Real acupuncture or acupressure is more effective than sham (placebo) acupuncture in preventing postsurgical nausea and reducing the need for anti-nausea medication.[40,41] Applying pressure to the P6 point intermittently following the initial acupuncture treatment prolongs its anti-emetic effect for up to 24 hours.[42] Acupressure bands are extremely safe. If your child requires surgery, ask about preventive acupuncture and use wristbands that stimulate the P6 points (SeaBands™) after surgery to help reduce anesthesia-related nausea.

Acupuncture and acupressure treatments at the P6 point have also proved useful in treating *morning sickness*. At least six comparison studies have shown that daily pressure (using wristbands) at the P6 point was more effective than sham or placebo acupressure (pressure at another point) in reducing nausea.[43] Acupressure wristbands reduce not just the nausea, but also the anxiety, depression, and other emotional discomforts that accompany morning sickness. Acupressure limited to the P6 points is safe during pregnancy.[44]

Stimulation of the P6 point has even proved effective in treating nausea due to *cancer chemotherapy in both adults and children*.[45,46] Acupressure using SeaBands™ or Reliefbands™ also effectively addresses chemotherapy-related nausea.[47] Like medications, acupuncture improves symptoms, but does not cure the underlying cause. Treatment may need to be repeated every few hours. Or children can use SeaBands for hours or even days to help reduce nausea.

A systematic analysis of over 142 randomized trials concluded that when treatment is limited to pressure at the P6 point or more extensive treatment is provided by a well-trained, licensed health professional, acupuncture is extremely safe for children and low in side effects compared with anti-emetic medications.[48] I recommend acupressure with Sea-Bands for most patients I see with ongoing nausea or vomiting such as children with cancer who need chemotherapy. Consider keeping a pair of SeaBands in your first aid kit.

Therapeutic Touch/Reiki/Hands on Healing

Healing Touch is a bio-field therapy similar to Therapeutic Touch and Reiki. In a study at Wake Forest, Healing Touch helped relieve fatigue and nausea in adults being treated in the hospital for acute leukemia.[49] Similarly, Healing Touch and Healing Harp music were helpful in reducing pain, anxiety, and nausea in hospitalized patients in a Minnesota study.[50] I have used Reiki and Therapeutic Touch several times with patients who had severe nausea, and patients told me it was very helpful; a few of them actually turned down their powerful nausea medicines, saying they were no longer necessary. Additional studies would be helpful in determining which kinds of hands-on healing are most helpful for different kinds of nausea.

Homeopathy

As expected from the homeopathic principle of "like cures like," all of these remedies actually *cause* vomiting if taken in higher, non-homeopathic doses: *Aconitum* (Aconite), *Antimonium tartaricum* (tartar emetic), *Arsenicum album* (arsenic), *Bryonia alba* (white bryony), *China officianalis* (Cinchona officianalis), *Ferrum metallicum* (iron), *Ipecac, Moonseed* (Cocculus) *Nux vomica* (poison nut), *Petroleum, Phosphorus, Pulsatilla,* and *Tabacum* (Nicotiana tabacum). In one study, a compound homeopathic formula, Cocculine™ (containing Cocculus, Nux Vomica, Petroleum, and Tabacum) was tested in a randomized trial of over 400 patients receiving chemotherapy for breast cancer who were also receiving anti-emetic medications; in this study, homeopathy offered no additional advantage over placebo.[51] Homeopathic remedies are safe and low cost, so I tolerate their use, but they are not a replacement for adequate hydration or other therapies with proven effectiveness, and I cannot recommend a specific brand or dose until more research has established their effectiveness in children.

WHAT I RECOMMEND TO TREAT VOMITING

Take Your Baby or Child to Be Evaluated If She

- Spits up more than a tablespoon of milk with each feeding
- Is losing weight or not gaining weight as well as predicted
- Is less than ten weeks old and has very forceful (projectile) vomiting
- Is vomiting when she drinks even small amounts (less than a tablespoon) of water
- Has violent retching
- Has a yellowish tinge to her skin or eyes (jaundice)
- Vomits up bile (greenish fluid) or blood or coffee-ground looking material
- Has a temperature over 100.5°F if she is less than two months old
- Has a temperature of over 103.9°F at any age
- Has severe abdominal pain, especially in the lower right side (could be appendicitis)
- Vomits more than three times an hour following a head injury
- Vomits continuously for more than 24 hours
- Seems uninterested in drinking or appears to be dehydrated
- Has symptoms or appearance that cause you concern

FOR BABIES WHO SPIT UP (OR HAVE REFLUX)

1. *Lifestyle—nutrition.* Small, frequent feedings, frequent burping, consider thickening feedings by adding 1–2 tablespoons of dry rice cereal to each ounce of formula.

2. *Lifestyle—environment.* Keep the baby's head elevated while feeding and for at least thirty minutes afterward.

FOR GASTROENTERITIS

1. *Lifestyle—nutrition.* Small amounts of rehydration fluid, ginger soda, miso soup, broth, rice water, or herbal tea containing ginger, chamomile, lemon balm, and mint sipped frequently; avoid solid foods and milk until the child has had at least two to four consecutive hours without vomiting; start with bananas, crackers, toast, and rice. Avoid very cold or very sweet foods.

2. *Biochemical—medications.* Avoid them unless you have consulted a health care professional.

FOR MOTION SICKNESS

1. *Lifestyle—environment.* Fresh air (keep the window open or stay on desk); ride in the front seat; focus on the horizon (count telephone poles or other distant objects); don't let the child read or focus on objects close up inside the car or boat.

2. *Biochemical—herbs.* Consider ginger supplements (ginger capsules, tea, candied ginger, or ginger soda).

3. *Biochemical—nutritional supplements.* Consider pyridoxine (B6) supplements—10 milligrams an hour or two before the ride.

4. *Lifestyle—nutrition.* Eat a light snack before traveling so the stomach isn't completely empty.

5. *Bioenergetic—acupuncture.* Consider acupressure or massage to the P6 point on the inside of the wrist. Elastic bands that rub this point are widely available under the trade name, SeaBands. I frequently recommend them.

6. *Biochemical—medications.* Consider medications such as Benadryl™ or Dramamine™ before the trip. If home remedies are unsuccessful, see your physician for stronger prescription medications.

7. *Lifestyle—mind-body.* If home remedies have not worked, consider professional hypnotherapy to prevent severe, recurrent motion sickness.

FOR MORNING SICKNESS

1. *Biochemical—nutritional supplements.* Vitamin B6 (pyridoxine): 10–25 milligrams one to three times daily.

2. *Biochemical—herbs.* Ginger: tablets, soda, or freshly grated.

3. *Bioenergetic—acupuncture.* Acupuncture or acupressure at the P6 point (one inch above the wrist crease between the tendons leading to the palm of the hand). Try SeaBands.

4. *Lifestyle—mind-body.* If home remedies have not worked, consider professional hypnotherapy, relaxation therapy, or biofeedback.

For vomiting due to other causes, including surgery and chemotherapy, ask your physician and consider non-pharmacological strategies like ginger, hypnosis, and acupressure.

27
WARTS

Nick Evans brought his six-year-old daughter, Amy, to be treated for warts. Amy had had a few warts on her knees for the last two months, but over the last week or so she had developed four new warts on her hands. The warts didn't hurt, but Amy kept picking at them and biting them, making them bleed. Nick had heard about having warts burned or frozen off, but he wondered if there wasn't some less drastic but still effective treatment.

WHAT ARE WARTS?

Warts are small, usually painless, dry, bumpy, firm growths in the skin. Most are raised and have a rough surface. Warts can be skin color, pink, tan, yellow, gray, black, or brown. Warts often contain brown dots and do not contain the whorls and ridges of fingerprints. They are usually harmless, but most people don't like the look of them and try to make them go away. Warts on the feet (plantar warts) grow into the skin on the soles of the feet and look flat; plantar warts can be painful. Warts can be itchy. If the top is cut off, warts bleed because tiny blood vessels grow into their cen-

ter. This chapter covers common warts on the hands and feet, but not genital warts, which are a sexually transmitted disease.

WHAT CAUSES WARTS?

Frogs and toads don't! There are several kinds of warts, and all are caused by human papilloma viruses (HPV). There are two kinds of warts that most often affect kids: *common warts* (usually found on the hands, face, knees, or elbows) and *plantar warts* (painful warts on the soles of the feet). Over the course of a lifetime, 75 percent of people develop warts. Common warts seldom appear before the age of two or after the age of forty. People with serious problems with their immune system can develop very extensive warts.

Warts are spread from person to person, but they are not very contagious. However, athletes who use communal showers or go barefoot in the locker room are slightly more likely to catch plantar warts. If your child takes a gym class or plays a team sport that involves locker rooms and showers, have her wear flip-flops, sandals, or other shower shoes to prevent direct contact between her feet and the floor, which may be harboring wart viruses.

WHAT IS THE BEST WAY TO TREAT WARTS?

Most warts go away by themselves without any therapy other than patience. In medical terms, patience is sometimes called "tincture of time." Interestingly, kids with eczema have fewer warts than other kids. A person with a healthy immune system gets rid of 80 percent of warts within two to three years.

Big warts and those that have persisted for at least twelve months are unlikely to go away by themselves. They might need treatment. Even when warts do go away, they often recur. With apologies to Tom Sawyer fans, spunk water (from an old tree stump) and dead cats have no proven efficacy in treating warts. Let's tour the Therapeutic Mountain to find out what does work, starting with a healthy lifestyle; if you want to skip to my bottom-line recommendations, flip to the end of the chapter.

LIFESTYLE THERAPIES: ENVIRONMENT, MIND-BODY

Environment

File it. For raised warts on the hands or elbows, you may want to try filing them with a nail file right after a bath or shower when the skin is soft. Do not use the same nail file on any other part of your body; you don't want to spread the wart virus.

Pad it. For plantar warts on the bottom of the feet, you may want to use special foot cushions to ease the pain while the wart heals. These are available over-the-counter in most grocery stores and pharmacies in the United States. Avoid tight shoes. Wear clean socks and wash the feet daily.

Keep warts in the *dark.* Cover the wart tightly with a piece of adhesive tape, duct tape, moleskin, or a Band-Aid. Keep the wart covered for a week. Remove the tape to check the wart, wash it well, and let it air out. Re-cover it for another week and take another peek. Many warts disappear in six weeks with this simple cover-up treatment. This technique seems to prompt the immune system to attack the wart, perhaps by increasing the temperature or humidity at the site. Several pediatricians I know say that this is their favorite wart remedy because it is so inexpensive and free of side effects. One study showed that it was more effective than cryotherapy (freezing the wart).[1] We like to call duct tape the MacGyver cure for warts.

A 1992 study showed that by *heating* warts to −45 to 50°C (about 110 to 120°F) for thirty to sixty seconds, warts disappeared twice as well as in an untreated group.[2] These results have been confirmed in more recent research using a slightly lower temperature (44°C) for a longer period (30 minutes).[3] Technically, this is called hyperthermia therapy. You can try soaking warts in hot water (careful not to scald the child!) for five to thirty minutes, three times a week; warts that respond to this treatment usually begin to melt within three weeks of the hot treatment.

Several severe environmental strategies are used as surgical approaches to eliminating warts—freezing, burning, lasers, zapping, and so on. (see the surgery section).

Mind/Body

Hypnosis can be very effective in getting rid of warts. In one study ten patients with warts on both the right and left sides of their body were given the hypnotic suggestion that the warts would disappear from just one side. Nine out of ten patients had warts disappear just on that one side (without any other treatment); the tenth lost warts on both sides.[4] Children have especially good imaginations, and hypnosis, guided imagery, or suggestion can work very well in healing their warts. Hypnosis can sometimes cure warts that have failed to respond to caustic conventional medical therapies. Involving the child in making the wart go away is more effective than simply telling him that a placebo will work.[5] Hypnosis is even effective in children whose immune systems are suppressed by illness or chemotherapy.[6]

I often use the technique my childhood doctor used with me: I tell the child that she can tell her warts to go away and that when she comes back in a month, I will give her a quarter for every wart she has made disappear. You can help your child imagine the wart disappearing, feeling the warmth or tingling as the blood carries it away. The more vivid the imagery, the more effective is the treatment. Hypnotherapy or suggestion work best if the child is *convinced* that the wart will go away, *involved* in the cure, and something is *done* to the wart.

No one knows whether hypnosis works by shrinking the blood vessels supplying the wart or by stimulating the immune system to fight the wart-causing virus. This would be a fascinating area for additional research because, either way, there are important implications for other, more serious illnesses such as cancer.

Nick planned to have Amy soak her warts in hot water for five minutes daily, file the top of the wart, and then cover it with an adhesive bandage. Amy received a sticker each day she soaked, filed using her special wart file, and covered it up. If she accumulated five stickers in a week, she could pick out her own movie to watch on the weekend. I asked Amy to return to the clinic in a month so we could count the warts. For every wart she got rid of, I'd pay her a quarter. We repeated this routine once. Within three months all of Amy's warts had disappeared, without a single scar, and she was very proud of herself. What might we have tried if a few warts remained?

BIOCHEMICAL THERAPIES: MEDICATIONS, HERBS, NUTRITIONAL SUPPLEMENTS

Medications

The British call topical wart remedies (those applied directly to the skin) "wart paint." The most commonly used topical medication to treat warts is *salicylic acid* (the active ingredient in Compound W® and Duofilm®). It is available as a liquid, gel, or in "plasters" that can be placed directly on the wart. Plasters contain higher concentrations of medication and are the least messy to apply.

For best results with topical wart medicine:

1. Soak the wart in warm water to soften it.
2. Peel away loose skin on top of the wart or pare down plantar warts until you get to the tender part; you don't need to make it bleed!
3. Apply the wart medicine. Don't let it get on healthy skin.
4. Cover the medication with a dressing such as a Band-Aid or piece of tape.
5. Repeat daily until the wart is gone—up to three months.

These medications may cause some burning or blistering on healthy skin. Be careful.

You may want to apply a protective emollient, such as Vaseline, to the healthy skin around the wart to prevent the wart paint from irritating the healthy skin.

Professionally applied medications may include other *acids, bleomycin, cantharidin* (a very irritating chemical extracted from the blister beetle known as "Spanish fly"), *podophyllum, tretinoin, liquid nitrogen* (to freeze the wart), or *silver nitrate*. Silver nitrate doesn't burn and is safe even on babies' tender skin. Most of the others on this list are pretty caustic and should only be applied by an experienced professional who is used to working with squirming children.

Some families and physicians use *acne remedies* on warts: *benzoyl peroxide, retinoic acid*, and other agents. These medications typically cause some irritation, which may help recruit the body's own immune system to eliminate the wart. There are not enough studies for me to recommend these therapies, but I have no objection to trying them if common sense precautions are used.

Because our own immune system rids us of wart viruses all the time, some clever immunologists are trying to boost the immune response to warts with a variety of medications. The most common immunologic approach to treat stubborn warts is injecting the wart with *interferon*. Others have tried applying agents that cause severe irritations, such as *squaric acid dibutylester* (SADBE) or *dinitrochlorobenzene* (DBCB) to the wart, which seems to help in about two-thirds of kids with resistant warts. The wart sufferer or the parent who applies the remedy may develop widespread allergic reactions if the application is not very precise, so it is definitely not a first-line therapy. I do not use these remedies, but your dermatologist might try them if all else fails. Another medication used by dermatologists for patients with extensive warts whose immune system is suppressed is *imiquimod* (Aldara®) cream, which is also used to treat actinic keratosis and basal cell carcinoma.[7]

There are a lot of weird wart remedies, and not all of them come from fictional characters. For example, someone noticed that patients who suffered from both peptic ulcers and warts and started taking *cimetidine* for their ulcers soon saw their warts melt away. A few dermatologists started recommending cimetidine (which is taken by mouth) to patients who had a variety of different kinds of warts all over their skin. Based on several case reports of remarkable success with cimetidine,[8] at least seven comparison studies have been done, but overall, the results are disappointing.[9] I do not recommend cimetidine as a wart remedy.

Nick decided to treat Amy's unresolved warts with Compound W®.

Herbs and Dietary Supplements

Some folks advocate the application of vitamin A, vitamin E oil, or a paste of baking soda and castor oil to the wart several times daily until the wart is gone. Some crush a vitamin C tablet or an aspirin, mix it with water to make a paste, and apply it to the wart. Others advise rubbing the wart with a cut raw potato and burying the potato under a tree in the backyard under a full moon. Vitamins A, C, and E, castor oil, crushed aspirin, and raw potatoes have not been scientifically studied as wart treatments, but they are inexpensive and safe. Similarly, common herbal remedies applied to warts include the juice of dandelion stalks, tincture of thuja (white cedar), milk of bitterroot herbs, milkweed juice, tea tree oil, and fresh elderberry juice. Based on scientific studies, I can't recommend them, but based on their safety, I tolerate them as long as common sense precautions are followed.

BIOMECHANICAL THERAPIES: SURGERY

Surgery

The most common surgical wart treatment is freezing the wart with liquid nitrogen. The treatment consists of soaking the wart (to soften it), paring it down (especially for warts on the feet), and then applying liquid nitrogen until the wart freezes. The liquid nitrogen is applied with a cotton swab or sprayed directly on the wart. Treatments are repeated every one to three weeks until the wart is gone. This works well for warts on the hand but is often painful and ineffective for warts on the sole of the foot.

Usually when parents ask to have the wart frozen, they have tried a nonprescription treatment for a week or so, and the wart is still there. Be patient. It typically takes at least a month for a home remedy to work. I generally reserve liquid nitrogen or other surgical treatments for children who have already tried a mind-body technique (suggestion or hypnosis), home therapy with wart plaster, and covering the wart with tape. If the warts survive three to six months of this treatment, I'll go ahead and use liquid nitrogen. It is inexpensive and has few side effects, but liquid nitrogen on normal skin (if the child moves during the application) can be quite painful.

Warts can also be electrically cauterized (zapped), lasered, or cut off. I don't recommend these treatments unless all else has failed, because they are painful and can leave scars.

BIOENERGETIC THERAPIES: ACUPUNCTURE, THERAPEUTIC TOUCH/REIKI, HOMEOPATHY

A few Chinese studies suggest that *acupuncture* is useful in treating large flat warts, but it is not widely used for this purpose in the United States;[10] acupuncture is safe, and if you'd like to try it, please let me know of your experience.

I haven't tried treating warts with *Therapeutic Touch* or *Reiki,* and I'm not aware of any reports that it helps. These kinds of healing therapies are unlikely to be harmful, and it might be worthwhile to combine them with guided imagery and a few simple home remedies.

In two controlled trials, *homeopathy* was no more effective than a placebo in treating warts.[11,12] On the other hand, homeopathy is very safe. I have no objection to trying homeopathy along with home remedies to treat warts, but I don't suggest you rush out and buy new remedies unless you are already a believer in homeopathy or you're a scientist studying this safe experiment.

WHAT I RECOMMEND FOR TREATING WARTS

Be patient. Most warts go away by themselves within 24 months.

Seek Professional Help If

- The wart becomes infected (red, painful, swollen).

- The wart is on the face or genital area.

- A wart on the foot is so painful that it interferes with walking.

- The wart doesn't disappear within six months with home therapy.

- You are concerned about the wart.

---

1. *Lifestyle—environment.* Clean it, soak it in hot water, file it, and cover it. Many warts go away in six to eight weeks with this remedy alone. For plantar warts, consider using an over-the-counter pad to cushion the wart while it heals.

2. *Lifestyle—mind-body.* Tell your child that her strong immune system can make the warts go away. Help her develop vivid imagery about her immune system fighting the warts, the warts melting away, and how the skin might feel tingly or warm as it's working. Consider an incentive for her success (a quarter, a sticker, or a book for every wart she eliminates).

3. *Biochemical—medications.* If environmental and mind-body therapies don't get rid of all the warts, try a non-prescription salicylic acid preparation such as Compound W® or Duofilm®. Apply the medication daily, and keep it covered with duct tape, adhesive tape, or a Band-Aid. If this doesn't work, your doctor may suggest other topical medicines. When all else fails …

4. *Biomechanical—surgery.* If warts have not disappeared with three to six months of home therapy, see your doctor about freezing the wart with liquid nitrogen or zapping it with a laser.

References

Chapter 3: Acne

1. Sardana K, Sharma RC, Sarkar R. Seasonal variation in acne vulgaris—myth or reality. *J Dermatol*, 2002;29:484–8.

2. Fulton JE, Jr., Plewig G, Kligman AM. Effect of chocolate on acne vulgaris. *JAMA* 1969;210:2071–4.

3. Hoehn GH. Acne and diet. *Cutis*, 1966;2:389–94.

4. Wortis J. Common acne and insulin hypoglycemia. *JAMA*, 1937;108:971.

5. Semon H. Some observations on the sugar metabolism In acne vulgaris, and its treatment by insulin. *British Journal of Dermatology*, 1940;52:123–8.

6. Bettley R. The treatment of acne vulgaris with tolbutamide. *British Journal of Dermatology*, 1961;73:149–51.

7. McCarty M. High-chromium yeast for acne? *Med Hypotheses*, 1984;14:307–10.

8. Reynolds RC, Lee S, Choi JY, et al. Effect of the glycemic index of carbohydrates on acne vulgaris. *Nutrients*, 2010;2:1060–72.

9. Mahmood SN, Bowe WP. Diet and acne update: carbohydrates emerge as the main culprit. *Journal of Drugs in Dermatology*, 2014;13:428–35.

10. Adebamowo CA, Spiegelman D, Berkey CS, et al. Milk consumption and acne in teenaged boys. *J Am Acad Dermatol*, 2008;58:787–93.

11. Grossi E, Cazzaniga S, Crotti S, et al. The constellation of dietary factors in adolescent acne: a semantic connectivity map approach. *J Eur Acad Dermatol Venereol*, 2014.

12. Motley RJ, Finlay AY. How much disability is caused by acne? *Clin Exp Dermatol*, 1989;14:194–8.

13. Hamilton FL, Car J, Lyons C, Car M, Layton A, Majeed A. Laser and other light therapies for the treatment of acne vulgaris: systematic review. *Br J Dermatol*, 2009;160:1273–85.

14. Snider BL, Dieteman DF. Letter: Pyridoxine therapy for premenstrual acne flare. *Archives of Dermatology*,1974;110:130–1.

15. Michaelsson G, Edqvist LE. Erythrocyte glutathione peroxidase activity in acne vulgaris and the effect of selenium and vitamin E treatment. *Acta Derm Venereol*, 1984;64:9–14.

16. Michaelsson G, Juhlin L, Vahlquist A. Effects of oral zinc and vitamin A in acne. *Archives of Dermatology*, 1977;113:31–6.

17. Weimar VM, Puhl SC, Smith WH, tenBroeke JE. Zinc sulfate in acne vulgaris. *Archives of Dermatology*, 1978;114:1776–8.

18. Sharquie KE, Noaimi AA, Al-Salih MM. Topical therapy of acne vulgaris using 2% tea lotion in comparison with 5% zinc sulphate solution. *Saudi Med J* 2008;29:1757–61.

19. Jung JY, Kwon HH, Hong JS, et al. Effect of dietary supplementation with omega-3 fatty acid and gamma-linolenic acid on acne vulgaris: a randomised, double-blind, controlled trial. *Acta Derm Venereol* 2014;94:521–5.

20. Kim J, Ko Y, Park YK, Kim NI, Ha WK, Cho Y. Dietary effect of lactoferrin-enriched fermented milk on skin surface lipid and clinical improvement of acne vulgaris. *Nutrition*, 2010;26:902–9.

21. Mueller EA, Trapp S, Frentzel A, Kirch W, Brantl V. Efficacy and tolerability of oral lactoferrin supplementation in mild to moderate acne vulgaris: an exploratory study. *Curr Med Res Opin*, 2011;27:793–7.

22. Elsaie ML, Abdelhamid MF, Elsaaiee LT, Emam HM. The efficacy of topical 2% green tea lotion in mild-to-moderate acne vul-

garis. *Journal of Drugs in Dermatology : JDD* 2009;8:358–64.

23. Fabbrocini G, Staibano S, De Rosa G, et al. Resveratrol-containing gel for the treatment of acne vulgaris: a single-blind, vehicle-controlled, pilot study. *Am J Clin Dermatol* 2011;12:133–41.

24. Hammer KA. Treatment of acne with tea tree oil (melaleuca) products: a review of efficacy, tolerability and potential modes of action. *International Journal of Antimicrobial Agents* 2015;45:106–10.

25. Bassett IB, Pannowitz DL, Barnetson RS. A comparative study of tea-tree oil versus benzoylperoxide in the treatment of acne. *Med J Aust* 1990;153:455–8.

26. Enshaieh S, Jooya A, Siadat AH, Iraji F. The efficacy of 5% topical tea tree oil gel in mild to moderate acne vulgaris: a randomized, double-blind placebo-controlled study. *Indian J Dermatol Venereol Leprol* 2007;73:22–5.

27. Stoughton RB, Leyden JJ. Efficacy of 4 percent chlorhexidine gluconate skin cleanser in the treatment of acne vulgaris. *Cutis* 1987;39:551–3.

28. Bowe W, Patel NB, Logan AC. Acne vulgaris, probiotics and the gut-brain-skin axis: from anecdote to translational medicine. *Beneficial Microbes* 2014;5:185–99.

29. Cao HJ, Yang GY, Wang YY, Liu JP. Acupoint stimulation for acne: a systematic review of randomized controlled trials. *Medical Acupuncture* 2013;25:173–94.

Chapter 4: ADHD

1. Elder TE. The importance of relative standards in ADHD diagnoses: evidence based on exact birth dates. *J Health Econ* 2010;29:641–56.

2. Sellers R, Maughan B, Pickles A, Thapar A, Collishaw S. Trends in parent- and teacher-rated emotional, conduct and ADHD problems and their impact in prepubertal children in Great Britain: 1999–2008. *J Child Psychol Psychiatry* 2015;56:49–57.

3. Feldman HM, Reiff MI. Clinical practice. Attention deficit-hyperactivity disorder in children and adolescents. *N Engl J Med* 2014;370:838–46.

4. Chirdkiatgumchai V, Xiao H, Fredstrom BK, et al. National trends in psychotropic medication use in young children: 1994–2009. *Pediatrics* 2013;132:615–23.

5. Owens JA, Maxim R, Nobile C, McGuinn M, Msall M. Parental and self-report of sleep in children with attention-deficit/hyperactivity disorder. *Arch Pediatr Adolesc Med* 2000;154:549–55.

6. Lee DH, Jacobs DR, Porta M. Association of serum concentrations of persistent organic pollutants with the prevalence of learning disability and attention deficit disorder. *J Epidemiol Community Health* 2007;61:591–596.

7. Ribas-Fito N, Cardo E, Sala M, et al. Breast-feeding, exposure to organochlorine compounds, and neurodevelopment in infants. *Pediatrics* 2003;111:e580–5.

8. Eskenazi B, Marks AR, Bradman A, et al. In utero exposure to dichlorodiphenyltrichloroethane (DDT) and dichlorodiphenyldichloroethylene (DDE) and neurodevelopment among young Mexican American children. *Pediatrics* 2006;118:233–41.

9. Bouchard MF, Bellinger DC, Wright RO, Weisskopf MG. Attention-deficit/hyperactivity disorder and urinary metabolites of organophosphate pesticides. *Pediatrics* 2010;125:e1270–7.

10. Fried PA, O'Connell CM, Watkinson B. 60- and 72-month follow-up of children prenatally exposed to marijuana, cigarettes, and alcohol: cognitive and language assessment. *J Dev Behav Pediatr* 1992;13:383–91.

11. Rydelius PA. Children of alcoholic fathers. Their social adjustment and their health status over 20 years. *Acta Paediatr Scand Suppl* 1981;286:1–89.

12. Byrd RS, Weitzman ML. Predictors of early grade retention among children in the United States. *Pediatrics* 1994;93(3):481–7.

13. Swing EL, Gentile DA, Anderson CA, Walsh DA. Television and video game exposure and the development of attention problems. *Pediatrics* 2010;126:214–21.

14. Bernstein GA, Carroll ME, Crosby RD, Perwien

AR, Go FS, Benowitz NL. Caffeine effects on learning, performance, and anxiety in normal school-age children. *J Am Acad Child Adolesc Psychiatry* 1994;33:407–15.

15. Rapoport JL. Diet and hyperactivity. *Nutr Rev* 1986;44 Suppl:158–62.

16. Harley JP, Ray RS, Tomasi L, et al. Hyperkinesis and food additives: testing the Feingold hypothesis. *Pediatrics* 1978;61:818–28.

17. Conners CK, Goyette CH, Southwick DA, Lees JM, Andrulonis PA. Food additives and hyperkinesis: a controlled double-blind experiment. *Pediatrics* 1976;58:154–66.

18. Egger J, Carter CM, Graham PJ, Gumley D, Soothill JF. Controlled trial of oligoantigenic treatment in the hyperkinetic syndrome. *Lancet* 1985;1:540–5.

19. Carter CM, Urbanowicz M, Hemsley R, et al. Effects of a few food diet in attention deficit disorder. *Arch Dis Child* 1993;69:564–8.

20. Bateman B, Warner JO, Hutchinson E, et al. The effects of a double blind, placebo controlled, artificial food colourings and benzoate preservative challenge on hyperactivity in a general population sample of preschool children. *Arch Dis Child* 2004;89:506–11.

21. McCann D, Barrett A, Cooper A, et al. Food additives and hyperactive behaviour in 3-year-old and 8/9-year-old children in the community: a randomised, double-blinded, placebo-controlled trial. *Lancet* 2007;370:1560–7.

22. Pelsser LM, Frankena K, Toorman J, et al. Effects of a restricted elimination diet on the behaviour of children with attention-deficit hyperactivity disorder (INCA study): a randomised controlled trial. *Lancet* 2011;377:494–503.

23. Biederman J, Faraone S, Milberger S, et al. Predictors of persistence and remission of ADHD into adolescence: results from a four-year prospective follow-up study. *J Am Acad Child Adolesc Psychiatry* 1996;35:343–51.

24. Stubberfield T, Parry T. Utilization of alternative therapies in attention-deficit hyperactivity disorder. *J Paediatr Child Health* 1999;35:450–3.

25. Glasser HE. *Transforming the difficult child: the nurtured heart approach*: Nurtured Heart Publications; 1999.

26. Edmonds CJ, Jeffes B. Does having a drink help you think? 6–7-year-old children show improvements in cognitive performance from baseline to test after having a drink of water. *Appetite* 2009;53:469–72.

27. Benton D, Maconie A, Williams C. The influence of the glycaemic load of breakfast on the behaviour of children in school. *Physiol Behav* 2007;92:717–24.

28. Cooper SB, Bandelow S, Nute ML, Morris JG, Nevill ME. Breakfast glycaemic index and cognitive function in adolescent school children. *Br J Nutr* 2012;107:1823–32.

29. Swanson JM, Kinsbourne M. Food dyes impair performance of hyperactive children on a laboratory learning test. *Science* 1980;207:1485–7.

30. Schuchardt JP, Huss M, Stauss-Grabo M, Hahn A. Significance of long-chain polyunsaturated fatty acids (PUFAs) for the development and behaviour of children. *Eur J Pediatr* 2010;169:149–64.

31. Cortese S, Angriman M, Lecendreux M, Konofal E. Iron and attention deficit/hyperactivity disorder: What is the empirical evidence so far? A systematic review of the literature. *Expert Rev Neurother* 2012;12:1227–40.

32. Kozielec T, Starobrat-Hermelin B. Assessment of magnesium levels in children with attention deficit hyperactivity disorder (ADHD). *Magnes Res* 1997;10:143–8.

33. Arnold LE, Bozzolo H, Hollway J, et al. Serum zinc correlates with parent- and teacher-rated inattention in children with attention-deficit/hyperactivity disorder. *J Child Adolesc Psychopharmacol* 2005;15:628–36.

34. Loffredo DA, Omizo M, Hammett VL. Group relaxation training and parental involvement with hyperactive boys. *J Learn Disabil* 1984;17:210–13.

35. Klein PS. Responses of hyperactive and normal children to variations in tempo of background of music. *Isr J Psychiatry Relat Sci* 1981;18:157–66.

36. Denkowski KM, Denkowski GC, Omizo MM. The effects of EMG-assisted relaxation training on the academic performance, locus of

control, and self-esteem of hyperactive boys. Biofeedback Self Regul 1983;8:363–75.

37. Potashkin BD, Beckles N. Relative efficacy of Ritalin and biofeedback treatments in the management of hyperactivity. *Biofeedback Self Regul* 1990;15:305–15.

38. Lee SW. Biofeedback as a treatment for childhood hyperactivity: a critical review of the literature. *Psychol Rep* 1991;68:163–92.

39. Monastra VJ, Lubar JF, Linden M, et al. Assessing attention deficit hyperactivity disorder via quantitative electroencephalography: an initial validation study. *Neuropsychology* 1999;13:424–33.

40. Steiner NJ, Frenette EC, Rene KM, Brennan RT, Perrin EC. In-school neurofeedback training for ADHD: sustained improvements from a randomized control trial. *Pediatrics* 2014;133:483–92.

41. Monastra VJ, Monastra DM, George S. The effects of stimulant therapy, EEG biofeedback, and parenting style on the primary symptoms of attention-deficit/hyperactivity disorder. *Appl Psychophysiol Biofeedback* 2002;27:231–49.

42. Micoulaud-Franchi JA, Geoffroy PA, Fond G, Lopez R, Bioulac S, Philip P. EEG neurofeedback treatments in children with ADHD: an updated meta-analysis of randomized controlled trials. *Frontiers in Human Neuroscience* 2014;8:906.

43. Shinaver CS, III, Entwistle PC, Soderqvist S. Cogmed WM training: reviewing the reviews. *Appl Neuropsychol Child* 2014;3:163–72.

44. Mehta S, Mehta V, Mehta S, et al. Multimodal behavior program for ADHD incorporating yoga and implemented by high school volunteers: a pilot study. *ISRN Pediatr* 2011;2011:780745.

45. Converse AK, Ahlers EO, Travers BG, Davidson RJ. Tai chi training reduces self-report of inattention in healthy young adults. *Frontiers in Human Neuroscience* 2014;8:13.

46. Grant JA, Duerden EG, Courtemanche J, Cherkasova M, Duncan GH, Rainville P. Cortical thickness, mental absorption and meditative practice: possible implications for disorders of attention. *Biol Psychol* 2013;92:275–81.

47. Black DS, Milam J, Sussman S. Sitting-meditation interventions among youth: a review of treatment efficacy. *Pediatrics* 2009;124:e532–41.

48. Flook S SS, et al. Effects of mindful awareness practices on executive functions in elementary school children. *Journal of Applied School Psychology* 2010;26:70–95.

49. Black DS, Fernando R. Mindfulness training and classroom behavior among lower-income and ethnic minority elementary school children. *J Child Fam Stud* 2014;23:1242–6.

50. Dunn FM, Howell RJ. Relaxation training and its relationship to hyperactivity in boys. *J Clin Psychol* 1982;38:92–100.

51. Barkley RA, Guevremont DC, Anastopoulos AD, Fletcher KE. A comparison of three family therapy programs for treating family conflicts in adolescents with attention-deficit hyperactivity disorder. *J Consult Clin Psychol* 1992;60:450–62.

52. Kemper KJ. *Mental Health, Naturally.* American Academy of Pediatrics; 2010.

53. Molina BS, Hinshaw SP, Swanson JM, et al. The MTA at 8 years: prospective follow-up of children treated for combined-type ADHD in a multisite study. *J Am Acad Child Adolesc Psychiatry* 2009;48:484–500.

54. Swanson JM, Elliott GR, Greenhill LL, et al. Effects of stimulant medication on growth rates across 3 years in the MTA follow-up. *J Am Acad Child Adolesc Psychiatry* 2007;46:1015–27.

55. McBride MC. An individual double-blind crossover trial for assessing methylphenidate response in children with attention deficit disorder. *J Pediatr* 1988;113:137–45.

56. White SR, Yadao CM. Characterization of methylphenidate exposures reported to a regional poison control center. *Arch Pediatr Adolesc Med* 2000;154:1199–203.

57. Molina BS, Hinshaw SP, Eugene Arnold L, et al. Adolescent substance use in the multimodal treatment study of attention-deficit/hyperactivity disorder (ADHD) (MTA) as a function of childhood ADHD, random assignment to childhood treatments, and subsequent medi-

cation. *J Am Acad Child Adolesc Psychiatry* 2013;52:250–63.

58. Savill NC, Buitelaar JK, Anand E, et al. The efficacy of atomoxetine for the treatment of children and adolescents with attention-deficit/hyperactivity disorder: a comprehensive review of over a decade of clinical research. *CNS Drugs* 2015;29:131–51.

59. Romano MJ, Dinh A. A 1000-fold overdose of clonidine caused by a compounding error in a 5-year-old child with attention-deficit/hyperactivity disorder. *Pediatrics* 2001;108(2):471–2.

60. Transler C, Eilander A, Mitchell S, van de Meer N. The impact of polyunsaturated fatty acids in reducing child attention deficit and hyperactivity disorders. *J Atten Disord* 2010;14:232–46.

61. Sorgi PJ, Hallowell EM, Hutchins HL, Sears B. Effects of an open-label pilot study with high-dose EPA/DHA concentrates on plasma phospholipids and behavior in children with attention deficit hyperactivity disorder. *Nutr J* 2007;6:16.

62. Gustafsson PA, Birberg-Thornberg U, Duchen K, et al. EPA supplementation improves teacher-rated behaviour and oppositional symptoms in children with ADHD. *Acta Paediatr* 2010;99:1540–9.

63. Widenhorn-Muller K, Schwanda S, Scholz E, Spitzer M, Bode H. Effect of supplementation with long-chain omega-3 polyunsaturated fatty acids on behavior and cognition in children with attention deficit/hyperactivity disorder (ADHD): a randomized placebo-controlled intervention trial. *Prostaglandins Leukot Essent Fatty Acids* 2014;91:49–60.

64. Richardson AJ, Burton JR, Sewell RP, Spreckelsen TF, Montgomery P. Docosahexaenoic acid for reading, cognition and behavior in children aged 7–9 years: a randomized, controlled trial (the DOLAB Study). *PLoS One* 2012;7:e43909.

65. Milte CM, Parletta N, Buckley JD, Coates AM, Young RM, Howe PR. Eicosapentaenoic and docosahexaenoic acids, cognition, and behavior in children with attention-deficit/hyperactivity disorder: a randomized controlled trial. *Nutrition* 2012;28:670–7.

66. Bloch MH, Qawasmi A. Omega-3 fatty acid supplementation for the treatment of children with attention-deficit/hyperactivity disorder symptomatology: systematic review and meta-analysis. *J Am Acad Child Adolesc Psychiatry* 2011;50:991–1000.

67. Villagomez A RU. Iron, magnesium, vitamin D, and zinc deficiencies in children presenting with symptoms of attention-deficit/hyperactivity disorder. *Children* 2014;1:261–79.

68. Konofal E, Lecendreux M, Deron J, et al. Effects of iron supplementation on attention deficit hyperactivity disorder in children. *Pediatr Neurol* 2008;38:20–6.

69. Low M, Farrell A, Biggs BA, Pasricha SR. Effects of daily iron supplementation in primary-school-aged children: systematic review and meta-analysis of randomized controlled trials. *CMAJ* 2013;185:E791–802.

70. Arnold LE, Votolato NA, Kleykamp D, Baker GB, Bornstein RA. Does hair zinc predict amphetamine improvement of ADD/hyperactivity? *Int J Neurosci* 1990;50:103–7.

71. Arnold LE, Disilvestro RA, Bozzolo D, et al. Zinc for attention-deficit/hyperactivity disorder: placebo-controlled double-blind pilot trial alone and combined with amphetamine. *J Child Adolesc Psychopharmacol* 2011;21:1–19.

72. Huss M, Volp A, Stauss-Grabo M. Supplementation of polyunsaturated fatty acids, magnesium and zinc in children seeking medical advice for attention-deficit/hyperactivity problems—an observational cohort study. *Lipids Health Dis* 2010;9:105.

73. Bendz LM, Scates AC. Melatonin treatment for insomnia in pediatric patients with attention-deficit/hyperactivity disorder. *Ann Pharmacother* 2010;44:185–91.

74. Hoebert M, van der Heijden KB, van Geijlswijk IM, Smits MG. Long-term follow-up of melatonin treatment in children with ADHD and chronic sleep onset insomnia. *J Pineal Res* 2009;47:1–7.

75. Abbasi SH, Heidari S, Mohammadi MR, Tabrizi M, Ghaleiha A, Akhondzadeh S. Acetyl-L-carnitine as an adjunctive therapy in the treatment of attention-deficit/hyperactivity disorder in children and adolescents: a

placebo-controlled trial. *Child Psychiatry Hum Dev* 2011;42:367–75.

76. Arnold LE, Amato A, Bozzolo H, et al. Acetyl-L-carnitine (ALC) in attention-deficit/hyperactivity disorder: a multi-site, placebo-controlled pilot trial. *J Child Adolesc Psychopharmacol* 2007;17:791–802.

77. CK C. A placebo-crossover study of caffeine treatment of hyperkinetic children. *Int J Mental Health* 1975;4:132–43.

78. Firestone P, Davey J, Goodman JT, Peters S. The effects of caffeine and methylphenidate on hyperactive children. *J Am Acad Child Psychiatry* 1978;17:445–56.

79. Huestis RD, Arnold LE, Smeltzer DJ. Caffeine versus methylphenidate and d-amphetamine in minimal brain dysfunction: a double-blind comparison. *Am J Psychiatry* 1975;132:868–70.

80. Garfinkel BD, Webster CD, Sloman L. Responses to methylphenidate and varied doses of caffeine in children with attention deficit disorder. *Can J Psychiatry* 1981;26:395–401.

81. Niederhofer H. Ginkgo biloba treating patients with attention-deficit disorder. *Phytother Res* 2010;24:26–7.

82. Lyon MR, Cline JC, Totosy de Zepetnek J, Shan JJ, Pang P, Benishin C. Effect of the herbal extract combination Panax quinquefolium and Ginkgo biloba on attention-deficit hyperactivity disorder: a pilot study. *J Psychiatry Neurosci* 2001;26:221–8.

83. Greenblatt J. Nutritional supplements in ADHD. *J Am Acad Child Adolesc Psychiatry* 1999;38:1209–11.

84. Heimann SW. Pycnogenol for ADHD? *J Am Acad Child Adolesc Psychiatry* 1999;38:357–8.

85. Katz M, Levine AA, Kol-Degani H, Kav-Venaki L. A compound herbal preparation (CHP) in the treatment of children with ADHD: a randomized controlled trial. *J Atten Disord* 2010;14:281–91.

86. Field TM, Quintino O, Hernandez-Reif M, Koslovsky G. Adolescents with attention deficit hyperactivity disorder benefit from massage therapy. *Adolescence* 1998;33:103–8.

87. Khilnani S, Field T, Hernandez-Reif M, Schanberg S. Massage therapy improves mood and behavior of students with attention-deficit/hyperactivity disorder. *Adolescence* 2003;38:623–38.

88. Karpouzis F, Bonello R, Pollard H. Chiropractic care for paediatric and adolescent attention-deficit/hyperactivity disorder: A systematic review. *Chiropr Osteopat* 2010;18:13.

89. Accorsi A, Lucci C, Di Mattia L, et al. Effect of osteopathic manipulative therapy in the attentive performance of children with attention-deficit/hyperactivity disorder. *J Am Osteopath Assoc* 2014;114:374–81.

90. Li S, Yu B, Lin Z, et al. Randomized-controlled study of treating attention deficit hyperactivity disorder of preschool children with combined electro-acupuncture and behavior therapy. *Complement Ther Med* 2010;18:175–83.

91. Lee MS, Choi TY, Kim JI, Kim L, Ernst E. Acupuncture for treating attention deficit hyperactivity disorder: a systematic review and meta-analysis. *Chin J Integr Med* 2011;17:257–60.

Chapter 5: Allergies

1. Asthma and Allergy Foundation of America. (website accessed 3/2/2015)

2. Salo PM, Arbes SJ, Jr., Jaramillo R, et al. Prevalence of allergic sensitization in the United States: results from the National Health and Nutrition Examination Survey (NHANES) 2005–2006. *J Allergy Clin Immunol* 2014;134:350–9.

3. Wegienka G, Havstad S, Zoratti EM, Kim H, Ownby DR, Johnson CC. Combined effects of prenatal medication use and delivery type are associated with eczema at age 2 years. *Clin Exp Allergy* 2015;45:660–8.

4. Hesselmar B, Hicke-Roberts A, Wennergren G. Allergy in children in hand versus machine dishwashing. *Pediatrics* 2015;135:e590–7.

5. Lack G. Update on risk factors for food allergy. *J Allergy Clin Immunol* 2012;129:1187–97.

6. Du Toit G, Roberts G, Sayre PH, et al. Randomized trial of peanut consumption in infants at risk for peanut allergy. *N Engl J Med* 2015;372:803–13.

7. Sansotta N, Piacentini GL, Mazzei F, Minniti F, Boner AL, Peroni DG. Timing of introduction of solid food and risk of allergic disease development: understanding the evidence. *Allergol Immunopathol (Madr)* 2013;41:337–45.

8. Allen KJ, Koplin JJ, Ponsonby AL, et al. Vitamin D insufficiency is associated with challenge-proven food allergy in infants. *J Allergy Clin Immunol* 2013;131:1109–16, 16 e1–6.

9. Untersmayr E, Jensen-Jarolim E. The role of protein digestibility and antacids on food allergy outcomes. *J Allergy Clin Immunol* 2008;121:1301–8; quiz 9–10.

10. Visness CM, London SJ, Daniels JL, et al. Association of obesity with IgE levels and allergy symptoms in children and adolescents: results from the National Health and Nutrition Examination Survey 2005–2006. *J Allergy Clin Immunol* 2009;123:1163–9, 9 e1–4.

11. Sampson HA, Ho DG. Relationship between food-specific IgE concentrations and the risk of positive food challenges in children and adolescents. *J Allergy Clin Immunol* 1997;100:444–51.

12. Panush RS. Food induced ("allergic") arthritis: clinical and serologic studies. *J Rheumatol* 1990;17:291–4.

13. Peters RL, Dharmage SC, Gurrin LC, et al. The natural history and clinical predictors of egg allergy in the first 2 years of life: a prospective, population-based cohort study. *J Allergy Clin Immunol* 2014;133:485–91.

14. van den Biggelaar AH, van Ree R, Rodrigues LC, et al. Decreased atopy in children infected with Schistosoma haematobium: a role for parasite-induced interleukin-10. *Lancet* 2000;356:1723–7.

15. Jalonen T. Identical intestinal permeability changes in children with different clinical manifestations of cow's milk allergy. *J Allergy Clin Immunol* 1991;88:737–42.

16. Steinman HA. "Hidden" allergens in foods. *J Allergy Clin Immunol* 1996;98:241–50.

17. Osborne NJ, Koplin JJ, Martin PE, et al. Prevalence of challenge-proven IgE-mediated food allergy using population-based sampling and predetermined challenge criteria in infants. *J Allergy Clin Immunol* 2011;127:668–76 e1–2.

18. Cavataio F, Carroccio A, Montalto G, Iacono G. Isolated rice intolerance: clinical and immunologic characteristics in four infants. *J Pediatr* 1996;128:558–60.

19. Saarinen UM, Kajosaari M. Breastfeeding as prophylaxis against atopic disease: prospective follow-up study until 17 years old. *Lancet* 1995;346:1065–9.

20. Lovegrove JA, Morgan JB, Hamptom SM. Dietary factors influencing levels of food antibodies and antigens in breast milk. *Acta Paediatr* 1996;85:778–84.

21. de Boissieu D, Matarazzo P, Rocchiccioli F, Dupont C. Multiple food allergy: a possible diagnosis in breastfed infants. *Acta Paediatr* 1997;86:1042–6.

22. Meydani SN, Ha WK. Immunologic effects of yogurt. *Am J Clin Nutr* 2000;71:861–72.

23. Fergusson DM, Horwood LJ, Shannon FT. Early solid feeding and recurrent childhood eczema: a 10-year longitudinal study. *Pediatrics* 1990;86:541–6.

24. Heiner DC, Sears JW, Kniker WT. Multiple precipitins to cow's milk in chronic respiratory disease. A syndrome including poor growth, gastrointestinal symptoms, evidence of allergy, iron deficiency anemia, and pulmonary hemosiderosis. *Am J Dis Child* 1962;103:634–54.

25. Sicherer SH. Food allergy: when and how to perform oral food challenges. *Pediatric Allergy and Immunology: Official Publication of the European Society of Pediatric Allergy and Immunology* 1999;10:226–34.

26. Caffarelli C, Terzi V, Perrone F, Cavagni G. Food related, exercise induced anaphylaxis. *Arch Dis Child* 1996;75:141–4.

27. Tilles S, Schocket A, Milgrom H. Exercise-induced anaphylaxis related to specific foods. *J Pediatr* 1995;127:587–9.

28. Hide DW, Matthews S, Tariq S, Arshad SH. Allergen avoidance in infancy and allergy at 4 years of age. *Allergy* 1996;51:89–93.

29. Platts-Mills T, Vaughan J, Squillace S, Woodfolk J, Sporik R. Sensitisation, asthma, and a modified Th2 response in children exposed

to cat allergen: a population-based cross-sectional study. *Lancet* 2001;357:752–6.

30. Avner DB, Perzanowski MS, Platts-Mills TA, Woodfolk JA. Evaluation of different techniques for washing cats: quantitation of allergen removed from the cat and the effect on airborne Fel d 1. *J Allergy Clin Immunol* 1997;100:307–12.

31. McDonald LG, Tovey E. The role of water temperature and laundry procedures in reducing house dust mite populations and allergen content of bedding. *J Allergy Clin Immunol* 1992;90:599–608.

32. Huang S-W. The effects of an air cleaner in the homes of children with perennial allergic rhinitis. *Pediatric Asthma, Allergy, & Immunology* 1993;7:111–17.

33. Tovey ER, McDonald LG. A simple washing procedure with eucalyptus oil for controlling house dust mites and their allergens in clothing and bedding. *J Allergy Clin Immunol* 1997;100:464–6.

34. Kemp TJ, Siebers RW, Fishwick D, et al. House dust mite allergen in pillows. *BMJ* 1996;313:916.

35. Georgitis JW. Local hyperthermia and nasal irrigation for perennial allergic rhinitis: effect on symptoms and nasal airflow. *Ann Allergy* 1993;71:385–9.

36. Zachariae R, Bjerring P. Increase and decrease of delayed cutaneous reactions obtained by hypnotic suggestions during sensitization. Studies on dinitrochlorobenzene and diphenylcyclopropenone. *Allergy* 1993;48:6–11.

37. Perloff MM, Spiegelman J. Hypnosis in the treatment of a child's allergy to dogs. *Am J Clin Hypn* 1973;15:269–72.

38. Anbar RD. Self-hypnosis for management of chronic dyspnea in pediatric patients. *Pediatrics* 2001;107:E21.

39. Locke SE, Ransil BJ, Zachariae R, et al. Effect of hypnotic suggestion on the delayed-type hypersensitivity response. *JAMA* 1994;272:47–52.

40. Langewitz W, Izakovic J, Wyler J, et al. Effect of self-hypnosis on hay fever symptoms—a randomised controlled intervention study. *Psychother Psychosom* 2005;74:165–72.

41. Freier S, Berger H. Disodium cromoglycate in gastrointestinal protein intolerance. *Lancet* 1973;1:913–15.

42. Schneider LC, Rachid R, LeBovidge J, et al. A pilot study of omalizumab to facilitate rapid oral desensitization in high-risk peanut-allergic patients. *J Allergy Clin Immunol* 2013;132:1368–74.

43. Vinuya RZ. Specific allergen immunotherapy for allergic rhinitis and asthma. *Pediatr Ann* 2000;29:425–32.

44. Anagnostou K, Clark A, King Y, et al. Efficacy and safety of high-dose peanut oral immunotherapy with factors predicting outcome. *Clin Exp Allergy* 2011;41:1273–81.

45. Tang ML, Ponsonby AL, Orsini F, et al. Administration of a probiotic with peanut oral immunotherapy: a randomized trial. *J Allergy Clin Immunol* 2015;135:737–44 e8.

46. Burks AW, Wood RA, Jones SM, et al. Sublingual immunotherapy for peanut allergy: long-term follow-up of a randomized multicenter trial. *J Allergy Clin Immunol* 2015;135:1240–8 e1–3.

47. Narisety SD, Frischmeyer-Guerrerio PA, Keet CA, et al. A randomized, double-blind, placebo-controlled pilot study of sublingual versus oral immunotherapy for the treatment of peanut allergy. *J Allergy Clin Immunol* 2015;135:1275–82 e1–6.

48. Mittman P. Randomized, double-blind study of freeze-dried Urtica dioica in the treatment of allergic rhinitis. *Planta Med* 1990;56:44–7.

49. Enomoto T, Sowa M, Nishimori K, et al. Effects of bifidobacterial supplementation to pregnant women and infants in the prevention of allergy development in infants and on fecal microbiota. *Allergology International* 2014;63:575–85.

50. Elazab N, Mendy A, Gasana J, et al. Probiotic administration in early life, atopy, and asthma: a meta-analysis of clinical trials. *Pediatrics* 2013;132:e666–76.

51. Bucca C, Rolla G, Oliva A, Farina JC. Effect of vitamin C on histamine bronchial responsiveness of patients with allergic rhinitis. *Ann Allergy* 1990;65:311–4.

52. Tecklenburg SL, Mickleborough TD, Fly AD,

Bai Y, Stager JM. Ascorbic acid supplementation attenuates exercise-induced bronchoconstriction in patients with asthma. *Respiratory Medicine* 2007;101:1770–8.

53. Cohen HA, Neuman I, Nahum H. Blocking effect of vitamin C in exercise-induced asthma. *Arch Pediatr Adolesc Med* 1997;151:367–70.

54. Duchen K, Casas R, Fageras-Bottcher M, Yu G, Bjorksten B. Human milk polyunsaturated long-chain fatty acids and secretory immunoglobulin A antibodies and early childhood allergy. *Pediatric Allergy and Immunology: Official Publication of the European Society of Pediatric Allergy and Immunology* 2000;11:29–39.

55. Lundberg JM, Saria A. Capsaicin-induced desensitization of airway mucosa to cigarette smoke, mechanical and chemical irritants. *Nature* 1983;302:251–3.

56. Andre C, Andre F, Colin L, Cavagna S. Measurement of intestinal permeability to mannitol and lactulose as a means of diagnosing food allergy and evaluating therapeutic effectiveness of disodium cromoglycate. *Ann Allergy* 1987;59:127–30.

57. Joos S, Schott C, Zou H, Daniel V, Martin E. Immunomodulatory effects of acupuncture in the treatment of allergic asthma: a randomized controlled study. *J Altern Complement Med* 2000;6:519–25.

58. Feng S, Han M, Fan Y, et al. Acupuncture for the treatment of allergic rhinitis: a systematic review and meta-analysis. *Am J Rhinol Allergy* 2015;29:57–62.

59. Weiser M, Gegenheimer LH, Klein P. A randomized equivalence trial comparing the efficacy and safety of Luffa comp.-Heel nasal spray with cromolyn sodium spray in the treatment of seasonal allergic rhinitis. *Forsch Komplementarmed* 1999;6:142–8.

60. Linde K, Clausius N, Ramirez G, et al. Are the clinical effects of homeopathy placebo effects? A meta-analysis of placebo-controlled trials. *Lancet* 1997;350:834–43.

61. Altunc U, Pittler MH, Ernst E. Homeopathy for childhood and adolescence ailments: systematic review of randomized clinical trials. *Mayo Clin Proc* 2007;82:69–75.

62. Roll S, Reinhold T, Pach D, et al. Comparative effectiveness of homoeopathic vs. conventional therapy in usual care of atopic eczema in children: long-term medical and economic outcomes. *PLoS One* 2013;8:e54973.

Chapter 6: Anxiety

1. Ask H, Torgersen S, Seglem KB, Waaktaar T. Genetic and environmental causes of variation in adolescent anxiety symptoms: a multiple-rater twin study. *J Anxiety Disord* 2014;28:363–71.

2. Foster JA, McVey Neufeld KA. Gut-brain axis: how the microbiome influences anxiety and depression. *Trends Neurosci* 2013;36:305–12.

3. Reinelt E, Aldinger M, Stopsack M, et al. High social support buffers the effects of 5-HTTLPR genotypes within social anxiety disorder. *Eur Arch Psychiatry Clin Neurosci* 2014;264:433–9.

4. Murray CJ, Atkinson C, Bhalla K, et al. The state of US health, 1990–2010: burden of diseases, injuries, and risk factors. *JAMA* 2013;310:591–608.

5. Skowronek IB, Mounsey A, Handler L. Clinical Inquiry: can yoga reduce symptoms of anxiety and depression? *J Fam Pract* 2014;63:398–407.

6. Field T, Diego M, Hernandez-Reif M. Tai chi/yoga effects on anxiety, heartrate, EEG and math computations. *Complement Ther Clin Pract* 2010;16:235–8.

7. Sharma M, Haider T. Tai chi as an alternative and complimentary therapy for anxiety: a systematic review. *Journal of Evidence-Based Complementary & Alternative Medicine* 2015;20:143–53.

8. Kemper KJ, Kelly EA. Treating children with therapeutic and healing touch. *Pediatr Ann* 2004;33:248–52.

9. Anderson JG, Taylor AG. Effects of healing touch in clinical practice: a systematic review of randomized clinical trials. *J Holist Nurs* 2011;29:221–8.

10. Butler LD, Symons BK, Henderson SL, Shortliffe LD, Spiegel D. Hypnosis reduces distress and duration of an invasive medical procedure for children. *Pediatrics* 2005;115:e77–85.

11. Huynh ME, Vandvik IH, Diseth TH. Hypnother-

apy in child psychiatry: the state of the art. *Clin Child Psychol Psychiatry* 2008;13:377–93.

12. Goldbeck L, Schmid K. Effectiveness of autogenic relaxation training on children and adolescents with behavioral and emotional problems. *J Am Acad Child Adolesc Psychiatry* 2003;42:1046–54.

13. Kanji N, White A, Ernst E. Autogenic training to reduce anxiety in nursing students: randomized controlled trial. *J Adv Nurs* 2006;53:729–35.

14. Bowden A, Lorenc A, Robinson N. Autogenic training as a behavioural approach to insomnia: a prospective cohort study. *Primary Health Care Research & Development* 2012;13:175–85.

15. Klassen JA, Liang Y, Tjosvold L, Klassen TP, Hartling L. Music for pain and anxiety in children undergoing medical procedures: a systematic review of randomized controlled trials. *Ambul Pediatr* 2008;8:117–28.

16. Wahbeh H, Calabrese C, Zwickey H, Zajdel D. Binaural beat technology in humans: a pilot study to assess neuropsychologic, physiologic, and electroencephalographic effects. *J Altern Complement Med* 2007;13:199–206.

17. Diehle J, Opmeer BC, Boer F, Mannarino AP, Lindauer RJ. Trauma-focused cognitive behavioral therapy or eye movement desensitization and reprocessing: what works in children with posttraumatic stress symptoms? A randomized controlled trial. *Eur Child Adolesc Psychiatry* 2015;24:227–36.

18. Lakhan SE, Vieira KF. Nutritional and herbal supplements for anxiety and anxiety-related disorders: systematic review. *Nutr J* 2010;9:42.

19. Carroll D, Ring C, Suter M, Willemsen G. The effects of an oral multivitamin combination with calcium, magnesium, and zinc on psychological well-being in healthy young male volunteers: a double-blind placebo-controlled trial. *Psychopharmacology (Berl)* 2000;150:220–5.

20. Carey PD, Warwick J, Harvey BH, Stein DJ, Seedat S. Single photon emission computed tomography (SPECT) in obsessive-compulsive disorder before and after treatment with inositol. *Metab Brain Dis* 2004;19:125–34.

21. Kiecolt-Glaser JK, Belury MA, Andridge R, Malarkey WB, Glaser R. Omega-3 supplementation lowers inflammation and anxiety in medical students: a randomized controlled trial. *Brain Behav Immun* 2011;25:1725–34.

22. Jezova D, Makatsori A, Smriga M, Morinaga Y, Duncko R. Subchronic treatment with amino acid mixture of L-lysine and L-arginine modifies neuroendocrine activation during psychosocial stress in subjects with high trait anxiety. *Nutr Neurosci* 2005;8:155–60.

23. Smriga M, Ghosh S, Mouneimne Y, Pellett PL, Scrimshaw NS. Lysine fortification reduces anxiety and lessens stress in family members in economically weak communities in Northwest Syria. *Proc Natl Acad Sci USA* 2004;101:8285–8.

24. Smriga M, Ando T, Akutsu M, Furukawa Y, Miwa K, Morinaga Y. Oral treatment with L-lysine and L-arginine reduces anxiety and basal cortisol levels in healthy humans. *Biomed Res* 2007;28:85–90.

25. Abdou AM, Higashiguchi S, Horie K, Kim M, Hatta H, Yokogoshi H. Relaxation and immunity enhancement effects of gamma-aminobutyric acid (GABA) administration in humans. *Biofactors* 2006;26:201–8.

26. Lu K, Gray MA, Oliver C, et al. The acute effects of L-theanine in comparison with alprazolam on anticipatory anxiety in humans. *Hum Psychopharmacol* 2004;19:457–65.

27. Unno K, Tanida N, Ishii N, et al. Anti-stress effect of theanine on students during pharmacy practice: positive correlation among salivary alpha-amylase activity, trait anxiety and subjective stress. *Pharmacol Biochem Behav* 2013;111:128–35.

28. Hofmann SG, Otto MW, Pollack MH, Smits JA. D-cycloserine augmentation of cognitive behavioral therapy for anxiety disorders: an update. *Curr Psychiatry Rep* 2015;17:532.

29. Bloch MH, Panza KE, Grant JE, Pittenger C, Leckman JF. N-acetylcysteine in the treatment of pediatric trichotillomania: a randomized, double-blind, placebo-controlled add-on trial. *J Am Acad Child Adolesc Psychiatry* 2013;52:231–40.

30. Oliver G, Dean O, Camfield D, et al. N-acetyl cysteine in the treatment of obsessive compulsive and related disorders: a systematic review. *Clin Psychopharmacol Neurosci* 2015;13:12–24.

31. McClure EA, Gipson CD, Malcolm RJ, Kalivas PW, Gray KM. Potential role of N-acetylcysteine in the management of substance use disorders. *CNS Drugs* 2014;28:95–106.

32. Mohajeri MH, Wittwer J, Vargas K, et al. Chronic treatment with a tryptophan-rich protein hydrolysate improves emotional processing, mental energy levels and reaction time in middle-aged women. *Br J Nutr* 2015:1–16.

33. Capello AE, Markus CR. Effect of sub chronic tryptophan supplementation on stress-induced cortisol and appetite in subjects differing in 5-HTTLPR genotype and trait neuroticism. *Psychoneuroendocrinology* 2014;45:96–107.

34. Schruers K, van Diest R, Overbeek T, Griez E. Acute L-5-hydroxytryptophan administration inhibits carbon dioxide-induced panic in panic disorder patients. *Psychiatry Res* 2002;113:237–43.

35. Maron E, Toru I, Vasar V, Shlik J. The effect of 5-hydroxytryptophan on cholecystokinin-4-induced panic attacks in healthy volunteers. *J Psychopharmacol* 2004;18:194–9.

36. Rasmussen SA. Lithium and tryptophan augmentation in clomipramine-resistant obsessive-compulsive disorder. *Am J Psychiatry* 1984;141:1283–5.

37. Blier P, Bergeron R. Sequential administration of augmentation strategies in treatment-resistant obsessive-compulsive disorder: preliminary findings. *Int Clin Psychopharmacol* 1996;11:37–44.

38. Messaoudi M, Violle N, Bisson JF, et al. Beneficial psychological effects of a probiotic formulation (Lactobacillus helveticus R0052 and Bifidobacterium longum R0175) in healthy human volunteers. *Gut Microbes* 2011;2:256–61.

39. Sathyanarayanan V, Thomas T, Einother SJ, et al. Brahmi for the better? New findings challenging cognition and anti-anxiety effects of Brahmi (Bacopa monniera) in healthy adults. *Psychopharmacology (Berl)* 2013;227:299–306.

40. Calabrese C, Gregory WL, Leo M, Kraemer D, Bone K, Oken B. Effects of a standardized Bacopa monnieri extract on cognitive performance, anxiety, and depression in the elderly: a randomized, double-blind, placebo-controlled trial. *J Altern Complement Med* 2008;14:707–13.

41. Benson S, Downey LA, Stough C, et al. An acute, double-blind, placebo-controlled cross-over study of 320 mg and 640 mg doses of Bacopa monnieri (CDRI 08) on multitasking stress reactivity and mood. *Phytother Res* 2014;28:551–9.

42. Woelk H, Arnoldt KH, Kieser M, Hoerr R. Ginkgo biloba special extract EGb 761 in generalized anxiety disorder and adjustment disorder with anxious mood: a randomized, double-blind, placebo-controlled trial. *J Psychiatr Res* 2007;41:472–80.

43. Gavrilova SI, Preuss UW, Wong JW, et al. Efficacy and safety of Ginkgo biloba extract EGb 761 in mild cognitive impairment with neuropsychiatric symptoms: a randomized, placebo-controlled, double-blind, multi-center trial. *Int J Geriatr Psychiatry* 2014;29:1087–95.

44. Sarris J, Stough C, Bousman CA, et al. Kava in the treatment of generalized anxiety disorder: a double-blind, randomized, placebo-controlled study. *J Clin Psychopharmacol* 2013;33:643–8.

45. Sarris J, LaPorte E, Schweitzer I. Kava: a comprehensive review of efficacy, safety, and psychopharmacology. *Aust NZ J Psychiatry* 2011;45:27–35.

46. Panossian A, Wikman G, Sarris J. Rosenroot (Rhodiola rosea): traditional use, chemical composition, pharmacology and clinical efficacy. *Phytomedicine* 2010;17:481–93.

47. Bystritsky A, Kerwin L, Feusner JD. A pilot study of Rhodiola rosea (Rhodax) for generalized anxiety disorder (GAD). *J Altern Complement Med* 2008;14:175–80.

48. Feijo L, Hernandez-Reif M, Field T, et al. Mothers' depressed mood and anxiety levels are reduced after massaging their preterm infants. *Infant Behav Dev* 2006;29:476–80.

49. Celebioglu A, Gurol A, Yildirim ZK, Buyukavci M. Effects of massage therapy on pain and anxiety arising from intrathecal therapy or bone marrow aspiration in children with cancer. *Int J Nurs Pract* 2014:2015;21(6):197–804.

50. O'Flaherty LA, van Dijk M, Albertyn R, Millar A, Rode H. Aromatherapy massage seems to enhance relaxation in children with burns: an observational pilot study. *Burns* 2012;38:840–5.

51. Hughes D, Ladas E, Rooney D, Kelly K. Massage therapy as a supportive care intervention for children with cancer. Oncol Nurs Forum 2008;35:431–42.

52. Scheewe S, Vogt L, Minakawa S, et al. Acupuncture in children and adolescents with bronchial asthma: a randomised controlled study. *Complement Ther Med* 2011;19:239–46.

53. Wang SM, Escalera S, Lin EC, Maranets I, Kain ZN. Extra-1 acupressure for children undergoing anesthesia. *Anesth Analg* 2008;107:811–16.

54. Zeltzer LK, Tsao JC, Stelling C, Powers M, Levy S, Waterhouse M. A phase I study on the feasibility and acceptability of an acupuncture/hypnosis intervention for chronic pediatric pain. *J Pain Symptom Manage* 2002;24:437–46.

55. Hollifield M. Acupuncture for posttraumatic stress disorder: conceptual, clinical, and biological data support further research. *CNS Neuroscience & Therapeutics* 2011;17:769–79.

56. Zhang ZJ, Wang XY, Tan QR, Jin GX, Yao SM. Electroacupuncture for refractory obsessive-compulsive disorder: a pilot waitlist-controlled trial. *J Nerv Ment Dis* 2009;197:619–22.

57. Cheuk DK, Yeung WF, Chung KF, Wong V. Acupuncture for insomnia. *Cochrane Database Syst Rev* 2012;9:CD005472.

58. Huang W, Kutner N, Bliwise DL. A systematic review of the effects of acupuncture in treating insomnia. *Sleep Med Rev* 2009;13:73–104.

59. Highfield ES, Kaptchuk TJ, Ott MJ, Barnes L, Kemper KJ. Availability of acupuncture in the hospitals of a major academic medical center: a pilot study. *Complement Ther Med* 2003;11:177–83.

60. Lin YC, Lee AC, Kemper KJ, Berde CB. Use of complementary and alternative medicine in pediatric pain management service: a survey. *Pain Med* 2005;6:452–8.

61. Feusner JD, Madsen S, Moody TD, et al. Effects of cranial electrotherapy stimulation on resting state brain activity. *Brain Behav* 2012;2:211–20.

62. Klawansky S, Yeung A, Berkey C, Shah N, Phan H, Chalmers TC. Meta-analysis of randomized controlled trials of cranial electrostimulation. Efficacy in treating selected psychological and physiological conditions. *J Nerv Ment Dis* 1995;183:478–84.

63. Wilson S, Molina Lde L, Preisch J, Weaver J. The effect of electronic dental anesthesia on behavior during local anesthetic injection in the young, sedated dental patient. *Pediatr Dent* 1999;21:12–17.

64. Kasen S, Wickramaratne P, Gameroff MJ. Religiosity and longitudinal change in psychosocial functioning in adult offspring of depressed parents at high risk for major depression. *Depress Anxiety* 2014;31:63–71.

65. Kemper KJ, Fletcher NB, Hamilton CA, McLean TW. Impact of healing touch on pediatric oncology outpatients: pilot study. *J Soc Integr Oncol* 2009;7:12–18.

Chapter 7: Asthma

1. Wang J, Visness CM, Sampson HA. Food allergen desensitization in inner-city children with asthma. *J Allergy Clin Immunol* 2005;115:5:1076–80.

2. Woods RK, Weiner J, Abramson M, et al. Patient's perceptions of food-induced asthma. *Aust NZ Med J* 1996;26(4):504–12.

3. Wilson NM. Bronchial hyper-reactivity in food and drink intolerance. *Ann Allergy* 1988;61(6 pt 2):75–9.

4. Unge G, Grubbstrom J, Olsson P, et al. Effects of dietary tryptophan restrictions on clinical symptoms in patients with endogenous asthma. *Allergy* 1983;38(3):211–12.

5. Woods RK, Weiner JM, Abramson M, et al. Do dairy products induce bronchoconstriction in adults with asthma? *J Allergy Clin Immunol* 1998;101(1):45–50.

6. Li J, Xun P, Zamora D. Intakes of long-chain omega-3 PUFAs and fish in relation to incidence of asthma among American young adults: the CARDIA study. *Am J Clin Nutr* 2013;97(1):173–8.

7. Netting MJ, Middleton PF, Makrides M. Does maternal diet during pregnancy and lactation affect outcomes in offspring? A systematic review of food-based approaches. *Nutrition* 2014;30(11–12):1225–41.

8. Dorsch W, Scharff J, Bayer T, et al. Antiasthmatic effects of onions. *Int Arch of All and Appl Immunol* 88(1989):228–30.

9. Wagner H, Dorsch W, Bayer Th, et al. Antiasthmatic effects of onions: inhibition of 5-lipoxygenase and cyclooxygenase in vitro by thiosulfinates and cepaenes. *Prostaglandin Leukotriene EFA* 1990;39:59–62.

10. Welsh EJ, Bara A, Barley E, et al. Caffeine for asthma. *Cochrane Database System Rev* 2010;Jan 20:1.

11. Wilson N, Vickers H., Taylor G., et al. Objective test for food sensitivity in asthmatic children: increased bronchial reactivity after cola drinks. *British Medical Journal* 284(1982):1226–8.

12. Bolte G, Winkler G, Holscher B. Margarine consumption, asthma, and allergy in young adults: results of the German National Health Survey 1998. *Ann Epidemiol* 2005;15(3):207–13.

13. Farchi S, Forastiere F, Agabiti N, et al. Dietary factors associated with wheezing and allergic rhinitis in children. *Eur Respir J* 2003;22(5):772–80.

14. Towns SJ, Mellis CM. Role of acetyl salicylic acid and sodium metabisulfite in chronic childhood asthma. *Pediatrics* 73(1984):631–7.

15. Friedman ME, Easton JG. Prevalence of positive metabisulphite challenges in children with asthma. *Pediatric Asthma, Allergy and Immunology* 1(1987):53–9.

16. Cramer H, Posadzki P, Dobos G, et al. Yoga for asthma: a systematic review and meta-analysis. *Ann Allergy Asthma Immunol* 2014;112(6):503–10.

17. Vempati R, Bijlani RL, Deepak KK. The efficacy of a comprehensive lifestyle modification programme based on yoga in the management of bronchial asthma: a randomized controlled trial. *BMC Pulm Med* 2009;9:37.

18. Singh V. Effect of respiratory exercises on asthma: the pink city lung exerciser. *Journal of Asthma* 24(1987):355–9.

19. Nagendra HR, Nagarantha R. An integrated approach of yoga therapy for bronchial asthma: a 3–54 month prospective study. *J Asthma* 1986;23(3):123–37.

20. Jain SC, Rai L, Valecha A, et al. Effect of yoga training on exercise tolerance in adolescents with childhood asthma. *J Asthma* 1991;28(6):437–42.

21. Chang YF, Yang YH, Chen CC, et al. Tai Chi Chuan improves the pulmonary function of asthmatic children. *J Microbiol Immunol Infect*, 2008;41(1):88–95.

22. Tiep BL, Burns M, Kao D, et al. Pursed lips breathing training using ear oximetry. *Chest* 90(1986):218–21.

23. Fluge T, Richter J, Fabel H, et al. Long-term effects of breathing exercises and yoga in patients with bronchial asthma. *Pneumologie* 1994;48:484–90.

24. Berlowitz D, Denehy L, Johns DP, Bish RM, Walters EH. The Buteyko asthma breathing technique. *Med J Aust* 1995;162:53.

25. Opat AJ, Cohen MM, Bailey MJ, et al. A clinical trial of the Buteyko breathing technique in asthma as taught by a video. *J Asthma* 2000;37(7):557–64.

26. Bowler SD, Green A, Mitchell CA. Buteyko breathing techniques in asthma: a blinded randomised controlled trial. *Med J Austr* 1998;169:575–8.

27. Cowie RL, Conley DP, Underwood MF, et al. A randomised controlled trial of the Buteyko technique as an adjunct to conventional management of asthma. *Respir Med* 2008;102(5):726–32.

28. Bar-On O, Inbar O. Swimming and asthma: benefits and deleterious effects. *Sports Medicine* 1992;14:397–405.

29. Schnall R, Ford P, Gillam I, Landau L. Swimming and dry land exercise in children with asthma. *Aust Paediatr J* 1982;18:23–7.

30. Tanizaki Y, Kitani H, Okazaki M, et al. Clinical effects of complex spa therapy on patients

with steroid-dependent intractable asthma (SDIA). *Aerugi* 1993;42:219–27.

31. Goldstein NA, Aronin C, Kantrowitz B, et al. The prevalence of sleep-disordered breathing in children with asthma and its behavioral effects. *Pediatr Pulmonol* 2015;50(11):1128–36.

32. Liu Y, Croft JB, Wheaton AG, et al. Association between perceived insufficient sleep, frequent mental distress, obesity, and chronic diseases. *BMC Public Health* 2013;13:84.

33. Bhattacharjee R, Choi BH, Gozal D, et al. Association of adenotonsillectomy with asthma outcomes in children: a longitudinal database analysis. *PLoS Medicine* 2014;11(11):e1001753.

34. Lanphear BP, Aligne A, Auinger P, et al. Residential exposures associated with asthma in US children. *Pediatrics* 2001;107(3):505–11.

35. Clarke D, Gormally M, Sheahan J, et al. Child car seats—a habitat for house dust mites and reservoir for harmful allergens. *Ann Agric Environ Med* 2015;22(1):17–22.

36. Beck AF, Huang B, Kercsmar CM, et al. Allergen sensitization profiles in a population-based cohort of children hospitalized for asthma. *Ann Am Thoracic Soc* 2015;12(3):376–84.

37. Kearney GD, Johnson LC, Xu X, et al. Eastern Carolina Asthma Prevention Program (ECAPP): an environmental intervention study among rural and underserved children with asthma in Eastern North Carolina. *Environ Health Insights* 2014;8:27–37.

38. Blackhall K, Appleton S, Cates CJ. Ionisers for chronic asthma. *Cochrane Database Syst Rev* 2012;Sept 12;9.

39. Hahn DL, Bukstein D, Luskin A, Zeitz H. Evidence for *Chlamydia pneumoniae* infection in steroid-dependent asthma. *Ann Allerg Asthma Immunol* 1998;80(1):45–9.

40. Wood BL, Miller BD, Lehman HK. Review of family relational stress and pediatric asthma: the value of biopsychosocial systemic models. *Fam Process* 2015;54(2):376–89.

41. McBride JJ, Vlieger AM, Anbar RD. Hypnosis in paediatric respiratory medicine. *Paediatr Respir Rev* 2014;15(1):82–5.

42. Henry M, DeRivera JLG, Gonzalez-Martin IJ, et al. Improvement of respiratory function in chronic asthmatic patients with autogenic therapy. *Journal of Psychosomatic Research* 37(1993):265–70.

43. Wilson AR, Honsberger R, Chiu TJ, Novey HS. Transcendental meditation and asthma. *Respiration* 1975;32:74–80.

44. Pbert L, Madison JM, Druker S, et al. Effect of mindfulness training on asthma quality of life and lung function: a randomised controlled trial. *Thorax* 2012;67(9):769–76.

45. Feldman GM. The effect of biofeedback training on respiratory resistance of asthmatic children. *Psychosomatic Medicine* 38(1976):27–34.

46. Coen BL, Conran PB, McGrady A, et al. Effects of biofeedback-assisted relaxation on asthma severity and immune function. *Ped Asthma, Allergy Immunol* 1996;10:71–8.

47. Scherr MS, Crawford PL. Three-year evaluation of biofeedback techniques in the treatment of children with chronic asthma in a summer camp environment. *Annals of Allergy* 41(1978):288–92.

48. Long KA, Ewing LJ, Cohen S, et al. Preliminary evidence for the feasibility of a stress management intervention for 7- to 12-year-olds with asthma. *J Asthma* 2011;48(2):162–70.

49. Smyth JM, et al. Effects of writing about stressful experiences on symptom reduction in patients with asthma or rheumatoid arthritis: a randomized trial. *JAMA* 1999;281:1304–9

50. Kemper KJ, Ritz RH, Benson MS, et al. Helium-oxygen mixture in the treatment of postextubation stridor in pediatric trauma patients. *Crit Care Med* 1991;19(3):356–9.

51. Wong JJ, Lee JH, Turner DA, et al. A review of the use of adjunctive therapies in severe acute asthma exacerbation in critically ill children. *Expert Rev Respir Med*, 2014;8(4):423–41.

52. Collipp PJ, Goldzier S, Weiss N, et al. Pyridoxine treatment of childhood bronchial asthma. *Ann Allergy* 1975;35:93–7.

53. Nakamura K, Wada K, Sahashi Y. Associations of intake of antioxidant vitamins and fatty acids with asthma in pre-school children. *Public Health Nutr* 2013;16(11):2040–5.

54. Soutar A, Seaton A, Brown K. Bronchial reactivity and dietary antioxidants. *Thorax* 1997;52:166–70.

55. Hemila H. Vitamin C may alleviate exercise-induced bronchoconstriction: a meta-analysis. *BMJ Open* 2013;3:e002416.

56. Grievink L, Zijlstra AG, Ke X, et al. Double-blind intervention trial on modulation of ozone effects on pulmonary function by antioxidant supplements. *Am J Epid* 1999;149(4):306–14

57. Romieu I, Meneses F, Ramirez M, et al. Anti-oxidant supplementation and respiratory functions among workers exposed to high levels of ozone. *Am J Respir Crit Care Med* 1998;158:226–32.

58. Hemila H. Vitamin C and common cold-induced asthma: a systematic review and statistical analysis. *Allergy Asthma Clin Immunol* 2013;9:46.

59. Rajabbik MH, Lotfi T, Alkhaled L, et al. Association between low vitamin D levels and the diagnosis of asthma in children: a systematic review of cohort studies. *Allergy Asthma Clin Immunol* 2014;10(1):31.

60. Kerley CP, Elnazir B, Faul J, et al. Vitamin D as an adjunctive therapy in asthma. Part 2: A review of human studies. *Pulm Pharmacol Ther* 2015;32:75–92.

61. Pojsupap S, Iliriani K, Sampaio TZ, et al. Efficacy of high-dose vitamin D in pediatric asthma: a systematic review and meta-analysis. *J Asthma*, 2014;Nov 21:1–9.

62. Soutar A, Seaton A, Brown K. Bronchial reactivity and dietary antioxidants. *Thorax* 1997;52(2):166–70.

63. Ciarallo L, Brousseau D, Reinert S. Higher-dose intravenous magnesium therapy for children with moderate to severe acute asthma. *Arch Pediatr Adolesc Med* 2000;154:979–83.

64. Gontijo-Amaral C, Ribeiro MA, Gontijo LS, et al. Oral magnesium supplementation in asthmatic children: a double-blind randomized, placebo-controlled trial. *Eur J Clin Nutr* 2007;61(1):54–60.

65. Broughton KS, Johnson CS, Pace BK, et al. Reduced asthma symptoms with n-3 fatty acid ingestion are related to 5-series leukotriene production. *Am J Clin Nutr* 1997;65:1011–17.

66. Yang H, Xun P, He K. Fish and fish oil intake in relation to the risk of asthma: a systematic review and meta-analysis. *PLoS One* 2013;8(11):e80048.

67. Nagakura T, Matsuda S, Shichijyo K, et al. Dietary supplementation with fish oil rich in omega-3 polyunsaturated fatty acids in children with bronchial asthma. *Eur J Respir J* 2000;16(5):861–5.

68. Villani F, Comazzi R, DeMaria P, et al. Effect of dietary supplementation with polyunsaturated fatty acids on bronchial hyperreactivity in subjects with seasonal asthma. *Respiration* 1998;65:265–9.

69. Klemens CM, Berman DR, Mozurkewich EL. The effect of perinatal omega-3 fatty acid supplementation on inflammatory markers and allergic diseases: a systematic review. *BJOG* 2011;118(8):916–25.

70. Ng TP, Niti M, Yap KB, et al. Curcumins-rich curry diet and pulmonary function in Asian older adults. *PLoS One* 2012;7(12):e51753.

71. Nakajima S, Tohda Y, Ohkawa K, et al. Effects of Saiboku-to (TJ-96) on bronchial asthma. *Ann NY Acad Sci* 1993;685:549–60.

72. Field T, Henteleff T, Hernandez-Rief M, et al. Children with asthma have improved pulmonary functions after massage therapy. *J Pediatr* 1998;132:854–8.

73. Fattah MA, Hamdy B. Pulmonary functions of children with asthma improve following massage therapy. *J Altern Complement Med* 2011;17(11):1065–8.

74. Kaminskyj A, Frazier M, Johnstone K, et al. Chiropractic care for patients with asthma: a systematic review of the literature. *J Can Chiropr Assoc* 2010;54(1):24–32.

75. Reinhold T, Brinkhaus B, Willich SN, et al. Acupuncture in patients suffering from allergic asthma: Is it worth the additional costs? *J Altern Complement Med* 2014;20(3):169–77.

76. Karlson G, Bennicke P. Acupuncture in asthmatic children: a prospective, randomized, controlled trial of efficacy. *Altern Ther Health Med* 2013;19(4):13–19

77. Elseify MY, Mohammed NH, Alsharkawy AA, et al. Laser acupuncture in treatment of childhood bronchial asthma. *J Complement Integr Med*, 2013; Jul 9:10.

78. Wacker von A. Healing in Asthma—a pilot study. *Erfahrungsheilkunde* 1996;July:428–33.

79. Shafei HF, AbdelDayem SM, Mohamed NH. Individualized homeopathy in a group of Egyptian asthmatic children. *Homeopathy* 2012;101(4):224–30.

80. McCarney RW, Linde K, Lasserson TJ. Homeopathy for chronic asthma. *Cochrane Database Syst Rev* 2004:1:CDC000353.

Chapter 8: Autism

1. McElhanon BO, McCracken C, Karpen S, Sharp WG. Gastrointestinal symptoms in autism spectrum disorder: a meta-analysis. *Pediatrics* 2014;133:872–3.

2. De Angelis M, Francavilla R, Piccolo M, De Giacomo A, Gobbetti M. Autism spectrum disorders and intestinal microbiota. *Gut Microbes* 2015;6(3):207–13.

3. Walker SJ, Fortunato J, Gonzalez LG, Krigsman A. Identification of unique gene expression profile in children with regressive autism spectrum disorder (ASD) and ileocolitis. *PLoS One* 2013;8:e58058.

4. Coury DL, Ashwood P, Fasano A, et al. Gastrointestinal conditions in children with autism spectrum disorder: developing a research agenda. *Pediatrics* 2012;130 Suppl 2:S160–8.

5. de Magistris L, Familiari V, Pascotto A, et al. Alterations of the intestinal barrier in patients with autism spectrum disorders and in their first-degree relatives. *Journal of Pediatric Gastroenterology and Nutrition* 2010;51:418–24.

6. Canitano R. Epilepsy in autism spectrum disorders. *Eur Child Adolesc Psychiatry* 2007;16:61–6.

7. Chen MH, Su TP, Chen YS, et al. Is atopy in early childhood a risk factor for ADHD and ASD? a longitudinal study. *J Psychosom Res* 2014;77:316–21.

8. Rossignol DA, Frye RE. Mitochondrial dysfunction in autism spectrum disorders: a systematic review and meta-analysis. *Mol Psychiatry* 2012;17:290–314.

9. Stein TP, Schluter MD, Steer RA, Guo L, Ming X. Bisphenol A exposure in children with autism spectrum disorders. *Autism Research:* 2015;8(3):272–83.

10. Yassa HA. Autism: a form of lead and mercury toxicity. *Environmental Toxicology and Pharmacology* 2014;38:1016–24.

11. Jo H, Schieve LA, Sharma AJ, Hinkle SN, Li R, Lind JN. Maternal prepregnancy body mass index and child psychosocial development at 6 years of age. *Pediatrics* 2015;135:e1198–209.

12. Krakowiak P, Walker CK, Bremer AA, et al. Maternal metabolic conditions and risk for autism and other neurodevelopmental disorders. *Pediatrics* 2012;129:e1121–8.

13. Xiang AH, Wang X, Martinez MP, et al. Association of maternal diabetes with autism in offspring. *JAMA* 2015;313:1425–34.

14. Sullivan EL, Riper KM, Lockard R, Valleau JC. Maternal high-fat diet programming of the neuroendocrine system and behavior. *Horm Behav* 2015;76:153–61.

15. Fernell E, Bejerot S, Westerlund J, et al. Autism spectrum disorder and low vitamin D at birth: a sibling control study. *Molecular Autism* 2015;6:3.

16. Raz R, Roberts AL, Lyall K, et al. Autism spectrum disorder and particulate matter air pollution before, during, and after pregnancy: a nested case-control analysis within the Nurses' Health Study II Cohort. *Environ Health Perspect* 2015;123:264–70.

17. Schmidt RJ, Tancredi DJ, Ozonoff S, et al. Maternal periconceptional folic acid intake and risk of autism spectrum disorders and developmental delay in the CHARGE (CHildhood Autism Risks from Genetics and Environment) case-control study. *Am J Clin Nutr* 2012;96:80–9.

18. Walker CK, Krakowiak P, Baker A, et al. Preeclampsia, placental insufficiency, and autism spectrum disorder or developmental delay. *JAMA Pediatrics* 2015;169:154–62.

19. Christensen J, Gronborg TK, Sorensen MJ, et al. Prenatal valproate exposure and risk of autism spectrum disorders and childhood autism. *JAMA* 2013;309:1696–703.

20. Man KK, Tong HH, Wong LY, Chan EW, Simonoff E, Wong IC. Exposure to selective serotonin reuptake inhibitors during preg-

nancy and risk of autism spectrum disorder in children: a systematic review and meta-analysis of observational studies. *Neurosci Biobehav Rev* 2015;49:82–9.

21. Lyall K, Schmidt RJ, Hertz-Picciotto I. Maternal lifestyle and environmental risk factors for autism spectrum disorders. *Int J Epidemiol* 2014;43:443–64.

22. Bauer AZ, Kriebel D. Prenatal and perinatal analgesic exposure and autism: an ecological link. *Environ Health* 2013;12:41.

23. Taylor LE, Swerdfeger AL, Eslick GD. Vaccines are not associated with autism: an evidence-based meta-analysis of case-control and cohort studies. *Vaccine* 2014;32:3623–9.

24. Jain A, Marshall J, Buikema A, et al. Autism occurrence by MMR vaccine status among US children with older siblings with and without autism. *JAMA* 2015;313:1534–40.

25. Macfabe D. Autism: metabolism, mitochondria, and the microbiome. *Global Advances in Health and Medicine* 2013;2:52–66.

26. Golla S, Sweeney JA. Corticosteroid therapy in regressive autism: Preliminary findings from a retrospective study. *BMC Med* 2014;12:79.

27. Duffy FH, Shankardass A, McAnulty GB, et al. Corticosteroid therapy in regressive autism: a retrospective study of effects on the Frequency Modulated Auditory Evoked Response (FMAER), language, and behavior. *BMC Neurology* 2014;14:70.

28. Pennesi CM, Klein LC. Effectiveness of the gluten-free, casein-free diet for children diagnosed with autism spectrum disorder: based on parental report. *Nutr Neurosci* 2012;15:85–91.

29. Pedersen L, Parlar S, Kvist K, Whiteley P, Shattock P. Data mining the ScanBrit study of a gluten- and casein-free dietary intervention for children with autism spectrum disorders: behavioural and psychometric measures of dietary response. *Nutr Neurosci* 2014;17:207–13.

30. Spilioti M, Evangeliou AE, Tramma D, et al. Evidence for treatable inborn errors of metabolism in a cohort of 187 Greek patients with autism spectrum disorder (ASD). *Frontiers in Human Neuroscience* 2013;7:858.

31. Neely L, Rispoli M, Gerow S, Ninci J. Effects of antecedent exercise on academic engagement and stereotypy during instruction. *Behav Modif* 2015;39:98–116.

32. Rosenblatt LE, Gorantla S, Torres JA, et al. Relaxation response-based yoga improves functioning in young children with autism: a pilot study. *J Altern Complement Med* 2011;17:1029–35.

33. Koenig KP, Buckley-Reen A, Garg S. Efficacy of the Get Ready to Learn yoga program among children with autism spectrum disorders: a pretest-posttest control group design. *Am J Occup Ther* 2012;66:538–46.

34. Ayyash HF, Preece P, Morton R, Cortese S. Melatonin for sleep disturbance in children with neurodevelopmental disorders: prospective observational naturalistic study. *Expert Rev Neurother* 2015:1–7.

35. Geretsegger M, Elefant C, Mossler KA, Gold C. Music therapy for people with autism spectrum disorder. *Cochrane Database Syst Rev* 2014;6:CD004381.

36. Section on Complementary and Integrative Medicine, Council on Children with Disabilities, American Academy of Pediatrics, Zimmer M, Desch L. Sensory integration therapies for children with developmental and behavioral disorders. *Pediatrics* 2012;129:1186–9.

37. Ghanizadeh A. Hyperbaric oxygen therapy for treatment of children with autism: a systematic review of randomized trials. *Medical Gas Research* 2012;2:13.

38. Oberman LM, Rotenberg A, Pascual-Leone A. Use of transcranial magnetic stimulation in autism spectrum disorders. *J Autism Dev Disord* 2015;45:524–36.

39. Braun JM, Kalkbrenner AE, Just AC, et al. Gestational exposure to endocrine-disrupting chemicals and reciprocal social, repetitive, and stereotypic behaviors in 4- and 5-year-old children: the HOME study. *Environ Health Perspect* 2014;122:513–20.

40. Kalkbrenner AE, Schmidt RJ, Penlesky AC. Environmental chemical exposures and autism spectrum disorders: a review of the epidemiological evidence. *Curr Probl Pediatr Adolesc Health Care* 2014;44:277–318.

41. Shelton JF, Geraghty EM, Tancredi DJ, et al. Neurodevelopmental disorders and prenatal residential proximity to agricultural pesticides: the CHARGE study. *Environ Health Perspect* 2014;122:1103–9.

42. Reichow B, Barton EE, Boyd BA, Hume K. Early intensive behavioral intervention (EIBI) for young children with autism spectrum disorders (ASD). *Cochrane Database Syst Rev* 2012;10:CD009260.

43. Rogers SJ, Vismara L, Wagner AL, McCormick C, Young G, Ozonoff S. Autism treatment in the first year of life: a pilot study of infant start, a parent-implemented intervention for symptomatic infants. *J Autism Dev Disord* 2014;44:2981–95.

44. Bradshaw J, Steiner AM, Gengoux G, Koegel LK. Feasibility and effectiveness of very early intervention for infants at-risk for autism spectrum disorder: a systematic review. *J Autism Dev Disord* 2015;45:778–94.

45. Orinstein AJ, Helt M, Troyb E, et al. Intervention for optimal outcome in children and adolescents with a history of autism. *J Dev Behav Pediatr* 2014;35:247–56.

46. Vivanti G, Paynter J, Duncan E, et al. Effectiveness and feasibility of the early start Denver model implemented in a group-based community childcare setting. *J Autism Dev Disord* 2014;44:3140–53.

47. Oono IP, Honey EJ, McConachie H. Parent-mediated early intervention for young children with autism spectrum disorders (ASD). *Cochrane Database Syst Rev* 2013;4:CD009774.

48. Bearss K, Johnson C, Smith T, et al. Effect of parent training vs parent education on behavioral problems in children with autism spectrum disorder: a randomized clinical trial. *JAMA* 2015;313:1524–33.

49. Reichow B, Steiner AM, Volkmar F. Cochrane review: social skills groups for people aged 6 to 21 with autism spectrum disorders (ASD). Evidence-based child health: a Cochrane review journal 2013;8:266–315.

50. Hurt E, Arnold LE, Lofthouse N. Quantitative EEG neurofeedback for the treatment of pediatric attention-deficit/hyperactivity disorder, autism spectrum disorders, learning disorders, and epilepsy. *Child Adolesc Psychiatr Clin N Am* 2014;23:465–86.

51. Cachia RL, Anderson A, Moore DW. Mindfulness, stress, and well-being in parents of children with autism spectrum disorder: a systematic review. *J Child Fam Stud* 2015;May 19:2016;25(1):1–14.

52. Holm MB, Baird JM, Kim YJ, et al. Therapeutic horseback riding outcomes of parent-identified goals for children with autism spectrum disorder: an ABA multiple case design examining dosing and generalization to the home and community. *J Autism Dev Disord* 2014;44:937–47.

53. Lanning BA, Baier ME, Ivey-Hatz J, Krenek N, Tubbs JD. Effects of equine assisted activities on autism spectrum disorder. *J Autism Dev Disord* 2014;44:1897–907.

54. Ajzenman HF, Standeven JW, Shurtleff TL. Effect of hippotherapy on motor control, adaptive behaviors, and participation in children with autism spectrum disorder: a pilot study. *Am J Occup Ther* 2013;67:653–63.

55. O'Haire ME, McKenzie SJ, McCune S, Slaughter V. Effects of classroom animal-assisted activities on social functioning in children with autism spectrum disorder. *J Altern Complement Med* 2014;20:162–8.

56. O'Haire ME. Animal-assisted intervention for autism spectrum disorder: a systematic literature review. *J Autism Dev Disord* 2013;43:1606–22.

57. Mortimer R, Privopoulos M, Kumar S. The effectiveness of hydrotherapy in the treatment of social and behavioral aspects of children with autism spectrum disorders: a systematic review. *Journal of Multidisciplinary Healthcare* 2014;7:93–104.

58. Coury D. Very little high-quality evidence to support most medications for children with autism spectrum disorders. *J Pediatr* 2011;159:872–3.

59. Sharma A, Shaw SR. Efficacy of risperidone in managing maladaptive behaviors for children with autistic spectrum disorder: a meta-analysis. *J Pediatr Health Care* 2012;26:291–9.

60. Ching H, Pringsheim T. Aripiprazole for

autism spectrum disorders (ASD). *Cochrane Database Syst Rev* 2012;5:CD009043.

61. McPheeters ML, Warren Z, Sathe N, et al. A systematic review of medical treatments for children with autism spectrum disorders. *Pediatrics* 2011;127:e1312–21.

62. Krishnaswami S, McPheeters ML, Veenstra-Vanderweele J. A systematic review of secretin for children with autism spectrum disorders. *Pediatrics* 2011;127:e1322–5.

63. James S, Stevenson SW, Silove N, Williams K. Chelation for autism spectrum disorder (ASD). *Cochrane Database Syst Rev* 2015;5:CD010766.

64. Blaurock-Busch E, Amin OR, Dessoki HH, Rabah T. Toxic metals and essential elements in hair and severity of symptoms among children with autism. *Maedica* 2012;7:38–48.

65. Adams JB, Audhya T, McDonough-Means S, et al. Toxicological status of children with autism vs. neurotypical children and the association with autism severity. *Biological Trace Element Research* 2013;151:171–80.

66. Brent J. Commentary on the abuse of metal chelation therapy in patients with autism spectrum disorders. *J of Med Toxicol* 2013;9:370–2.

67. Adams JB, Holloway C. Pilot study of a moderate dose multivitamin/mineral supplement for children with autistic spectrum disorder. *J Altern Complement Med* 2004;10:1033–9.

68. Adams JB, Audhya T, McDonough-Means S, et al. Effect of a vitamin/mineral supplement on children and adults with autism. *BMC Pediatr* 2011;11:111.

69. Mehl-Madrona L, Leung B, Kennedy C, Paul S, Kaplan BJ. Micronutrients versus standard medication management in autism: a naturalistic case-control study. *J Child Adolesc Psychopharmacol* 2010;20:95–103.

70. Kaluzna-Czaplinska J, Zurawicz E, Michalska M, Rynkowski J. A focus on homocysteine in autism. *Acta Biochimica Polonica* 2013;60:137–42.

71. Adams JB, George F, Audhya T. Abnormally high plasma levels of vitamin B6 in children with autism not taking supplements compared to controls not taking supplements. *J Altern Complement Med* 2006;12:59–63.

72. Mousain-Bosc M, Roche M, Polge A, et al. Improvement of neurobehavioral disorders in children supplemented with magnesium-vitamin B6. II. Pervasive developmental disorder-autism. *Magnes Res* 2006;19:53–62.

73. Suren P, Roth C, Bresnahan M, et al. Association between maternal use of folic acid supplements and risk of autism spectrum disorders in children. *JAMA* 2013;309:570–7.

74. Bertoglio K, Jill James S, Deprey L, Brule N, Hendren RL. Pilot study of the effect of methyl B12 treatment on behavioral and biomarker measures in children with autism. *J Altern Complement Med* 2010;16:555–60.

75. Kumar J, Muntner P, Kaskel FJ, Hailpern SM, Melamed ML. Prevalence and associations of 25-hydroxyvitamin D deficiency in US children: NHANES 2001–2004. *Pediatrics* 2009;124:e362–70.

76. Mostafa GA, Al-Ayadhi LY. Reduced serum concentrations of 25-hydroxy vitamin D in children with autism: relation to autoimmunity. *Journal of Neuroinflammation* 2012;9:201.

77. Saad K, Abdel-Rahman AA, Elserogy YM, et al. Vitamin D status in autism spectrum disorders and the efficacy of vitamin D supplementation in autistic children. *Nutr Neurosci* 2015;Apr 15;epub ahead of print.

78. Jia F, Wang B, Shan L, Xu Z, Staal WG, Du L. Core symptoms of autism improved after vitamin D supplementation. *Pediatrics* 2015;135:e196–8.

79. Patrick RP, Ames BN. Vitamin D hormone regulates serotonin synthesis. Part 1: relevance for autism. *FASEB J* 2014;28:2398–413.

80. Mayer EA, Padua D, Tillisch K. Altered brain-gut axis in autism: comorbidity or causative mechanisms? *BioEssays* 2014;36:933–9.

81. Hardan AY, Fung LK, Libove RA, et al. A randomized controlled pilot trial of oral N-acetylcysteine in children with autism. *Biol Psychiatry* 2012;71:956–61.

82. Ghanizadeh A, Moghimi-Sarani E. A randomized double blind placebo controlled clinical trial of N-Acetylcysteine added to risperidone for treating autistic disorders. *BMC Psychiatry* 2013;13:196.

83. Nikoo M, Radnia H, Farokhnia M, Mohammadi

MR, Akhondzadeh S. N-acetylcysteine as an adjunctive therapy to risperidone for treatment of irritability in autism: a randomized, double-blind, placebo-controlled clinical trial of efficacy and safety. *Clin Neuropharmacol* 2015;38:11–17.

84. Mankad D, Dupuis A, Smile S, et al. A randomized, placebo controlled trial of omega-3 fatty acids in the treatment of young children with autism. *Molecular Autism* 2015;6:18.

85. James S, Montgomery P, Williams K. Omega-3 fatty acids supplementation for autism spectrum disorders (ASD). *Cochrane Database Syst Rev* 2011:CD007992.

86. Yui K, Koshiba M, Nakamura S, Kobayashi Y. Effects of large doses of arachidonic acid added to docosahexaenoic acid on social impairment in individuals with autism spectrum disorders: a double-blind, placebo-controlled, randomized trial. *J Clin Psychopharmacol* 2012;32:200–6.

87. Silva LM, Schalock M, Gabrielsen KR, Budden SS, Buenrostro M, Horton G. Early intervention with a parent-delivered massage protocol directed at tactile abnormalities decreases severity of autism and improves child-to-parent interactions: a replication study. *Autism Research and Treatment* 2015;2015:904585.

88. Silva L, Schalock M. Treatment of tactile impairment in young children with autism: results with QiGong massage. *International Journal of Therapeutic Massage & Bodywork* 2013;6:12–20.

89. Lee MS, Kim JI, Ernst E. Massage therapy for children with autism spectrum disorders: a systematic review. *J Clin Psychiatry* 2011;72:406–11.

90. Ming X, Chen X, Wang XT, et al. Acupuncture for treatment of autism spectrum disorders. *Evid Based Complement Alternat Med* 2012;2012:679845.

91. Lee MS, Choi TY, Shin BC, Ernst E. Acupuncture for children with autism spectrum disorders: a systematic review of randomized clinical trials. *J Autism Dev Disord* 2012;42:1671–83.

Chapter 9: Burns

1. Montemarano AD, Gupta RK, Burge JR, Klein K. Insect repellents and the efficacy of sunscreens. *Lancet* 1997;349:1670–1.

2. Sies H, Stahl W. Nutritional protection against skin damage from sunlight. *Annu Rev Nutr* 2004;24:173–200.

3. Eberlein-Konig B, Placzek M, Przybilla B. Protective effect against sunburn of combined systemic ascorbic acid (vitamin C) and d-alpha-tocopherol (vitamin E). *J Am Acad Dermatol* 1998;38:45–8.

4. Greul AK, Grundmann JU, Heinrich F, et al. Photoprotection of UV-irradiated human skin: an antioxidative combination of vitamins E and C, carotenoids, selenium and proanthocyanidins. *Skin Pharmacol Appl Skin Physiol* 2002;15:307–15.

5. Fortes C, Mastroeni S, Melchi F, et al. A protective effect of the Mediterranean diet for cutaneous melanoma. *Int J Epidemiol* 2008;37:1018–29.

6. Kopcke W, Krutmann J. Protection from sunburn with beta-carotene—a meta-analysis. *Photochem Photobiol* 2008;84:284–8.

7. Darvin ME, Sterry W, Lademann J, Patzelt A. Alcohol consumption decreases the protection efficiency of the antioxidant network and increases the risk of sunburn in human skin. *Skin Pharmacol Physiol* 2013;26:45–51.

8. Mogollon JA, Boivin C, Lemieux S, Blanchet C, Claveau J, Dodin S. Chocolate flavanols and skin photoprotection: a parallel, double-blind, randomized clinical trial. *Nutr J* 2014;13:66.

9. Alexander JW, MacMillan BG, Stinnett JD, et al. Beneficial effects of aggressive protein feeding in severely burned children. *Ann Surg* 1980;192:505–17.

10. Faber AW, Patterson DR, Bremer M. Repeated use of immersive virtual reality therapy to control pain during wound dressing changes in pediatric and adult burn patients. *J Burn Care Res* 2013;34:563–8.

11. Berger MM, Davadant M, Marin C, et al. Impact of a pain protocol including hypnosis in major burns. *Burns* 2010;36:639–46.

12. Kipping B, Rodger S, Miller K, Kimble RM. Virtual reality for acute pain reduction in adolescents undergoing burn wound care: a prospective randomized controlled trial. *Burns* 2012;38:650–7.

13. Kavanagh C, Freeman R. Should children participate in burn care? *Am J Nurs* 1984;84:601.

14. Genuino GA, Baluyut-Angeles KV, Espiritu AP, Lapitan MC, Buckley BS. Topical petrolatum gel alone versus topical silver sulfadiazine with standard gauze dressings for the treatment of superficial partial thickness burns in adults: a randomized controlled trial. *Burns* 2014;40:1267–73.

15. Kaplan JZ. Acceleration of wound healing by a live yeast cell derivative. *Arch Surg* 1984;119:1005–8.

16. Crowe MJ, McNeill RB, Schlemm DJ, Greenhalgh DG, Keller SJ. Topical application of yeast extract accelerates the wound healing of diabetic mice. *J Burn Care Rehabil* 1999;20:155–62.

17. Grindlay D, Reynolds T. The aloe vera phenomenon: a review of the properties and modern uses of the leaf parenchyma gel. *J Ethnopharmacol* 1986;16:117–51.

18. Shahzad MN, Ahmed N. Effectiveness of Aloe Vera gel compared with 1% silver sulphadiazine cream as burn wound dressing in second degree burns. *J Pak Med Assoc* 2013;63:225–30.

19. Gegova G, Manolova N, Serkedzhieva I, et al. Combined effect of selected antiviral substances of natural and synthetic origin. II. Anti-influenza activity of a combination of a polyphenolic complex isolated from Geranium sanguineum L. and rimantadine in vivo. *Acta Microbiol Bulg* 1993;30:37–40.

20. Sienkiewicz M, Poznanska-Kurowska K, Kaszuba A, Kowalczyk E. The antibacterial activity of geranium oil against gram-negative bacteria isolated from difficult-to-heal wounds. *Burns* 2014;40:1046–51.

21. Mehrabani D, Farjam M, Geramizadeh B, Tanideh N, Amini M, Panjehshahin MR. The healing effect of curcumin on burn wounds in rat. *World J Plast Surg* 2015;4:29–35.

22. Amish Burn Study Group, Kolacz NM, Jaroch MT, et al. The effect of Burns & Wounds (B&W)/burdock leaf therapy on burn-injured Amish patients: a pilot study measuring pain levels, infection rates, and healing times. *J Holist Nurs* 2014;32:327–40.

23. Jull AB, Cullum N, Dumville JC, Westby MJ, Deshpande S, Walker N. Honey as a topical treatment for wounds. *Cochrane Database Syst Rev* 2015;3:CD005083.

24. Garty BZ. Garlic burns. *Pediatrics* 1993;91:658–9.

25. Barbosa E, Faintuch J, Machado Moreira EA, et al. Supplementation of vitamin E, vitamin C, and zinc attenuates oxidative stress in burned children: a randomized, double-blind, placebo-controlled pilot study. *J Burn Care Res* 2009;30:859–66.

26. Berger MM, Cavadini C, Chiolero R, Guinchard S, Krupp S, Dirren H. Influence of large intakes of trace elements on recovery after major burns. *Nutrition* 1994;10:327–34; discussion 52.

27. Berger MM, Spertini F, Shenkin A, et al. Trace element supplementation modulates pulmonary infection rates after major burns: a double-blind, placebo-controlled trial. *Am J Clin Nutr* 1998;68:365–71.

28. Klein GL, Nicolai M, Langman CB, Cuneo BF, Sailer DE, Herndon DN. Dysregulation of calcium homeostasis after severe burn injury in children: possible role of magnesium depletion. *J Pediatr* 1997;131:246–51.

29. Saffle JR, Wiebke G, Jennings K, Morris SE, Barton RG. Randomized trial of immune-enhancing enteral nutrition in burn patients. *J Trauma* 1997;42:793–800; discussion 2.

30. Gottschlich MM, Warden GD, Michel M, et al. Diarrhea in tube-fed burn patients: incidence, etiology, nutritional impact, and prevention. *JPEN J Parenter Enteral Nutr* 1988;12:338–45.

31. Ghanayem H. Towards evidence based emergency medicine: Best BETs from the Manchester Royal Infirmary. BET 3: Vitamin C in severe burns. *Emerg Med J* 2012;29:1017–18.

32. Haberal M, Hamaloglu E, Bora S, Oner G, Bilgin N. The effects of vitamin E on immune regulation after thermal injury. *Burns Incl Therm Inj* 1988;14:388–93.

33. Shewmake KB, Talbert GE, Bowser-Wallace BH, Caldwell FT, Jr., Cone JB. Alterations in plasma copper, zinc, and ceruloplasmin levels in patients with thermal trauma. *J Burn Care Rehabil* 1988;9:13–17.

34. Caldis-Coutris N, Gawaziuk JP, Logsetty S. Zinc supplementation in burn patients. *J Burn Care Res* 2012;33:678–82.

35. Lin JJ, Chung XJ, Yang CY, Lau HL. A meta-analysis of trials using the intention to treat principle for glutamine supplementation in critically ill patients with burn. *Burns* 2013;39:565–70.

36. Field T, Peck M, Krugman S, et al. Burn injuries benefit from massage therapy. *J Burn Care Rehabil* 1998;19:241–4.

37. Field T, Peck M, Scd, et al. Postburn itching, pain, and psychological symptoms are reduced with massage therapy. *J Burn Care Rehabil* 2000;21:189–93.

38. Lewis SM, Clelland JA, Knowles CJ, Jackson JR, Dimick AR. Effects of auricular acupuncture-like transcutaneous electric nerve stimulation on pain levels following wound care in patients with burns: a pilot study. *J Burn Care Rehabil* 1990;11:322–9.

39. Sumano H, Mateos G. The use of acupuncture-like electrical stimulation for wound healing of lesions unresponsive to conventional treatment. Am J Acupunct 1999;27:5–14.

40. Turner JG, Clark AJ, Gauthier DK, Williams M. The effect of therapeutic touch on pain and anxiety in burn patients. *J Adv Nurs* 1998;28:10–20.

41. Leaman AM, Gorman D. Cantharis in the early treatment of minor burns. *Arch Emerg Med* 1989;6:259–61.

Chapter 10: Colds

1. Cohen S, Doyle WJ, Skoner DP, Rabin BS, Gwaltney JM, Jr. Social ties and susceptibility to the common cold. *JAMA* 1997;277:1940–4.

2. Saketkhoo K, Januszkiewicz A, Sackner MA. Effects of drinking hot water, cold water, and chicken soup on nasal mucus velocity and nasal airflow resistance. *Chest* 1978;74:408–10.

3. Rennard BO, Ertl RF, Gossman GL, Robbins RA, Rennard SI. Chicken soup inhibits neutrophil chemotaxis in vitro. *Chest* 2000;118:1150–7.

4. Sanu A, Eccles R. The effects of a hot drink on nasal airflow and symptoms of common cold and flu. *Rhinology* 2008;46:271–5.

5. Nieman DC, Henson DA, Dumke CL, et al. Relationship between salivary IgA secretion and upper respiratory tract infection following a 160-km race. *J Sports Med Phys Fitness* 2006;46:158–62.

6. Orzech KM, Acebo C, Seifer R, Barker D, Carskadon MA. Sleep patterns are associated with common illness in adolescents. *J Sleep Res* 2014;23:133–42.

7. Singh M, Singh M. Heated, humidified air for the common cold. *Cochrane Database Syst Rev* 2013;6:CD001728.

8. Pach D, Knochel B, Ludtke R, et al. Visiting a sauna: does inhaling hot dry air reduce common cold symptoms? A randomised controlled trial. *Med J Aust* 2010;193:730–4.

9. Smith SM, Schroeder K, Fahey T. Over-the-counter (OTC) medications for acute cough in children and adults in community settings. *Cochrane Database Syst Rev* 2014;11:CD001831.

10. Klein-Schwartz W, Sorkin JD, Doyon S. Impact of the voluntary withdrawal of over-the-counter cough and cold medications on pediatric ingestions reported to poison centers. *Pharmacoepidemiology and Drug Safety* 2010;19:819–24.

11. Gaffey MJ, Gwaltney JM, Jr., Sastre A, Dressler WE, Sorrentino JV, Hayden FG. Intranasally and orally administered antihistamine treatment of experimental rhinovirus colds. *Am Rev Respir Dis* 1987;136:556–60.

12. Taylor JA, Novack AH, Almquist JR, Rogers JE. Efficacy of cough suppressants in children. *J Pediatr* 1993;122:799–802.

13. Oduwole O, Meremikwu MM, Oyo-Ita A, Udoh EE. Honey for acute cough in children. *Cochrane Database Syst Rev* 2014;12:CD007094.

14. Hoffer-Schaefer A, Rozycki HJ, Yopp MA, Rubin BK. Guaifenesin has no effect on sputum volume or sputum properties in adoles-

cents and adults with acute respiratory tract infections. *Respir Care* 2014;59:631–6.

15. Kuhn JJ, Hendley JO, Adams KF, Clark JW, Gwaltney JM, Jr. Antitussive effect of guaifenesin in young adults with natural colds. Objective and subjective assessment. *Chest* 1982;82:713–18.

16. Hirsch SR, Viernes PF, Kory RC. The expectorant effect of glyceryl guaiacolate in patients with chronic bronchitis. A controlled in vitro and in vivo study. *Chest* 1973;63:9–14.

17. Graham NM, Burrell CJ, Douglas RM, Debelle P, Davies L. Adverse effects of aspirin, acetaminophen, and ibuprofen on immune function, viral shedding, and clinical status in rhinovirus-infected volunteers. *The Journal of Infectious Diseases* 1990;162:1277–82.

18. Dart RC, Paul IM, Bond GR, et al. Pediatric fatalities associated with over the counter (nonprescription) cough and cold medications. *Ann Emerg Med* 2009;53:411–17.

19. Eccles R, Jawad MS, Morris S. The effects of oral administration of (-)-menthol on nasal resistance to airflow and nasal sensation of airflow in subjects suffering from nasal congestion associated with the common cold. *J Pharm Pharmacol* 1990;42:652–4.

20. AlBalawi ZH, Othman SS, Alfaleh K. Intranasal ipratropium bromide for the common cold. *Cochrane Database Syst Rev* 2013;6:CD008231.

21. Stansfield SK, Pierre-Louis M, Lerebours G, Augustin A. Vitamin A supplementation and increased prevalence of childhood diarrhoea and acute respiratory infections. *Lancet* 1993;342:578–82.

22. Kartasasmita CB, Rosmayudi O, Soemantri ES, Deville W, Demedts M. Vitamin A and acute respiratory infections. *Paediatr Indones* 1991;31:41–9.

23. Hunter DC, Skinner MA, Wolber FM, et al. Consumption of gold kiwifruit reduces severity and duration of selected upper respiratory tract infection symptoms and increases plasma vitamin C concentration in healthy older adults. *Br J Nutr* 2012;108:1235–45.

24. Johnston CS, Barkyoumb GM, Schumacher SS. Vitamin C supplementation slightly improves physical activity levels and reduces cold incidence in men with marginal vitamin C status: a randomized controlled trial. *Nutrients* 2014;6:2572–83.

25. Fondell E, Balter O, Rothman KJ, Balter K. Dietary intake and supplement use of vitamins C and E and upper respiratory tract infection. *J Am Coll Nutr* 2011;30:248–58.

26. Chalmers TC. Effects of ascorbic acid on the common cold. An evaluation of the evidence. *Am J Med* 1975;58:532–6.

27. Douglas RM, Hemila H, Chalker E, Treacy B. Vitamin C for preventing and treating the common cold. *Cochrane Database Syst Rev* 2007:CD000980.

28. Bergman P, Lindh AU, Bjorkhem-Bergman L, Lindh JD. Vitamin D and respiratory tract infections: a systematic review and meta-analysis of randomized controlled trials. *PLoS One* 2013;8:e65835.

29. McFarlin BK, Carpenter KC, Davidson T, McFarlin MA. Baker's yeast beta glucan supplementation increases salivary IgA and decreases cold/flu symptomatic days after intense exercise. *J Diet Suppl* 2013;10:171–83.

30. Talbott SM, Talbott JA. Baker's yeast beta-glucan supplement reduces upper respiratory symptoms and improves mood state in stressed women. *J Am Coll Nutr* 2012;31:295–300.

31. Hao Q, Lu Z, Dong BR, Huang CQ, Wu T. Probiotics for preventing acute upper respiratory tract infections. *Cochrane Database Syst Rev* 2011:CD006895.

32. Ozen M, Kocabas Sandal G, Dinleyici EC. Probiotics for the prevention of pediatric upper respiratory tract infections: a systematic review. *Expert Opin Biol Ther* 2015;15:9–20.

33. Garaiova I, Muchova J, Nagyova Z, et al. Probiotics and vitamin C for the prevention of respiratory tract infections in children attending preschool: a randomised controlled pilot study. *Eur J Clin Nutr* 2015;69:373–9.

34. Zhang JS, Tian Z, Lou ZC. [Quality evaluation of twelve species of Chinese Ephedra (ma huang)]. *Yao Xue Xue Bao* 1989;24:865–71.

35. Schapowal A, Klein P, Johnston SL. Echinacea reduces the risk of recurrent respiratory tract infections and complications: a meta-analysis

of randomized controlled trials. *Adv Ther* 2015;32:187–200.

36. Weber W, Taylor JA, Stoep AV, Weiss NS, Standish LJ, Calabrese C. Echinacea purpurea for prevention of upper respiratory tract infections in children. *J Altern Complement Med* 2005;11:1021–6.

37. Minetti AM, Forti S, Tassone G, Torretta S, Pignataro L. Efficacy of complex herbal compound of Echinacea angustifolia (Imoviral(R) Junior) in recurrent upper respiratory tract infections during pediatric age: preliminary results. *Minerva Pediatr* 2011;63:177–82.

38. Saxena RC, Singh R, Kumar P, et al. A randomized double blind placebo controlled clinical evaluation of extract of Andrographis paniculata (KalmCold) in patients with uncomplicated upper respiratory tract infection. *Phytomedicine* 2010;17:178–85.

39. Kligler B, Ulbricht C, Basch E, et al. Andrographis paniculata for the treatment of upper respiratory infection: a systematic review by the natural standard research collaboration. *Explore (NY)* 2006;2:25–9.

40. Zakay-Rones Z, Thom E, Wollan T, Wadstein J. Randomized study of the efficacy and safety of oral elderberry extract in the treatment of influenza A and B virus infections. *The Journal of International Medical Research* 2004;32:132–40.

41. Zakay-Rones Z, Varsano N, Zlotnik M, et al. Inhibition of several strains of influenza virus in vitro and reduction of symptoms by an elderberry extract (Sambucus nigra L.) during an outbreak of influenza B Panama. *J Altern Complement Med* 1995;1:361–9.

42. Nantz MP, Rowe CA, Muller CE, Creasy RA, Stanilka JM, Percival SS. Supplementation with aged garlic extract improves both NK and gammadelta-T cell function and reduces the severity of cold and flu symptoms: a randomized, double-blind, placebo-controlled nutrition intervention. *Clinical Nutrition* 2012;31:337–44.

43. Seida JK, Durec T, Kuhle S. North American (Panax quinquefolius) and Asian Ginseng (Panax ginseng) preparations for prevention of the common cold in healthy adults: a systematic review. *Evid Based Complement Alternat Med* 2011;2011:282151.

44. Timmer A, Gunther J, Motschall E, Rucker G, Antes G, Kern WV. Pelargonium sidoides extract for treating acute respiratory tract infections. *Cochrane Database Syst Rev* 2013;10:CD006323.

Chapter 11: Colic

1. Barr RG, Kramer MS, Pless IB, Boisjoly C, Leduc D. Feeding and temperament as determinants of early infant crying/fussing behavior. *Pediatrics* 1989;84:514–21.

2. Fillion L, Duval S, Dumont S, et al. Impact of a meaning-centered intervention on job satisfaction and on quality of life among palliative care nurses. *Psychooncology* 2009;18:1300–10.

3. Zeskind PS, Barr RG. Acoustic characteristics of naturally occurring cries of infants with "colic." *Child Dev* 1997;68:394–403.

4. Ahmad K, Fatemeh F, Mehri N, Maryam S. Probiotics for the treatment of pediatric Helicobacter pylori infection: a randomized double blind clinical trial. *Iranian Journal of Pediatrics* 2013;23:79–84.

5. Canivet C, Hagander B, Jakobsson I, Lanke J. Infantile colic—less common than previously estimated? *Acta Paediatr* 1996;85:454–8.

6. Rautava P, Lehtonen L, Helenius H, Sillanpaa M. Infantile colic: child and family three years later. *Pediatrics* 1995;96:43–7.

7. Sillanpaa M, Saarinen M. Infantile colic associated with childhood migraine: A prospective cohort study. *Cephalalgia* 2015;35(14):1240–51.

8. Karp H. The fourth trimester and the calming reflex: novel ideas for nurturing young infants. *Midwifery Today with International Midwife* 2012:25–6, 67.

9. Lothe L, Lindberg T. Cow's milk whey protein elicits symptoms of infantile colic in colicky formula-fed infants: a double-blind crossover study. *Pediatrics* 1989;83:262–6.

10. Clyne PS, Kulczycki A, Jr. Human breast milk contains bovine IgG. Relationship to infant colic? *Pediatrics* 1991;87:439–44.

11. Estep DC, Kulczycki A, Jr. Treatment of

infant colic with amino acid-based infant formula: a preliminary study. *Acta Paediatr* 2000;89:22–7.

12. Campbell JP. Dietary treatment of infant colic: a double-blind study. *J Royal Coll of Gen Prac* 1989;39:11–14.

13. Swadling C, Griffiths P. Is modified cow's milk formula effective in reducing symptoms of infant colic? *Br J Community Nurs* 2003;8:24–7.

14. Evans RW, Fergusson DM, Allardyce RA, Taylor B. Maternal diet and infantile colic in breast-fed infants. *Lancet* 1981;1:1340–2.

15. Duro D, Rising R, Cedillo M, Lifshitz F. Association between infantile colic and carbohydrate malabsorption from fruit juices in infancy. *Pediatrics* 2002;109:797–805.

16. Barr RG, McMullan SJ, Spiess H, et al. Carrying as colic "therapy": a randomized controlled trial. *Pediatrics* 1991;87:623–30.

17. Savino F, Palumeri E, Castagno E, et al. Reduction of crying episodes owing to infantile colic: A randomized controlled study on the efficacy of a new infant formula. *Eur J Clin Nutr* 2006;60:1304–10.

18. Forsyth BW. Colic and the effect of changing formulas: a double-blind, multiple-crossover study. *J Pediatr* 1989;115:521–6.

19. Barr RG, Wooldridge J, Hanley J. Effects of formula change on intestinal hydrogen production and crying and fussing behavior. *J Dev Behav Pediatr* 1991;12:248–53.

20. Arikan D, Alp H, Gozum S, Orbak Z, Cifci EK. Effectiveness of massage, sucrose solution, herbal tea or hydrolysed formula in the treatment of infantile colic. *J Clin Nurs* 2008;17:1754–61.

21. Keane V, Straus J, Roberts K. Do solids help baby sleep through the night? *Am J Dis Child* 1988;142:404–5.

22. Treem WR, Hyams JS, Blankschen E, et al. Evaluation of the effect of a fiber-enriched formula on infant colic. *J Pediatr* 1991;119:695–701.

23. Larson K, Ayllon T. The effects of contingent music and differential reinforcement on infantile colic. *Behav Res Ther* 1990;28:119–25.

24. Savino F, Ceratto S, De Marco A, Cordero di Montezemolo L. Looking for new treatments of infantile colic. *Italian Journal of Pediatrics* 2014;40:53.

25. Dobson D, Lucassen PL, Miller JJ, et al. Manipulative therapies for infantile colic. *Cochrane Database Syst Rev* 2012;12:CD004796.

26. Kanabar D, Randhawa M, Clayton P. Improvement of symptoms in infant colic following reduction of lactose load with lactase. *J Hum Nutr Diet* 2001;14:359–63.

27. Weissbluth M, Christoffel KK, Davis AT. Treatment of infantile colic with dicyclomine hydrochloride. *J Pediatr* 1984;104:951–5.

28. Savino F, Cordisco L, Tarasco V, et al. Lactobacillus reuteri DSM 17938 in infantile colic: a randomized, double-blind, placebo-controlled trial. *Pediatrics* 2010;126:e526–33.

29. Szajewska H, Gyrczuk E, Horvath A. Lactobacillus reuteri DSM 17938 for the management of infantile colic in breastfed infants: a randomized, double-blind, placebo-controlled trial. *J Pediatr* 2013;162:257–62.

30. Chau K, Lau E, Greenberg S, et al. Probiotics for infantile colic: a randomized, double-blind, placebo-controlled trial investigating Lactobacillus reuteri DSM 17938. *J Pediatr* 2015;166:74–8.

31. Savino F, Pelle E, Palumeri E, Oggero R, Miniero R. Lactobacillus reuteri (American Type Culture Collection Strain 55730) versus simethicone in the treatment of infantile colic: a prospective randomized study. *Pediatrics* 2007;119:e124–30.

32. Indrio F, Di Mauro A, Riezzo G, et al. Prophylactic use of a probiotic in the prevention of colic, regurgitation, and functional constipation: a randomized clinical trial. *JAMA Pediatrics* 2014;168:228–33.

33. Savino F, Ceratto S, Poggi E, Cartosio ME, Cordero di Montezemolo L, Giannattasio A. Preventive effects of oral probiotic on infantile colic: a prospective, randomised, blinded, controlled trial using Lactobacillus reuteri DSM 17938. *Beneficial Microbes* 2015;6:245–51.

34. Giovannini M, Verduci E, Gregori D, et al. Prebiotic effect of an infant formula supplemented with galacto-oligosaccharides: randomized multicenter trial. *J Am Coll Nutr* 2014;33:385–93.

35. Weizman Z, Alkrinawi S, Goldfarb D, Bitran C. Efficacy of herbal tea preparation in infantile colic. *Journal of Pediatrics* 1993;122:650–2.

36. Savino F, Cresi F. A randomized double-blind placebo-controlled trial of a standardized extract of chamomile, fennel, and lemon balm in the treatment of colicky infants. *Phytother Res* 2005;19(4):335–40.

37. Alexandrovich I, Rakovitskaya O, Kolmo E, Sidorova T, Shushunov S. The effect of fennel (Foeniculum Vulgare) seed oil emulsion in infantile colic: a randomized, placebo-controlled study. *Altern Ther Health Med* 2003;9:58–61.

38. Madden GR, Schmitz KH, Fullerton K. A case of infantile star anise toxicity. *Pediatr Emerg Care* 2012;28:284–5.

39. Cetinkaya B, Basbakkal Z. The effectiveness of aromatherapy massage using lavender oil as a treatment for infantile colic. *Int J Nurs Pract* 2012;18:164–9.

40. Klougart N, Nilsson N, Jacobsen J. Infantile colic treated by chiropractors: a prospective study of 316 cases. *J Manipulative Physiol Ther* 1989;12:281–8.

41. Wiberg JM, Nordsteen J, Nilsson N. The short-term effect of spinal manipulation in the treatment of infantile colic: a randomized controlled clinical trial with a blinded observer. *J Manipulative Physiol Ther* 1999;22:517–22.

42. Miller JE, Newell D, Bolton JE. Efficacy of chiropractic manual therapy on infant colic: a pragmatic single-blind, randomized controlled trial. *J Manipulative Physiol Ther* 2012;35:600–7.

43. Hayden C, Mullinger B. A preliminary assessment of the impact of cranial osteopathy for the relief of infantile colic. *Complement Ther Clin Pract* 2006;12:83–90.

44. Gustafsson PA, Birberg-Thornberg U, Duchen K, et al. EPA supplementation improves teacher-rated behaviour and oppositional symptoms in children with ADHD. *Acta Paediatr* 2010;99:1540–9.

45. Landgren K, Kvorning N, Hallstrom I. Feeding, stooling and sleeping patterns in infants with colic—a randomized controlled trial of minimal acupuncture. *BMC Complement Altern Med* 2011;11:93.

46. Skjeie H, Skonnord T, Fetveit A, Brekke M. Acupuncture for infantile colic: a blinding-validated, randomized controlled multicentre trial in general practice. *Scandinavian Journal of Primary Health Care* 2013;31:190–6.

Chapter 12: Constipation

1. Egger G, Wolfenden K, Pares J, Mowbray G. "Bread: it's a great way to go." Increasing bread consumption decreases laxative sales in an elderly community. *Med J Aust* 1991;155:820–1.

2. Dupont C, Campagne A, Constant F. Efficacy and safety of a magnesium sulfate-rich natural mineral water for patients with functional constipation. *Clin Gastroenterol Hepatol* 2014;12:1280–7.

3. Miller LE, Ouwehand AC. Probiotic supplementation decreases intestinal transit time: meta-analysis of randomized controlled trials. *World J Gastroenterol* 2013;19:4718–25.

4. Korterink JJ, Ockeloen L, Benninga MA, Tabbers MM, Hilbink M, Deckers-Kocken JM. Probiotics for childhood functional gastrointestinal disorders: a systematic review and meta-analysis. *Acta Paediatr* 2014;103:365–72.

5. Gordon M, Naidoo K, Akobeng AK, Thomas AG. Cochrane Review: Osmotic and stimulant laxatives for the management of childhood constipation (Review). *Evidence-Based Child Health: A Cochrane Review Journal* 2013;8:57–109.

6. Chen SL, Cai SR, Deng L, et al. Efficacy and complications of polyethylene glycols for treatment of constipation in children: a meta-analysis. *Medicine (Baltimore)* 2014;93:e65.

7. Clark JH, Russell GJ, Fitzgerald JF, Nagamori KE. Serum beta-carotene, retinol, and alpha-tocopherol levels during mineral oil therapy for constipation. *Am J Dis Child* 1987;141:1210–12.

8. Siegers CP, von Hertzberg-Lottin E, Otte M, Schneider B. Anthranoid laxative abuse—a risk for colorectal cancer? *Gut* 1993;34:1099101.

9. Klauser AG, Flaschentrager J, Gehrke A, Muller-Lissner SA. Abdominal wall massage: effect on colonic function in healthy volunteers and in patients with chronic constipation. *Z Gastroenterol* 1992;30:247–51.

10. Lamas K, Graneheim UH, Jacobsson C. Experiences of abdominal massage for constipation. *J Clin Nurs* 2012;21:757–65.

11. Silva CA, Motta ME. The use of abdominal muscle training, breathing exercises and abdominal massage to treat paediatric chronic functional constipation. *Colorectal Dis* 2013;15:e250–5.

Chapter 13: Cough

1. Johnston ID, Strachan DP, Anderson HR. Effect of pneumonia and whooping cough in childhood on adult lung function. *N Engl J Med* 1998;338:581–7.

2. Arroll B. Non-antibiotic treatments for upper-respiratory tract infections (common cold). *Respiratory Medicine* 2005;99:1477–84.

3. Larkin A, Lassetter J. Vitamin D deficiency and acute lower respiratory infections in children younger than 5 years: identification and treatment. *J Pediatr Health Care* 2014;28:572–82.

4. Camargo CA, Jr., Ganmaa D, Sidbury R, et al. Randomized trial of vitamin D supplementation for winter-related atopic dermatitis in children. *J Allergy Clin Immunol* 2014;134:831–5 e1.

5. Imhoff-Kunsch B, Stein AD, Martorell R, et al. Prenatal docosahexacnoic acid supplementation and infant morbidity: randomized controlled trial. *Pediatrics* 2011;128:e505–12.

6. Lapillonne A, Pastor N, Zhuang W, Scalabrin DM. Infants fed formula with added long chain polyunsaturated fatty acids have reduced incidence of respiratory illnesses and diarrhea during the first year of life. *BMC Pediatr* 2014;14:168.

7. Hageman JH, Hooyenga P, Diersen-Schade DA, Scalabrin DM, Wichers HJ, Birch EE. The impact of dietary long-chain polyunsaturated fatty acids on respiratory illness in infants and children. *Curr Allergy Asthma Rep* 2012;12:564–73.

8. Pinnock CB, Arney WK. The milk-mucus belief: sensory analysis comparing cow's milk and a soy placebo. *Appetite* 1993;20:61–70.

9. Fluge O, Omenaas E, Eide GE, Gulsvik A. Fish consumption and respiratory symptoms among young adults in a Norwegian community. *The European Respiratory Journal* 1998;12:336–40.

10. Cohen HA, Rozen J, Kristal H, et al. Effect of honey on nocturnal cough and sleep quality: a double-blind, randomized, placebo-controlled study. *Pediatrics* 2012;130:465–71.

11. Oduwole O, Meremikwu MM, Oyo-Ita A, Udoh EE. Honey for acute cough in children. *Cochrane Database Syst Rev* 2014;12:CD007094.

12. Mortensen J, Falk M, Groth S, Jensen C. The effects of postural drainage and positive expiratory pressure physiotherapy on tracheobronchial clearance in cystic fibrosis. *Chest* 1991;100:1350–7.

13. Mamolen M, Lewis DM, Blanchet MA, Satink FJ, Vogt RL. Investigation of an outbreak of "humidifier fever" in a print shop. *Am J Ind Med* 1993;23:483–90.

14. Stein MT, Harper G, Chen J. Persistent cough in an adolescent. *J Dev Behav Pediatr* 1999;20:434–6.

15. Elkins GR, Carter BD. Hypnotherapy in the treatment of childhood psychogenic coughing: a case report. *Am J Clin Hypn* 1986;29:59–63.

16. Lavigne JV, Davis AT, Fauber R. Behavioral management of psychogenic cough: alternative to the "bedsheet" and other aversive techniques. *Pediatrics* 1991;87:532–7.

17. Liu X. 41 Cases of cough treated with cupping. *International Journal of Clinical Acupuncture* 1991;2:319–22.

18. Cui X, Wang SM, Wu LQ. Sixty-eight cases of child chronic cough treated by moxibustion. *J Tradit Chin Med* 2009;29:9–10.

Chapter 14: Diaper Rash

1. Mendling W, Brasch J, German Society for G, et al. Guideline vulvovaginal candidosis (2010) of the German Society for Gynecol-

ogy and Obstetrics, the Working Group for Infections and Infectimmunology in Gynecology and Obstetrics, the German Society of Dermatology, the Board of German Dermatologists and the German Speaking Mycological Society. *Mycoses* 2012;55 Suppl 3:1–13.

2. De Seta F, Parazzini F, De Leo R, et al. Lactobacillus plantarum P17630 for preventing candida vaginitis recurrence: a retrospective comparative study. *Eur J Obstet Gynecol Reprod Biol* 2014;182:136–9.

3. Oncel MY, Arayici S, Sari FN, et al. Comparison of Lactobacillus reuteri and nystatin prophylaxis on candida colonization and infection in very low birth weight infants. *J Matern Fetal Neonatal Med* 2014:1–5.

4. Klunk C, Domingues E, Wiss K. An update on diaper dermatitis. *Clin Dermatol* 2014;32:477–87.

5. Coughlin CC, Eichenfield LF, Frieden IJ. Diaper dermatitis: clinical characteristics and differential diagnosis. *Pediatr Dermatol* 2014;31 Suppl 1:19–24.

6. Mohamadi J, Motaghi M, Panahi J, et al. Antifungal resistance in candida isolated from oral and diaper rash candidiasis in neonates. *Bioinformation* 2014;10:667–70.

7. Healy CE. Precocious puberty due to a diaper ointment. *Indiana Med* 1984;77:610.

8. Panahi Y, Sharif MR, Sharif A, et al. A randomized comparative trial on the therapeutic efficacy of topical aloe vera and calendula officinalis on diaper dermatitis in children. *ScientificWorldJournal* 2012;2012:810234.

9. Despard C. Diaper dermatitis: another simple remedy. *CMAJ* 1988;139:706.

10. Collipp PJ, Kuo B, Castro-Magana M, Chen SY, Salvatore S. Hair zinc, scalp hair quantity, and diaper rash in normal infants. *Cutis* 1985;35:66–70.

Chapter 15: Diarrhea

1. Jarvinen KM, Nowak-Wegrzyn A. Food protein-induced enterocolitis syndrome (FPIES): current management strategies and review of the literature. *The Journal of Allergy and Clinical Immunology in Practice* 2013;1:317–22.

2. Sopo SM, Giorgio V, Dello Iacono I, Novembre E, Mori F, Onesimo R. A multicentre retrospective study of 66 Italian children with food protein-induced enterocolitis syndrome: different management for different phenotypes. *Clin Exp Allergy* 2012;42:1257–65.

3. Caubet JC, Ford LS, Sickles L, et al. Clinical features and resolution of food protein-induced enterocolitis syndrome: 10-year experience. *J Allergy Clin Immunol* 2014;134:382–9.

4. Goldenberg JZ, Ma SS, Saxton JD, et al. Probiotics for the prevention of Clostridium difficile-associated diarrhea in adults and children. *Cochrane Database Syst Rev* 2013;5:CD006095.

5. Avery ME, Snyder JD. Oral therapy for acute diarrhea. The underused simple solution. *N Engl J Med* 1990;323:891–4.

6. Simakachorn N, Tongpenyai Y, Tongtan O, Varavithya W. Randomized, double-blind clinical trial of a lactose-free and a lactose-containing formula in dietary management of acute childhood diarrhea. *J Med Assoc Thai* 2004;87:641–9.

7. Brown KH, Peerson JM, Fontaine O. Use of nonhuman milks in the dietary management of young children with acute diarrhea: a meta-analysis of clinical trials. *Pediatrics* 1994;93:17–27.

8. Lamberti LM, Walker CL, Chan KY, Jian WY, Black RE. Oral zinc supplementation for the treatment of acute diarrhea in children: a systematic review and meta-analysis. *Nutrients* 2013;5:4715–40.

9. Crisinel PA, Verga ME, Kouame KS, et al. Demonstration of the effectiveness of zinc in diarrhoea of children living in Switzerland. *Eur J Pediatr* 2015;174:1061–7.

10. Applegate JA, Fischer Walker CL, Ambikapathi R, Black RE. Systematic review of probiotics for the treatment of community-acquired acute diarrhea in children. *BMC Public Health* 2013;13 Suppl 3:S16.

11. Szajewska H, Urbanska M, Chmielewska A, Weizman Z, Shamir R. Meta-analysis: Lactobacillus reuteri strain DSM 17938 (and the original strain ATCC 55730) for treating acute

gastroenteritis in children. *Beneficial Microbes* 2014;5:285–93.

12. Patro-Golab B, Shamir R, Szajewska H. Yogurt for treating acute gastroenteritis in children: systematic review and meta-analysis. *Clinical Nutrition* 2015;34(5):818–24.

13. Ciccarelli S, Stolfi I, Caramia G. Management strategies in the treatment of neonatal and pediatric gastroenteritis. *Infection and Drug Resistance* 2013;6:133–61.

14. Guarino A, Ashkenazi S, Gendrel D, et al. European Society for Pediatric Gastroenterology, Hepatology, and Nutrition/European Society for Pediatric Infectious Diseases evidence-based guidelines for the management of acute gastroenteritis in children in Europe: update 2014. *Journal of Pediatric Gastroenterology and Nutrition* 2014;59:132–52.

15. Mittra D, Bukutu C, Vohra S. American Academy of Pediatrics Section on C, Integrative M. Complementary, holistic, and integrative medicine: a review of therapies for diarrhea. *Pediatr Rev* 2008;29:349–53.

16. Yalcin SS, Yurdakok K, Tezcan I, Oner L. Effect of glutamine supplementation on diarrhea, interleukin-8 and secretory immunoglobulin A in children with acute diarrhea. *Journal of Pediatric Gastroenterology and Nutrition* 2004;38:494–501.

17. Kamuchaki JM, Kiguli S, Wobudeya E, Bortolussi R. No benefit of glutamine supplementation on persistent diarrhea in Ugandan children. *Pediatr Infect Dis J* 2013;32:573–6.

18. Aksit S, Caglayan S, Cukan R, Yaprak I. Carob bean juice: a powerful adjunct to oral rehydration solution treatment in diarrhoea. *Paediatric and Perinatal Epidemiology* 1998;12:176–81.

19. Loeb H, Vandenplas Y, Wursch P, Guesry P. Tannin-rich carob pod for the treatment of acute-onset diarrhea. *Journal of Pediatric Gastroenterology and Nutrition* 1989;8:480–5.

20. Becker B, Kuhn U, Hardewig-Budny B. Double-blind, randomized evaluation of clinical efficacy and tolerability of an apple pectin-chamomile extract in children with unspecific diarrhea. *Arzneimittelforschung* 2006;56:387–93.

21. Jacobs J, Jonas WB, Jimenez-Perez M, Crothers D. Homeopathy for childhood diarrhea: combined results and metaanalysis from three randomized, controlled clinical trials. *Pediatr Infect Dis J* 2003;22:229–34.

22. Jacobs J, Guthrie BL, Montes GA, et al. Homeopathic combination remedy in the treatment of acute childhood diarrhea in Honduras. *J Altern Complement Med* 2006;12:723–32.

Chapter 16: Ear Infections

1. Zemek R, Szyszkowicz M, Rowe BH. Air pollution and emergency department visits for otitis media: a case-crossover study in Edmonton, Canada. *Environ Health Perspect* 2010;118:1631–6.

2. Fortanier AC, Venekamp RP, Boonacker CW, et al. Pneumococcal conjugate vaccines for preventing otitis media. *Cochrane Database Syst Rev* 2014;4:CD001480.

3. Norhayati MN, Ho JJ, Azman MY. Influenza vaccines for preventing acute otitis media in infants and children. *Cochrane Database Syst Rev* 2015;3:CD010089.

4. Cayir A, Turan MI, Ozkan O, et al. Serum vitamin D levels in children with recurrent otitis media. *European Archives of Oto-Rhino-Laryngology* 2014;271:689–93.

5. Marchisio P, Consonni D, Baggi E, et al. Vitamin D supplementation reduces the risk of acute otitis media in otitis-prone children. *Pediatr Infect Dis J* 2013;32:1055–60.

6. Roberts JE, Sanyal MA, Burchinal MR, Collier AM, Ramey CT, Henderson FW. Otitis media in early childhood and its relationship to later verbal and academic performance. *Pediatrics* 1986;78:423–30.

7. Roberts JE, Burchinal MR, Clarke-Klein SM. Otitis media in early childhood and cognitive, academic, and behavior outcomes at 12 years of age. *J Pediatr Psychol* 1995;20:645–60.

8. Paradise JL, Feldman HM, Campbell TF, et al. Tympanostomy tubes and developmental outcomes at 9 to 11 years of age. *N Engl J Med* 2007;356:248–61.

9. Lapillonne A, Pastor N, Zhuang W, Scalabrin

DM. Infants fed formula with added long chain polyunsaturated fatty acids have reduced incidence of respiratory illnesses and diarrhea during the first year of life. *BMC Pediatr* 2014;14:168.

10. Linday LA, Dolitsky JN, Shindledecker RD, Pippenger CE. Lemon-flavored cod liver oil and a multivitamin-mineral supplement for the secondary prevention of otitis media in young children: pilot research. *The Annals of Otology, Rhinology, and Laryngology* 2002;111:642–52.

11. Nsouli TM, Nsouli SM, Linde RE, O'Mara F, Scanlon RT, Bellanti JA. Role of food allergy in serous otitis media. *Ann Allergy* 1994;73:215–19.

12. Gisselsson-Solen M. Acute otitis media in children-current treatment and prevention. *Current Infectious Disease Reports* 2015;17:476.

13. Spurling GK, Del Mar CB, Dooley L, Foxlee R, Farley R. Delayed antibiotics for respiratory infections. *Cochrane Database Syst Rev* 2013;4:CD004417.

14. Lieberthal AS, Carroll AE, Chonmaitree T, et al. The diagnosis and management of acute otitis media. *Pediatrics* 2013;131:e964–99.

15. Kozyrskyj A, Klassen TP, Moffatt M, Harvey K. Short-course antibiotics for acute otitis media. *Cochrane Database Syst Rev* 2010:CD001095.

16. Berman S, Roark R, Luckey D. Theoretical cost effectiveness of management options for children with persisting middle ear effusions. Pediatrics 1994;93:353–63.

17. Petrou S, Dakin H, Abangma G, Benge S, Williamson I. Cost-utility analysis of topical intranasal steroids for otitis media with effusion based on evidence from the GNOME trial. *Value in Health* 2010;13:543–51.

18. Cantekin EI, Mandel EM, Bluestone CD, et al. Lack of efficacy of a decongestant-antihistamine combination for otitis media with effusion ("secretory" otitis media) in children. Results of a double-blind, randomized trial. *N Engl J Med* 1983;308:297–301.

19. Griffin G, Flynn CA. Antihistamines and/or decongestants for otitis media with effusion (OME) in children. *Cochrane Database Syst Rev* 2011:CD003423.

20. Elemraid MA, Mackenzie IJ, Fraser WD, Brabin BJ. Nutritional factors in the pathogenesis of ear disease in children: a systematic review. *Annals of Tropical Paediatrics* 2009;29:85–99.

21. Kumar J, Muntner P, Kaskel FJ, Hailpern SM, Melamed ML. Prevalence and associations of 25-hydroxyvitamin D deficiency in US children: NHANES 2001–2004. *Pediatrics* 2009;124:e362–70.

22. Uhari M, Kontiokari T, Koskela M, Niemela M. Xylitol chewing gum in prevention of acute otitis media: double blind randomised trial. *BMJ* 1996;313:1180–4.

23. Uhari M, Kontiokari T, Niemela M. A novel use of xylitol sugar in preventing acute otitis media. *Pediatrics* 1998;102:879–84.

24. Azarpazhooh A, Limeback H, Lawrence HP, Shah PS. Xylitol for preventing acute otitis media in children up to 12 years of age. *Cochrane Database Syst Rev* 2011:CD007095.

25. Vernacchio L, Corwin MJ, Vezina RM, et al. Xylitol syrup for the prevention of acute otitis media. *Pediatrics* 2014;133:289–95.

26. Adams JB, Audhya T, McDonough-Means S, et al. Toxicological status of children with autism vs. neurotypical children and the association with autism severity. *Biological Trace Element Research* 2013;151:171–80.

27. Boettcher J, Astrom V, Pahlsson D, Schenstrom O, Andersson G, Carlbring P. Internet-based mindfulness treatment for anxiety disorders: a randomized controlled trial. *Behav Ther* 2014;45:241–53.

28. Sarrell EM, Mandelberg A, Cohen HA. Efficacy of naturopathic extracts in the management of ear pain associated with acute otitis media. *Arch Pediatr Adolesc Med* 2001;155:796–9.

29. Sarrell EM, Cohen HA, Kahan E. Naturopathic treatment for ear pain in children. *Pediatrics* 2003;111:e574–9.

30. Marchisio P, Esposito S, Bianchini S, et al. Effectiveness of a propolis and zinc solution in preventing acute otitis media in children with a history of recurrent acute otitis media. *International Journal of Immunopathology and Pharmacology* 2010;23:567–75.

31. Pohlman KA, Holton-Brown MS. Otitis media and spinal manipulative therapy: a literature review. *Journal of Chiropractic Medicine* 2012;11:160–9.

32. Berkman ND, Wallace IF, Steiner MJ, et al. *Otitis Media With Effusion: Comparative Effectiveness of Treatments.* Rockville (MD) 2013.

33. Friese KH, Kruse S, Ludtke R, Moeller H. The homoeopathic treatment of otitis media in children—comparisons with conventional therapy. *Int J Clin Pharmacol Ther* 1997;35:296–301.

34. Harrison H, Fixsen A, Vickers A. A randomized comparison of homoeopathic and standard care for the treatment of glue ear in children. *Complement Ther Med* 1999;7:132–5.

35. Antman K, Benson MC, Chabot J, et al. Complementary and alternative medicine: the role of the cancer center. *J Clin Oncol* 2001;19:55S–60S.

36. Anderson JG, Taylor AG. Effects of healing touch in clinical practice: a systematic review of randomized clinical trials. *J Holist Nurs* 2011;29:221–8.

37. Sinha MN, Siddiqui VA, Nayak C, et al. Randomized controlled pilot study to compare homeopathy and conventional therapy in acute otitis media. *Homeopathy* 2012;101:5–12.

38. Grimaldi-Bensouda L, Begaud B, Rossignol M, et al. Management of upper respiratory tract infections by different medical practices, including homeopathy, and consumption of antibiotics in primary care: the EPI3 cohort study in France 2007–2008. *PLoS One* 2014;9:e89990.

39. Bell IR, Boyer NN. Homeopathic medications as clinical alternatives for symptomatic care of acute otitis media and upper respiratory infections in children. *Global Advances in Health and Medicine: Improving Healthcare Outcomes Worldwide* 2013;2:32–43.

Chapter 17: Eczema

1. Coughlin CC, Eichenfield LF, Frieden IJ. Diaper dermatitis: clinical characteristics and differential diagnosis. *Pediatr Dermatol* 2014;31 Suppl 1:19–24.

2. Fergusson DM, Horwood LJ. Early solid food diet and eczema in childhood: a 10-year longitudinal study. *Pediatric Allergy and Immunology: Official Publication of the European Society of Pediatric Allergy and Immunology* 1994;5:44–7.

3. Kanny G, Hatahet R, Moneret-Vautrin DA, Kohler C, Bellut A. Allergy and intolerance to flavouring agents in atopic dermatitis in young children. *Allergie et immunologie* 1994;26:204–6, 9–10.

4. Mertens J, Stock S, Lungen M, et al. Is prevention of atopic eczema with hydrolyzed formulas cost-effective? A health economic evaluation from *Pediatric Allergy and Immunology: Official Publication of the European Society of Pediatric Allergy and Immunology* 2012;23:597–604.

5. Yang C, Hao Z, Zhang LL, Guo Q. Efficacy and safety of acupuncture in children: an overview of systematic reviews. *Pediatr Res* 2015;78(2):112–19.

6. Al Rashoud AS, Abboud RJ, Wang W, Wigderowitz C. Efficacy of low-level laser therapy applied at acupuncture points in knee osteoarthritis: a randomised double-blind comparative trial. *Physiotherapy* 2014;100:242–8.

7. Schmitt J, Schmitt NM, Kirch W, Meurer M. Early exposure to antibiotics and infections and the incidence of atopic eczema: a population-based cohort study. *Pediatric Allergy and Immunology* 2010;21:292–300.

8. Saarinen UM, Kajosaari M. Breastfeeding as prophylaxis against atopic disease: prospective follow-up study until 17 years old. *Lancet* 1995;346:1065–9.

9. Hoppu U, Rinne M, Salo-Vaananen P, Lampi AM, Piironen V, Isolauri E. Vitamin C in breast milk may reduce the risk of atopy in the infant. *Eur J Clin Nutr* 2005;59:123–8.

10. Blomqvist S, Wester A, Sundelin G, Rehn B. Test-retest reliability, smallest real difference and concurrent validity of six different balance tests on young people with mild to moderate intellectual disability. *Physiotherapy* 2012;98:313–19.

11. Enomoto T, Sowa M, Nishimori K, et al.

Effects of bifidobacterial supplementation to pregnant women and infants in the prevention of allergy development in infants and on fecal microbiota. *Allergology International* 2014;63:575–85.

12. Bertelsen RJ, Brantsaeter AL, Magnus MC, et al. Probiotic milk consumption in pregnancy and infancy and subsequent childhood allergic diseases. *J Allergy Clin Immunol* 2014;133:165–71, e1–8.

13. Kremmyda LS, Vlachava M, Noakes PS, Diaper ND, Miles EA, Calder PC. Atopy risk in infants and children in relation to early exposure to fish, oily fish, or long-chain omega-3 fatty acids: a systematic review. *Clinical Reviews in Allergy & Immunology* 2011;41:36–66.

14. Bergmann MM, Caubet JC, Boguniewicz M, Eigenmann PA. Evaluation of food allergy in patients with atopic dermatitis. *The Journal of Allergy and Clinical Immunology in Practice* 2013;1:22–8.

15. Berkman ND, Wallace IF, Steiner MJ, et al. *Otitis Media With Effusion: Comparative Effectiveness of Treatments.* Rockville (MD) 2013.

16. Silverberg JI, Hanifin J, Simpson EL. Climatic factors are associated with childhood eczema prevalence in the United States. *The Journal of Investigative Dermatology* 2013;133:1752–9.

17. Casimir GJ, Duchateau J, Gossart B, Cuvelier P, Vandaele F, Vis HL. Atopic dermatitis: role of food and house dust mite allergens. *Pediatrics* 1993;92:252–6.

18. Ricci G, Patrizi A, Specchia F, et al. Effect of house dust mite avoidance measures in children with atopic dermatitis. *Br J Dermatol* 2000;143:379–84.

19. Ersser SJ, Cowdell F, Latter S, et al. Psychological and educational interventions for atopic eczema in children. *Cochrane Database Syst Rev* 2014;1:CD004054.

20. McMenamy CJ, Katz RC, Gipson M. Treatment of eczema by EMG biofeedback and relaxation training: a multiple baseline analysis. *Journal of Behavior Therapy and Experimental Psychiatry* 1988;19:221–7.

21. Fredriksson T, Gip L. Urea creams in the treatment of dry skin and hand dermatitis. *International Journal of Dermatology* 1975;14:442–4.

22. Munkvad M. A comparative trial of Clinitar versus hydrocortisone cream in the treatment of atopic eczema. *Br J Dermatol* 1989;121:763–6.

23. Sigurgeirsson B, Boznanski A, Todd G, et al. Safety and efficacy of pimecrolimus in atopic dermatitis: a 5-year randomized trial. *Pediatrics* 2015;135:597–606.

24. Foolad N, Brezinski EA, Chase EP, Armstrong AW. Effect of nutrient supplementation on atopic dermatitis in children: a systematic review of probiotics, prebiotics, formula, and fatty acids. *JAMA Dermatology* 2013;149:350–5.

25. Biagi PL, Bordoni A, Masi M, et al. A long-term study on the use of evening primrose oil (Efamol) in atopic children. *Drugs Exp Clin Res* 1988;14:285–90.

26. Morse NL, Clough PM. A meta-analysis of randomized, placebo-controlled clinical trials of Efamol evening primrose oil in atopic eczema. Where do we go from here in light of more recent discoveries? *Current Pharmaceutical Biotechnology* 2006;7:503–24.

27. Palmer DJ, Sullivan T, Gold MS, et al. Effect of n-3 long chain polyunsaturated fatty acid supplementation in pregnancy on infants' allergies in first year of life: randomised controlled trial. *BMJ* 2012;344:e184.

28. Camargo CA, Jr., Ganmaa D, Sidbury R, et al. Randomized trial of vitamin D supplementation for winter-related atopic dermatitis in children. *J Allergy Clin Immunol* 2014;134:831–5, e1.

29. Di Filippo P, Scaparrotta A, Rapino D, et al. Vitamin D supplementation modulates the immune system and improves atopic dermatitis in children. *International Archives of Allergy and Immunology* 2015;166:91–6.

30. Vestita M, Filoni A, Congedo M, Foti C, Bonamonte D. Vitamin D and atopic dermatitis in childhood. *Journal of Immunology Research* 2015;2015:257879.

31. Javanbakht MH, Keshavarz SA, Djalali M, et al. Randomized controlled trial using vitamins E and D supplementation in atopic dermatitis. *The Journal of Dermatological Treatment* 2011;22:144–50.

32. Allen KJ, Koplin JJ, Ponsonby AL, et al.

Vitamin D insufficiency is associated with challenge-proven food allergy in infants. *J Allergy Clin Immunol* 2013;131:1109–16, 16 e1–6.

33. Uehara M, Sugiura H, Sakurai K. A trial of oolong tea in the management of recalcitrant atopic dermatitis. *Archives of Dermatology* 2001;137:42–3.

34. Schachner L, Field T, Hernandez-Reif M, Duarte AM, Krasnegor J. Atopic dermatitis symptoms decreased in children following massage therapy. *Pediatr Dermatol* 1998;15:390–5.

35. Rossi E, Bartoli P, Bianchi A, Da Fre M. Homeopathy in paediatric atopic diseases: long-term results in children with atopic dermatitis. *Homeopathy* 2012;101:13–20.

36. Irwin DF, Gross HE, Stucky BD, et al. Development of six PROMIS pediatrics proxy-report item banks. *Health Qual Life Outcomes* 2012;10:22.

37. Roll S, Reinhold T, Pach D, et al. Comparative effectiveness of homeopathic vs. conventional therapy for atopic eczema. *PLoS One* 2013;8(1):e54973.

Chapter 18: Fatigue

1. Chapenko S, Krumina A, Kozireva S, et al. Activation of human herpesviruses 6 and 7 in patients with chronic fatigue syndrome. *Journal of Clinical Virology: The Official Publication of the Pan American Society for Clinical Virology* 2006;37 Suppl 1:S47–51.

2. Allen JJ, Schnyer RN, Chambers AS, Hitt SK, Moreno FA, Manber R. Acupuncture for depression: a randomized controlled trial. *J Clin Psychiatry* 2006;67:1665–73.

3. Larun L, Brurberg KG, Odgaard-Jensen J, Price JR. Exercise therapy for chronic fatigue syndrome. *Cochrane Database Syst Rev* 2015;2:CD003200.

4. Lauche R, Cramer H, Hauser W, Dobos G, Langhorst J. A systematic review and meta-analysis of qigong for the fibromyalgia syndrome. *Evid Based Complement Alternat Med* 2013;2013:635182.

5. Ades PA, Keteyian SJ, Balady GJ, et al. Cardiac rehabilitation exercise and self-care for chronic heart failure. *JACC Heart Failure* 2013;1:540–7.

6. Langhorst J, Klose P, Dobos GJ, Bernardy K, Hauser W. Efficacy and safety of meditative movement therapies in fibromyalgia syndrome: a systematic review and meta-analysis of randomized controlled trials. *Rheumatol Int* 2013;33:193–207.

7. Cachia D, Kamiya-Matsuoka C, Pinnix CC, et al. Myelopathy following intrathecal chemotherapy in adults: a single institution experience. *Journal of Neuro-Oncology* 2015;122:391–8.

8. Burschka JM, Keune PM, Oy UH, Oschmann P, Kuhn P. Mindfulness-based interventions in multiple sclerosis: beneficial effects of Tai Chi on balance, coordination, fatigue and depression. *BMC Neurology* 2014;14:165.

9. Larkey LK, Roe DJ, Weihs KL, et al. Randomized controlled trial of Qigong/Tai Chi Easy on cancer-related fatigue in breast cancer survivors. *Ann Behav Med* 2015;49:165–76.

10. Masuda A, Kihara T, Fukudome T, Shinsato T, Minagoe S, Tei C. The effects of repeated thermal therapy for two patients with chronic fatigue syndrome. *J Psychosom Res* 2005;58:383–7.

11. Soejima Y, Munemoto T, Masuda A, Uwatoko Y, Miyata M, Tei C. Effects of Waon therapy on chronic fatigue syndrome: a pilot study. *Internal Medicine* 2015;54:333–8.

12. Fraioli A, Grassi M, Mennuni G, et al. Clinical researches on the efficacy of spa therapy in fibromyalgia. A systematic review. *Annali dell'Istituto Superiore di Sanita* 2013;49:219–29.

13. Herttua K, Makela P, Martikainen P. Changes in alcohol-related mortality and its socioeconomic differences after a large reduction in alcohol prices: a natural experiment based on register data. *Am J Epidemiol* 2008;168:1110–18; discussion 26–31.

14. Tummers M, Knoop H, van Dam A, Bleijenberg G. Implementing a minimal intervention for chronic fatigue syndrome in a mental health centre: a randomized controlled trial. *Psychol Med* 2012;42:2205–15.

15. Knight SJ, Scheinberg A, Harvey AR. Interventions in pediatric chronic fatigue syndrome/

myalgic encephalomyelitis: a systematic review. *J Adolesc Health* 2013;53:154–65.

16. Rimes KA, Wingrove J. Mindfulness-based cognitive therapy for people with chronic fatigue syndrome still experiencing excessive fatigue after cognitive behaviour therapy: a pilot randomized study. *Clin Psychol Psychother* 2013;20:107–17.

17. James LC, Folen RA. EEG biofeedback as a treatment for chronic fatigue syndrome: a controlled case report. *Behav Med* 1996;22:77–81.

18. Hauser W, Bernardy K, Uceyler N, Sommer C. Treatment of fibromyalgia syndrome with antidepressants: a meta-analysis. *JAMA* 2009;301:198–209.

19. Baschetti R. Chronic fatigue syndrome and Addison's disease. *J Pediatr* 2003;142:217.

20. Baschetti R. Chronic fatigue syndrome: a form of Addison's disease. *J Intern Med* 2000;247:737–9.

21. Anbar RD. Self-hypnosis for management of chronic dyspnea in pediatric patients. *Pediatrics* 2001;107:E21.

22. Nijhof SL, Rutten JM, Uiterwaal CS, Bleijenberg G, Kimpen JL, Putte EM. The role of hypocortisolism in chronic fatigue syndrome. *Psychoneuroendocrinology* 2014;42:199–206.

23. Blockmans D, Persoons P, Van Houdenhove B, Lejeune M, Bobbaers H. Combination therapy with hydrocortisone and fludrocortisone does not improve symptoms in chronic fatigue syndrome: a randomized, placebo-controlled, double-blind, crossover study. *Am J Med* 2003;114:736–41.

24. Sulheim D, Fagermoen E, Winger A, et al. Disease mechanisms and clonidine treatment in adolescent chronic fatigue syndrome: a combined cross-sectional and randomized clinical trial. *JAMA Pediatrics* 2014;168:351–60.

25. Regland B, Forsmark S, Halaouate L, et al. Response to vitamin B12 and folic acid in myalgic encephalomyelitis and fibromyalgia. *PLoS One* 2015;10:e0124648.

26. Witham MD, Adams F, McSwiggan S, et al. Effect of intermittent vitamin D3 on vascular function and symptoms in chronic fatigue syndrome—a randomised controlled trial. *Nutrition, Metabolism, and Cardiovascular Diseases: NMCD* 2015;25:287–94.

27. Clark CE, Horvath IA, Taylor RS, Campbell JL. Doctors record higher blood pressures than nurses: systematic review and meta-analysis. *Br J Gen Pract* 2014;64:e223–32.

28. Malaguarnera M, Gargante MP, Cristaldi E, et al. Acetyl L-carnitine (ALC) treatment in elderly patients with fatigue. *Arch Gerontol Geriatr* 2008;46:181–90.

29. Miyamae T, Seki M, Naga T, et al. Increased oxidative stress and coenzyme Q10 deficiency in juvenile fibromyalgia: amelioration of hypercholesterolemia and fatigue by ubiquinol-10 supplementation. *Redox Rep* 2013;18:12–9.

30. Castro-Marrero J, Cordero MD, Segundo MJ, et al. Does oral coenzyme Q10 plus NADH supplementation improve fatigue and biochemical parameters in chronic fatigue syndrome? *Antioxidants & Redox Signaling* 2015;22:67985.

31. Alraek T, Lee MS, Choi TY, Cao H, Liu J. Complementary and alternative medicine for patients with chronic fatigue syndrome: a systematic review. *BMC Complement Altern Med* 2011;11:87.

32. Datieva VK, Rosinskaia AV, Levin OS. [The use of melatonin in the treatment of chronic fatigue syndrome and circadian rhythm disorders in Parkinson's disease]. *Zhurnal nevrologii i psikhiatrii imeni SS Korsakova* 2013;113:77–81.

33. van Heukelom RO, Prins JB, Smits MG, Bleijenberg G. Influence of melatonin on fatigue severity in patients with chronic fatigue syndrome and late melatonin secretion. *Eur J Neurol* 2006;13:55–60.

34. Forsyth LM, Preuss HG, MacDowell AL, et al. Therapeutic effects of oral NADH on the symptoms of patients with chronic fatigue syndrome. *Ann Allergy Asthma Immunol* 1999;82:185–91.

35. Santaella ML, Font I, Disdier OM. Comparison of oral nicotinamide adenine dinucleotide (NADH) versus conventional therapy for

chronic fatigue syndrome. *Puerto Rico Health Sciences Journal* 2004;23:89–93.

36. Das YT, Bagchi M, Bagchi D, Preuss HG. Safety of 5-hydroxy-L-tryptophan. *Toxicol Lett* 2004;150:111–22.

37. Berkman ND, Wallace IF, Steiner MJ, et al. *Otitis Media with Effusion: Comparative Effectiveness of Treatments.* Rockville (MD) 2013.

38. Hartz AJ, Bentler S, Noyes R, et al. Randomized controlled trial of Siberian ginseng for chronic fatigue. *Psychol Med* 2004;34:51–61.

39. Belcaro G, Cornelli U, Luzzi R, et al. Robuvit(R) (Quercus robur extract) supplementation in subjects with chronic fatigue syndrome and increased oxidative stress. A pilot registry study. *Journal of Neurosurgical Sciences* 2015;59:105–17.

40. Burrows TL, Martin RJ, Collins CE. A systematic review of the validity of dietary assessment methods in children when compared with the method of doubly labeled water. *J Am Diet Assoc* 2010;110:1501–10.

41. Yuemei L, Hongping L, Shulan F, Dongfang G. The therapeutic effects of electrical acupuncture and auricular-plaster in 32 cases of chronic fatigue syndrome. *J Tradit Chin Med* 2006;26:163–4.

42. *Group Care for Chronic Disease Management: A Review of the Clinical Effectiveness, Cost-effectiveness, and Guidelines.* Ottawa (ON) 2013.

43. Al Rashoud AS, Abboud RJ, Wang W, Wigderowitz C. Efficacy of low-level laser therapy applied at acupuncture points in knee osteoarthritis: a randomised double-blind comparative trial. *Physiotherapy* 2014;100:242–8.

44. Dean ME, Karsandas R, Bland JM, Gooch D, MacPherson H. Homeopathy for mental fatigue: lessons from a randomized, triple blind, placebo-controlled cross-over clinical trial. *BMC Complement Altern Med* 2012;12:167.

45. Weatherley-Jones E, Nicholl JP, Thomas KJ, et al. A randomised, controlled, triple-blind trial of the efficacy of homeopathic treatment for chronic fatigue syndrome. *J Psychosom Res* 2004;56:189–97.

Chapter 20: Headache

1. Moffett AM, Swash M, Scott DF. Effect of chocolate in migraine: a double-blind study. *J Neurol Neurosurg Psychiatry* 1974;37:445–8.

2. Marcus DA, Scharff L, Turk D, Gourley LM. A double-blind provocative study of chocolate as a trigger of headache. *Cephalalgia* 1997;17:855–62.

3. Forsythe WI, Redmond A. Two controlled trials of tyramine in children with migraine. *Dev Med Child Neurol* 1974;16:794–9.

4. Salfield SA, Wardley BL, Houlsby WT, et al. Controlled study of exclusion of dietary vasoactive amines in migraine. *Arch Dis Child* 1987;62:458–60.

5. Bigal ME, Krymchantowski AV. Migraine triggered by sucralose—a case report. *Headache* 2006;46:515–17.

6. Jacob SE, Stechschulte S. Formaldehyde, aspartame, and migraines: a possible connection. *Dermatitis* 2008;19:E10–11.

7. Suez J, Korem T, Zeevi D, et al. Artificial sweeteners induce glucose intolerance by altering the gut microbiota. *Nature* 2014;514:181–6.

8. Sigmon SC, Herning RI, Better W, Cadet JL, Griffiths RR. Caffeine withdrawal, acute effects, tolerance, and absence of net beneficial effects of chronic administration: cerebral blood flow velocity, quantitative EEG, and subjective effects. *Psychopharmacology (Berl)* 2009;204:573–85.

9. Ferrari A, Baraldi C, Sternieri E. Medication overuse and chronic migraine: a critical review according to clinical pharmacology. *Expert Opinion on Drug Metabolism & Toxicology* 2015;11:1127–44.

10. Cooke LJ, Rose MS, Becker WJ. Chinook winds and migraine headache. *Neurology* 2000;54:302–7.

11. Solomon GD. Circadian rhythms and migraine. *Cleve Clin J Med* 1992;59:326–9.

12. Egger J, Carter CM, Wilson J, Turner MW, Soothill JF. Is migraine food allergy? A double-blind controlled trial of oligoantigenic diet treatment. *Lancet* 1983;2:865–9.

13. Egger J, Carter CM, Soothill JF, Wilson J. Oligo-

antigenic diet treatment of children with epilepsy and migraine. *J Pediatr* 1989;114:51–8.

14. Alpay K, Ertas M, Orhan EK, Ustay DK, Lieners C, Baykan B. Diet restriction in migraine, based on IgG against foods: a clinical double-blind, randomised, cross-over trial. *Cephalalgia* 2010;30:829–37.

15. Monro J, Brostoff J, Carini C, Zilkha K. Food allergy in migraine. Study of dietary exclusion and RAST. *Lancet* 1980;2:1–4.

16. Mansfield LE. The role of food allergy in migraine: a review. *Ann Allergy* 1987;58:313–17.

17. Carter CM, Egger J, Soothill JF. A dietary management of severe childhood migraine. *Human Nutrition Applied Nutrition* 1985;39:294–303.

18. Mansfield LE, Vaughan TR, Waller SF, Haverly RW, Ting S. Food allergy and adult migraine: double-blind and mediator confirmation of an allergic etiology. *Ann Allergy* 1985;55:126–9.

19. Guarnieri P, Radnitz CL, Blanchard EB. Assessment of dietary risk factors in chronic headache. *Biofeedback Self Regul* 1990;15:15–25.

20. Narin SO, Pinar L, Erbas D, Ozturk V, Idiman F. The effects of exercise and exercise-related changes in blood nitric oxide level on migraine headache. *Clin Rehabil* 2003;17:624–30.

21. Lockett DM, Campbell JF. The effects of aerobic exercise on migraine. *Headache* 1992;32:50–4.

22. Landy SH, Griffin B. Pressure, heat, and cold help relieve headache pain. *Arch Fam Med* 2000;9:792–3.

23. Blumenfeld AM, Varon SF, Wilcox TK, et al. Disability, HRQoL and resource use among chronic and episodic migraineurs: results from the International Burden of Migraine Study (IBMS). *Cephalalgia* 2011;31:301–15.

24. Bhola R, Kinsella E, Giffin N, et al. Single-pulse transcranial magnetic stimulation (sTMS) for the acute treatment of migraine: evaluation of outcome data for the UK post market pilot program. *The Journal of Headache and Pain* 2015;16:535.

25. Holden EW, Deichmann MM, Levy JD. Empirically supported treatments in pediatric psychology: recurrent pediatric headache.

[comment]. *Journal of Pediatric Psychology* 1999;24:91–109.

26. Holroyd KA, Nash JM, Pingel JD, Cordingley GE, Jerome A. A comparison of pharmacological (amitriptyline HCL) and nonpharmacological (cognitive-behavioral) therapies for chronic tension headaches. *J Consult Clin Psychol* 1991;59:387–93.

27. D'Anci K E, Vibhakar A, Kanter JH, Mahoney CR, Taylor HA. Voluntary dehydration and cognitive performance in trained college athletes. *Percept Mot Skills* 2009;109:251–69.

28. Fentress D, Mehegan J, Benson H. Biofeedback and relaxation response training in the treatment of pediatric migraine. *Developmental Medicine and Child Neurology* 1986;28:139–46.

29. Engel JM. Relaxation training: a self-help approach for children with headaches. *Am J Occup Ther* 1992;46:591–6.

30. Wallbaum AB, Rzewnicki R, Steele H, Suedfeld P. Progressive muscle relaxation and restricted environmental stimulation therapy for chronic tension headache: a pilot study. *Int J Psychosom* 1991;38:33–9.

31. Richter IL, McGrath PJ, Humphreys PJ, Goodman JT, Firestone P, Keene D. Cognitive and relaxation treatment of paediatric migraine. *Pain* 1986;25:195–203.

32. Penzien DB, Irby MB, Smitherman TA, Rains JC, Houle TT. Well-established and empirically supported behavioral treatments for migraine. *Curr Pain Headache Rep* 2015;19:500.

33. Primavera JP, III, Kaiser RS. Non-pharmacological treatment of headache: is less more? *Headache* 1992;32:393–5.

34. Olness K, MacDonald JT, Uden DL. Comparison of self-hypnosis and propranolol in the treatment of juvenile classic migraine. *Pediatrics* 1987;79:593–7.

35. Sartory G, Muller B, Metsch J, Pothmann R. A comparison of psychological and pharmacological treatment of pediatric migraine. *Behav Res Ther* 1998;36:1155–70.

36. Lisspers J, Ost LG. Long-term follow-up of migraine treatment: Do the effects remain up to six years? *Behav Res Ther* 1990;28:313–22.

37. Powers SW, Mitchell MJ, Byars KC, Bentti AL, LeCates SL, Hershey AD. A pilot study of one-session biofeedback training in pediatric headache. *Neurology* 2001;56:133.

38. Andrasik F. Biofeedback in headache: an overview of approaches and evidence. *Cleve Clin J Med* 2010;77 Suppl 3:S72–6.

39. Grazzi L, Leone M, Frediani F, Bussone G. A therapeutic alternative for tension headache in children: treatment and 1-year follow-up results. *Biofeedback Self Regul* 1990;15:1–6.

40. Nuechterlein KH, Holroyd JC. Biofeedback in the treatment of tension headache. Current status. *Arch Gen Psychiatry* 1980;37:866–73.

41. Wells RE, Burch R, Paulsen RH, et al. Meditation for migraines: a pilot randomized controlled trial. *Headache* 2014;54:1484–95.

42. Hershey AD, Powers SW, Coffey CS, et al. Childhood and Adolescent Migraine Prevention (CHAMP) study: a double-blinded, placebo-controlled, comparative effectiveness study of amitriptyline, topiramate, and placebo in the prevention of childhood and adolescent migraine. *Headache* 2013;53:799–816.

43. Holroyd KA, Penzien DB. Pharmacological versus non-pharmacological prophylaxis of recurrent migraine headache: a meta-analytic review of clinical trials. *Pain* 1990;42:1–13.

44. Shamliyan TA, Kane RL, Ramakrishnan R, Taylor FR. *Migraine in Children: Preventive Pharmacologic Treatments.* Rockville (MD) 2013.

45. Schrader H, Stovner LJ, Helde G, Sand T, Bovim G. Prophylactic treatment of migraine with angiotensin converting enzyme inhibitor (lisinopril): randomised, placebo controlled, crossover study. *BMJ* 2001;322:19–22.

46. El-Chammas K, Keyes J, Thompson N, et al. Pharmacologic treatment of pediatric headaches: a meta-analysis. *JAMA Pediatrics* 2013;167:250–8.

47. Bille B, Ludvigsson J, Sanner G. Prophylaxis of migraine in children. *Headache* 1977;17:61–3.

48. Bearss K, Johnson C, Smith T, et al. Effect of parent training vs parent education on behavioral problems in children with autism spectrum disorder: a randomized clinical trial. *JAMA* 2015;313:1524–33.

49. Hirfanoglu T, Serdaroglu A, Gulbahar O, Cansu A. Prophylactic drugs and cytokine and leptin levels in children with migraine. *Pediatr Neurol* 2009;41:281–7.

50. Bidabadi E, Mashouf M. A randomized trial of propranolol versus sodium valproate for the prophylaxis of migraine in pediatric patients. *Paediatr Drugs* 2010;12:269–75.

51. Lewis D, Winner P, Saper J, et al. Randomized, double-blind, placebo-controlled study to evaluate the efficacy and safety of topiramate for migraine prevention in pediatric subjects 12 to 17 years of age. *Pediatrics* 2009;123:924–34.

52. O'Brien HL, Kabbouche MA, Kacperski J, Hershey AD. Treatment of pediatric migraine. *Curr Treat Options Neurol* 2015;17:326.

53. Ahn AC, Kaptchuk TJ. Advancing acupuncture research. *Altern Ther Health Med* 2005;11:40–5.

54. Leung S, Bulloch B, Young C, Yonker M, Hostetler M. Effectiveness of standardized combination therapy for migraine treatment in the pediatric emergency department. *Headache* 2013;53:491–7.

55. Maizels M, Geiger AM. Intranasal lidocaine for migraine: a randomized trial and open-label follow-up. *Headache* 1999;39:543–51.

56. Ching H, Pringsheim T. Aripiprazole for autism spectrum disorders (ASD). *Cochrane Database Syst Rev* 2012;5:CD009043.

57. Condo M, Posar A, Arbizzani A, Parmeggiani A. Riboflavin prophylaxis in pediatric and adolescent migraine. *The Journal of Headache and Pain* 2009;10:361–5.

58. Bruijn J, Duivenvoorden H, Passchier J, Locher H, Dijkstra N, Arts WF. Medium-dose riboflavin as a prophylactic agent in children with migraine: a preliminary placebo-controlled, randomised, double-blind, cross-over trial. *Cephalalgia* 2010;30:1426–34.

59. Sun-Edelstein C, Mauskop A. Foods and supplements in the management of migraine headaches. *Clin J Pain* 2009;25:446–52.

60. Hershey AD, Powers SW, Vockell AL, et al. Coenzyme Q10 deficiency and response to supplementation in pediatric and adolescent migraine. *Headache* 2007;47:73–80.

61. Menon S, Lea RA, Ingle S, et al. Effects of dietary folate intake on migraine disability and frequency. *Headache* 2015;55:301–9.

62. Utterback G, Zacharias R, Timraz S, Mershman D. Butterbur extract: prophylactic treatment for childhood migraines. *Complement Ther Clin Pract* 2014;20:61–4.

63. Grossman W, Schmidramsl H. An extract of Petasites hybridus is effective in the prophylaxis of migraine. *Altern Med Rev* 2001;6:303–10.

64. Buse D, Manack A, Serrano D, et al. Headache impact of chronic and episodic migraine: results from the American Migraine Prevalence and Prevention study. *Headache* 2012;52:3–17.

65. Agosti R, Duke RK, Chrubasik JE, Chrubasik S. Effectiveness of Petasites hybridus preparations in the prophylaxis of migraine: a systematic review. *Phytomedicine* 2006;13:743–6.

66. Murphy JJ, Heptinstall S, Mitchell JR. Randomised double-blind placebo-controlled trial of feverfew in migraine prevention. *Lancet* 1988;2:189–92.

67. Pfaffenrath V, Diener HC, Fischer M, Friede M, Henneicke-von Zepelin HH, Investigators. The efficacy and safety of Tanacetum parthenium (feverfew) in migraine prophylaxis—a double-blind, multicentre, randomized placebo-controlled dose-response study. *Cephalalgia* 2002;22:523–32.

68. Mustafa T, Srivastava KC. Ginger (Zingiber officinale) in migraine headache. *J Ethnopharmacol* 1990;29:267–73.

69. Maghbooli M, Golipour F, Moghimi Esfandabadi A, Yousefi M. Comparison between the efficacy of ginger and sumatriptan in the ablative treatment of the common migraine. *Phytother Res* 2014;28:412–15.

70. Schattner P, Randerson D. Tiger Balm as a treatment of tension headache. A clinical trial in general practice. *Aust Fam Physician* 1996;25:216, 8, 20 passim.

71. Shrivastava R, Pechadre JC, John GW. Tanacetum parthenium and Salix alba (Mig-RL) combination in migraine prophylaxis: a prospective, open-label study. *Clin Drug Investig* 2006;26:287–96.

72. Maizels M, Blumenfeld A, Burchette R. A combination of riboflavin, magnesium, and feverfew for migraine prophylaxis: a randomized trial. *Headache* 2004;44:885–90.

73. Cady RK, Goldstein J, Nett R, Mitchell R, Beach ME, Browning R. A double-blind placebo-controlled pilot study of sublingual feverfew and ginger (LipiGesic M) in the treatment of migraine. *Headache* 2011;51:1078–86.

74. Ferro EC, Biagini AP, da Silva IE, Silva ML, Silva JR. The combined effect of acupuncture and Tanacetum parthenium on quality of life in women with headache: randomised study. *Acupunct Med* 2012;30:252–7.

75. Moraska AF, Stenerson L, Butryn N, Krutsch JP, Schmiege SJ, Mann JD. Myofascial trigger point-focused head and neck massage for recurrent tension-type headache: a randomized, placebo-controlled clinical trial. *Clin J Pain* 2015;31:159–68.

76. Chatchawan U, Eungpinichpong W, Sooktho S, Tiamkao S, Yamauchi J. Effects of Thai traditional massage on pressure pain threshold and headache intensity in patients with chronic tension-type and migraine headaches. *J Altern Complement Med* 2014;20:486–92.

77. Wylie KR, Jackson C, Crawford PM. Does psychological testing help to predict the response to acupuncture or massage/relaxation therapy in patients presenting to a general neurology clinic with headache? *Journal of Traditional Chinese Medicine* 1997;17:130–9.

78. von Stulpnagel C, Reilich P, Straube A, et al. Myofascial trigger points in children with tension-type headache: a new diagnostic and therapeutic option. *J Child Neurol* 2009;24:406–9.

79. Launso L, Brendstrup E, Arnberg S. An exploratory study of reflexological treatment for headache.[comment]. *Alternative Therapies in Health & Medicine* 1999;5:57–65.

80. Chaibi A, Tuchin PJ, Russell MB. Manual therapies for migraine: a systematic review. *The Journal of Headache and Pain* 2011;12:127–33.

81. Bowing G, Zhou J, Endres HG, Coeytaux RR, Diener HC, Molsberger AF. Differences in Chinese diagnoses for migraine and tension-

type headache: an analysis of the German acupuncture trials (GERAC) for headache. *Cephalalgia* 2010;30:224–32.

82. Schetzek S, Heinen F, Kruse S, et al. Headache in children: update on complementary treatments. *Neuropediatrics* 2013;44:25–33.

83. Keller E, Bzdek VM. Effects of therapeutic touch on tension headache pain. *Nurs Res* 1986;35:101–6.

84. Mesa-Jimenez JA, Lozano-Lopez C, Angulo-Diaz-Parreno S, et al. Multimodal manual therapy vs. pharmacological care for management of tension type headache: A meta-analysis of randomized trials. *Cephalalgia* 2015;35:1323–32.

85. Paemeleire K, Louis P, Magis D, et al. Diagnosis, pathophysiology and management of chronic migraine: a proposal of the Belgian Headache Society. *Acta neurologica Belgica* 2015;115:1–17.

Chapter 21: Jaundice

1. Maisels MJ. Managing the jaundiced newborn: a persistent challenge. *CMAJ* 2015;187:335–43.

2. Nagar G, Vandermeer B, Campbell S, Kumar M. Reliability of transcutaneous bilirubin devices in preterm infants: a systematic review. *Pediatrics* 2013;132:871–81.

3. Gourley GR, Kreamer B, Cohnen M, Kosorok MR. Neonatal jaundice and diet. *Arch Pediatr Adolesc Med* 1999;153:184–8.

4. Maisels MJ, McDonagh AF. Phototherapy for neonatal jaundice. *N Engl J Med* 2008;358:920–8.

5. Vandborg PK, Hansen BM, Greisen G, Ebbesen F. Dose-response relationship of phototherapy for hyperbilirubinemia. *Pediatrics* 2012;130:e352–7.

6. Sherbiny HS, Youssef DM, Sherbini AS, El-Behedy R, Sherief LM. High-intensity light-emitting diode vs fluorescent tubes for intensive phototherapy in neonates. *Paediatrics and International Child Health* 2015:2046905515Y0000000006.

7. Malwade US, Jardine LA. Home- versus hospital-based phototherapy for the treatment of non-haemolytic jaundice in infants at more than 37 weeks' gestation. *Cochrane Database Syst Rev* 2014;6:CD010212.

8. Usatin D, Liljestrand P, Kuzniewicz MW, Escobar GJ, Newman TB. Effect of neonatal jaundice and phototherapy on the frequency of first-year outpatient visits. *Pediatrics* 2010;125:729–34.

9. Seyyedrasooli A, Valizadeh L, Hosseini MB, Asgari Jafarabadi M, Mohammadzad M. Effect of vimala massage on physiological jaundice in infants: a randomized controlled trial. *Journal of Caring Sciences* 2014;3:165–73.

Chapter 22: Pain (Abdominal Pain, Back Pain, Joint Pain)

1. Anand KJ, Hickey PR. Pain and its effects in the human neonate and fetus. *N Engl J Med* 1987;317:1321–9.

2. King S, Chambers CT, Huguet A, et al. The epidemiology of chronic pain in children and adolescents revisited: a systematic review. *Pain* 2011;152:2729–38.

3. Roth-Isigkeit A, Thyen U, Stoven H, Schwarzenberger J, Schmucker P. Pain among children and adolescents: restrictions in daily living and triggering factors. *Pediatrics* 2005;115:e152–62.

4. Tsai WS, Yang YH, Wang LC, Chiang BL. Abrupt temperature change triggers arthralgia in patients with juvenile rheumatoid arthritis. *J Microbiol Immunol Infect* 2006;39:465–70.

5. Chumpitazi BP, Cope JL, Hollister EB, et al. Randomised clinical trial: gut microbiome biomarkers are associated with clinical response to a low FODMAP diet in children with the irritable bowel syndrome. *Aliment Pharmacol Ther* 2015;42:418–27.

6. Ong DK, Mitchell SB, Barrett JS, et al. Manipulation of dietary short chain carbohydrates alters the pattern of gas production and genesis of symptoms in irritable bowel syndrome. *J Gastroenterol Hepatol* 2010;25:1366–73.

7. Chatzi L, Apostolaki G, Bibakis I, et al. Protective effect of fruits, vegetables and the Mediterranean diet on asthma and allergies among children in Crete. *Thorax* 2007;62:677–83.

8. Long AC, Krishnamurthy V, Palermo TM. Sleep disturbances in school-age children with chronic pain. *J Pediatr Psychol* 2008;33:258–68.

9. Song R, Lee EO, Lam P, Bae SC. Effects of tai chi exercise on pain, balance, muscle strength, and perceived difficulties in physical functioning in older women with osteoarthritis: a randomized clinical trial. *J Rheumatol* 2003;30:2039–44.

10. Wall RB. Tai Chi and mindfulness-based stress reduction in a Boston Public Middle School. *J Pediatr Health Care* 2005;19:230–7.

11. Brands MM, Purperhart H, Deckers-Kocken JM. A pilot study of yoga treatment in children with functional abdominal pain and irritable bowel syndrome. *Complement Ther Med* 2011;19:109–14.

12. Evans S, Sternlieb B, Zeltzer L, Tsao J. Iyengar yoga and the use of props for pediatric chronic pain: a case study. *Altern Ther Health Med* 2013;19:66–70.

13. Gemelli F, Cattani R. Carbon monoxide poisoning in childhood. *Br Med J (Clin Res Ed)* 1985;291:1197.

14. Caprilli S, Anastasi F, Grotto RP, Scollo Abeti M, Messeri A. Interactive music as a treatment for pain and stress in children during venipuncture: a randomized prospective study. *J Dev Behav Pediatr* 2007;28:399–403.

15. Whitehead-Pleaux AM, Zebrowski N, Baryza MJ, Sheridan RL. Exploring the effects of music therapy on pediatric pain: phase 1. *J Music Ther* 2007;44:217–41.

16. Daveson BA, Kennelly J. Music therapy in palliative care for hospitalized children and adolescents. *J Palliat Care* 2000;16:35–8.

17. McCurdy LE, Winterbottom KE, Mehta SS, Roberts JR. Using nature and outdoor activity to improve children's health. *Curr Probl Pediatr Adolesc Health* Care 2010;40:102–17.

18. Diette GB, Lechtzin N, Haponik E, Devrotes A, Rubin HR. Distraction therapy with nature sights and sounds reduces pain during flexible bronchoscopy: a complementary approach to routine analgesia. *Chest* 2003;123:941–8.

19. Al Rashoud AS, Abboud RJ, Wang W, Wigderowitz C. Efficacy of low-level laser therapy applied at acupuncture points in knee osteoarthritis: a randomised double-blind comparative trial. *Physiotherapy* 2014;100:242–8.

20. Haslerud S, Magnussen LH, Joensen J, Lopes-Martins RA, Bjordal JM. The efficacy of low-level laser therapy for shoulder tendinopathy: a systematic review and meta-analysis of randomized controlled trials. *Physiotherapy Research International* 2015;20:108–25.

21. Takahashi H, Okuni I, Ushigome N, et al. Low level laser therapy for patients with cervical disk hernia. *Laser Therapy* 2012;21:193–7.

22. Vallone F, Benedicenti S, Sorrenti E, Schiavetti I, Angiero F. Effect of diode laser in the treatment of patients with nonspecific chronic low back pain: a randomized controlled trial. *Photomed Laser Surg* 2014;32:490–4.

23. Afifi M. Positive health practices and depressive symptoms among high school adolescents in Oman. *Singapore Med J* 2006;47:960–6.

24. Foley-Nolan D, Barry C, Coughlan RJ, O'Connor P, Roden D. Pulsed high frequency (27MHz) electromagnetic therapy for persistent neck pain. A double blind, placebo-controlled study of 20 patients. *Orthopedics* 1990;13:445–51.

25. Eccles NK. A critical review of randomized controlled trials of static magnets for pain relief. *J Altern Complement Med* 2005;11:495–509.

26. Johnson MI, Paley CA, Howe TE, Sluka KA. Transcutaneous electrical nerve stimulation for acute pain. *Cochrane Database Syst Rev* 2015;6:CD006142.

27. Harvey M, Elliott M. Transcutaneous electrical nerve stimulation (TENS) for pain management during cavity preparations in pediatric patients. *ASDC Journal of Dentistry for Children* 1995;62:49–51.

28. Casimiro L, Brosseau L, Robinson V, et al. Therapeutic ultrasound for the treatment of rheumatoid arthritis. *Cochrane Database Syst Rev* 2002:CD003787.

29. Ruiz-Molinero C, Jimenez-Rejano JJ, Chillon-Martinez R, Suarez-Serrano C, Rebollo-Roldan J, Perez-Cabezas V. Efficacy of therapeutic

ultrasound in pain and joint mobility in whiplash traumatic acute and subacute phases. *Ultrasound Med Biol* 2014;40:2089–95.

30. Rutjes AW, Nuesch E, Sterchi R, Juni P. Therapeutic ultrasound for osteoarthritis of the knee or hip. *Cochrane Database Syst Rev* 2010:CD003132.

31. Anthony KK, Schanberg LE. Assessment and management of pain syndromes and arthritis pain in children and adolescents. *Rheum Dis Clin North Am* 2007;33:625–60.

32. Folks DG. The interface of psychiatry and irritable bowel syndrome. *Curr Psychiatry Rep* 2004;6:210–15.

33. Prasko J, Jelenova D, Mihal V. Psychological aspects and psychotherapy of inflammatory bowel diseases and irritable bowel syndrome in children. *Biomed Pap Med Fac Univ Palacky Olomouc Czech Repub* 2010;154:307–14.

34. Gerik SM. Pain management in children: developmental considerations and mind-body therapies. *South Med J* 2005;98:295–302.

35. Weydert JA, Shapiro DE, Acra SA, et al. Evaluation of guided imagery as treatment for recurrent abdominal pain in children: a randomized controlled trial. *BMC Pediatr* 2006;6:29.

36. Palermo TM, Eccleston C, Lewandowski AS, Williams AC, Morley S. Randomized controlled trials of psychological therapies for management of chronic pain in children and adolescents: an updated meta-analytic review. *Pain* 2010;148:387–97.

37. Vlieger AM, Menko-Frankenhuis C, Wolfkamp SC, Tromp E, Benninga MA. Hypnotherapy for children with functional abdominal pain or irritable bowel syndrome: a randomized controlled trial. *Gastroenterology* 2007;133:1430–6.

38. Rutten JM, Reitsma JB, Vlieger AM, Benninga MA. Gut-directed hypnotherapy for functional abdominal pain or irritable bowel syndrome in children: a systematic review. *Arch Dis Child* 2013;98:252–7.

39. Rutten JM, Vlieger AM, Frankenhuis C, et al. Gut-directed hypnotherapy in children with irritable bowel syndrome or functional abdominal pain (syndrome): a randomized

controlled trial on self exercises at home using CD versus individual therapy by qualified therapists. *BMC Pediatr* 2014;14:140.

40. Sibinga EM, Kemper KJ. Complementary, holistic, and integrative medicine: meditation practices for pediatric health. *Pediatr Rev* 2010;31:e91–103.

41. Ott MJ. Mindfulness meditation in pediatric clinical practice. *Pediatr Nurs* 2002;28:487–90.

42. Levy RL, Langer SL, Walker LS, et al. Cognitive-behavioral therapy for children with functional abdominal pain and their parents decreases pain and other symptoms. *Am J Gastroenterol* 2010;105:946–56.

43. Palermo TM, Wilson AC, Peters M, Lewandowski A, Somhegyi H. Randomized controlled trial of an Internet-delivered family cognitive-behavioral therapy intervention for children and adolescents with chronic pain. *Pain* 2009;146:205–13.

44. Rutten JM, Korterink JJ, Venmans LM, Benninga MA, Tabbers MM. Nonpharmacologic treatment of functional abdominal pain disorders: a systematic review. *Pediatrics* 2015;135:522–35.

45. Sanders MR, Shepherd RW, Cleghorn G, Woolford H. The treatment of recurrent abdominal pain in children: a controlled comparison of cognitive-behavioral family intervention and standard pediatric care. *J Consult Clin Psychol* 1994;62:306–14.

46. Lightdale JR, Gremse DA, Section on Gastroenterology H, Nutrition. Gastroesophageal reflux: management guidance for the pediatrician. *Pediatrics* 2013;131:e1684–95.

47. Teitelbaum JE, Arora R. Long-term efficacy of low-dose tricyclic antidepressants for children with functional gastrointestinal disorders. *Journal of Pediatric Gastroenterology and Nutrition* 2011;53:260–4.

48. Kaminski A, Kamper A, Thaler K, Chapman A, Gartlehner G. Antidepressants for the treatment of abdominal pain-related functional gastrointestinal disorders in children and adolescents. *Cochrane Database Syst Rev* 2011:CD008013.

49. Rufo PA, Bousvaros A. Current therapy of

inflammatory bowel disease in children. *Paediatr Drugs* 2006;8:279–302.

50. Breda L, Del Torto M, De Sanctis S, Chiarelli F. Biologics in children's autoimmune disorders: efficacy and safety. *Eur J Pediatr* 2011;170:157–67.

51. Pierce CA, Voss B. Efficacy and safety of ibuprofen and acetaminophen in children and adults: a meta-analysis and qualitative review. *Ann Pharmacother* 2010;44:489–506.

52. Krause I, Cleper R, Eisenstein B, Davidovits M. Acute renal failure, associated with nonsteroidal anti-inflammatory drugs in healthy children. *Pediatr Nephrol* 2005;20:1295–8.

53. Weiss JE, Luca NJ, Boneparth A, Stinson J. Assessment and management of pain in juvenile idiopathic arthritis. *Paediatr Drugs* 2014;16:473–81.

54. Bekkali NL, Bongers ME, Van den Berg MM, Liem O, Benninga MA. The role of a probiotics mixture in the treatment of childhood constipation: a pilot study. *Nutr J* 2007;6:17.

55. Bu LN, Chang MH, Ni YH, Chen HL, Cheng CC. Lactobacillus casei rhamnosus Lcr35 in children with chronic constipation. *Pediatr Int* 2007;49:485–90.

56. Gawronska A, Dziechciarz P, Horvath A, Szajewska H. A randomized double-blind placebo-controlled trial of Lactobacillus GG for abdominal pain disorders in children. *Aliment Pharmacol Ther* 2007;25:177–84.

57. Guandalini S, Magazzu G, Chiaro A, et al. VSL#3 improves symptoms in children with irritable bowel syndrome: a multicenter, randomized, placebo-controlled, double-blind, crossover study. *Journal of Pediatric Gastroenterology and Nutrition* 2010;51:24–30.

58. Horvath A, Dziechciarz P, Szajewska H. Meta-analysis: Lactobacillus rhamnosus GG for abdominal pain-related functional gastrointestinal disorders in childhood. *Aliment Pharmacol Ther* 2011;33:1302–10.

59. Guandalini S. Are probiotics or prebiotics useful in pediatric irritable bowel syndrome or inflammatory bowel disease? *Frontiers in Medicine* 2014;1:23.

60. Romano C, Ferrau V, Cavataio F, et al. Lactobacillus reuteri in children with functional abdominal pain (FAP). *J Paediatr Child Health* 2014;50:E68–71.

61. Plotnikoff GA, Quigley JM. Prevalence of severe hypovitaminosis D in patients with persistent, nonspecific musculoskeletal pain. *Mayo Clin Proc* 2003;78:1463–70.

62. Kremer JM, Lawrence DA, Petrillo GF, et al. Effects of high-dose fish oil on rheumatoid arthritis after stopping nonsteroidal antiinflammatory drugs. Clinical and immune correlates. *Arthritis Rheum* 1995;38:1107–14.

63. Nagakura T, Matsuda S, Shichijyo K, Sugimoto H, Hata K. Dietary supplementation with fish oil rich in omega-3 polyunsaturated fatty acids in children with bronchial asthma. *The European Respiratory Journal: Official Journal of the European Society for Clinical Respiratory Physiology* 2000;16:861–5.

64. De Silva V, El-Metwally A, Ernst E, et al. Evidence for the efficacy of complementary and alternative medicines in the management of osteoarthritis: a systematic review. *Rheumatology (Oxford)* 2011;50:911–20.

65. Gardiner P. Complementary, holistic, and integrative medicine: chamomile. *Pediatr Rev* 2007;28:e16–8.

66. Wigler I, Grotto I, Caspi D, Yaron M. The effects of Zintona EC (a ginger extract) on symptomatic gonarthritis. *Osteoarthritis Cartilage* 2003;11:783–9.

67. Cappello G, Spezzaferro M, Grossi L, Manzoli L, Marzio L. Peppermint oil (Mintoil) in the treatment of irritable bowel syndrome: a prospective double blind placebo-controlled randomized trial. *Dig Liver Dis* 2007;39:530–6.

68. Ford AC, Talley NJ, Spiegel BM, et al. Effect of fibre, antispasmodics, and peppermint oil in the treatment of irritable bowel syndrome: systematic review and meta-analysis. *BMJ* 2008;337:a2313.

69. Gardiner P, Kemper KJ. Herbs in pediatric and adolescent medicine. *Pediatr Rev* 2000;21:44–57.

70. Kline RM, Kline JJ, Di Palma J, Barbero GJ. Enteric-coated, pH-dependent peppermint oil capsules for the treatment of irritable bowel syndrome in children. *Journal of Pediatrics* 2001;138:125–8.

71. van Tilburg MA, Felix CT. Diet and functional abdominal pain in children and adolescents. *Journal of Pediatric Gastroenterology and Nutrition* 2013;57:141–8.

72. Grigoleit HG, Grigoleit P. Peppermint oil in irritable bowel syndrome. *Phytomedicine* 2005;12:601–6.

73. Ottillinger B, Storr M, Malfertheiner P, Allescher HD. STW 5 (Iberogast(R))—a safe and effective standard in the treatment of functional gastrointestinal disorders. *Wien Med Wochenschr* 2013;163:65–72.

74. Miranda A, Saps M. The use of non-narcotic pain medication in pediatric gastroenterology. *Paediatr Drugs* 2014;16:293–307.

75. Maroon JC, Bost JW, Maroon A. Natural anti-inflammatory agents for pain relief. *Surg Neurol Int* 2010;1:80.

76. Kuptniratsaikul V, Thanakhumtorn S, Chinswangwatanakul P, Wattanamongkonsil L, Thamlikitkul V. Efficacy and safety of Curcuma domestica extracts in patients with knee osteoarthritis. *J Altern Complement Med* 2009;15:891–7.

77. Hanai H, Iida T, Takeuchi K, et al. Curcumin maintenance therapy for ulcerative colitis: randomized, multicenter, double-blind, placebo-controlled trial. *Clin Gastroenterol Hepatol* 2006;4:1502–6.

78. Cropley M, Cave Z, Ellis J, Middleton RW. Effect of kava and valerian on human physiological and psychological responses to mental stress assessed under laboratory conditions. *Phytother Res* 2002;16:23–7.

79. Stanos SP. Topical agents for the management of musculoskeletal pain. *J Pain Symptom Manage* 2007;33:342–55.

80. Mason L, Moore RA, Edwards JE, McQuay HJ, Derry S, Wiffen PJ. Systematic review of efficacy of topical rubefacients containing salicylates for the treatment of acute and chronic pain. *BMJ* 2004;328:995.

81. Field T, Hernandez-Reif M, Seligman S, et al. Juvenile rheumatoid arthritis: benefits from massage therapy. *J Pediatr Psychol* 1997;22:607–17.

82. Alcantara J, Davis J. The chiropractic care of children with "growing pains": a case series

and systematic review of the literature. *Complement Ther Clin Pract* 2011;17:28–32.

83. Zeltzer LK, Tsao JC, Stelling C, Powers M, Levy S, Waterhouse M. A phase I study on the feasibility and acceptability of an acupuncture/hypnosis intervention for chronic pediatric pain. *J Pain Symptom Manage* 2002;24:437–46.

84. Kemper KJ, Sarah R, Silver-Highfield E, Xiarhos E, Barnes L, Berde C. On pins and needles? Pediatric pain patients' experience with acupuncture. *Pediatrics* 2000;105:941–7.

85. Kemper KJ, Kelly EA. Treating children with therapeutic and healing touch. *Pediatr Ann* 2004;33:248–52.

86. vanderVaart S, Gijsen VM, de Wildt SN, Koren G. A systematic review of the therapeutic effects of Reiki. *J Altern Complement Med* 2009;15:1157–69.

Chapter 23: Ringworm and Other Fungal Infections

1. El-Gohary M, van Zuuren EJ, Fedorowicz Z, et al. Topical antifungal treatments for tinea cruris and tinea corporis. *Cochrane Database Syst Rev* 2014;8:CD009992.

2. Lebwohl M, Elewski B, Eisen D, Savin RC. Efficacy and safety of terbinafine 1% solution in the treatment of interdigital tinea pedis and tinea corporis or tinea cruris. *Cutis* 2001;67:261–6.

3. Savin RC. Oral terbinafine versus griseofulvin in the treatment of moccasin-type tinea pedis. *J Am Acad Dermatol* 1990;23:807–9.

4. Gupta AK, Drummond-Main C. Meta-analysis of randomized, controlled trials comparing particular doses of griseofulvin and terbinafine for the treatment of tinea capitis. *Pediatr Dermatol* 2013;30:1–6.

5. Stein Gold LF, Parish LC, Vlahovic T, et al. Efficacy and safety of naftifine HCl Gel 2% in the treatment of interdigital and moccasin type tinea pedis: pooled results from two multicenter, randomized, double-blind, vehicle-controlled trials. *Journal of Drugs in Dermatology: JDD* 2013;12:911–18.

6. Chren MM, Landefeld CS. A cost analysis of

topical drug regimens for dermatophyte infections. *JAMA* 1994;272:1922–5.

7. Yuen CW, Yip J, Cheung HC. Treatment of interdigital type tinea pedis with a 2-week regimen of wearing hygienic socks loaded with antifungal microcapsules. *J Am Acad Dermatol* 2013;69(3):495–6.

8. Maissa BJ, Walid H. Antifungal activity of chemically different essential oils from wild Tunisian Thymus spp. *Natural Product Research* 2015;29:869–73.

9. Szweda P, Gucwa K, Kurzyk E, et al. Essential oils, silver nanoparticles and propolis as alternative agents against fluconazole resistant Candida albicans, Candida glabrata and Candida krusei clinical isolates. *Indian Journal of Microbiology* 2015;55:175–83.

10. Boukhatem MN, Ferhat MA, Kameli A, Saidi F, Kebir HT. Lemon grass (Cymbopogon citratus) essential oil as a potent anti-inflammatory and antifungal drugs. *The Libyan Journal of Medicine* 2014;9:25431.

11. Saxena S, Uniyal V, Bhatt RP. Inhibitory effect of essential oils against Trichosporon ovoides causing Piedra Hair Infection. *Brazilian Journal of Microbiology* [publication of the Brazilian Society for Microbiology] 2012;43:1347–54.

12. Sanguinetti M, Posteraro B, Romano L, et al. In vitro activity of Citrus bergamia (bergamot) oil against clinical isolates of dermatophytes. *The Journal of Antimicrobial Chemotherapy* 2007;59:305–8.

13. van Vuuren SF, Suliman S, Viljoen AM. The antimicrobial activity of four commercial essential oils in combination with conventional antimicrobials. *Letters in Applied Microbiology* 2009;48:440–6.

14. Ledezma E, Lopez JC, Marin P, et al. Ajoene in the topical short-term treatment of tinea cruris and tinea corporis in humans. Randomized comparative study with terbinafine. *Arzneimittelforschung* 1999;49:544–7.

15. Ledezma E, Marcano K, Jorquera A, et al. Efficacy of ajoene in the treatment of tinea pedis: a double-blind and comparative study with terbinafine. *J Am Acad Dermatol* 2000;43:829–32.

16. Ogita A, Fujita K, Taniguchi M, Tanaka T.

Enhancement of the fungicidal activity of amphotericin B by allicin, an allyl-sulfur compound from garlic, against the yeast Saccharomyces cerevisiae as a model system. *Planta Med* 2006;72:1247–50.

17. Ikeda S, Kanoya Y, Nagata S. Effects of a foot bath containing green tea polyphenols on interdigital tinea pedis. *Foot* 2013;23:58–62.

18. Pisseri F, Bertoli A, Nardoni S, et al. Antifungal activity of tea tree oil from Melaleuca alternifolia against Trichophyton equinum: an in vivo assay. *Phytomedicine* 2009;16:1056–8.

19. Flores FC, de Lima JA, Ribeiro RF, et al. Antifungal activity of nanocapsule suspensions containing tea tree oil on the growth of Trichophyton rubrum. *Mycopathologia* 2013;175:281–6.

20. Homeyer DC, Sanchez CJ, Mende K, et al. In vitro activity of Melaleuca alternifolia (tea tree) oil on filamentous fungi and toxicity to human cells. *Medical Mycology* 2015;53:285–94.

21. Tong MM, Altman PM, Barnetson RS. Tea tree oil in the treatment of tinea pedis. *Australas J Dermatol* 1992;33:145–9.

22. Garg AP, Muller J. Inhibition of growth of dermatophytes by Indian hair oils. *Mycoses* 1992;35:363–9.

Chapter 24: Sleep Problems

1. Keyes KM, Maslowsky J, Hamilton A, Schulenberg J. The great sleep recession: changes in sleep duration among US adolescents, 1991–2012. *Pediatrics* 2015;135:460–8.

2. Petit D, Pennestri MH, Paquet J, et al. Childhood Sleepwalking and Sleep Terrors: A Longitudinal Study of Prevalence and Familial Aggregation. *JAMA Pediatrics* 2015;169:653–8.

3. Kahn A, Mozin MJ, Rebuffat E, Sottiaux M, Muller MF. Milk intolerance in children with persistent sleeplessness: a prospective double-blind crossover evaluation. *Pediatrics* 1989;84:595–603.

4. Pajno GB, Barberio F, Vita D, et al. Diagnosis of cow's milk allergy avoided melatonin intake in infant with insomnia. *Sleep* 2004;27:1420–1.

5. Kahn A, Francois G, Sottiaux M, et al. Sleep

characteristics in milk-intolerant infants. *Sleep* 1988;11:291–7.

6. Cain N, Gradisar M. Electronic media use and sleep in school-aged children and adolescents: A review. *Sleep Med* 2010;11:735–42.

7. Pinilla T, Birch LL. Help me make it through the night: behavioral entrainment of breast-fed infants' sleep patterns. *Pediatrics* 1993;91:436–44.

8. Wolfson A, Lacks P, Futterman A. Effects of parent training on infant sleeping patterns, parents' stress, and perceived parental competence. *J Consult Clin Psychol* 1992;60:41–8.

9. Spadafora A, Hunt HT. The multiplicity of dreams: cognitive-affective correlates of lucid, archetypal, and nightmare dreaming. *Percept Mot Skills* 1990;71:627–44.

10. Galbally M, Lewis AJ, McEgan K, Scalzo K, Islam FA. Breastfeeding and infant sleep patterns: an Australian population study. *J Paediatr Child Health* 2013;49:E147–52.

11. Brown A, Harries V. Infant sleep and night feeding patterns during later infancy: association with breastfeeding frequency, daytime complementary food intake, and infant weight. *Breastfeeding Medicine: The Official Journal of the Academy of Breastfeeding Medicine* 2015;10:246–52.

12. Macknin ML, Medendorp SV, Maier MC. Infant sleep and bedtime cereal. *Am J Dis Child* 1989;143:1066–8.

13. Castel LD, Williams KA, Bosworth HB, et al. Content validity in the PROMIS social-health domain: a qualitative analysis of focus-group data. *Qual Life Res* 2008;17:737–49.

14. Sakuma Y, Sasaki-Otomaru A, Ishida S, et al. Effect of a home-based simple yoga program in child-care workers: a randomized controlled trial. *J Altern Complement Med* 2012;18:769–76.

15. Halpern J, Cohen M, Kennedy G, Reece J, Cahan C, Baharav A. Yoga for improving sleep quality and quality of life for older adults. *Altern Ther Health Med* 2014;20:37–46.

16. Carroll JE, Seeman TE, Olmstead R, et al. Improved sleep quality in older adults with insomnia reduces biomarkers of disease risk: pilot results from a randomized controlled comparative efficacy trial. *Psychoneuroendocrinology* 2015;55:184–92.

17. Fong SS, Ng SS, Lee HW, et al. The effects of a 6-month Tai Chi Qigong training program on temporomandibular, cervical, and shoulder joint mobility and sleep problems in nasopharyngeal cancer survivors. *Integr Cancer Ther* 2015;14:16–25.

18. Accorsi A, Lucci C, Di Mattia L, et al. Effect of osteopathic manipulative therapy in the attentive performance of children with attention-deficit/hyperactivity disorder. *J Am Osteopath Assoc* 2014;114:374–81.

19. Blair PS, Mitchell EA, Heckstall-Smith EM, Fleming PJ. Head covering—a major modifiable risk factor for sudden infant death syndrome: a systematic review. *Arch Dis Child* 2008;93:778–83.

20. Blair PS, Sidebotham P, Pease A, Fleming PJ. Bed-sharing in the absence of hazardous circumstances: Is there a risk of sudden infant death syndrome? An analysis from two case-control studies conducted in the UK. *PLoS One* 2014;9:e107799.

21. Wolf AW, Lozoff B. Object attachment, thumb-sucking, and the passage to sleep. *J Am Acad Child Adolesc Psychiatry* 1989;28:287–92.

22. Seymour FW, Brock P, During M, Poole G. Reducing sleep disruptions in young children: evaluation of therapist-guided and written information approaches: a brief report. *J Child Psychol Psychiatry* 1989;30:913–18.

23. Montgomery P, Stores G, Wiggs L. The relative efficacy of two brief treatments for sleep problems in young learning disabled (mentally retarded) children: a randomised controlled trial. *Arch Dis Child* 2004;89:125–30.

24. Weiskop S, Richdale A, Matthews J. Behavioural treatment to reduce sleep problems in children with autism or fragile X syndrome. *Dev Med Child Neurol* 2005;47:94–104.

25. Adams LA, Rickert VI. Reducing bedtime tantrums: comparison between positive routines and graduated extinction. *Pediatrics* 1989;84:756–61.

26. Reid MJ, Walter AL, O'Leary SG. Treatment of young children's bedtime refusal and nighttime wakings: a comparison of "standard"

and graduated ignoring procedures. *J Abnorm Child Psychol* 1999;27:5–16.

27. de Bruin EJ, Oort FJ, Bogels SM, Meijer AM. Efficacy of internet and group-administered cognitive behavioral therapy for insomnia in adolescents: a pilot study. *Behav Sleep Med* 2014;12:235–54.

28. Trauer JM, Qian MY, Doyle JS, Rajaratnam SM, Cunnington D. Cognitive behavioral therapy for chronic insomnia: a systematic review and meta-analysis. *Ann Intern Med* 2015;163(3):191–204.

29. Kohen DP, Mahowald MW, Rosen GM. Sleep-terror disorder in children: the role of self-hypnosis in management. *Am J Clin Hypn* 1992;34:233–44.

30. Palace EM, Johnston C. Treatment of recurrent nightmares by the dream reorganization approach. *Journal of Behavior Therapy and Experimental Psychiatry* 1989;20:219–26.

31. Ong JC, Manber R, Segal Z, Xia Y, Shapiro S, Wyatt JK. A randomized controlled trial of mindfulness meditation for chronic insomnia. *Sleep* 2014;37:1553–63.

32. Black DS, O'Reilly GA, Olmstead R, Breen EC, Irwin MR. Mindfulness meditation and improvement in sleep quality and daytime impairment among older adults with sleep disturbances: a randomized clinical trial. *JAMA Internal Medicine* 2015;175:494–501.

33. Bowden A, Lorenc A, Robinson N. Autogenic Training as a behavioural approach to insomnia: a prospective cohort study. *Primary Health Care Research & Development* 2012;13:175–85.

34. Barowsky EI, Moskowitz J, Zweig JB. Biofeedback for disorders of initiating and maintaining sleep. Ann N Y Acad Sci 1990;602:97–103.

35. Griffiths WJ, Lester BK, Coulter JD, Williams HL. Tryptophan and sleep in young adults. *Psychophysiology* 1972;9:345–56.

36. Yogman MW, Zeisel SH. Nutrients, neurotransmitters and infant behavior. *Am J Clin Nutr* 1985;42:352–60.

37. Steinberg LA, O'Connell NC, Hatch TF, Picciano MF, Birch LL. Tryptophan intake influences infants' sleep latency. *J Nutr* 1992;122:1781–91.

38. Cubero J, Chanclon B, Sanchez S, Rivero M, Rodriguez AB, Barriga C. Improving the quality of infant sleep through the inclusion at supper of cereals enriched with tryptophan, adenosine-5'-phosphate, and uridine-5'-phosphate. *Nutr Neurosci* 2009;12:272–80.

39. Okawa M, Mishima K, Nanami T, et al. Vitamin B12 treatment for sleep-wake rhythm disorders. *Sleep* 1990;13:15–23.

40. Ohta T, Ando K, Iwata T, et al. Treatment of persistent sleep-wake schedule disorders in adolescents with methylcobalamin (vitamin B12). *Sleep* 1991;14:414–18.

41. Liira J, Verbeek JH, Costa G, et al. Pharmacological interventions for sleepiness and sleep disturbances caused by shift work. *Cochrane Database Syst Rev* 2014;8:CD009776.

42. Shamseer L, Vohra S. Complementary, holistic, and integrative medicine: melatonin. *Pediatr Rev* 2009;30:223–8.

43. Lehrl S. Clinical efficacy of kava extract WS 1490 in sleep disturbances associated with anxiety disorders. Results of a multicenter, randomized, placebo-controlled, double-blind clinical trial. *J Affect Disord* 2004;78:101–10.

44. Boerner RJ, Sommer H, Berger W, Kuhn U, Schmidt U, Mannel M. Kava-Kava extract LI 150 is as effective as Opipramol and Buspirone in Generalised Anxiety Disorder—an 8-week randomized, double-blind multi-centre clinical trial in 129 out-patients. *Phytomedicine* 2003;10 Suppl 4:38–49.

45. Lillehei AS, Halcon LL, Savik K, Reis R. Effect of inhaled lavender and sleep hygiene on self-reported sleep issues: A Randomized Controlled Trial. *J Altern Complement Med* 2015;21:430–8.

46. Lillehei AS, Halcon LL. A systematic review of the effect of inhaled essential oils on sleep. *J Altern Complement Med* 2014;20:441–51.

47. Leathwood PD, Chauffard F. Aqueous extract of valerian reduces latency to fall asleep in man. *Planta Med* 1985:144–8.

48. Lindahl O, Lindwall L. Double blind study of a valerian preparation. *Pharmacol Biochem Behav* 1989;32:1065–6.

49. Koetter U, Schrader E, Kaufeler R, Brattstrom A. A randomized, double blind, placebo-

controlled, prospective clinical study to demonstrate clinical efficacy of a fixed valerian hops extract combination (Ze 91019) in patients suffering from non-organic sleep disorder. *Phytother Res* 2007;21:847–51.

50. Chrousos GP, Gold PW. The concepts of stress and stress system disorders. Overview of physical and behavioral homeostasis. *JAMA* 1992;267:1244–52.

51. Yates CC, Mitchell AJ, Booth MY, Williams DK, Lowe LM, Whit Hall R. The effects of massage therapy to induce sleep in infants born preterm. *Pediatr Phys Ther* 2014;26:405–10.

52. Vickers A, Ohlsson A, Lacy JB, Horsley A. Massage for promoting growth and development of preterm and/or low birth-weight infants. *Cochrane Database Syst Rev* 2004:CD000390.

53. Cheuk DK, Yeung WF, Chung KF, Wong V. Acupuncture for insomnia. *Cochrane Database Syst Rev* 2012;9:CD005472.

54. Krieger D, Peper E, Ancoli S. Therapeutic touch: searching for evidence of physiological change. *Am J Nurs* 1979;79:660–2.

55. Heidt PR. Helping patients to rest: clinical studies in therapeutic touch. *Holist Nurs Pract* 1991;5:57–66.

56. Braun C, Layton J, Braun J. Therapeutic touch improves residents' sleep. *J Am Health Care Assoc* 1986;12:48–9.

57. Bukowski EL, Berardi D. Reiki brief report: using Reiki to reduce stress levels in a nine-year-old child. *Explore (NY)* 2014;10:253–5.

58. Wood AM, Joseph S, Lloyd J, Atkins S. Gratitude influences sleep through the mechanism of pre-sleep cognitions. *J Psychosom Res* 2009;66:43–8.

59. Bell IR, Howerter A, Jackson N, Aickin M, Baldwin CM, Bootzin RR. Effects of homeopathic medicines on polysomnographic sleep of young adults with histories of coffee-related insomnia. *Sleep Med* 2011;12:505–11.

Chapter 25: Sore Throats

1. Cingi C, Songu M, Ural A, et al. Effect of chlorhexidine gluconate and benzydamine hydrochloride mouth spray on clinical signs and quality of life of patients with streptococcal tonsillopharyngitis: multicentre, prospective, randomised, double-blinded, placebo-controlled study. *J Laryngol Otol* 2011;125:620–5.

2. Wethington JF. Double-blind study of benzydamine hydrochloride, a new treatment for sore throat. *Clin Ther* 1985;7:641–6.

3. Wonnemann M, Helm I, Stauss-Grabo M, et al. Lidocaine 8 mg sore throat lozenges in the treatment of acute pharyngitis. A new therapeutic option investigated in comparison to placebo treatment. *Arzneimittelforschung* 2007;57:689–97.

4. Annuk H, Hirmo S, Turi E, Mikelsaar M, Arak E, Wadstrom T. Effect on cell surface hydrophobicity and susceptibility of Helicobacter pylori to medicinal plant extracts. *FEMS Microbiology Letters* 1999;172:41–5.

5. Lennon DR, Farrell E, Martin DR, Stewart JM. Once-daily amoxicillin versus twice-daily penicillin V in group A beta-haemolytic streptococcal pharyngitis. *Arch Dis Child* 2008;93:474–8.

6. Richter SS, Heilmann KP, Dohrn CL, Riahi F, Beekmann SE, Doern GV. Changing epidemiology of antimicrobial-resistant Streptococcus pneumoniae in the United States, 2004–2005. *Clinical Infectious Diseases: An Official Publication of the Infectious Diseases Society of America* 2009;48:e23–33.

7. Farrell DJ, Klugman KP, Pichichero M. Increased antimicrobial resistance among nonvaccine serotypes of Streptococcus pneumoniae in the pediatric population after the introduction of 7-valent pneumococcal vaccine in the United States. *Pediatr Infect Dis J* 2007;26:123–8.

8. Pichichero ME, Disney FA, Talpey WB, et al. Adverse and beneficial effects of immediate treatment of Group A beta-hemolytic streptococcal pharyngitis with penicillin. *Pediatr Infect Dis J* 1987;6:635–43.

9. Gerber MA, Randolph MF, DeMeo KK, Kaplan EL. Lack of impact of early antibiotic therapy for streptococcal pharyngitis on recurrence rates. *J Pediatr* 1990;117:853–8.

10. Johnston BC, Goldenberg JZ, Vandvik PO, Sun X, Guyatt GH. Probiotics for the prevention

of pediatric antibiotic-associated diarrhea. *Cochrane Database Syst Rev* 2011:CD004827.

11. Amir J, Harel L, Smetana Z, Varsano I. Treatment of herpes simplex gingivostomatitis with aciclovir in children: a randomised double blind placebo controlled study. *BMJ* 1997;314:1800–3.

12. Afman CE, Welge JA, Steward DL. Steroids for post-tonsillectomy pain reduction: meta-analysis of randomized controlled trials. *Otolaryngol Head Neck Surg* 2006;134:181–6.

13. Bender CE. The value of corticosteroids in the treatment of infectious mononucleosis. *JAMA* 1967;199:529–31.

14. Tynell E, Aurelius E, Brandell A, et al. Acyclovir and prednisolone treatment of acute infectious mononucleosis: a multicenter, double-blind, placebo-controlled study. *J Infect Dis* 1996;174:324–31.

15. Stelter K. Tonsillitis and sore throat in children. *GMS Current Topics in Otorhinolaryngology, Head and Neck Surgery* 2014;13:Doc07.

16. Adams JB, Audhya T, McDonough-Means S, et al. Toxicological status of children with autism vs. neurotypical children and the association with autism severity. *Biological Trace Element Research* 2013;151:171–80.

17. Hemila H, Chalker E. The effectiveness of high dose zinc acetate lozenges on various common cold symptoms: a meta-analysis. *BMC Family Practice* 2015;16:24.

18. Macknin ML, Piedmonte M, Calendine C, Janosky J, Wald E. Zinc gluconate lozenges for treating the common cold in children: a randomized controlled trial. *JAMA* 1998;279:1962–7.

19. Mangaiyarkarasi SP, Manigandan T, Elumalai M, Cholan PK, Kaur RP. Benefits of aloe vera in dentistry. *Journal of Pharmacy & Bioallied Sciences* 2015;7:S255–9.

20. Sahebjamee M, Mansourian A, Mohammad MH, et al. Comparative efficacy of aloe vera and benzydamine mouthwashes on radiation-induced oral mucositis: a triple-blind, randomised, controlled clinical trial. *Oral Health & Preventive Dentistry* 2014.

21. Partridge M, Poswillo DE. Topical carbenoxolone sodium in the management of herpes simplex infection. *The British Journal of Oral & Maxillofacial Surgery* 1984;22:138–45.

22. Schapowal A, Berger D, Klein P, Suter A. Echinacea/sage or chlorhexidine/lidocaine for treating acute sore throats: a randomized double-blind trial. *Eur J Med Res* 2009;14:406–12.

23. Schapowal A, Klein P, Johnston SL. Echinacea reduces risk of recurrent respiratory infections. *Adv Ther* 2015;32(3):187–200.

24. Minetti AM, Forti S, Tassone G, Torretta S, Pignataro L. Efficacy of complex herbal compound of Echinacea angustifolia (Imoviral(R) Junior) in recurrent upper respiratory tract infections during pediatric age: preliminary results. *Minerva Pediatr* 2011;63:177–82.

25. Cohen HA, Varsano I, Kahan E, Sarrell EM, Uziel Y. Effectiveness of an herbal preparation containing echinacea, propolis, and vitamin C in preventing respiratory tract infections in children: a randomized, double-blind, placebo-controlled, multicenter study. *Arch Pediatr Adolesc Med* 2004;158:217–21.

26. Bereznoy VV, Riley DS, Wassmer G, Heger M. Efficacy of extract of Pelargonium sidoides in children with acute non-group A beta-hemolytic streptococcus tonsillopharyngitis: a randomized, double-blind, placebo-controlled trial. *Altern Ther Health Med* 2003;9:68–79.

27. Hubbert M, Sievers H, Lehnfeld R, Kehrl W. Efficacy and tolerability of a spray with Salvia officinalis in the treatment of acute pharyngitis—a randomised, double-blind, placebo-controlled study with adaptive design and interim analysis. *Eur J Med Res* 2006;11:20–6.

28. Hwang SH, Song JN, Jeong YM, Lee YJ, Kang JM. The efficacy of honey for ameliorating pain after tonsillectomy: a meta-analysis. *European Archives of Oto-Rhino-Laryngology* 2016;273(4):811–18.

29. Mohebbi S, Nia FH, Kelantari F, Nejad SE, Hamedi Y, Abd R. Efficacy of honey in reduction of post tonsillectomy pain, randomized clinical trial. *International Journal of Pediatric Otorhinolaryngology* 2014;78:1886–9.

30. Paradise JL, Bluestone CD, Bachman RZ, et al. Efficacy of tonsillectomy for recurrent throat infection in severely affected children. Results of parallel randomized and nonrandomized clinical trials. *N Engl J Med* 1984;310:674–83.

31. Burton MJ, Glasziou PP, Chong LY, Venekamp RP. Tonsillectomy or adenotonsillectomy versus non-surgical treatment for chronic/ recurrent acute tonsillitis. *Cochrane Database Syst Rev* 2014;11:CD001802.

32. Fleckenstein J, Lill C, Ludtke R, Gleditsch J, Rasp G, Irnich D. A single point acupuncture treatment at large intestine meridian: a randomized controlled trial in acute tonsillitis and pharyngitis. *Clin J Pain* 2009;25:624–31.

33. Malapane E, Solomon EM, Pellow J. Efficacy of a homeopathic complex on acute viral tonsillitis. *J Altern Complement Med* 2014;20:868–73.

34. Haidvogl M, Riley DS, Heger M, et al. Homeopathic and conventional treatment for acute respiratory and ear complaints: a comparative study on outcome in the primary care setting. *BMC Complement Altern Med* 2007;7:7.

Chapter 26: Vomiting and Nausea

1. Niebyl JR. Clinical practice. Nausea and vomiting in pregnancy. *N Engl J Med* 2010;363:1544–50.

2. Boekema PJ, Samsom M, van Berge Henegouwen GP, Smout AJ. Coffee and gastrointestinal function: facts and fiction. A review. *Scandinavian Journal of Gastroenterology Supplement* 1999;230:35–9.

3. Vanderhoof JA, Moran JR, Harris CL, Merkel KL, Orenstein SR. Efficacy of a pre-thickened infant formula: a multicenter, double-blind, randomized, placebo-controlled parallel group trial in 104 infants with symptomatic gastroesophageal reflux. *Clin Pediatr (Phila)* 2003;42:483–95.

4. Chao HC, Vandenplas Y. Effect of cereal-thickened formula and upright positioning on regurgitation, gastric emptying, and weight gain in infants with regurgitation. *Nutrition* 2007;23:23–8.

5. Horvath A, Dziechciarz P, Szajewska H. The effect of thickened-feed interventions on gastroesophageal reflux in infants: systematic review and meta-analysis of randomized, controlled trials. *Pediatrics* 2008;122:e1268–77.

6. Wenzl TG, Schneider S, Scheele F, Silny J, Heimann G, Skopnik H. Effects of thickened feeding on gastroesophageal reflux in infants: a placebo-controlled crossover study using intraluminal impedance. *Pediatrics* 2003;111:e355–9.

7. Iacono G, Carroccio A, Cavataio F, et al. Gastroesophageal reflux and cow's milk allergy in infants: a prospective study. *J Allergy Clin Immunol* 1996;97:822–7.

8. Farahmand F, Najafi M, Ataee P, Modarresi V, Shahraki T, Rezaei N. Cow's milk allergy among children with gastroesophageal reflux disease. *Gut and Liver* 2011;5:298–301.

9. Borrelli O, Mancini V, Thapar N, et al. Cow's milk challenge increases weakly acidic reflux in children with cow's milk allergy and gastroesophageal reflux disease. *J Pediatr* 2012;161:476–81 e1.

10. Moses FM. The effect of exercise on the gastrointestinal tract. *Sports Med* 1990;9:159–72.

11. Hines S, Steels E, Chang A, Gibbons K. Aromatherapy for treatment of postoperative nausea and vomiting. *Cochrane Database Syst Rev* 2012;4:CD007598.

12. Hodge NS, McCarthy MS, Pierce RM. A prospective randomized study of the effectiveness of aromatherapy for relief of postoperative nausea and vomiting. *J Perianesth Nurs* 2014;29:5–11.

13. Sites DS, Johnson NT, Miller JA, et al. Controlled breathing with or without peppermint aromatherapy for postoperative nausea and/or vomiting symptom relief: a randomized controlled trial. *J Perianesth Nurs* 2014;29:12–19.

14. Ferruggiari L, Ragione B, Rich ER, Lock K. The effect of aromatherapy on postoperative nausea in women undergoing surgical procedures. *J Perianesth Nurs* 2012;27:246–51.

15. Lua PL, Salihah N, Mazlan N. Effects of inhaled ginger aromatherapy on chemotherapy-induced nausea and vomiting and health-

related quality of life in women with breast cancer. *Complement Ther Med* 2015;23:396–404.

16. Cadranel JF, Tarbe de Saint Hardouin C, Elouaer-Blanc L, Ruszniewski P, Mignon M, Bonfils S. Hypnosis for intractable vomiting. *Lancet* 1987;1:1140.

17. Sokel BS, Devane SP, Bentovim A, Milla PJ. Self hypnotherapeutic treatment of habitual reflex vomiting. *Arch Dis Child* 1990;65:626–7.

18. Zeltzer LK, Dolgin MJ, LeBaron S, LeBaron C. A randomized, controlled study of behavioral intervention for chemotherapy distress in children with cancer. *Pediatrics* 1991;88:34–42.

19. LaGrone RG. Hypnobehavioral therapy to reduce gag and emesis with a 10-year-old pill swallower. *Am J Clin Hypn* 1993;36:132–6.

20. Jacknow DS, Tschann JM, Link MP, Boyce WT. Hypnosis in the prevention of chemotherapy-related nausea and vomiting in children: a prospective study. *J Dev Behav Pediatr* 1994;15:258–64.

21. Burish TG, Shartner CD, Lyles JN. Effectiveness of multiple muscle-site EMG biofeedback and relaxation training in reducing the aversiveness of cancer chemotherapy. *Biofeedback Self Regul* 1981;6:523–35.

22. Banks RD Salisbury DA, Ceresia PJ. The Canadian Forces Airsickness Rehabilitation Program, 1981–1991. *Aviat Space Environ Med* 1992;63:1098–101.

23. Spinks A, Wasiak J. Scopolamine (hyoscine) for preventing and treating motion sickness. *Cochrane Database Syst Rev* 2011:CD002851.

24. Flank J, Thackray J, Nielson D, et al. Olanzapine for treatment and prevention of acute chemotherapy-induced vomiting in children: a retrospective, multi-center review. *Pediatr Blood Cancer* 2015;62:496–501.

25. Derakhshanfar H, Amree AH, Alimohammadi H, Shojahe M, Sharami A. Results of double blind placebo controlled trial to assess the effect of vitamin B6 on managing of nausea and vomiting in pediatrics with acute gastroenteritis. *Global Journal of Health Science* 2013;5:197–201.

26. Diop L, Guillou S, Durand H. Probiotic food supplement reduces stress-induced gastrointestinal symptoms in volunteers: a double-blind, placebo-controlled, randomized trial. *Nutr Res* 2008;28:1–5.

27. Tolone S, Pellino V, Vitaliti G, Lanzafame A, Tolone C. Evaluation of Helicobacter pylori eradication in pediatric patients by triple therapy plus lactoferrin and probiotics compared to triple therapy alone. *Italian Journal of Pediatrics* 2012;38:63.

28. Ahmad K, Fatemeh F, Mehri N, Maryam S. Probiotics for the treatment of pediatric Helicobacter pylori infection: a randomized double blind clinical trial. *Iranian Journal of Pediatrics* 2013;23:79–84.

29. Gong Y, Li Y, Sun Q. Probiotics improve efficacy and tolerability of triple therapy to eradicate Helicobacter pylori: a meta-analysis of randomized controlled trials. *International Journal of Clinical and Experimental Medicine* 2015;8:6530–43.

30. Chaiyakunapruk N, Kitikannakorn N, Nathisuwan S, Leeprakobboon K, Leelasettagool C. The efficacy of ginger for the prevention of postoperative nausea and vomiting: a meta-analysis. *Am J Obstet Gynecol* 2006;194:95–9.

31. Thomson M, Corbin R, Leung L. Effects of ginger for nausea and vomiting in early pregnancy: a meta-analysis. *J Am Board Fam Med* 2014;27:115–22.

32. Bone ME, Wilkinson DJ, Young JR, McNeil J, Charlton S. Ginger root—a new antiemetic. The effect of ginger root on postoperative nausea and vomiting after major gynaecological surgery. *Anaesthesia* 1990;45:669–71.

33. Weizman Z, Alkrinawi S, Goldfarb D, Bitran C. Efficacy of herbal tea preparation in infantile colic. *Journal of Pediatrics* 1993;122:650–2.

34. Otake K, Uchida K, Mori K, et al. Efficacy of the Japanese herbal medicine rikkunshito in infants with gastroesophageal reflux disease. *Pediatr Int* 2015;57(4):673–6.

35. Neu M, Pan Z, Workman R, et al. Benefits of massage therapy for infants with symptoms of gastroesophageal reflux disease. *Biol Res Nurs* 2014;16:387–97.

36. Post-White J, Kinney ME, Savik K, et al. Therapeutic massage and healing touch improve

symptoms in cancer. *Integr Cancer Ther* 2003;2:332–44.

37. Mazlum S, Chaharsoughi NT, Banihashem A, Vashani HB. The effect of massage therapy on chemotherapy-induced nausea and vomiting in pediatric cancer. *Iranian Journal of Nursing and Midwifery Research* 2013;18:280–4.

38. Ezzo J, Streitberger K, Schneider A. Cochrane systematic reviews examine P6 acupuncture-point stimulation for nausea and vomiting. *J Altern Complement Med* 2006;12:489–95.

39. Lee A, Fan LT. Stimulation of the wrist acupuncture point P6 for preventing postoperative nausea and vomiting. *Cochrane Database Syst Rev* 2009:CD003281.

40. Dundee JW, Ghaly RG, Bill KM, Chestnutt WN, Fitzpatrick KT, Lynas AG. Effect of stimulation of the P6 antiemetic point on postoperative nausea and vomiting. *Br J Anaesth* 1989;63:612–18.

41. Zhang GL, Yang SY, Zhu ZL, Mu PX. Meta-analysis on postoperative complications of wristband acupoint pressure therapy. *Journal of Biological Regulators and Homeostatic Agents* 2015;29:187–93.

42. Dundee JW, Yang J. Prolongation of the antiemetic action of P6 acupuncture by acupressure in patients having cancer chemotherapy. *J R Soc Med* 1990;83:360–2.

43. Helmreich RJ, Shiao SY, Dune LS. Meta-analysis of acustimulation effects on nausea and vomiting in pregnant women. *Explore (NY)* 2006;2:412–21.

44. Park J, Sohn Y, White AR, Lee H. The safety of acupuncture during pregnancy: a systematic review. *Acupunct Med* 2014;32:257–66.

45. Gottschling S, Reindl TK, Meyer S, et al. Acupuncture to alleviate chemotherapy-induced nausea and vomiting in pediatric oncology—a randomized multicenter crossover pilot trial. *Klinische Padiatrie* 2008;220:365–70.

46. Dibble SL, Luce J, Cooper BA, et al. Acupressure for chemotherapy-induced nausea and vomiting: a randomized clinical trial. *Oncol Nurs Forum* 2007;34:813–20.

47. Treish I, Shord S, Valgus J, et al. Randomized double-blind study of the Reliefband as an adjunct to standard antiemetics in patients receiving moderately-high to highly emetogenic chemotherapy. *Support Care Cancer* 2003;11:516–21.

48. Yang C, Hao Z, Zhang LL, Guo Q. Efficacy and safety of acupuncture in children: an overview of systematic reviews. *Pediatr Res* 2015;78(2):112–19.

49. Danhauer S, Tooze J, Holder P, et al. Healing touch as a supportive intervention for adult acute leukemia patients: a pilot investigation of effects on distress and symptoms. *JSIO* 2008;6:89–97.

50. Lincoln V, Nowak EW, Schommer B, Briggs T, Fehrer A, Wax G. Impact of healing touch with healing harp on inpatient acute care pain: a retrospective analysis. *Holist Nurs Pract* 2014;28:164–70.

51. Perol D, Provencal J, Hardy-Bessard AC, et al. Can treatment with Cocculine improve the control of chemotherapy-induced emesis in early breast cancer patients? A randomized, multi-centered, double-blind, placebo-controlled Phase III trial. *BMC Cancer* 2012;12:603.

Chapter 27: Warts

1. Focht DR, III, Spicer C, Fairchok MP. The efficacy of duct tape vs cryotherapy in the treatment of verruca vulgaris (the common wart). *Arch Pediatr Adolesc Med* 2002;156:971–4.

2. Stern P, Levine N. Controlled localized heat therapy in cutaneous warts. *Archives of Dermatology* 1992;128:945–8.

3. Huo W, Gao XH, Sun XP, et al. Local hyperthermia at 44 degrees C for the treatment of plantar warts: a randomized, patient-blinded, placebo-controlled trial. *J Infect Dis* 2010;201:1169–72.

4. Spanos NP, Stenstrom RJ, Johnston JC. Hypnosis, placebo, and suggestion in the treatment of warts. *Psychosom Med* 1988;50:245–60.

5. Noll R. Hypnotherapy for warts in children and adolescents. *Developmental and Behavioral Pediatrics* 1994;15:170–3.

6. Tasini MF, Hackett TP. Hypnosis in the treatment of warts in immunodeficient children. *Am J Clin Hypn* 1977;19:152–4.

References

7. Harwood CA, Perrett CM, Brown VL, et al. Imiquimod cream 5% for recalcitrant cutaneous warts in immunosuppressed individuals. *Br J Dermatol* 2005;152:122–9.

8. Glass AT, Solomon BA. Cimetidine therapy for recalcitrant warts in adults. *Archives of Dermatology* 1996;132:680–2.

9. Fit KE, Williams PC. Use of histamine2-antagonists for the treatment of verruca vulgaris. *Ann Pharmacother* 2007;41:1222–6.

10. Ning S, Li F, Qian L, et al. The successful treatment of flat warts with auricular acupuncture. *International Journal of Dermatology* 2012;51:211–15.

11. Labrecque M, Audet D, Latulippe LG, Drouin J. Homeopathic treatment of plantar warts. *CMAJ* 1992;146:1749–53.

12. Smolle J, Prause G, Kerl H. A double-blind, controlled clinical trial of homeopathy and an analysis of lunar phases and postoperative outcome. *Archives of Dermatology* 1998;134:1368–70.

INDEX